Introduction to
Modern Biochemistry

Fourth Edition

Introduction to Modern Biochemistry

Fourth Edition

By

P. Karlson

Institute of Physiological Chemistry
University of Marburg
Marburg/Lahn, Germany

Translated by

Charles H. Doering

Department of Psychiatry
Stanford University School of Medicine
Stanford, California

Academic Press *New York San Francisco London*

A Subsidiary of Harcourt Brace Jovanovich, Publishers

First published in the German language under the title
Kurzes Lehrbuch der Biochemie für Mediziner und Naturwissenschaftler
and copyrighted in 1974 by GEORG THIEME VERLAG.
Stuttgart, Germany

ACADEMIC PRESS, INC.
111 Fifth Avenue, New York, New York 10003

United Kingdom Edition published by
ACADEMIC PRESS, INC. (LONDON) LTD.
24/28 Oval Road, London NW1

Library of Congress Cataloging in Publication Data

Karlson, Peter, Date
Introduction to modern biochemistry.

Translation of Kurzes Lehrbuch der Biochemie für Mediziner und Naturwissenschaftler.
1. Biological chemistry.
I. Title. [DNLM: 1. Biochemistry.
QU4 K17k 1973]
QH345.K313 1974 574.1'92 73-9429
ISBN 0-12-399764-X

PRINTED IN THE UNITED STATES OF AMERICA

CONTENTS

CHAPTER IV *Proteins*

CHAPTER V *Enzymes and Biocatalysis*

CHAPTER VI *Coenzymes*

CHAPTER VII *Nucleic Acids and Protein Biosynthesis*

CHAPTER VIII *Metabolism of Proteins*

CHAPTER IX *Porphyrins and Hemins*

CHAPTER XVII *Glycosides, Oligosaccharides, Polysaccharides*

CHAPTER XVIII *Topochemistry of the Cell*

CHAPTER XIX *Regulatory Mechanisms and Correlations in Intermediary Metabolism*

CHAPTER XX *Hormones*

Appendix

PREFACE

Since modern biology has extended its analysis of life to molecular dimensions, biochemistry can be regarded as the basis for many disciplines of biology and medicine. However, the vast amount of detailed information available can prove confusing to the student, indicating a need for a book in which an introduction to biochemistry and a survey of the existing facts and concepts are presented. It is hoped that this textbook, now in its fourth edition, will fulfill its purpose, to provide a better understanding of the chemical background of biological phenomena.

The material has been arranged according to didactic needs. No distinction has been made between "descriptive biochemistry," the chemistry of natural products, and "dynamic biochemistry" or metabolism. In some chapters the chemical structures of the natural compounds have been deduced from the description of their biosyntheses.

In the selection of material, biochemistry has been regarded as part of the science of life. Concepts of general importance, such as the generation and utilization of free energy, the role of genes in protein and enzyme synthesis, and the importance of the submicroscopic structure of the cell in biochemical reactions, have been emphasized. Clinical problems have been treated only insofar as they have contributed to the knowledge of normal biochemistry.

Because progress in biochemistry is so rapid a revision of the third edition, translated from the sixth German edition (1967), has long been due. This text is based largely on the ninth German edition and contains all the changes made in the seventh and eighth German editions. Several chapters have been completely rewritten; more emphasis has been placed on regulation of metabolism and structure and function of membranes. A section on normal and malignant growth has been added.

I thank the translator, Dr. Charles H. Doering, for incorporating in his translation the latest changes in the German text.

P. Karlson

CHAPTER I

Organic Chemistry and Biochemistry

Biochemistry deals with carbon compounds and the reactions they undergo in living organisms. Organic chemistry, the chemistry of carbon compounds, therefore, is basic to biochemistry. The chemistry of natural products and biochemistry overlap extensively and lack a clearly defined border. Numerous compounds that have only recently been recognized as intermediates of metabolism had long been known in organic chemistry.

A good knowledge of organic chemistry, in particular, a thorough appreciation of structural formulas, therefore, is a prerequisite for an intelligent study of biochemistry. The following brief outline is not intended to replace a course in organic chemistry, but rather is to serve as a handy reference for selected compounds and for reactions of special significance in biochemistry.

1. Hydrocarbons as Parent Substances

The great variety of organic compounds arises from the ease with which carbon atoms can attach to each other to form chain compounds. Since carbon is tetravalent, a great many possibilities for branching arise, leading to a staggering number of carbon skeletons. If the remaining valence bonds are filled with hydrogen, then we speak of the resulting compound as a hydrocarbon. From a systematic viewpoint, hydrocarbons are the parent substances of all organic compounds; in practice, however, they are very rarely used to prepare other classes of compounds.

The molecular formulas of all saturated open-chain hydrocarbons are represented by the general formula C_nH_{2n+2}. Every time an open chain closes to a ring, irrespective of size of the ring, two hydrogen atoms are lost with a corresponding change in the molecular formula. Thus it becomes possible to deduce from the molecular formula of a saturated hydrocarbon such as cholestane, $C_{27}H_{48}$ (parent substance of cholesterol), that there must be four rings in its carbon skeleton.

In biochemistry saturated or alicyclic ring systems are at least as important as aromatic structures derived from benzene, which dominate such a large segment of classical organic chemistry. It is well known that the aromatic ring is a resonating system which may be represented only imperfectly by the traditional hexagon and the three double bonds. Nevertheless we shall adopt this symbol to distinguish aromatic compounds from the saturated ring systems; these, in turn, will be represented by simple pentagons and hexagons. The most important ring compounds and their formulas are listed in Table I-1.

Besides the all-carbon rings, we will frequently come across so-called heterocyclic compounds, in which one (or more) carbon atom is exchanged for one (or more) of nitrogen, oxygen, or sulfur. Ring systems of this type have usually received trivial names (see Table I-1). Heterocyclic compounds possess chemical properties quite different from analogous carbocyclic compounds; nitrogen, for example, usually imparts basic properties (but occasionally is acidic). This is an important feature of heterocyclic compounds.

Most saturated hydrocarbons are rather inert. By contrast, unsaturated compounds, designated by

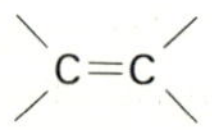 (double bond)

or by

$-C\equiv C-$ (triple bond)

are more reactive. They may easily add other atoms or groups of atoms, such as hydrogen or the elements of water:

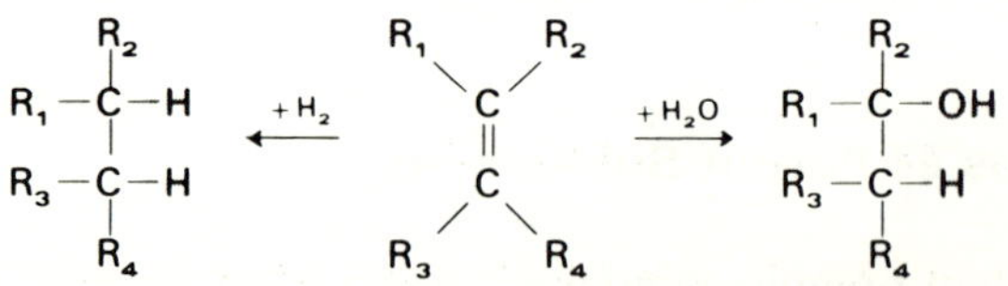

For further addition reactions, the reader should consult textbooks of organic chemistry. Unsaturated compounds can be cleaved by oxidation at their double bonds.

Presentation of Structural Formulas. Strictly speaking, every valence bond in a formula should be represented by a dash, the symbol for an electron pair. For convenience, however, groups saturated with hydrogen are usually written as $—CH_3$ or $—CH_2—$ instead of the cumbersome

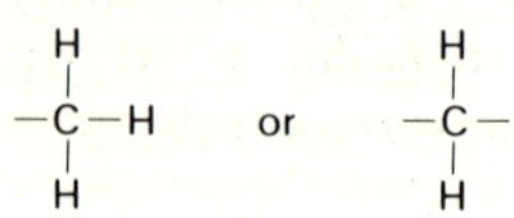

TABLE I-1

Ring Systems

Number of atoms in ring	Carbocyclic rings		Heterocyclic rings	
	Saturated	Unsaturated (aromatic)	N-containing	O-containing
5	 Cyclopentane	 Cyclopentadiene	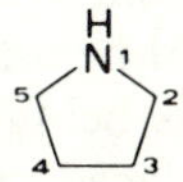 Pyrrolidine 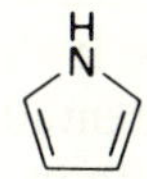 Pyrrole 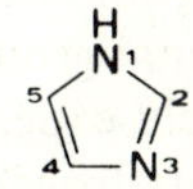 Imidazole	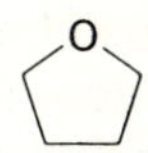 Tetrahydrofuran 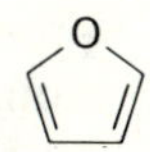 Furan
6	 Cyclohexane	 Benzene	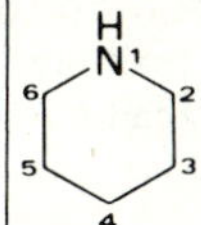 Piperidine 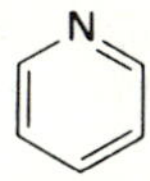 Pyridine 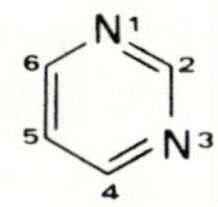 Pyrimidine	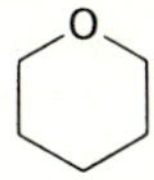 Tetrahydropyran 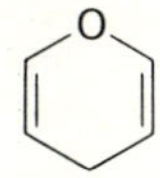 Pyran
6 + 5	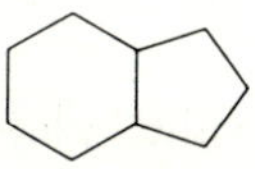 Hydrindane	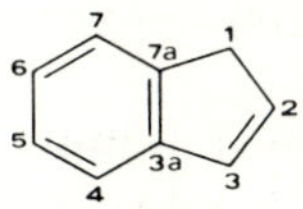 Indene	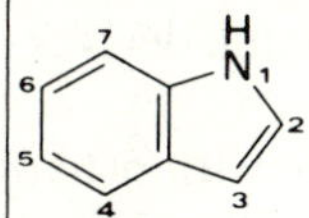 Indole 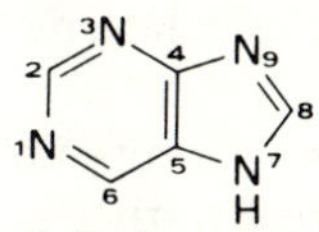 Purine	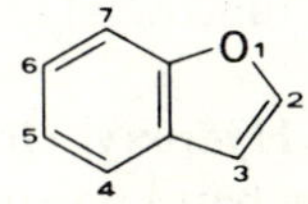 Coumarone
6 + 6	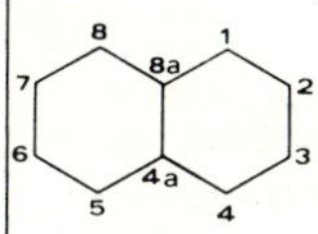 Decalin	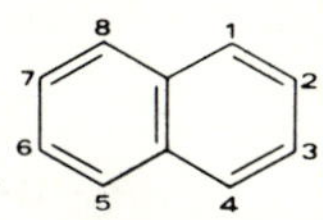 Naphthalene	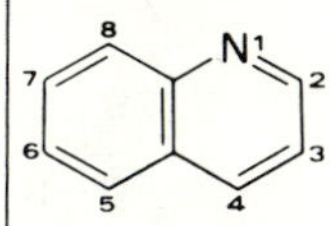 Quinoline 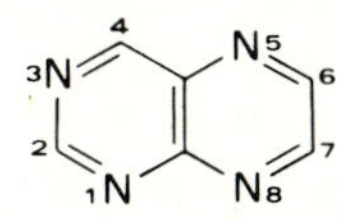 Pteridine	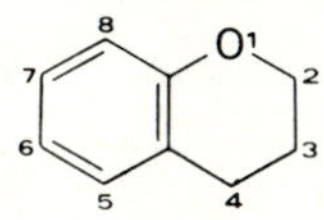 Chroman
Higher systems	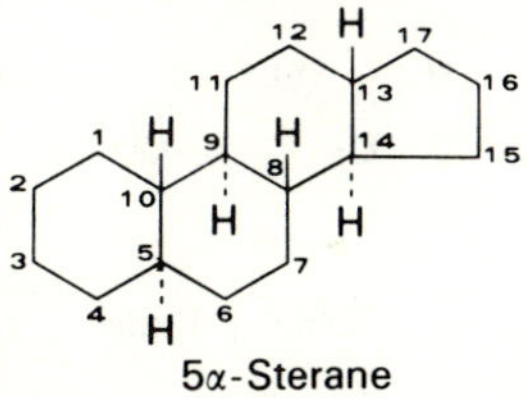 5α-Sterane	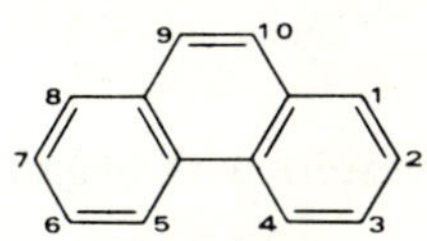 Phenanthrene	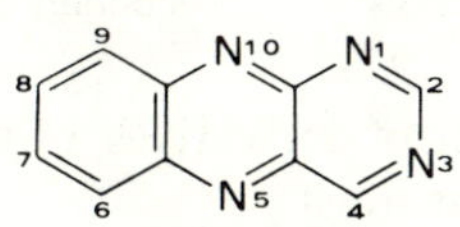 Alloxazine	

We will simplify further in many cases and let a system of angled lines represent a hydrocarbon chain in which each corner or kink indicates a CH_2 group. The same holds for ring compounds where the abbreviated form is more common. Double bonds are always shown by a double line; a hexagon with single lines hence means a cyclohexane ring, i.e., a saturated compound and not a benzene ring.

Groups of atoms of lesser importance in any particular context, or repeated several times, usually are designated simply by R (= radical). Often R is used at a position where different groups may be substituted.

2. Functional Groups

A further reason for the great variability of organic compounds is found in the presence of functional groups. It should be understood that a strictly formal derivation, seldom feasible experimentally, is being described in this section. Depending on the number of hydrogen atoms replaced on a carbon, we distinguish between

monovalent functional group:	—OH (hydroxyl group),	$-NH_2$ (amino group)
divalent functional group:	=O (carbonyl group),	=NH (imino group)
trivalent functional group:	=O, —OH (carboxyl group),	≡N (nitrile)

Chemical reactions of the different classes of substances ordinarily are reactions of functional groups; the latter's name derives from this fact.

Hydroxyl Groups. Compounds with a hydroxyl group are called alcohols after the best known representative, namely, ordinary or ethyl alcohol. It is convenient to distinguish the following:

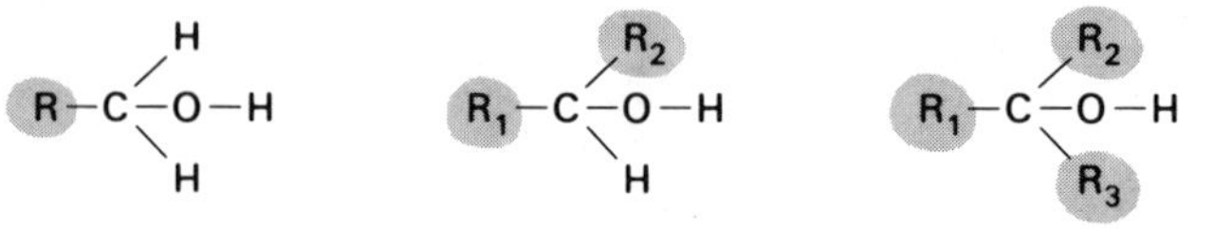

Primary alcohols Secondary alcohols Tertiary alcohols

Alcohols form a number of derivatives, of which the esters with organic acids and phosphoric acid interest us most:

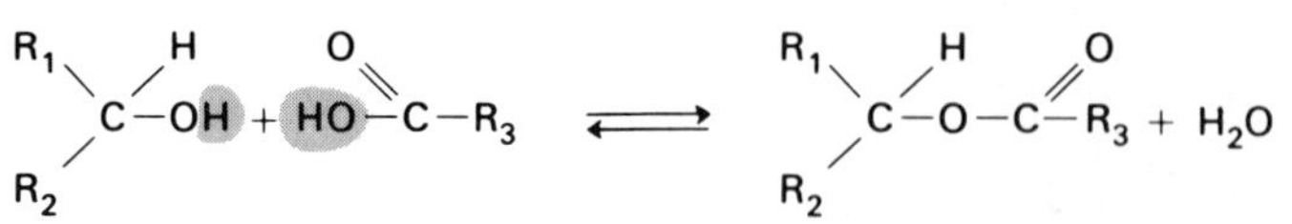

Alcohol + Carboxylic acid ⇌ Ester + Water

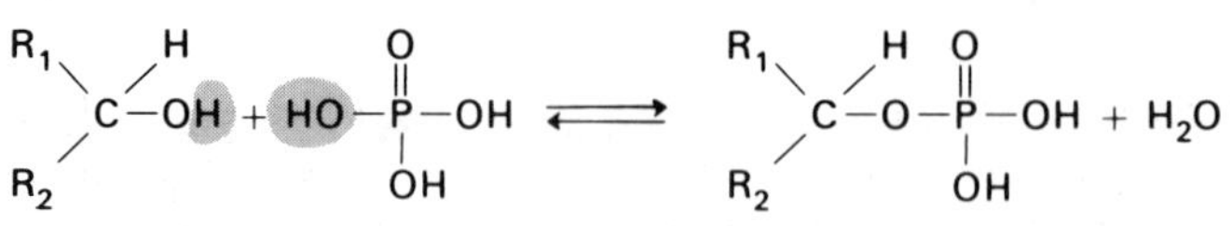

Alcohol + Phosphoric acid

Ester of phosphoric acid or alcohol phosphate

The formation of esters is a reversible reaction. The final equilibrium may be approached from either side of the equation (cf. Chapter V,2).

The removal of one molecule of water from two alcohol molecules produces an ether:

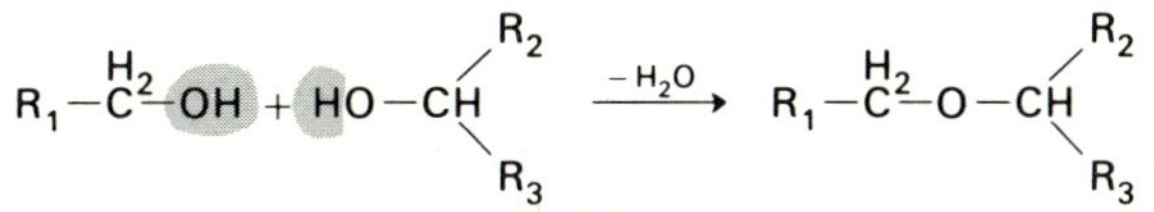

Many natural products contain ether groups.

One more example of the reactions of the hydroxyl group is the dehydrogenation to the carbonyl group:

(a) (b)

Dehydrogenation of alcohols

In reaction (a), a primary alcohol yields an aldehyde; in (b), a secondary alcohol yields a ketone. Tertiary alcohols cannot be dehydrogenated without destroying the carbon skeleton. Instead, they split off one molecule of water rather easily to form the unsaturated compound. The other alcohols may also be dehydrated.

Amines. In former days (of "type theories" in organic chemistry), alcohols had been conceived of as monoalkylated water, and amines as alkylated ammonia. Many properties of these two classes of compounds can indeed be explained that way.

The nomenclature of amines differs from that of the alcohols. The designations *primary*, *secondary*, and *tertiary* refer to the number of substituents on the ammonia nitrogen:

Primary amine — Secondary amine — Tertiary amine — Quaternary ammonium salt

The fourth formula shows a quaternary ammonium salt, derived formally by replacing all four hydrogen atoms of the salt NH_4Cl with organic radicals. The hydroxides of the quaternary ammonium ion, the "free bases," exist in addition to the neutral salts.

The nitrogen compounds of this series are organic bases. Salt formation, written analogously to ammonium ion formation, is the most important functional reaction of amino compounds (cf. Chapter II,1).

$$NH_3 + H^{\oplus} \rightleftharpoons NH_4^{\oplus}; \qquad R{-}NH_2 + H^{\oplus} \rightleftharpoons R{-}NH_3^{\oplus}$$

Cyclic bases actually are secondary or tertiary amines. The R_2 and R_3 groups have joined to close a ring. Examples of this are seen in Table I-1. Several of the cyclic bases (pyridine, imidazole, etc.) have a double bond between a C and an N atom (aromatic systems).

Another reaction of amines, comparable to ester formation of alcohols, is the formation of amides. In practice, the preparation of amides is achieved only over some detours.

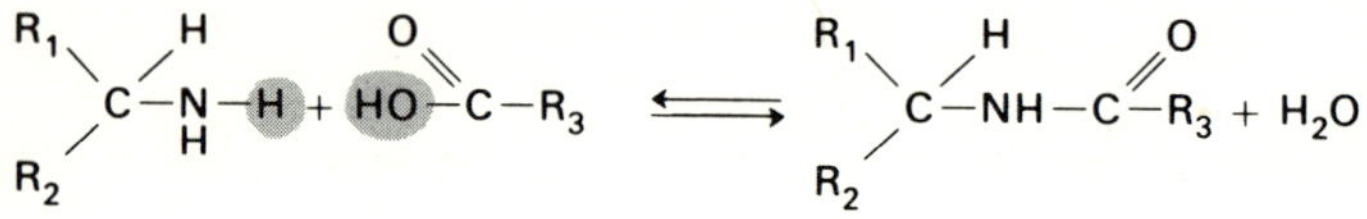

Dehydrogenation of a primary amino group introduces, again in analogy to alcohols, a double bond,

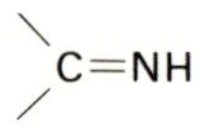 imino group

changing it to an imino group. Imines are unstable and are hydrolyzed to ketones, (or aldehydes, if $R_1 = H$) and ammonia. The oxidative metabolism of amino acids, for example, follows this reaction scheme.

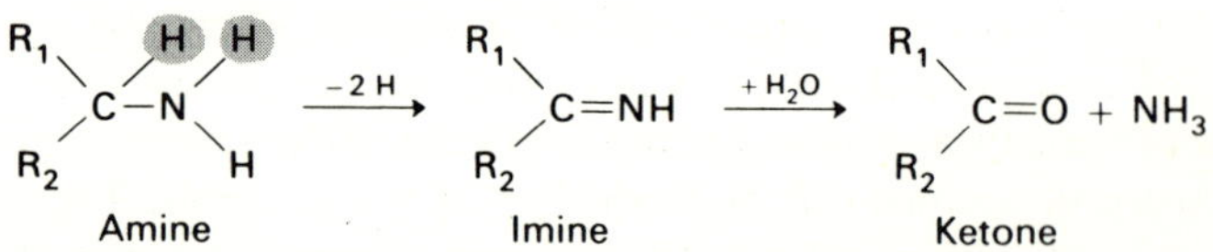

Carbonyl Compounds: Aldehydes and Ketones. The general formulas of these groups of compounds are

Aldehyde Ketone

Common to both is the carbonyl group,

which undergoes certain reactions. By reduction (hydrogenation), i.e., addition of hydrogen to the

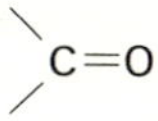 double bond

alcohols are formed. The reaction is reversible: Primary or secondary alcohols can be dehydrated to yield carbonyl compounds as illustrated above in the discussion on alcohols. This reversible reaction is particularly important in biochemistry.

In another addition reaction, a molecule of water is introduced. The resulting hydrates are quite unstable but may serve as intermediates. Addition of an alcohol instead of a water molecule gives rise to the analogous hemiacetals. By further reaction with alcohol and removal of water, acetals are formed. The name acetal applies only to derivatives of aldehydes; ketals are derived from ketones.

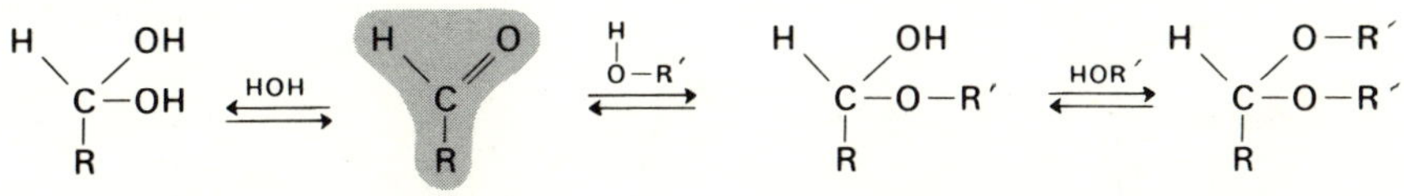

Aldehyde hydrate — Aldehyde — Hemiacetal — (Full-)Acetal

Acetal formation is facilitated greatly by reagents that bind the released water; acetals are also easily hydrolyzed.

Simple dehydrogenation of an aldehyde hydrate, i.e., without addition of oxygen, produces a carboxylic acid:

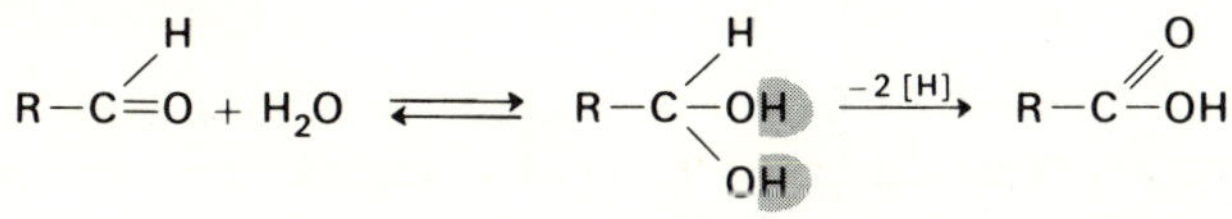

This reaction was the starting point for H. Wieland's "dehydrogenation theory" of biological oxidation, the subject of bitter controversy in the 1920's. The dispute over "activation of hydrogen" (dehydrogenation) and O. Warburg's "activation of oxygen" (oxidation) seems meaningless today. Now we are able to cite numerous examples of both types of reactions; cell respiration employs both types.

For the purpose of chemical identification, the reaction of the carbonyl compound with derivatives of hydrazine is often used. The example chosen here is the reaction with phenylhydrazine:

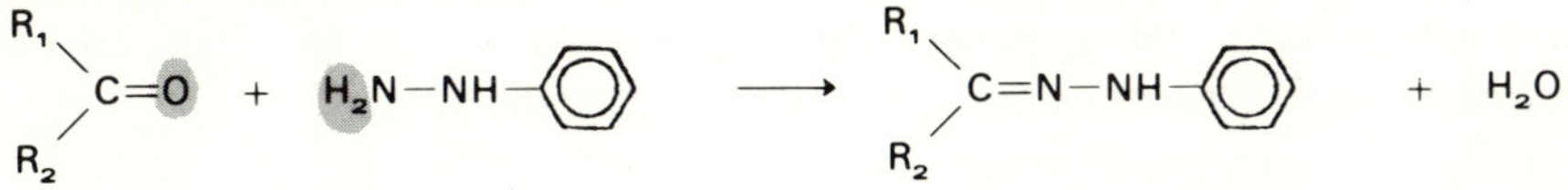

Ketone + Phenylhydrazine — Phenylhydrazone of the ketone

The hydrazone formed in this reaction usually crystallizes easily and is only sparingly soluble. A variation of this reaction is used in the chemistry of sugars. Hydroxylamine analogously forms oximes.

Carbonyl compounds react not only with H_2N groups, but with reactive H_2C groups as well. In the aldol condensation there is first an addition to the carbonyl double bond and then water is split off. These steps are taken separately [see examples in Chapters VI,7, reaction (2) of coenzyme A, and XV,5]:

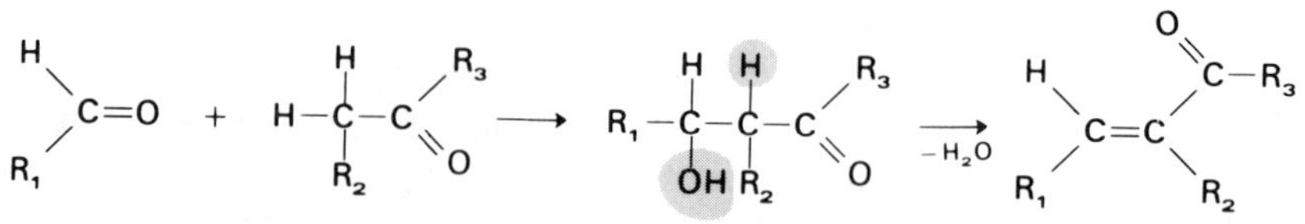

Aldehydes and ketones can change to their *enol forms*; a hydrogen atom wanders to the oxygen atom and the double bond shifts:

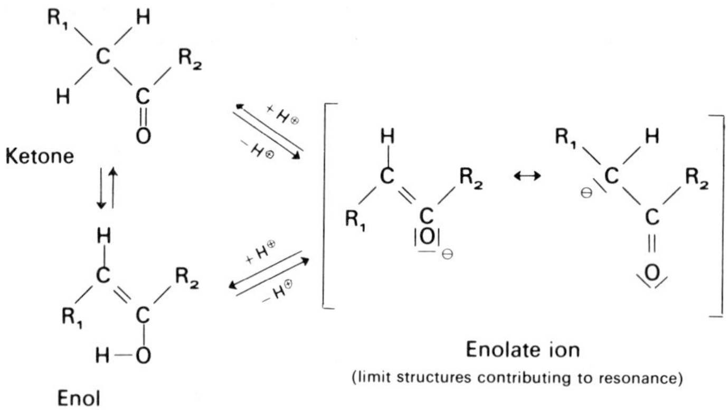

The name "enol" is derived from "en" for double bond and "ol" for alcohol group. The transition from carbonyl to enol is called "tautomerism"; it occurs especially easily whenever other unsaturated groups are in a potentially conjugated position nearby. As seen above, an enol can readily donate a proton, because the resulting enolate ion is stabilized by resonance.

Phenols. The phenols are closely related to the enols and are characterized by a hydroxyl group attached directly to an aromatic nucleus. The phenolic OH group is always next to a double bond which is part of the aromatic system. Many structures contributing to the resonance can be written for the phenolate ion; phenols, for this reason, are weak acids, and the simplest representative, phenol itself, originally was called "carbolic acid." Nevertheless, the phenolic OH group can form ethers and esters, and in these properties more nearly resembles the alcohols.

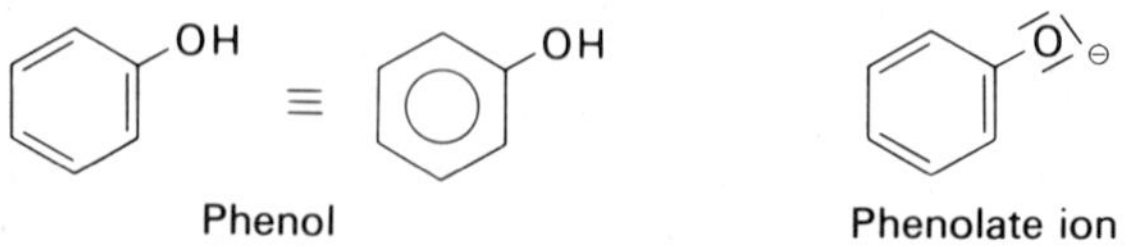

Imines. The imines are substances with a C=N double bond. They hydrolyze easily and are stable only in special combinations. Ring compounds and the guanidino

group

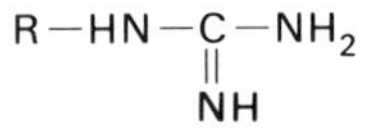

Guanidine, R = H

are among such combinations (see Table I-1 and Chapter VII,1). Guanidine is stable in water.

The older chemical nomenclature also calls secondary amines, R—NH—R′, "imines." In some instances such naming has survived stubbornly. We will avoid it whenever possible and call them "secondary amines."

Carboxylic Acids. These contain the carboxyl group —COOH; three valence bonds of the carbon atom have been replaced by oxygen. The name indicates their acidic nature. Carboxylic acids dissociate in water to give one proton (more properly a hydronium ion, H_3O^+) and the negatively charged anion.

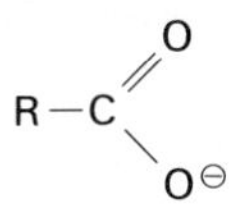

Carboxylate ion

The negative charge, however, is not localized on one oxygen atom. Both oxygens share this charge by resonance. Nevertheless, we will continue to represent the carboxylate anion in this manner for convenience.

Salts. At physiological pH most of the organic acids exist in their anionic form, i.e., in their salt form. Acids occurring frequently in biochemistry are listed along with their formulas in Table I-2.

For the preparation and purification of organic acids, their salts are occasionally used. The silver salts, for example, of most organic acids are only slightly soluble.

Dissociation of Carboxylic Acids. Within cells and in other body fluids the carboxylic acids exist almost entirely in their dissociated anionic form; only an infinitesimal fraction retains the group COOH. We may represent the dissociation, somewhat simplified, by the equation

$$\text{R—COOH} \rightleftharpoons \text{R—COO}^{\ominus} + \text{H}^{\oplus} \qquad (1)$$

and apply it to the law of mass action. The law states that the ratio of the concentrations is a constant:

$$\frac{[\text{R—COO}^{\ominus}] \cdot [\text{H}^{\oplus}]}{[\text{R—COOH}]} = K \qquad (2)$$

Conventionally, the brackets signify *concentration* of the individual entities (either molecules or ions). For simplicity we write $[H^{\oplus}]$ in these equations, although the ion is $[H_3O^{\oplus}]$ in reality. With more concentrated solutions, furthermore, the simple stoichiometric concentrations should be replaced by "activities." The interaction among ions seems to lower their effective concentration, i.e., their activity. This phenomenon is expressed by the activity coefficient f (which is always less than unity and is used as a factor for concentrations). For the "physiological NaCl" solution (0.154 M), for example, $f = 0.76$.

TABLE I-2

Important Acids

Monocarboxylic acids	Dicarboxylic acids	Hydroxy or keto acids
$HCOOH$ Formic acid	$HOOC-COOH$ Oxalic acid	$H_3C-C(H)(OH)-COOH$ Lactic acid
$H_3C-COOH$ Acetic acid	$HOOC-CH_2-COOH$ Malonic acid	$H_3C-C(=O)-COOH$ Pyruvic acid
H_3C-CH_2-COOH Propionic acid	$H_2C(COOH)-CH_2-COOH$ Succinic acid	$H_2C(OH)-C(H)(OH)-COOH$ Glyceric acid
$H_3C-CH_2-CH_2-COOH$ Butyric acid	$(H)(HOOC)C=C(COOH)(H)$ Fumaric acid	$H_2C(COOH)-C(H)(OH)-COOH$ Malic acid
$(H_3C)_2C(H)-CH_2-COOH$ Isovaleric acid	$H_2C(COOH)-CH_2-CH_2-COOH$ Glutaric acid	$HO-C(H)(COOH)-C(H)(OH)-COOH$ Tartaric acid

Sörensen has introduced the term pH of a solution as a scale for the concentration (or rather activity) of $H^{\oplus}$ ions appearing in Eq. (2). The pH is defined as the negative common logarithm (base of 10) of the $H^{\oplus}$ concentration, or

$$pH = -\log [H^{\oplus}] \tag{3}$$

Because of the dissociation to $H^{\oplus}$ and $OH^{\ominus}$, even pure water has a $H^{\oplus}$ ion concentration of $10^{-7} M$, or a pH of 7; this is also the neutral pH.[1] The concept of the pH is very important in practical biochemistry, because most cellular processes are affected strongly by the $H^{\oplus}$ ion concentration, i.e., the pH.

[1] The temperature dependence of the pH of neutrality, at which the $H^{\oplus}$ concentration equals the $OH^{\ominus}$ concentration, is often overlooked. Neutrality is pH 7.0 at 22°, pH 6.75 at 37°, and pH 6.5 at 57°.

The use of pH values follows, of course, the rules of logarithmic calculations; every change of the pH by one unit corresponds to a change in $H^{\oplus}$ ion concentration by the factor of 10. For example, the $H^{\oplus}$ ion concentration of 2×10^{-5} *M* can be expressed as a pH of

$$-\log(2 \times 10^{-5}) = -(\log 2 + \log 10^{-5}) = -(0.3 - 5.0) = 4.7$$

Let us return once more to the law of mass action. We may rewrite Eq. (2) in the following way ($A^{\ominus}$ = anion, HA = undissociated acid):

$$\frac{[A^{\ominus}]}{[HA]} \cdot [H^{\oplus}] = K \tag{4}$$

If the quotient equals unity, i.e., the amount of undissociated acid equals the amount of anion, then the $H^{\oplus}$ ion concentration becomes the dissociation constant of the acid. If $[H^{\oplus}]$ is expressed as the pH, it is expedient to express also the dissociation constant as the negative logarithm. This value is then called the pK; pK = $-\log K$.

The conversion of Eq. (4) into its logarithmic form yields:

$$\log\frac{[A^{\ominus}]}{[HA]} + \log[H^{\oplus}] = \log K$$

$$\log\frac{[A^{\ominus}]}{[HA]} - \log K = -\log[H^{\oplus}]$$

$$\log\frac{[A^{\ominus}]}{[HA]} + pK = pH$$

This last equation, generally known as the Henderson-Hasselbalch equation, is useful particularly for the calculation of pH values arising from mixtures of salts and acids. Mixtures of this kind are called buffer solutions.

Buffers. Buffer solutions are capable of absorbing $H^{\oplus}$ and $OH^{\ominus}$ ions and in this way can resist changes of pH which would otherwise occur with the additions of acids or bases. This phenomenon can be explained by the law of mass action. If acid is added to the equilibrium mixture represented by Eq. (4), the $[H^{\oplus}]$ will rise. However, in order to reach equilibrium again, [HA] must also rise; the latter can be achieved only by the recombination of $H^{\oplus}$ and $A^{\ominus}$ to form HA. This reaction consumes nearly all the added $H^{\oplus}$ ions, and the consequent change in pH is negligible. All this presupposes, of course, that not too many $H^{\oplus}$ ions, or not too much acid, have been added. The capacity of the buffer is limited by the number of $A^{\ominus}$ ions available to trap $H^{\oplus}$ ions.

Weak acids or weak bases in combination with their salts are, in general, adaptable for setting up buffers; their greatest capacity to buffer is around their pK value. Buffers are indispensable in most biochemical work. Commonly used are buffers of phosphate, citrate, glycine, tris(hydroxymethyl)aminomethane (or simply "tris buffer"), and others. The exact compositions of different buffer mixtures are listed in biochemical handbooks.

The physiological buffer system in blood is described in Chapter XXI,4.

Esters. Further functional derivatives of the acids may be derived formally from the acyl residue. (The group R—CO— is called the acyl residue.) Esters, as the first example, are formed from acids and alcohols by an equilibrium reaction:

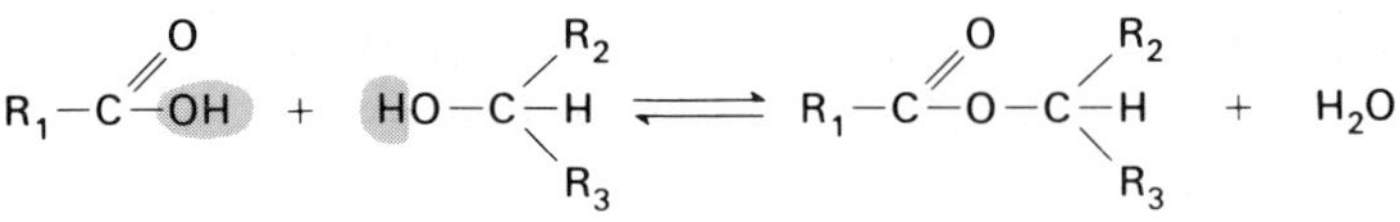

In preparative chemistry, conditions can be chosen to obtain the ester in nearly quantitative yields. Under physiological conditions—i.e., in aqueous medium at neutral pH—the equilibrium favors the reactants, acid and alcohol. Yet many esters are found in cells. For biological ester formation and cleavage, see Chapters VI,5, reaction type (a); VI,7, reaction (1) of coenzyme A; and XII,6.

Amides. Acid amides also play an important role in biochemistry. They are compounds in which the hydroxyl group of an acid has been replaced with $—NH_2$ or —NH—R. In contrast to esters, amides cannot be prepared from acid and ammonia (or amine), but rather have to be made via reactive or "activated" derivatives of the acids. They are very stable in aqueous solution, despite the fact that the reaction equilibrium of hydrolysis is far on the side of cleavage:

$$R-CO-NH_2 + H_2O \rightleftharpoons R-COO^{\ominus} + NH_4^{\oplus}$$

"Activated" Acid Derivatives. These are very reactive and interact with other substances exergonically (Chapter V,2). Even esters belong to this class, because they undergo many reactions that are impossible for free acids. In organic chemistry, acid chlorides and acid anhydrides are preferred.

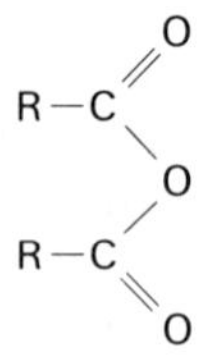

Acid anhydride

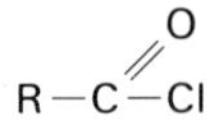

Acid chloride

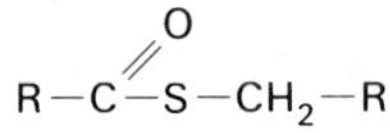

Thioester

When acid chlorides react with other substances, the chloride is released as HCl (in the case of acid anhydrides, the acyl group is released as the free acid). Thus esters, amides, etc., are prepared conveniently. Such "activated" acids are also known in biochemistry. Activated acids are of the anhydride type or of the thioester type. Thioesters differ from simple esters in that the alcohol component R—OH is replaced by an R—SH compound. They are highly reactive (cf. also group transfer, Chapter VI,7, coenzyme A).

3. Polymeric Compounds

Compounds can combine to form chainlike larger molecules, which, in turn, are susceptible to hydrolytic breakdown. They contain C—O—C or N—C bonds that may be thought to have arisen by condensation of two parent molecules with loss of one water molecule. Their hydrolysis, therefore, appears to be a reversal of their formation. The list of such polymeric compounds is headed by the acid derivatives, esters and amides. Esters and acetals also belong to this group.

Nearly all macromolecular natural products are combinations of smaller compounds.[2] Molecular weights of monomeric compounds rarely exceed 600. Polysaccharides (starch, cellulose) are linked by hemiacetal bonds; proteins, by acid–amide bonds, usually called peptide bonds (see Table I-3). Ester linkages are found in fats and lipids (not of very high molecular weights) as well as in the macromolecular nucleic acids, which are phosphoric acids of exceedingly high molecular weights.

TABLE I-3
Macromolecular Natural Products

Type of linkage	Natural product	See Chapter
Polyacetals	Cellulose, starch, etc.	XVII
Polyamides	Proteins, polypeptides	III, IV
Polyesters	Nucleic acids (phosphate esters)	VII
	Fats (carboxylate esters)	XII

Macromolecules usually consist of subunits with two functional groups, both of which contribute to the linking; in the case of proteins, they are the amino and carboxyl group. Biosynthesis, however, does not occur simply by splitting out water between components.

4. Isomerisms

Another phenomenon contributing to the great variability of organic compounds is isomerism. Isomers are compounds that have exactly the same molecular formula but differ in at least one of their chemical or physical properties.

The difference in properties may stem from differing functional groups. Both lactic acid and dihydroxyacetone have the same molecular formula, $C_3H_6O_3$; the

[2] An exception is natural rubber, which is a polymer of isoprene units with C—C linkages only. The molecular weights may range as high as 350,000. Highly polymerized synthetic products frequently follow the same pattern of composition, i.e., of the very stable C—C linkage. Synthetic products of the polyamide (nylon), polyester, and polyurethane types are other familiar examples of polymers.

former has a carboxyl and a hydroxyl group, while the latter has two hydroxyl groups and one carbonyl group and is not an acid:

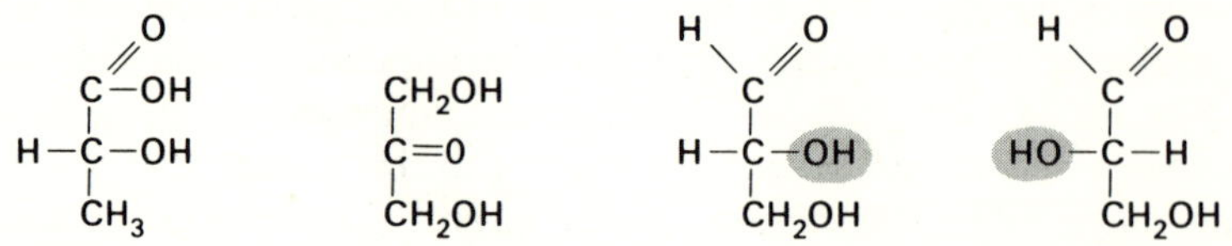

Lactic acid Dihydroxyacetone Glyceraldehyde

D-Form L-Form

Such differences are rather obvious. For such compounds to have identical molecular formulas appears to be merely an accident. The difference is smaller between dihydroxyacetone and glyceraldehyde (second and third formula, respectively); the carbonyl and the hydroxyl groups have simply been exchanged. In the case of the D- and L-forms of glyceraldehyde, only the hydroxyl group and the H atom on the middle carbon have changed places, which results in mirror-image isomerism (see below).

Positional Isomerisms. When one or more groups assume different positions on a carbon skeleton, as with dihydroxyacetone and glyceraldehyde in the above series of formulas, positional isomerism occurs. Further examples are methylmalonic acid and succinic acids as well as citric acid and isocitric acid:

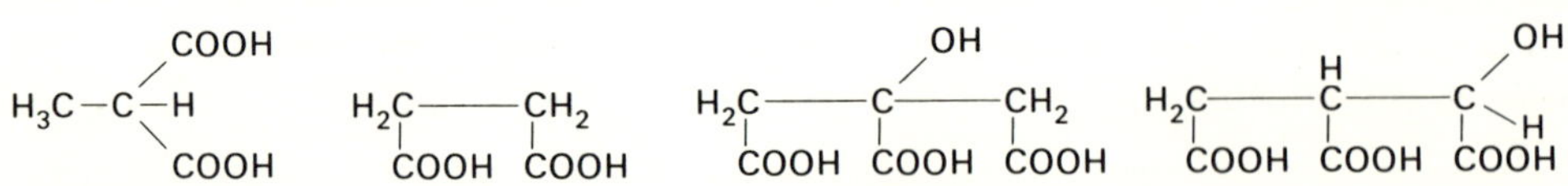

Methylmalonic acid Succinic acid Citric acid Isocitric acid

The structural differences are evident without further explanation. The properties of positional isomers generally differ more than those of homologs. A homolog is a compound whose carbon chain has been extended by one extra CH_2 group.

Geometric Isomerism, (*cis-trans*-Isomerism). To determine this type of isomerism requires that an imaginary plane be drawn through a molecule. In the four-membered ring, for example, all four C atoms lie in one plane. Two COOH groups at positions 1 and 3 may either both be on the same side of the plane (*cis*-compound) or be on opposite sides (*trans*-compound).

The drawing clearly shows the difference between the two model molecules. This difference is reflected in some of their physical properties, for example, their melting points:

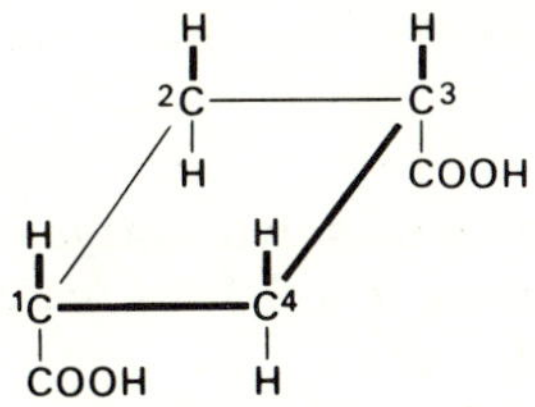

m.p. 131°
cis-

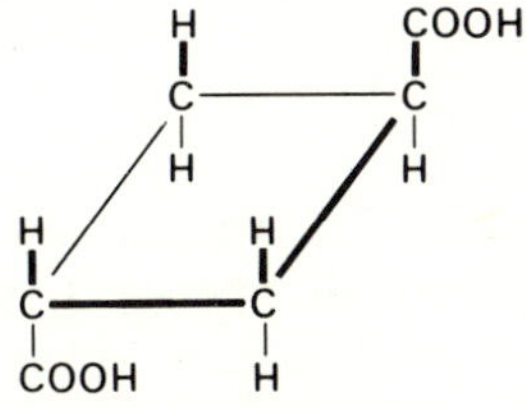

m.p. 190°
trans-

Cyclobutanedicarboxylic acid

A double bond, in a sense a two-membered ring, presents the simplest case of geometric isomerism. Here a plane may be imagined to run through the two C atoms perpendicular to the surface of the paper. A well-known pair of isomers is maleic and fumaric acids. Stable isomers of this type are not possible with single bonds because of the generally unhindered rotation around the C—C axis. The two formulas of succinic acid to the right are just two equivalent modes of presenting the same substance.

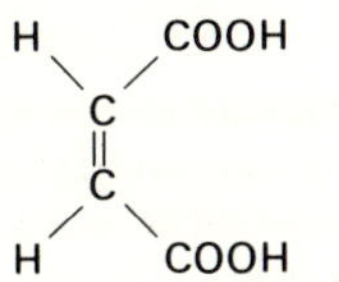

Maleic acid
(*cis*-compound)

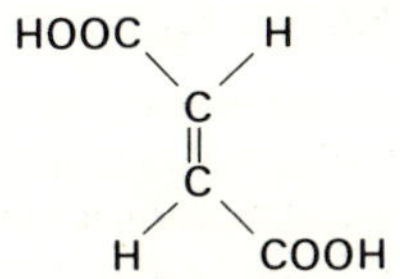

Fumaric acid
(*trans*-compound)

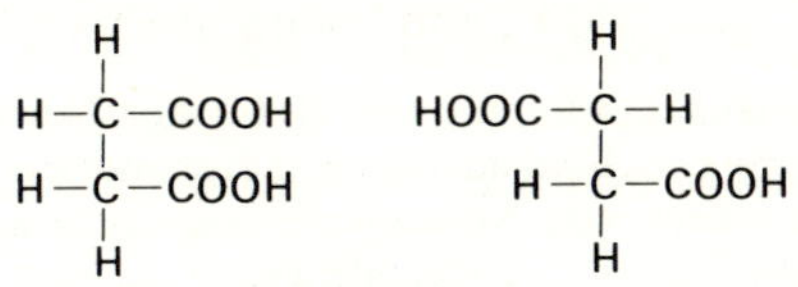

Succinic acid
(only two manners of presentation, no isomerism)

Geometric isomerism with respect to the plane of the ring also figures prominently with five- and six-membered ring compounds. In many cases it has been convenient to let the designations *cis* and *trans* refer to one particular substituent rather than to indicate relative positions of two groups. According to the conventions of steroid chemistry (cf. Chapter XIV,2), *trans* in the expression "3-*trans*-hydroxy" establishes the orientation of the hydroxyl group at C atom 3 with respect to the methyl group at C atom 10 as reference point.

Mirror-Image Isomerism (Chirality). This type of isomerism appears in asymmetric molecules, of which the simplest member is a C atom with four different substituents. The C atom thus substituted is usually—but somewhat incorrectly—called the asymmetric C atom. Two substances related by this mirror-image isomerism are called enantiomers. They possess the same chemical and physical properties (melting point, boiling point, solubility, etc.), the only difference being in the direction of optical rotation, from which the term "optical antipodes" is derived.

Solutions of such isomers rotate the plane of plane-polarized light. The direction of rotation is indicated by + (dextrorotatory) and − (levorotatory). The degree of rotation is based on a light path of 10 cm through a solution of the substance with a concentration of 1 gm per 1 ml of a specified solvent. Comprehensive tables list this "specific rotation" along with wavelength and temperature. Experimental determinations on far less concentrated solutions may be converted to specific rotation, since the rotation is proportional to concentration.

Projection Formulas. It is possible to represent mirror-image isomerism in the plane of the paper through a perspective drawing (Fig. I-1a). Clearer, however, is E. Fischer's projection formula now widely used in print.

The following conventions apply to the planar projection of a three-dimensional model. If a plane is imagined to pass through the asymmetric carbon atom, then all other carbon atoms of the carbon chain (in vertical position) must lie behind that plane and the other substituents of the asymmetric carbon atom must appear in front of the plane. The arrangement is then projected onto the plane as shown in Fig. I-1b.

D- and L-Series. The compound shown in the planar projection formula is L-serine. All amino acids of the L-series have their NH_2 groups to the left (if the —COOH is on top) which is represented by projection formulas. The mirror-image forms (D-series) have their NH_2 groups pointed to the right. The same conventions hold for sugars, whose simplest member is glyceraldehyde. The projection formulas given above at the beginning of Section 4 show both dextrorotatory D-glyceraldehyde,[3] $\alpha_D = +14°$, and levoratatory L-glyceraldehyde, $\alpha_D = -14°$.

When projection formulas are used, it should be remembered that they must not be moved out of the plane of the paper (which would result in the enantiomeric forms). Within the plane, however, they may be pushed or turned about freely. Normally the formulas are written such that the carbon atom in the highest oxidation state stands on top; D-glyceraldehyde, in that case, has its OH group pointing to the right.

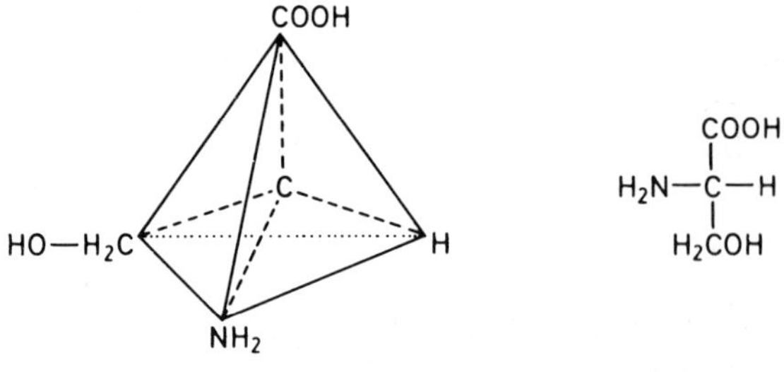

Fig. I-1a. Tetrahedral structure and projection formula of L-serine.

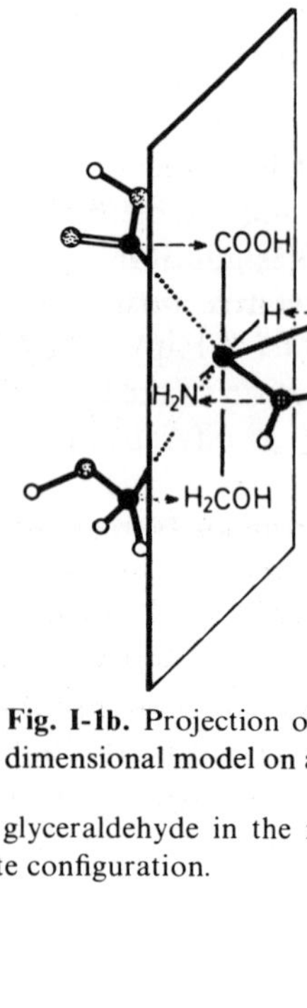

Fig. I-1b. Projection of three-dimensional model on a plane.

[3] The arbitrarily adopted convention to represent dextrorotatory glyceraldehyde in the manner indicated has now been substantiated by the determination of its absolute configuration.

Only very laborious chemical conversions and comparisons can establish whether a compound belongs to the D- or the L-series. The assignment of configuration, unfortunately, is not related to the direction of rotation of a compound, which may easily be determined experimentally. One example, D-glyceraldehyde, is dextrorotatory; oxidation affords D-(−)glyceric acid, which is levoratatory. By reduction of the CH_2OH group to the CH_3 group, levoratatory D-(−)-lactic acid is obtained, whose ester is again dextrorotatory:

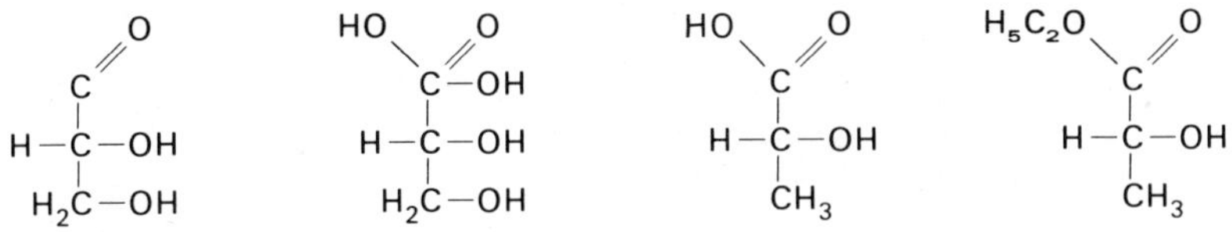

D-Glyceraldehyde — $\alpha_D = +14°$

D-Glyceric acid — $\alpha_D = -2°$

D-Lactic acid — $\alpha_D = -3.8°$

D-Lactic ethyl ester (D-ethyl lactate) — $\alpha_D = +11.5°$

During all these reactions, the configuration at the asymmetric center has remained unchanged.

During chemical synthesis inactive (symmetrical) substances yield a mixture of equal amounts of both enantiomers. A molecular complex often formed from equal parts of the L- and D-isomer is called a *racemate*. Methods are available to prepare the pure optical antipodes from racemate mixtures.

The D- and L-forms differ in their chemical reactivity only when interacting with other optically active compounds. Since all enzymes are optically active, the D- and L-forms are converted enzymatically at very different rates; in many instances, one form is converted immeasurably slowly.

The R—S System. The unequivocal assignment to either the D- or L-series is not always possible, even when the formula of a compound is known completely. A new system of nomenclature has therefore been devised[4] for the description of the configuration of individual asymmetric atoms. The new system called sequence rule or R—S system, is based directly on the tetrahedral model (Fig. I-2).

The four groups on the asymmetric center are placed in a pecking order that depends on the atomic number of the substituting atom. If the substituting atom itself does not permit ranking, then its substituents enter into the decision. For example, in serine (Fig. I-2) the central atom is surrounded by H (atomic number 1), by N (atomic number 7), and two C atoms (atomic number 6). One of the latter bears a hydroxyl group (oxygen, atomic number 8); the other is a carboxyl group (doubly bound oxygen is counted as 2×8 because of the double bond, and a singly bound oxygen as 8). The amino group consequently ranks highest, followed by the COOH group, the CH_2OH group, and finally by H. The tetrahedral model is now oriented in such a way that the substituting atom with the lowest rank (hydrogen in this example) points to the rear. The other three substituents then form a regular star in front. If the *direction from highest to second to third ranking substituent* is counterclockwise, as in the model for serine shown above, then the central atom is called "S" (from Latin *sinister* = left). If the direction is found to be clockwise, the designation is "R" (from Latin *rectus* = right). By definition, this nomenclature specifies the absolute configuration, and hence can be used only for compounds whose absolute configuration is established. The advantage is that even with complicated compounds the steric arrangement of each individual asymmetric center can be described uniquely without any further points of reference.

---

[4] R. S. Cahn, C. K. Ingold, and V. Prelog, *Experientia* **12,** 81–94 (1956); R. S. Cahn, An introduction to the sequence rule. *J. Chem. Ed.* **41,** 116–125 (1964).

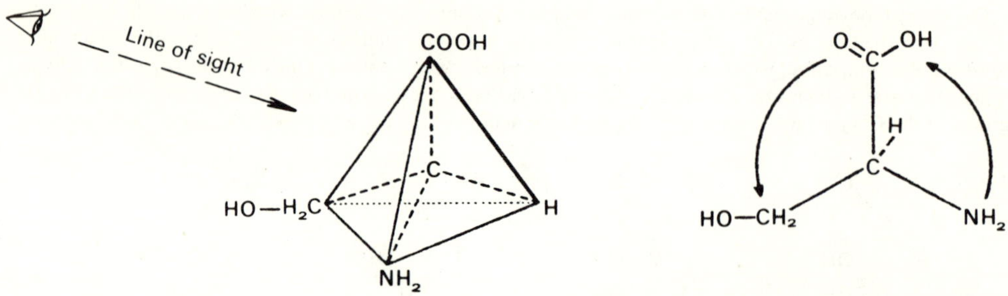

Fig. I-2. Designation of absolute configuration by the R—S sequence rule. At left the orientation of the tetrahedron; at right, the diagrammatic representation for determining the direction from highest to lowest ranking substituent. The arrow points to the left, and thus the central atom is described as "S."

If a molecule possesses two (generally, n) different asymmetric C atoms, then there must be four (generally, 2^n) different forms. The number of isomers (always a multiple of 2) is grouped into pairs of *enantiomers* (mirror-image isomers), which agree in their melting points, solubility, etc. These physical properties distinguish one enantiomeric pair from another. An illustration can be found in threonine and allothreonine.

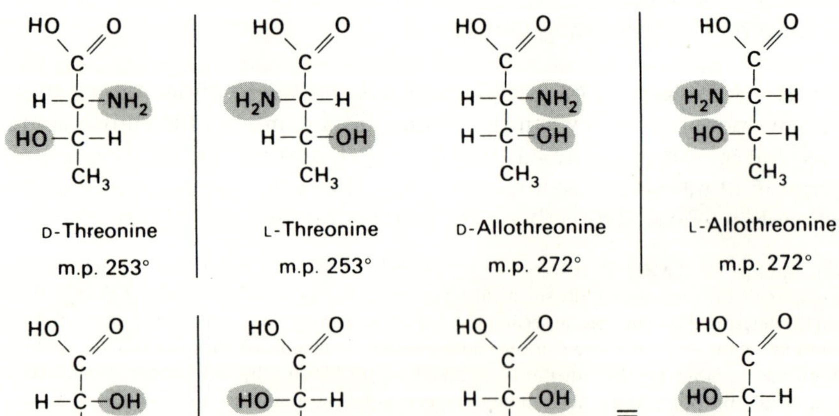

D-Threonine m.p. 253° | L-Threonine m.p. 253° | D-Allothreonine m.p. 272° | L-Allothreonine m.p. 272°

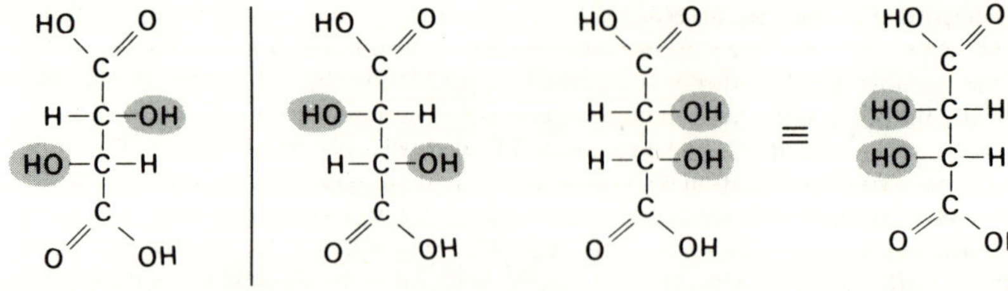

R,R(+)-Tartaric acid | S,S(−)-Tartaric acid | R,S-*meso*-Tartaric acid (identical with its mirror image)

With two identical asymmetric centers (e.g., as in tartaric acid), one of the two forms is constructed symmetrically, i.e., its mirror images can be superimposed perfectly. Such compounds are optically inactive and are called *meso*-forms.

With more complicated molecules, particularly with ring compounds, it is often difficult to predict whether a compound will be optically active. One reliable rule states that any molecular structure will exist in two optical antipodes if it cannot be superimposed completely with its mirror image. This decision can be reached most easily by means of models. In general, the various stereochemical relations can be appreciated fully only if one constructs models of the molecules, which may be achieved with rather simple means. For example, corks and pins may help to improvise models adequate for many considerations.

Geometric and mirror-image isomerism frequently appear together. Thus, 1,3-dihydroxycyclopentane exists both in a *cis*- and a *trans*-form; the *cis*-compound is a *meso*-form, i.e., possesses symmetry; (cf., also *myo*-inositol, Chapter XIII,2). The *trans*-compound, on the other hand, cannot be brought to complete superimposition with its mirror image and, hence, displays two optical antipodes:

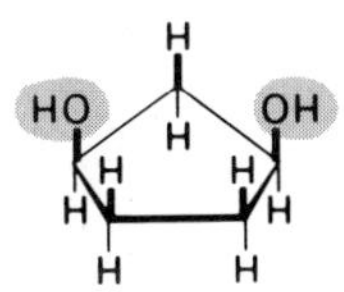

cis-1,3-Dihydroxy-cyclopentane

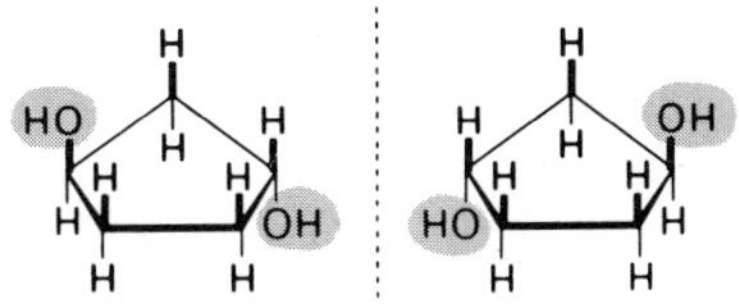

trans-1,3-Dihydroxycyclopentane

5. Reactions with Biochemical Significance

The living cell utilizes a much smaller number of different reactions than a synthetic chemist does. One very prominent type of reaction is the transfer of groups. Certain groups, e.g., an acetyl or methyl group or even an entire glucose molecule, are transferred to another molecule. The various group transfers are discussed in Chapter VI,3 (Table VI-1). Similarly, hydrogen may be transferred, very often reversibly.

The following reactions are of great importance in the organism:

1. *Formation and Cleavage of Macromolecules.* Cleavage is mostly hydrolytic (occasionally phosphorolytic). Synthesis, however, employs group transfer.
2. *Dehydrogenation and Hydrogenation.* Either of special redox systems (see respiratory chain, Chapter X,4), where, in effect, oxygen is reduced (hydrogenated) to water, or of any intermediate product of metabolism:

   a. Alcohols to carbonyl groups (reversible)
   b. Saturated to unsaturated compounds (reversible)
   c. Aldehydes to carboxylic acids (not readily reversible)
   d. Amines to imines, which in turn are hydrolyzed to carbonyl compounds (reversible only over detours)

3. *Formation and Cleavage of C—C Bonds*

   a. By decarboxylation of keto acids or, in reverse, by carboxylation of ketones and acids (to malonic acid derivatives)

b. By condensation of activated carboxylic acids to β-keto esters and its reverse, according to the principle of ester condensation
c. By reactions analogous to aldol condensations (Chapter XV,5) or condensations of ketones with activated CH_2 groups (frequently reversible)
d. By addition to double bonds (establishing a C—C bond)

4. *Loss of Water* (or ammonia) leaving a double bond and *Addition of Water* to a double bond.

The number of reaction types in cells evidently is small. Specificity, on the other hand, is extraordinarily great. From the multitude of potential reactants present in the cell, usually one particular substance is chosen and its specific reaction guided in a definite way. Maybe one of the most striking examples is the direct introduction of oxygen into the steroid molecule by an enzyme which permits the insertion of a hydroxyl group at one and only one position and with a certain steric orientation. Many more examples of specific reactions will be seen later.

BIBLIOGRAPHY

E. E. Bittar, "Cell pH." Butterworths, Washington, D.C., 1964.
R. Breslow, "Organic Reaction Mechanisms." 2nd ed. Benjamin, New York, 1969; available as paperback.
T. C. Bruice and S. J. Benkovic, "Bioorganic Mechanisms," 2 vols. Benjamin, New York, 1966.
P. Sykes, "A Guidebook to Mechanism in Organic Chemistry," 3rd ed. Wiley, New York, 1970.

CHAPTER II

Amino Acids

Proteins are essential components of all living cells. During hydrolysis proteins break down to amino acids, whereas in the organism proteins are built up from amino acids. The significance of amino acids, however, is not limited to their being components of proteins. Amino acids undergo many metabolic reactions and supply precursors for other endogenous substances (e.g., hemoglobin of blood).

This chapter will describe chemical properties of amino acids, chiefly of those contained in proteins. The metabolism of amino acids is discussed in Chapter VIII along with that of proteins.

1. Chemical Properties

Their name indicates that amino acids possess two characteristic functional groups: the amino group, $-NH_2$, and the carboxyl group, $-COOH$. All amino acids that occur as natural components of proteins have their amino group in α-position to the carboxyl group. A list of the 20 amino acids found regularly in proteins is presented in Table II-1. They have trivial names that are abbreviated to the first three letters, with a few exceptions.

The general formula of all amino acids is:

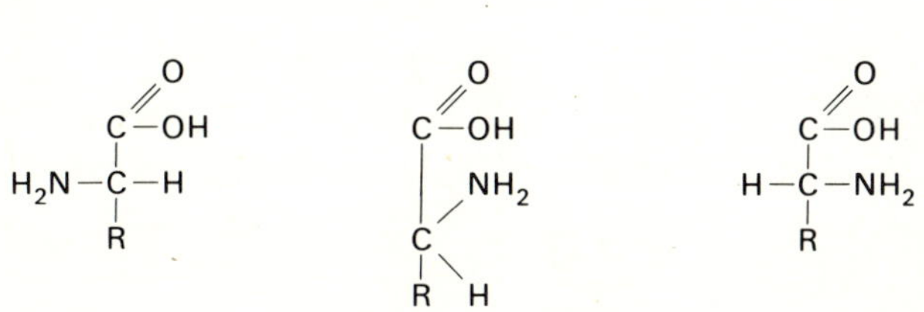

L-Amino acid General formula D-Amino acid

TABLE II-1

Important Amino Acids

(I)

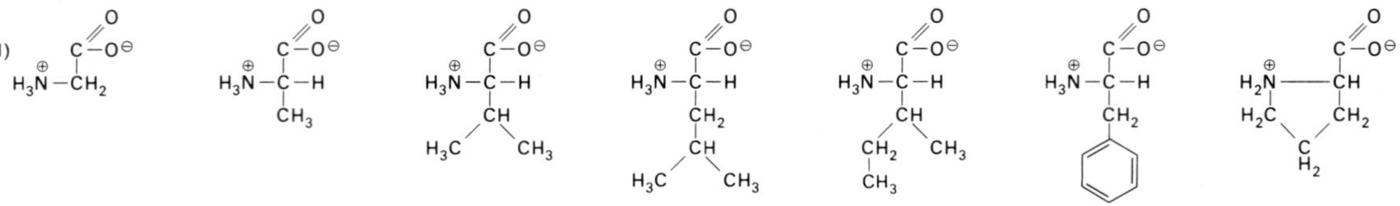

Glycine (Gly) | L-Alanine (Ala) | L-Valine (Val) | L-Leucine (Leu) | L-Isoleucine (Ile) | L-Phenylalanine (Phe) | L-Proline (Pro)

(II)

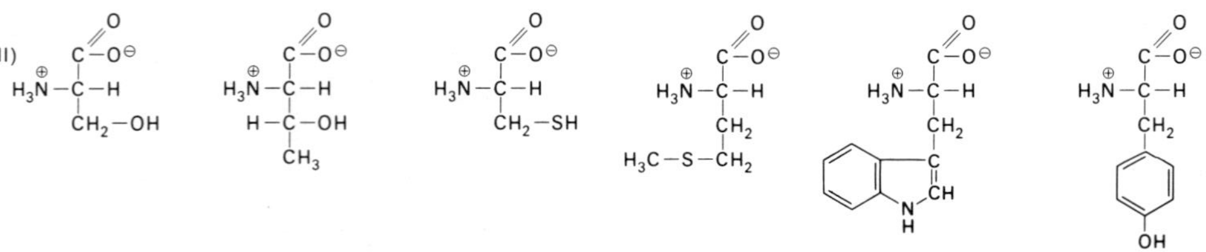

L-Serine (Ser) | L-Threonine (Thr) | L-Cysteine (Cys) | L-Methionine (Met) | L-Tryptophan (Trp) | L-Tyrosine (Tyr)

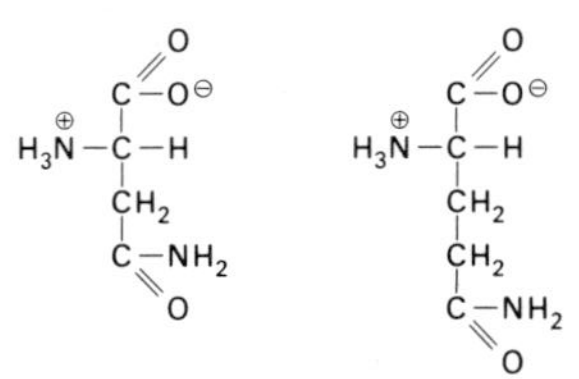

L-Asparagine (Asn or Asp NH$_2$) | L-Glutamine (Gln or Glu·NH$_2$)

(III)

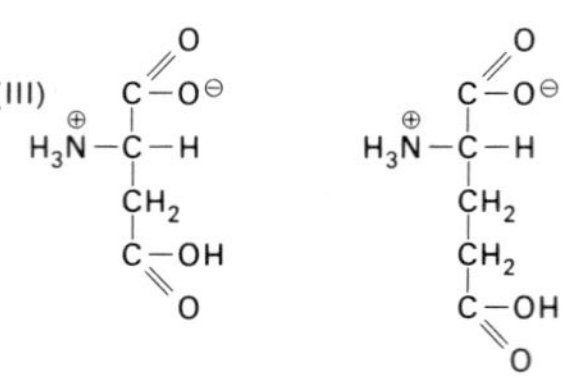

L-Aspartate (Asp) | L-Glutamate (Glu)

(IV)

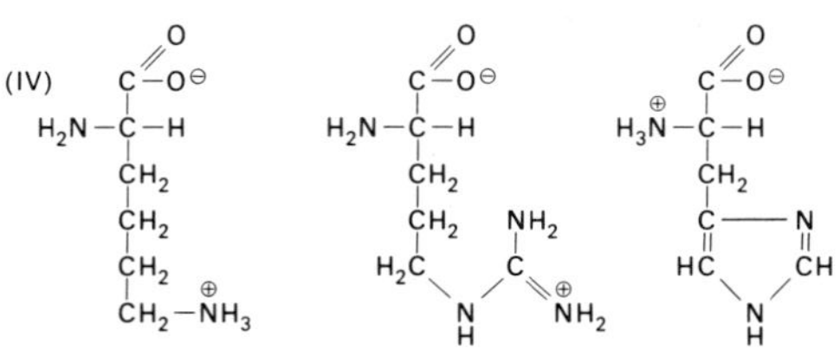

L-Lysine (Lys) | L-Arginine (Arg) | L-Histidine (His)

What distinguishes the individual amino acids is the residue R, an aliphatic or aromatic carbon skeleton which may itself carry further functional groups. Glycine, $H_2N—CH_2—COOH$, is the only amino acid in which the α-carbon substituent R is a hydrogen atom. In addition, the nitrogen can be substituted or be part of a ring structure.

In all cases of the general formula, except when R = H, the α-carbon is substituted asymmetrically, resulting in optical activity, since the model cannot be superimposed on its mirror image (cf. Chapter I,4). We recognize two steric configurations, the D- and the L-series. The amino acids appearing in proteins are members of the L-series which means their α-carbons are substituted with the same orientation as those in L-alanine and L-serine (R = CH_3 and CH_2OH, respectively). The direction of optical rotation, however, is independent of the configuration.

Zwitterions. The carboxyl group, being acidic, can lose a $H^{\oplus}$ by dissociation. The basic amino group can pick up a proton, in analogy to ammonia: $NH_3 + H^{\oplus} \longrightarrow NH_4^{\oplus}$. If both groups are ionized, the result is the so-called zwitterion (German, "ion of both kinds"; amphoteric ion) shown in Formula I. Amino acids in aqueous solution are in zwitterionic form:

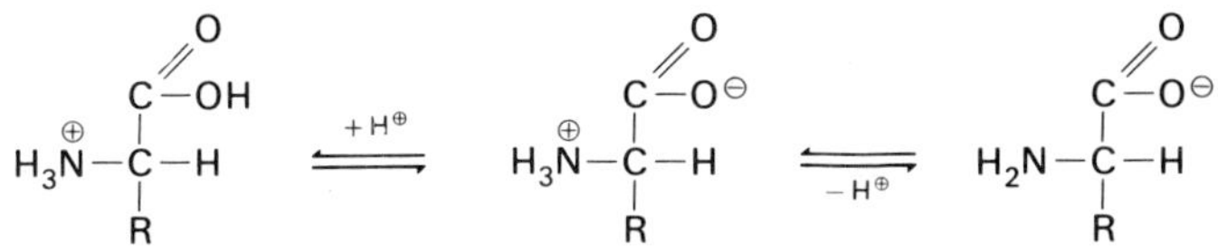

(II) (I) (III)

The addition of $H^{\oplus}$ ions (acid) will neutralize the negatively charged groups, as shown in Formula II. Addition of $OH^{\ominus}$ ions (base) will remove protons attached to the amino groups (cf. Formula III) and form water. A single amino acid molecule in aqueous solution can exist only in one of these three forms (the uncharged variation is never possible). A solution of an amino acid with many quintillions (10^{18}) of molecules adjusts itself to a mixture of Formulas I and II or I and III with one form predominating. The dissociation constants of the carboxyl and the amino group and the pH of the solution, i.e., the $H^{\oplus}$ ion concentration, determine which form will predominate.

The dissociation of the acidic and basic groups of amino acids is governed by the law of mass action, as with all weak acids and bases. To each group is ascribed a dissociation constant, K, whose negative logarithm is called the pK value (cf. Chapter I,2, Dissociation of Carboxylic Acids). The pK of the carboxyl group is usually around 2.2; evidently in amino acids the carboxyl group is dissociated to a much greater extent than in acetic acid (pK = 4.65). The amino group, which may also be considered to be an acid according to the following equation, has a pK ranging between 9.0 and 9.8:

$$R—NH_3^{\oplus} = R—NH_2 + H^{\oplus}$$

It is often valuable to express this dissociation in the form of a titration curve where pH is plotted as a function of the amount of added acid ($H^{\oplus}$ ions) and base. With the addition of 0.5 mole the pK is attained. Figure II-1 shows the titration curve of histidine which with one acidic and two basic groups has three pK values. In the vicinity of its pK value an amino acid behaves as a buffer (cf. Chapter I,2).

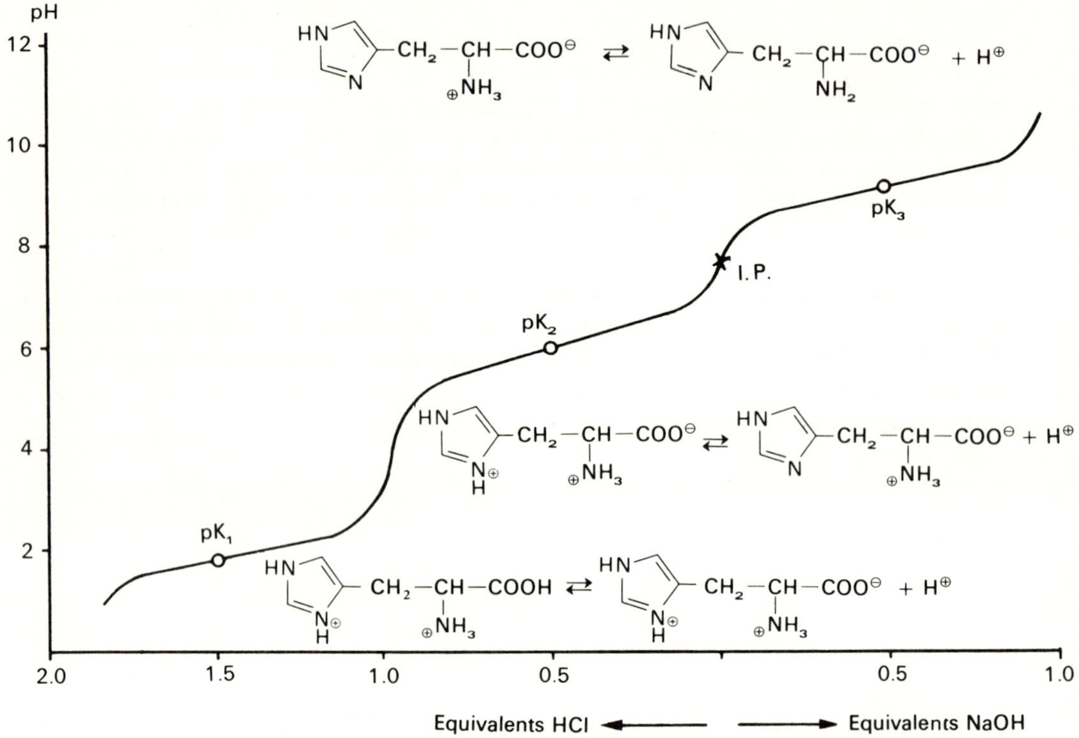

Fig. II-1. Titration curve of histidine. I.P. = isoelectric point.

At a certain pH, an amino acid will be present almost exclusively in its zwitterionic form (Formula I) with little, but exactly equal, proportions of Formulas II and III. This particular pH value is called the *isoelectric point* (I.P.), and it can be calculated from the acidic and basic groups:

$$\frac{pK_1 + pK_2}{2} = \text{I.P.}$$

The amino acid appears to be without charge, although each molecule carries one positive and one negative charge.

General Reactions. The usual derivatives can be prepared from the two functional groups of amino acids (cf. Chapter I,2).

Salt formation of the carboxyl group has already been discussed in principle; formula III (of the preceding page) depicts the anion of a salt.

Of biochemical significance is the formation of amides. Ammonia affords unsubstituted amides,

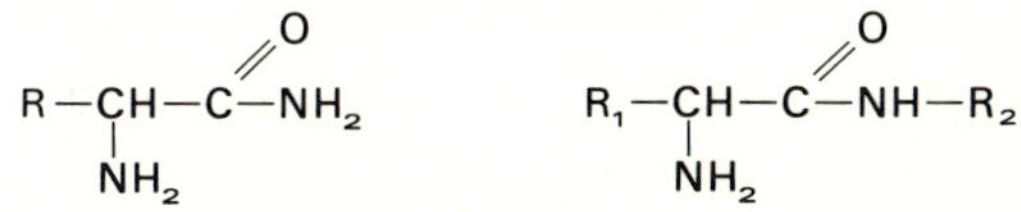

while amines produce substituted amides, better known as peptides (cf. Chapter III).

Salt formation, again, is also one of the reactions of the amino group. As with ammonia, it consists of an addition of a proton from another acid:

$$R{-}NH_2 + H^{\oplus} \rightleftharpoons R{-}NH_3^{\oplus}$$

Another important reaction is acylation, which yields again acid amides. The acetyl and benzoyl derivatives of glycine (the latter is called hippuric acid) occur naturally, in addition to a few other compounds of similar composition. If the acylating agent is another amino acid, the result is again a peptide.

Amino groups may also be alkylated, e.g., with methyl groups. Methyl derivatives of glycine play a physiological role.

One important color reaction of amino acids is the reaction with *ninhydrin*. It is the most widely used method for detecting and determining amino acids and follows the scheme shown here:

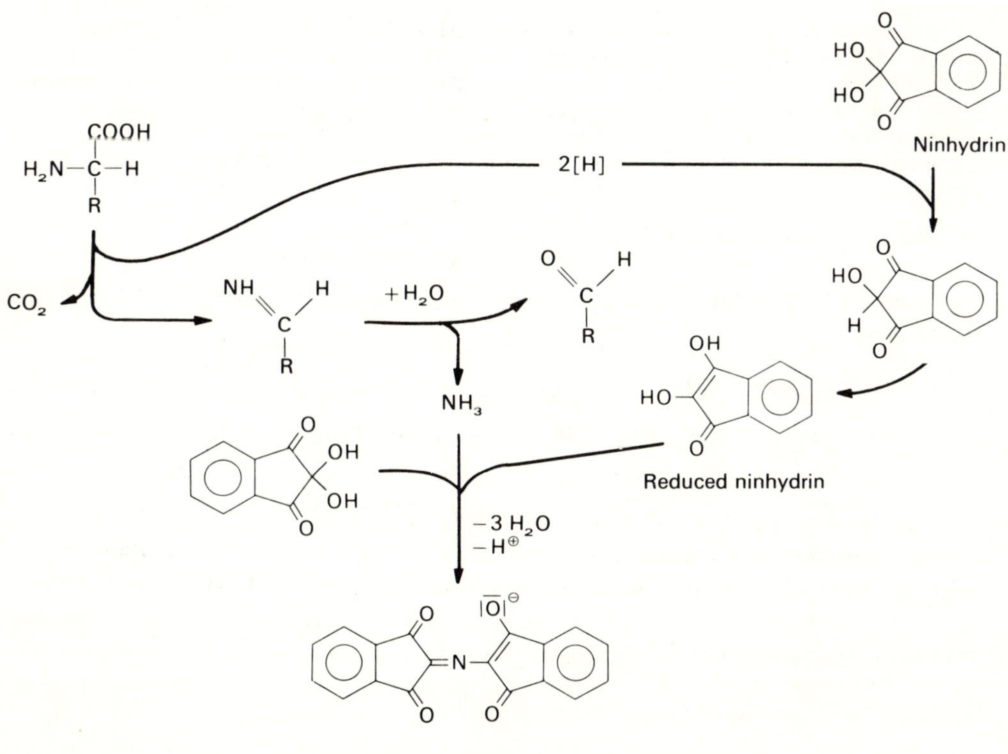

The amino acid is degraded to the next lower aldehyde and CO_2. The reagent combines with the liberated ammonia and produces a blue color.

2. Individual Amino Acids

Twenty different amino acids regularly appear in proteins. As is evident from the general formula (cf. Section 1), they differ only in the substituent R. The formulas of these amino acids are included in Table II-1 above. One system of classification groups the amino acids into five divisions:

I. Amino acids with a nonpolar R, i.e., with an unsubstituted hydrocarbon side chain: glycine, alanine, valine, leucine, isoleucine, phenylalanine, and proline.

II. Amino acids with nonionized,[1] but polar, substituents, such as —OH, —SH, —CO · NH_2, and a few heterocyclicones: tyrosine, tryptophan, serine, threonine, cysteine, cystine, and methionine. To this group belong also asparagine and glutamine, the acid amides of the acidic amino acids (Group III). The acidic properties of the carboxyl group are lost through amide formation.

III. Acidic amino acids, or monoaminodicarboxylic acids, containing an additional carboxyl group: glutamic acid and aspartic acid.

IV. Basic amino acids, or diaminomonocarboxylic acids, containing an additional amino group: lysine, arginine, and histidine.

V. Less common amino acids.

This classification emphasizes the significance of the amino acid side chains for the three-dimensional structure of proteins. The side chains of Group I enter hydrophobic bonds (see Chapter IV,3), those of Group II form hydrogen bonds between side chains, and the amino acids of Groups III and IV contribute the electrical charges determining the electrophoretic mobility of the protein molecule.

The organism of man and most animals cannot synthesize every amino. Some must be supplied in the diet, and, consequently, they are known as essential amino acids (see the list in Chapter VIII,6).

Group I. Nonpolar Side Chains.

Glycine (aminoacetate). The simplest amino acid and the only one lacking an asymmetric atom.

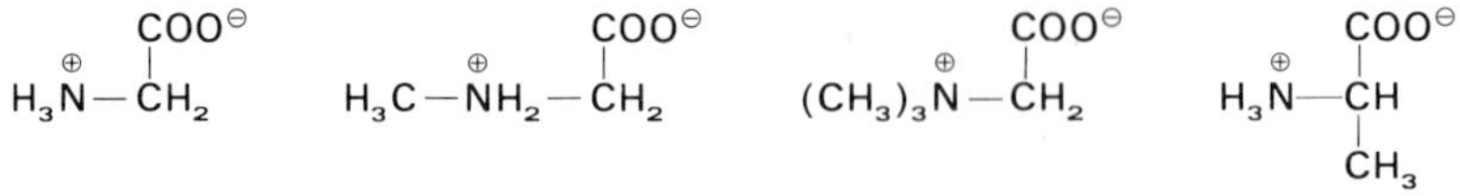

Glycine Sarcosine Betaine L-Alanine

Glycine is present abundantly in scleroproteins (fibrous protein; cf. Chapter IV,3). Methylation yields sarcosine, which occurs in several peptides. Further methylation

---

[1] In a more alkaline environment the SH group of cysteine and the phenolic OH group of tyrosine can dissociate their proton.

yields betaine. This substance exists exclusively in the zwitterionic or salt form (zwitterionic formulas sometimes are called betaine structures).

The glycine conjugates acetylglycine and benzoylglycine (hippuric acid) have already been mentioned. Bile acids form similar derivatives (conjugates) with glycine (cf. Chapter XIV,5).

Alanine (α-aminopropionate). All amino acids, except glycine, may be regarded as derivatives of alanine; by replacement of one or two H atoms of the methyl group, all the other amino acids can be constructed.

Valine (α-aminoisovalerate), ***leucine*** (α-aminoisocaproate), and ***isoleucine*** (α-amino-β-methylvalerate). Each possesses a branched carbon chain. Chemically they are quite alike. Leucine and isoleucine are particularly difficult to separate. Most animals cannot synthetize the branched carbon chain. These amino acids, therefore, are indispensable or essential in the diet (cf. list in Chapter VIII,6).

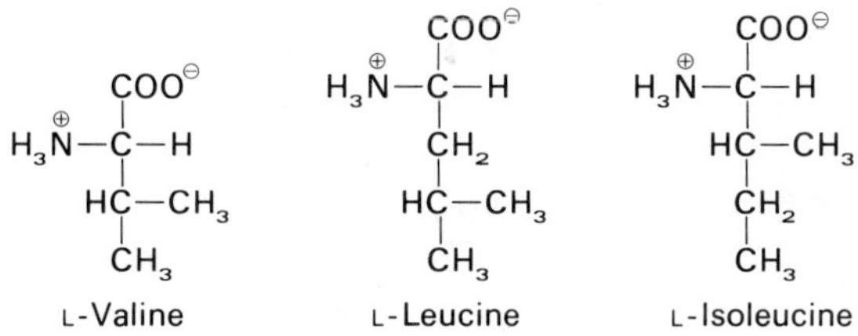

Proline (2-pyrrolidinecarboxylate). Frequently found in protein hydrolyzates, proline is a cyclic amino acid. The amino group in α-position has become part of the ring and consequently is a secondary amino group instead of being the usual primary. The same is true for hydroxyproline, which occurs only in structural proteins. Collagen is rich in both proline and hydroxyproline. These secondary amino acids occasionally are called "imino acids" after some archaic chemical nomenclature. Compounds containing the group C—NH—C and formerly called "imines" must not be confused with the true imines which contain the imino group, $R_1R_2C{=}NH$.

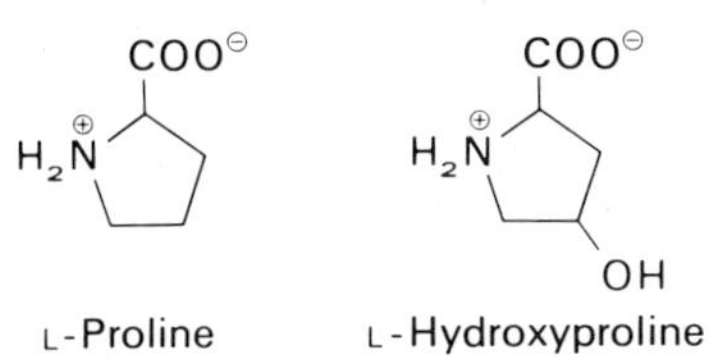

Phenylalanine (α-amino-β-phenylpropionate). Because of its aromatic ring, phenylalanine cannot be formed in the animal organism. It is closely related to tyrosine, which belongs to Group II with its phenolic —OH group.

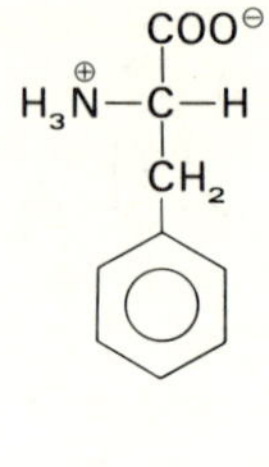

L-Phenylalanine

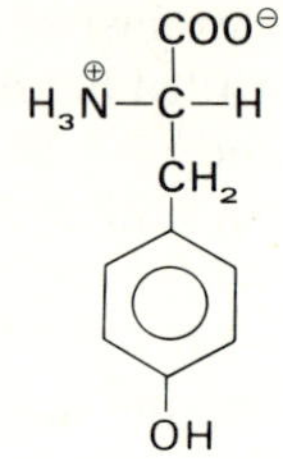

L-Tyrosine

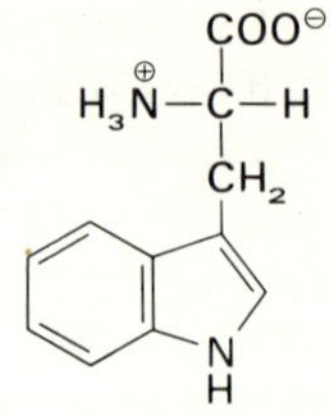

L-Tryptophan

Group II. Nonionized Polar Side Chains.

Tyrosine [α-amino-β-(*p*-hydroxyphenyl)propionate]. The phenolic group of tyrosine is weakly acidic and loses its proton above pH 9 (pK = 10.1).

Tryptophan (α-amino-β-3-indolepropionate). A heterocyclic amino acid, tryptophan is a derivative of indole (with one N atoms).

The three amino acids phenylalanine, tyrosine, and tryptophan may be grouped together on the basis of their aromatic side chains.

Serine (α-amino-β-hydroxypropionate). This amino acid contains an alcoholic hydroxyl group that may engage in the usual reactions, such as ester formation. Esters with phosphoric acid have physiological significance as components of nucleotides and some proteins. The unesterified OH group of the serine residue possesses special functions in several enzymes.

Threonine (α-amino-β-hydroxybutyrate). The next higher homolog to serine is threonine. Its name points to the relationship with the sugar threose. Threonine has two asymmetric centers (cf. Chapter I,4) manifested by four stereoisomeric forms, of which two are always mirror images of the other two, namely, D- and L-threonine and D- and L-allothreonine. Threonine was discovered in an interesting manner: it was identified as a necessary supplementary factor of a synthetic diet mixture (Rose, 1935) and thus became the first amino acid to be recognized as indispensable.

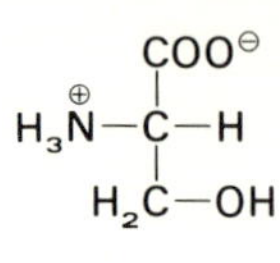

L-Serine

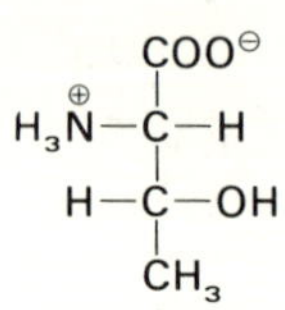

L-Threonine

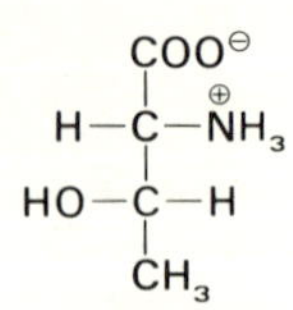

D-Threonine

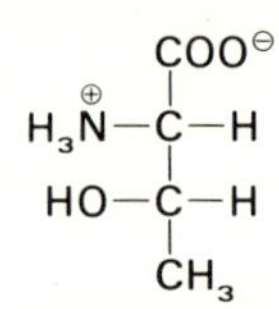

L-Allothreonine

Cysteine (α-amino-β-mercaptopropionate). This amino acids contains sulfur. The SH group is quite reactive and readily dehydrogenated (even while part of a peptide chain). When it is dehydrogenated (i.e., oxidized), two molecules will join to form the amino acid cystine (compare with reaction of glutathione; cf. Chapter III,3).

Cystine [di-(α-amino-β-mercaptopropionate)]. Being a diaminodicarboxylic acid, cystine may be part of two different peptide chains at the same time, or be part of the same chain at widely separated places; such S—S linkages occur in many proteins.

Methionine (α-amino-γ-methylmercaptobutyrate). Methionine may be looked upon as a methylation product of homocysteine and aside from its role as a protein constituent and as an essential amino acid, is also important as a donor of active methyl groups.

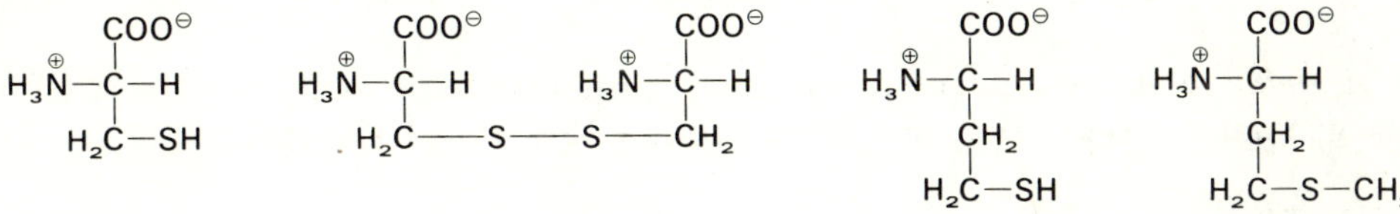

L-Cysteine L-Cystine L-Homocysteine L-Methionine

Asparagine and Glutamine. The amides of the "acidic amino acids," are grouped here. The amide group bestows hydrophilic properties to these two amino acids. Acid or alkaline hydrolysis of proteins also splits amide groups, and ammonia and aspartate or glutamate are released.

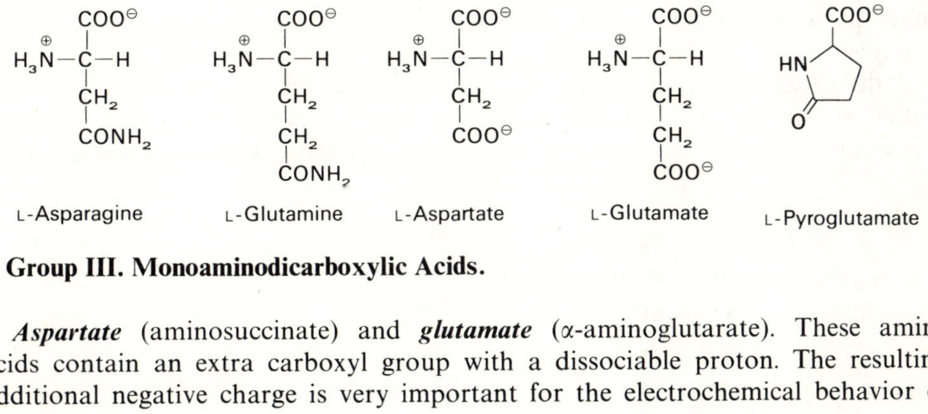

L-Asparagine L-Glutamine L-Aspartate L-Glutamate L-Pyroglutamate

Group III. Monoaminodicarboxylic Acids.

Aspartate (aminosuccinate) and ***glutamate*** (α-aminoglutarate). These amino acids contain an extra carboxyl group with a dissociable proton. The resulting additional negative charge is very important for the electrochemical behavior of proteins.

Intramolecular amide formation between the amino and the γ-carboxyl groups of glutamate results in pyrrolidone α-carboxylate or pyroglutamate, a ring compound which in some peptide chains constitutes the N-terminus.

Group IV. Diaminomonocarboxylic Acids.

Lysine (α,ε-diaminocaproate). Lysine is distinguished by the extra amino group in its side chain which imparts basic properties to it. Depending on the pH of the solution, lysine may bear various charges. Collagen also contains *hydroxylysine.*

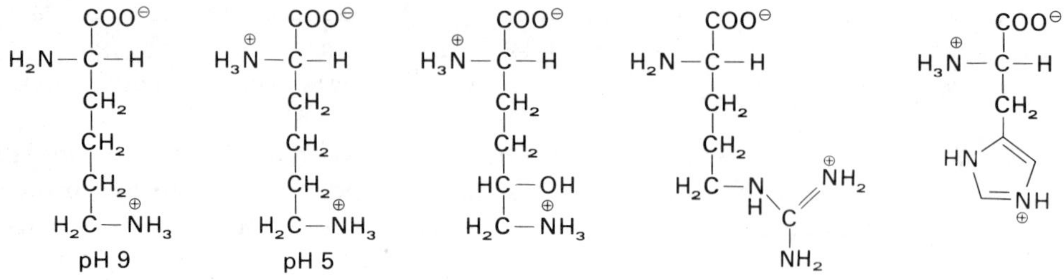

L-Lysine L-Hydroxylysine L-Arginine L-Histidine

Arginine (α-amino-δ-guanidinovalerate). Another basic amino acid because of the guanidino group, arginine is more strongly basic than lysine ($pK_3 = 12.5$).

Histidine (α-amino-β-imidazolylpropionate). The last entry in this list, histidine contains the weakly basic imidazole system ($pK = 6.1$). In many enzyme proteins it functions as a proton donor or acceptor.

Group V. Less Common Amino Acids. Besides the 20 amino acids regularly present in proteins, several others occur in natural material. Some are found free in plant or animal tissue, some are bound to simple peptides and more rarely to special proteins, while others function as intermediates in metabolism. From over a hundred different amino acids so far isolated from plants and microorganisms, we have selected a few examples.

Both *hydroxyproline* and *hydroxylysine* mentioned above have been found only in protein of connective tissue. In protein from corn, *α-aminoadipate* (one CH_2 group more than glutamate) was detected, and in bacterial protein, *α,ε-diaminopimelate*:

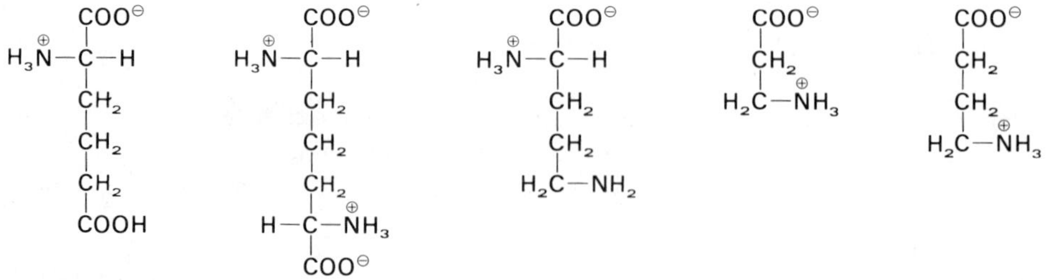

L-α-Amino-adipate L,L-α,ε-Diamino-pimelate L-Ornithine β-Alanine γ-Amino-butyrate

There are some amino acids that occur in several peptides, but not in proteins. They arise by ordinary metabolic processes from the common amino acids. Examples are *ornithine* (from arginine, cf. Chapter VIII,8); *β-alanine* (from aspartate, cf. Chapter VIII,5), a component among other things, of the vitamin pantothenic acid (cf. Chapter VI,7); and last, *γ-aminobutyrate* found in the brain in free form.

Sulfur-containing amino acids related to cysteine and methionine are *cysteic acid* and its decarboxylation product *taurine*. *Lanthionine*, the thioester of two molecules of cysteine, has been isolated from wool hydrolyzates. *Homocysteine* is the demethylation product of methionine (formula above under Group II).

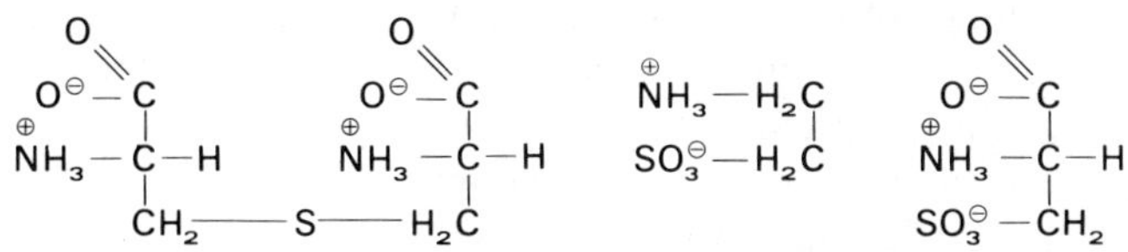

L-Lanthionine Taurine L-Cysteic acid

Two derivatives of serine produced by streptomyces have become recognized as antibiotics: *azaserine* with an unusual diazoacetyl group and cycloserine. Further, *chloramphenicol* (Chloromycetin) should be mentioned, because it is structurally similar to *phenylserine* and is employed therapeutically as an antibiotic. With the last few examples, it should be noted that apparently microorganisms readily synthesize functional groups that seem entirely "unphysiological" such as nitro, hydroxylamino, dichloroacetyl, and diazoacetyl groups.

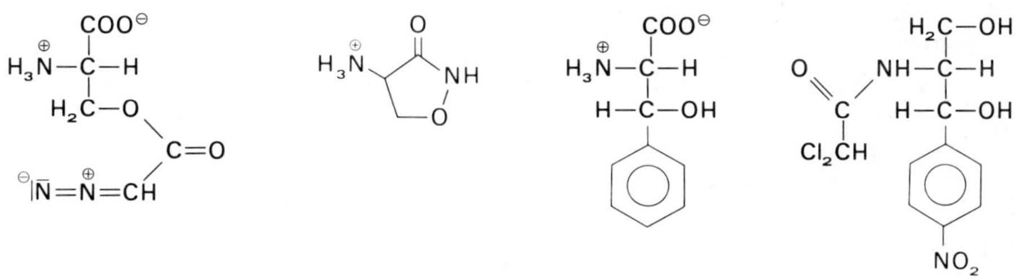

Azaserine Cycloserine L-Phenylserine L-Chloramphenicol

3. Separation of Amino Acids

The components of proteins may be freed by hydrolysis. The boiling of protein in 6 *N* hydrochloric acid for some time results in a mixture of amino acids. The first step in the analysis of proteins is to determine the content of amino acids both qualitatively and quantitatively. This approach has met with great difficulty, since amino acids are chemically rather similar and often have identical functional groups. Progress has been held up for decades because of these analytical problems. Two methods have been developed since 1940, with outstanding results, namely, paper chromatography[2] and ion-exchange chromatography.

Paper Chromatography. Its primary purpose is identification of substances, as very small amounts (5 μg and less) may be detected. The technique is simple. The

[2] Chromatography was defined by Tswett (1906) as a procedure for separating dyes based on repeated adsorption and desorption in successive layers of some absorption medium. Later the term was applied to all techniques of separation in which a selective process is repeated a great number of times on carrier material. Minute differences in properties are thus magnified for effective separation.

material is applied to filter paper and allowed to come into equilibrium with the vapors of a solvent. The paper is then dipped into the organic solvent system, which is drawn up by capillary action and carries the mixture of substances with it part of the way. Easily soluble substances are carried farther than less soluble ones. Substances are separated according to their relative solubilities in water (bound to the cellulose of paper) and the organic solvent (advancing by capillary action). This separation of substances according to their "partition coefficients" between two immiscible solvents is generally called *partition chromatography*. The distance traveled by a substance relative to the distance the solvent rises is called the R_f value (cf. Fig. II-2).

With minimal technical effort, paper chromatography provides excellent separation not only of amino acids, but also of many other substances, as long as the correct solvent system is chosen. Particularly, its use for qualitative fractionation and identification has become valuable. Monographs should be consulted for the various applications and modifications of this method.

Thin-Layer Chromatography. Traditionally this method involves a thin layer (usually $\frac{1}{4}$ mm thick) of adsorbent material spread on a glass plate. Silica gel, aluminum oxide, cellulose powder, etc., have been used as adsorbents. The procedure and apparatus are similar to those used in paper chromatography. The chief advantages are greater speed and wider choice of adsorbents. As originally developed for the separation of lipids on silica gel, it is a typical example of so-called *adsorption*

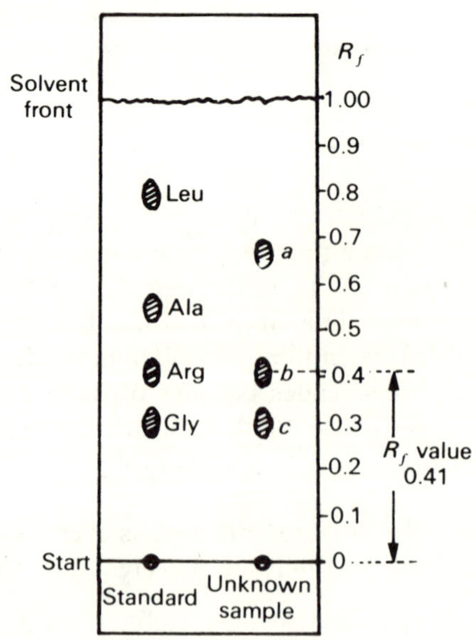

Fig. II-2. Separation of amino acids by paper chromatography. Substance *a* is unknown; *b* is identical with arginine, and *c* with glycine.

chromatography, where substances adhering directly to the inert adsorbent material are eluted, and consequently moved forward, to different degrees by the organic solvent.

Ion-Exchange Chromatography. For the quantitative separation and determination of amino acids from hydrolyzates, the preferred method is now ion-exchange chromatography. Ion-exchange resins are synthetic polymers with many acidic or

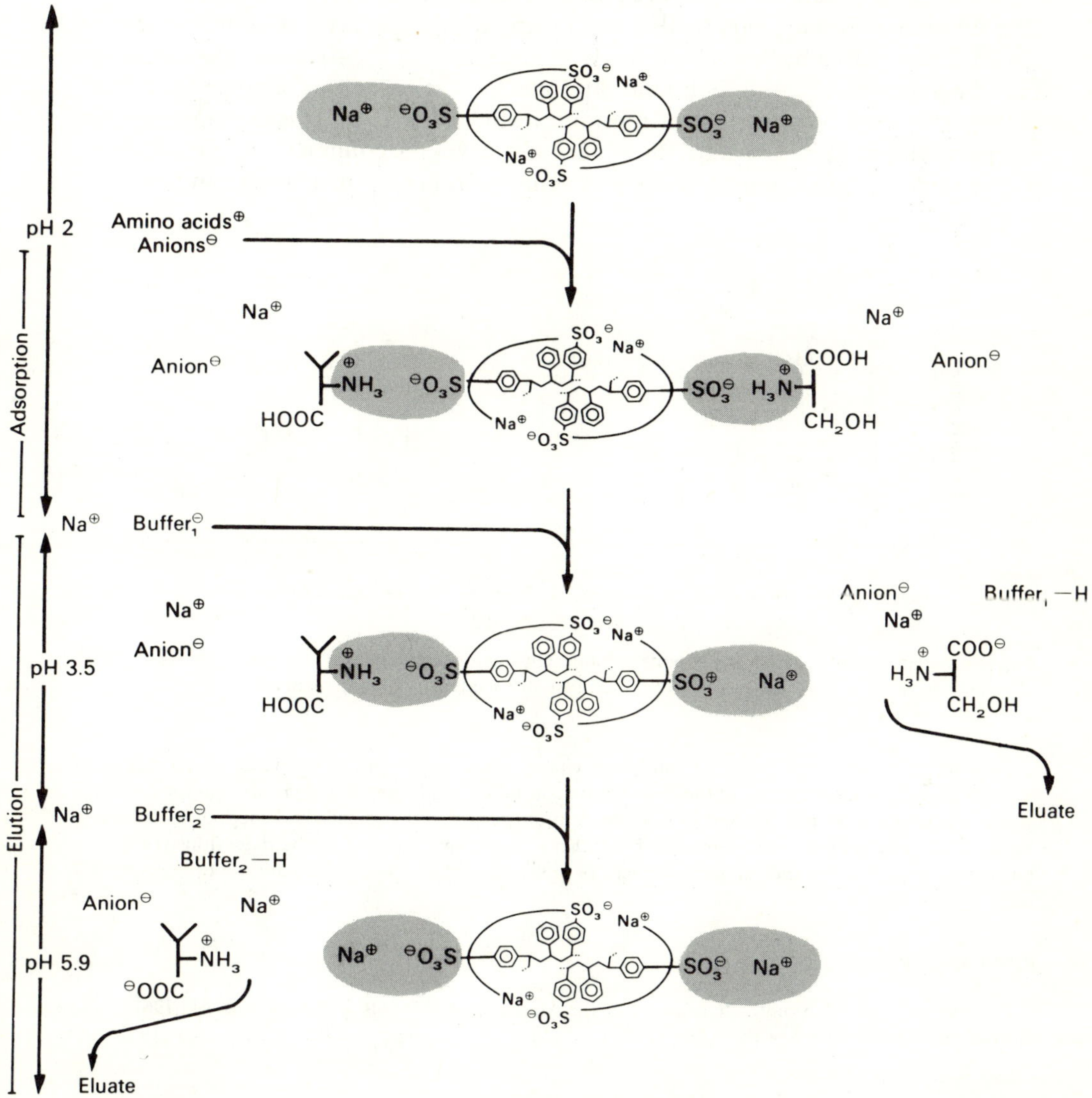

Fig. II-3. Mechanism of ion-exchange chromatography of amino acids.

basic groups. Each of the acidic groups (—SO_3H or —COOH) can dissociate one proton which, however, for electrostatic reasons, remains in the vicinity. The proton may move away only when another positively charged ion (cation) takes its place. This is the mechanism of ion exchange. For example, $Na^{\oplus}$ or $NH_4^{\oplus}$ ions in solution are loosely bound to the resin and $H^{\oplus}$ ions are released into solution in exchange until equilibrium concentrations are reestablished. Analogously, the $OH^{\ominus}$ of resin —$NH_3^{\oplus}OH^{\ominus}$ may exchange for $Cl^{\ominus}$ or R—$COO^{\ominus}$.

In practice, a "column" is prepared by filling a glass tube with resin in its $Na^{\oplus}$ form. The solution of amino acids is allowed to seep into the top layer of resin and then is eluted slowly with a buffer to hold the $H^{\oplus}$ concentration approximately constant. At first the amino acid cation is bound, but later it is released in exchange for a $Na^{\oplus}$ of the buffer when it converts to zwitterionic form (cf. Fig. II-3). This exchange process is repeated layer after layer until the amino acid leaves the column in the eluate, where it may be determined quantitatively by the ninhydrin reaction. Curves can be drawn as shown in Fig. II-4.

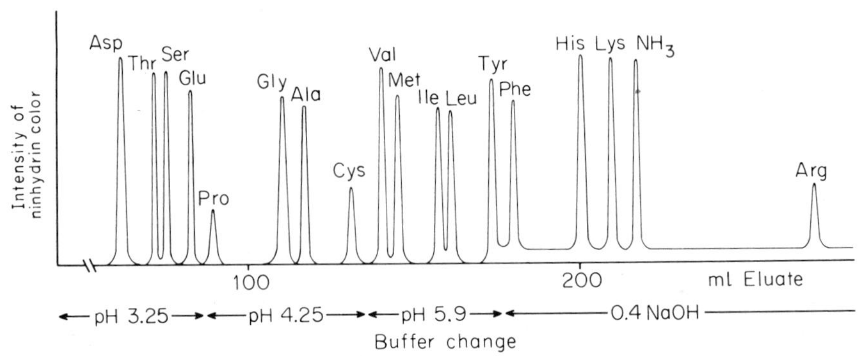

Fig. II-4. Separation of amino acids by ion-exchange chromatography (partial diagram). Abscissa, volume of eluant; ordinate, concentration of amino acid in eluate (= intensity of ninhydrin color).

To what extent an amino acid is bound by exchange resins is, of course, a function of the degree of ionization of the amino acid, i.e., of the dissociation constant K of the individual groups and the pH of the eluant. In other words, this method of fractionation utilizes the rather small differences of ionization of amino or carboxyl groups. Fully automatic but rather intricate apparatuses have been developed that carry out the chromatography and the measurement of ninhydrin color in the eluates and plot the result as a curve ("amino acid analyzer").

BIBLIOGRAPHY

J. P. Greenstein and M. Winitz, "Chemistry of the Amino Acids," 3 vols. Wiley, New York, 1961.
A. Meister, "Biochemistry of the Amino Acids," 2nd ed., 2 vols. Academic Press, New York, 1965.

CHAPTER III

Peptides

1. Structure and Nomenclature

Chemically peptides are acid amides. According to the definition in Chapter I,3, they are polymeric compounds that can be decomposed by hydrolysis to amino acids:

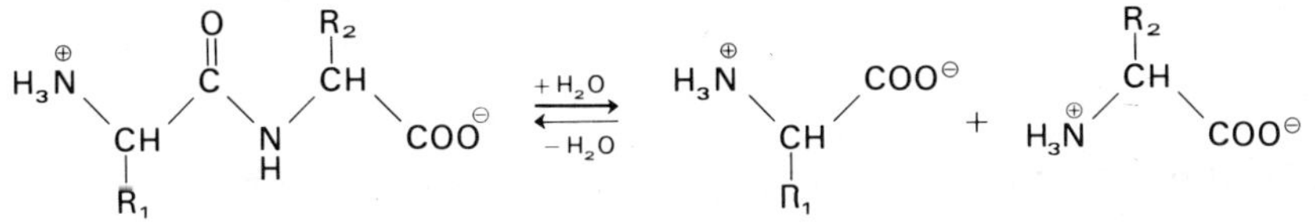

The above reaction comes to equilibrium, as every chemical reaction does, but strongly favors the direction of hydrolysis. The major reason is that peptide links arise only from the nonionized nonpolar forms of amino acids, which really do not exist in aqueous solution.

The bond

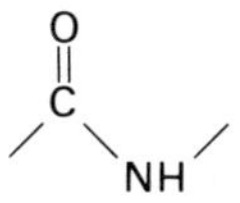

is called the peptide bond. In a typical peptide both components are amino acids. However, in some substances, only the amine component is an amino acid. Although such substances should more properly be called acylamino acids, occasionally they are included among the peptides.

Two amino acids form a dipeptide, three a tripeptide, eight an octapeptide, etc. If a peptide is made up of not more than ten amino acids, it is called an oligopeptide; beyond that it is a polypeptide. Polypeptides become proteins when they are made up of over a hundred amino acids; sometimes they are also called "macropeptides."

Since in systematic nomenclature peptides are acylamino acids, the specific name for peptides is obtained by attaching the ending *-yl* to that amino acid whose carboxyl group has undergone reaction; e.g., glycylalanine and alanylleucyltyrosine.

Merely knowing the kind and number of the constituent amino acids of polypeptides is not sufficient. The sequence in which the amino acids are linked must also be known. Notice that even the dipeptide glycylalanine is not the same as alanylglycine.

To avoid lengthy names a suggestion by Brand and Edsall will be followed here. Many amino acid residues are abbreviated by the first three letters of the names of the amino acid. Therefore, Ala stands for the alanyl-, Glu for the glutamyl-, His for the histidyl residue, etc. (summarized in Table IV). The amide group (as in glutamine) and the cystine link are expressed especially by adding $-NH_2$ and $-S-S$; end groups are indicated by $-H$ and $-OH$. The accompanying figure compares a structural formula with the

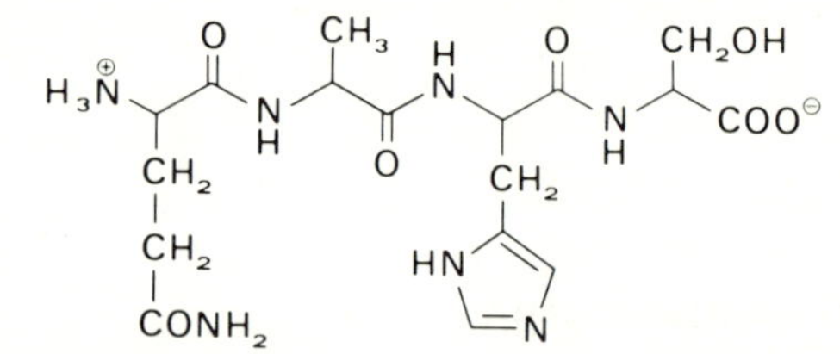

Glutaminyl—Alanyl—Histidyl—Serine
= Gln—Ala—His—Ser

systematic name and its abbrevated form. In writing the formulas of peptides it is conventional to write the terminal amino group always to the left. If this proves impractical, then an arrow indicates the direction: CO → NH. We will frequently use these conventions, since they are used widely in describing sequences of amino acids.

2. Determination of Amino Acid Sequence in Peptides

It is essential to be able to ascertain the sequence of amino acids in naturally occurring peptides as well as in peptides obtained from the fragmentation of proteins (cf. Chapter IV,2). The *Edman degradation* has been particularly useful for this purpose.

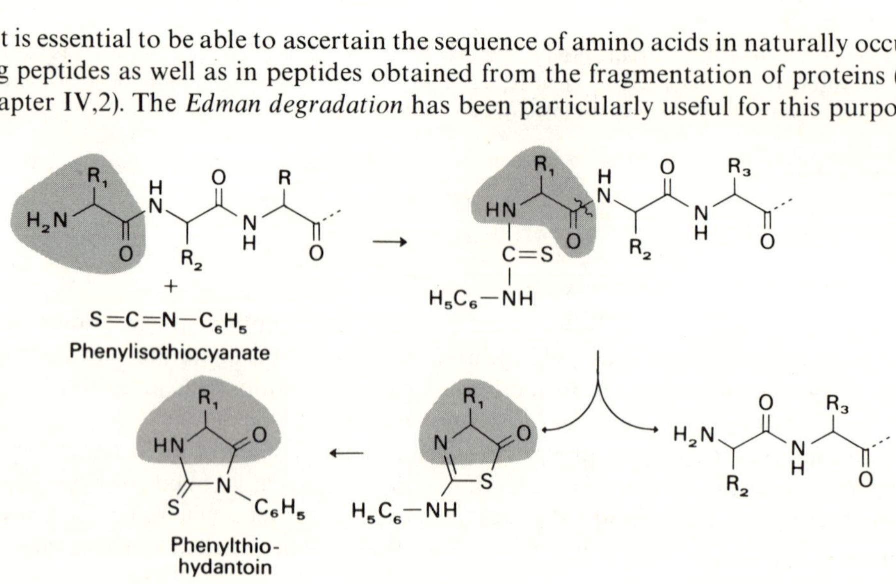

The amino group of the peptide undergoes reaction at pH 8–9 with phenylisothiocyanate. With acid-catalyzed cleavage, the newly formed peptide derivative loses its terminal residue as a substituted thiohydantoin, which can be identified, for example, by paper chromatography. The reaction can then be repeated on the remaining peptide. One amino acid unit after another can be removed from the amino end of the peptide in this way. This procedure has been automated and can be carried out on commerically available apparatus.

The N-terminal amino acid is identified by the reaction with *dinitrofluorobenzene*, according to Sanger ("end group analysis").

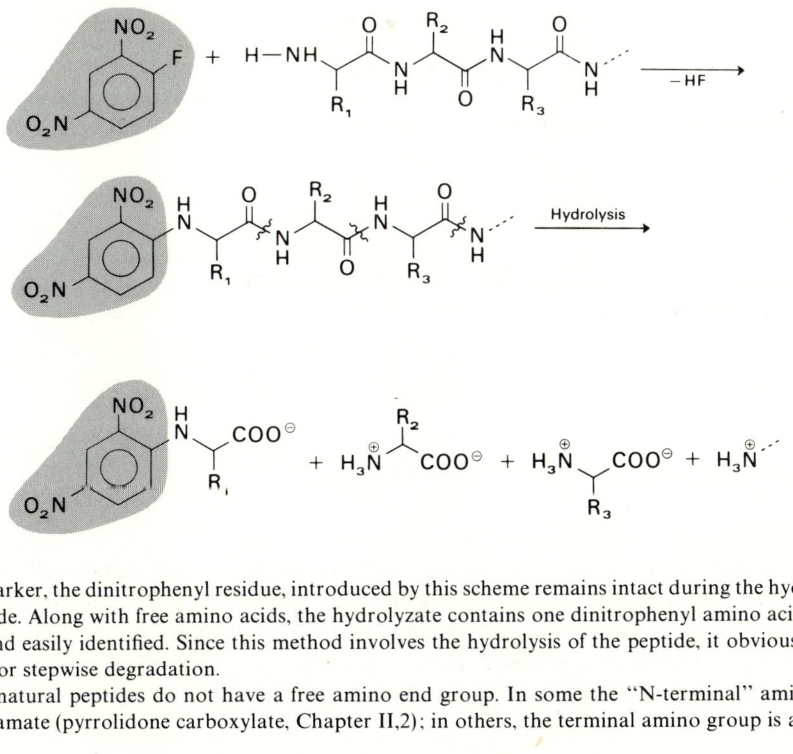

The marker, the dinitrophenyl residue, introduced by this scheme remains intact during the hydrolysis of the peptide. Along with free amino acids, the hydrolyzate contains one dinitrophenyl amino acid, colored yellow and easily identified. Since this method involves the hydrolysis of the peptide, it obviously cannot be used for stepwise degradation.

Some natural peptides do not have a free amino end group. In some the "N-terminal" amino acid is pyroglutamate (pyrrolidone carboxylate, Chapter II,2); in others, the terminal amino group is acetylated.

Finally, enzymatic methods are also available for the cleavage of larger peptides and for the identification of terminal groups (cf. Chapter VIII,3). Long sequences have been established by combining these methods.

Peptide Synthesis. Any synthesis of peptides must resort to some activation (cf. Chapter I,2) of the carboxyl or the amino group. One of the oldest methods of organic chemistry involves acid chlorides. In that case, the amino group of the one amino acid must be protected, e.g., by a carbobenzoxy residue, $C_6H_5—CH_2—O—CO—$. The protective carbobenzoxy group can subsequently be removed by hydrogenolysis (cleavage by catalytically activated hydrogen).

An ingenious variation of the usual methods of peptide synthesis involves attachment of a protected amino acid (*t*-butyloxycarbonyl group) to a solid polystyrene resin, removal of the amino protecting group, condensation with a second protected amino acid, and so on. The entire stepwise process takes place in one reaction vessel and has been fully automated. Cumbersome purification between steps is replaced by mere washing of the insoluble resin. In the last step, the finished peptide is cleaved from the resin. Since each reaction is essentially quantitative, very pure peptides are obtained (Merrifield).[1] Larger peptides can, of course, be prepared by condensing smaller peptide fragments. Recently the enzyme ribonuclease consisting of 124 amino acids has been synthesized; it was enzymatically active.

Chemical synthesis of peptides has been important for the development of peptide chemistry. We cannot describe all the newer methods which employ various other derivatives for activation. One important feature has been the prevention of racemization of amino acids. Numerous longer peptides, including peptide hormones, have recently been synthesized. The complete synthesis of insulin with 51 amino acid residues has been accomplished independently by at least three different laboratories, as well as the synthesis of other small proteins and enzymes.

3. Naturally Occurring Peptides

Glutathione (γ-Glutamylcysteylglycine). Glutathione is a natural and widely distributed tripeptide.

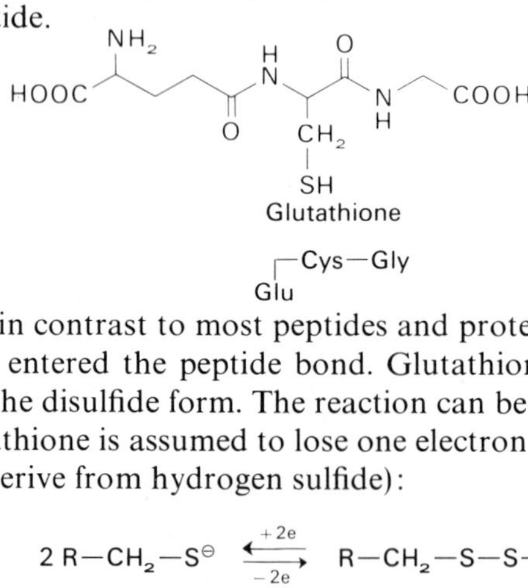

Note that here, in contrast to most peptides and proteins, the γ-carboxyl group of glutamic acid has entered the peptide bond. Glutathione is easily dehydrogenated and converted to the disulfide form. The reaction can be taken as an oxidation if the ionic form of glutathione is assumed to lose one electron pair (SH groups are weakly acidic since they derive from hydrogen sulfide):

$$2\,R{-}CH_2{-}S^{\ominus} \underset{-2e}{\overset{+2e}{\rightleftarrows}} R{-}CH_2{-}S{-}S{-}CH_2{-}R$$

The reaction is reversible, and glutathione has been thought to participate in a biological redox system. It is also coenzyme to glyoxalase.

Carnosine and Anserine. Found in muscle, both carnosine and anserine are dipeptides of β-alanine and histidine. Their biological role is still unknown (cf.

[1] R. B. Merrifield, Solid-phase peptide synthesis. III. An improved synthesis of bradykinin. *Biochemistry* **3**, 1385–1390 (1964). R. B. Merrifield and J. M. Stewart, Automated peptide synthesis. *Nature* (*London*) **207**, 522–523 (1965).

pantothenic acid, Chapter VI,7).

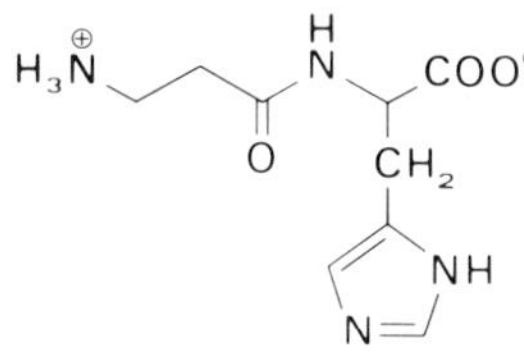

Carnosine

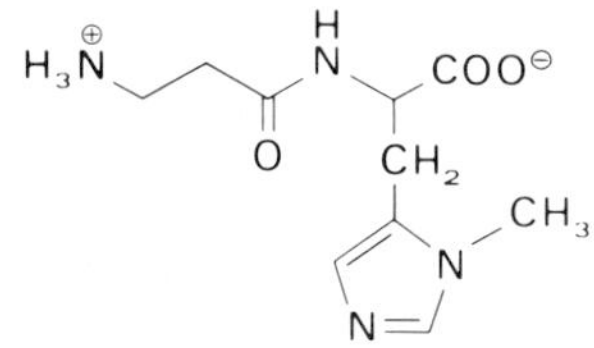

Anserine

Protamines. These are strongly basic polypeptides with molecular weights of 1 to 5000. They are isolated from fish sperm, where they occur together with nucleic acids. Their arginine content is very high; clupein, for example, consists of 22 moles of arginine and only 11 moles of all other amino acids (K. Felix). It is assumed presently that because of their polycationic nature protamines interact with nucleic acids (polyanionic) and thereby facilitate the tight packing of nucleic acids in sperm cells.

Peptide Hormones. Several hormones of the hypophyseal (pituitary) gland and the pancreas are polypeptides.

The structure of the hypothalamic hormone *thyrotropin releasing factor* is relatively simple, but it should be noted that neither the terminal H_2N group nor the carboxyl group are free. The physiological action of this hormone is discussed in Chapter XX,9.

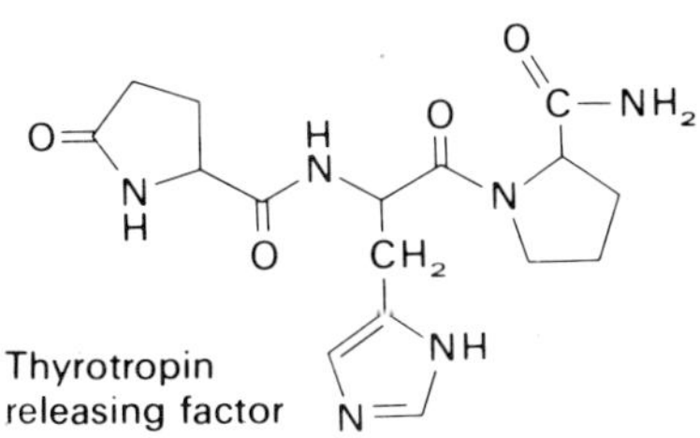

Thyrotropin releasing factor

Ocytocin[2] and Vasopressin. These are hormones of the posterior lobe of the hypophysis (pituitary) and are very similar to each other:

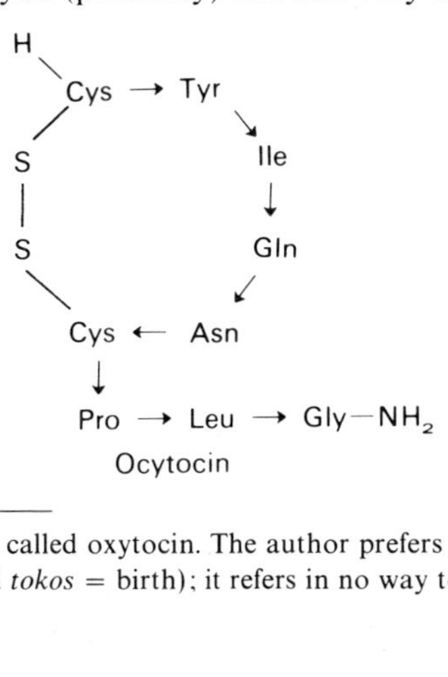

Ocytocin

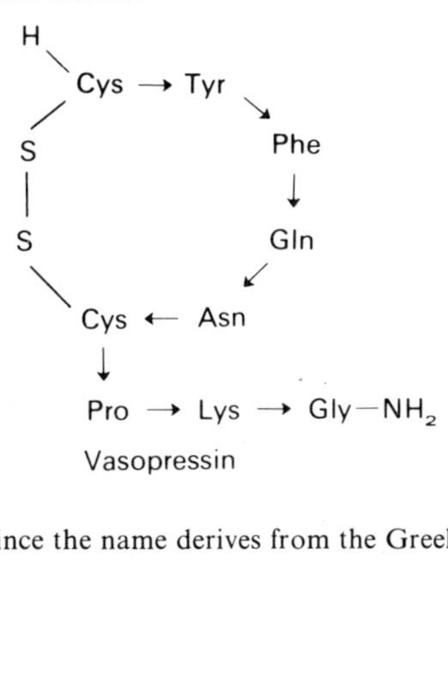

Vasopressin

---

[2] Often called oxytocin. The author prefers "ocytocin," since the name derives from the Greek (*oky* = quick and *tokos* = birth); it refers in no way to oxygen.

Insulin:

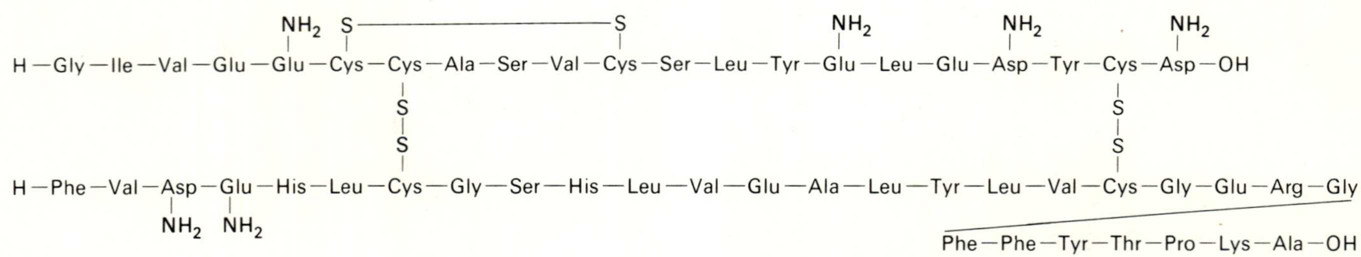
H—Gly—Ile—Val—Glu—Glu—Cys—Cys—Ala—Ser—Val—Cys—Ser—Leu—Tyr—Glu—Leu—Glu—Asp—Tyr—Cys—Asp—OH
H—Phe—Val—Asp—Glu—His—Leu—Cys—Gly—Ser—His—Leu—Val—Glu—Ala—Leu—Tyr—Leu—Val—Cys—Gly—Glu—Arg—Gly
Phe—Phe—Tyr—Thr—Pro—Lys—Ala—OH

The occurrence of cystine, one of whose amino groups is terminal, enables these peptides to assume a ring structure. Lysine may be replaced by arginine. The C-terminal glycine is present in the form of its amide. *Vasotocin* is the designation for an analogous hormone, which has isoleucine in position 3 (like ocytocin) and arginine in position 8 (like vasopressin). First produced synthetically, it was subsequently identified as a hypophyseal hormone of frogs. The physiological roles differ for these chemically similar hormones (cf. Chapter XX,9).

Corticotropin, or *a*drenocorticotropic *h*ormone (ACTH). This is a rather complicated peptide of the hypophyseal anterior lobe. The sequence of its 39 component amino acids has been established. The molecular weight is 4500.

In corticotropins of different animal origin (sheep, pigs, and cattle) minor changes occur in positions 30–33. The same is true for the melanocyte-stimulating hormone (MSH), or *melanotropin*, of the intermediate lobe of the pituitary gland. In this case, species specificity is also reflected in the length of the peptide chain. The sequence of melanotropin is the same as that of corticotropin for a major portion of the peptide chain.

Insulin. This hormone is secreted by the pancreas and causes a lowering of blood sugar (cf. Chapter XX,8). Its molecular weight is around 6000 (more exactly, 5750), and for that reason should be classified as a peptide. However, in solution, particularly in the presence of metal ions, it readily aggregates to form molecules with relative weights of 12,000, 36,000, or 48,000. It is evident that there is no clear distinction between peptides and proteins. Insulin, hence, is considered to be the first protein whose structure has been determined completely. The establishment of the sequence of the 51 amino acid residues through Sanger's efforts (1953) was an immensely important step for biology in general and provided great impetus to protein research in particular.

Sanger and his co-workers broke insulin into small peptides by partial hydrolysis and separated them by paper chromatography. The sequences in these peptide fragments were determined by the dinitrophenyl method and compared with the composition of larger fragments resulting from enzymatic cleavage of the insulin molecule. The partial formulas were finally integrated to the complete formula of insulin shown above.

The sequence of another pancreatic hormone *glucagon*, composed of 29 amino acid residues, is also known. *Parathormone*, secreted by the parathyroid gland has a molecular weight of 8500, approaching that of proteins.

Antibiotics and Toxins. Fungi and microorganisms produce various substances with characteristic chemical features. Either they contain amino acids that are not found in any other proteins or peptides or they have highly unusual linkages.

One of the best known representatives is *penicillin*:

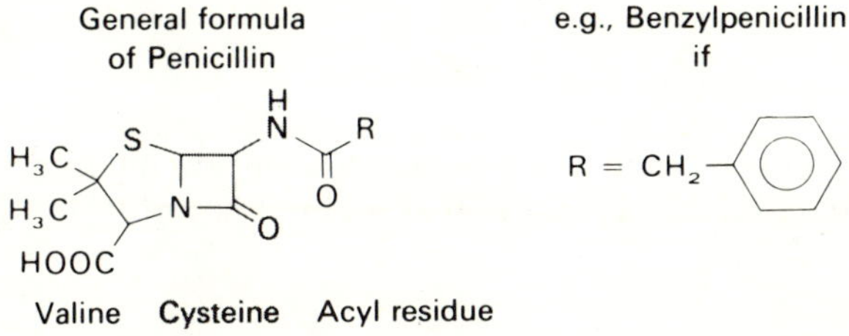

It is produced by the mold *Penicillium notatum* and excreted into the culture medium. Its biogenesis begins with the amino acids valine and cysteine; the formula can easily be derived, especially if one remembers that the molecule also contains a highly strained four-membered β-lactam ring and another S-containing ring. The amino group of the cysteine component is acylated; several different radicals may be attached in place of R. At present the bulk of manufactured penicillin carries a benzyl radical, $C_6H_5CH_2$. Recently, it has become possible to induce the mold to produce penicillanic acid (lacking the entire group labeled "acyl residue" in the general formula above) and to manufacture from it all the different penicillin derivatives.

Penicillin is the prototype of an *antibiotic*, which is generally defined as a substance that inhibits the growth and multiplication of bacteria and other microorganisms and is of biological origin. These were the properties responsible for the discovery of penicillin by Fleming in 1928. Antibiotics affect only certain groups of bacteria; they exhibit a definite spectrum of action. Antibiotics are an invaluable aid for the control of infectious diseases because they prevent rapid growth of the causative agents. Excessive doses occasionally cause undesirable side effects, such as destroying the natural intestinal flora and permitting a colony of resistant strains of bacteria or fungi to lodge there. Many antibiotics in high doses are toxic, and only a few have proved generally useful in therapy.

Antibiotics belong to chemically very diverse classes of compounds. *Azaserine* and *chloramphenicol* have already been mentioned (Chapter II,2). Another peptide antibiotic is *gramicidin*, a cyclic peptide consisting of ten aminoacyl residues, including two of D-phenylalanine. *Actinomycin* has a more complicated structure; a chromophoric group (a phenoxazone dicarboxylic acid) is linked to the terminal amino group of each of two identical pentapeptides by acid amide bond. The terminal carboxyl group of each peptide has formed a macrocyclic lactone ring with the hydroxyl group of threonine, as shown in the formula (Sar = Sarcosine; MeVal = *N*-methylvaline).

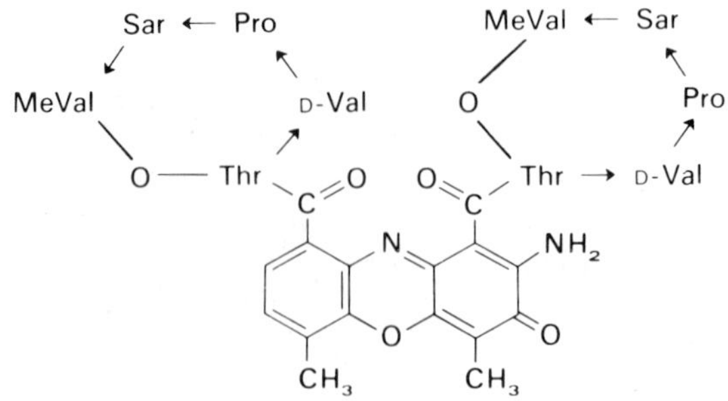

Actinomycin D

Actinomycin is an antibiotic and a cytostatic, which means it inhibits cellular division. It has become a standard tool in research as an inhibitor of RNA synthesis (Chapter VII,6).

Phalloidin, one of the poisons of the mushroom *Amanita phalloides*, is highly toxic. As little as 50 μg are lethal for a mouse. Seven different amino acids are involved in its structure, these include such unusual hydroxy amino acids as hydroxyleucine and allothreonine (T. Wieland). There is a bridge between tryptophan and the SH group of cysteine which yields hydroxytryptophan upon hydrolysis. The formula is not shown here. Closely related *amanitin* has recently been applied as a specific inhibitor of RNA polymerase (Chapter VII,5).

Valinomycin is also a cyclic molecule in which six amino acids and six hydroxy acids are alternatively linked by ester and acid amide bonds. The antibiotic is able to complex $K^{\oplus}$ but not $Na^{\oplus}$. This differentiation is intriguing in connection with the postulated carrier mechanisms of active transport (Chapter XXI).

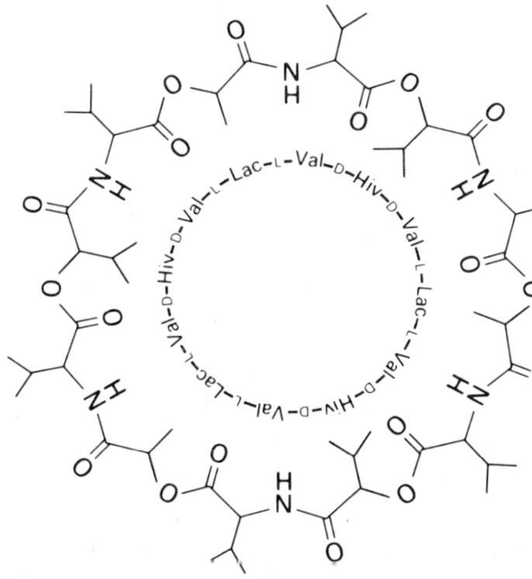

Lac = Lactate
Hiv = Hydroxyisovalerate

Valinomycin

The biosynthesis of peptide antibiotics does not proceed by the same mechanism as that of proteins. It generally takes place on multienzyme complexes, and the sequence of amino acids in the peptide is determined by the particular arrangement of enzyme proteins in the complex. Protein-bound thioesters of amino acids frequently are the "activated" intermediate products of the biosynthesis.

BIBLIOGRAPHY

R. E. Canfield and C. B. Anfinsen, Determination of the primary structure of proteins. *In* "The Proteins" (H. Neurath, ed.), Vol. I, pp. 311–378. Academic Press, New York, 1963.

K. Hofmann and P. G. Katsoyannis, Synthesis and function of peptides. *In* "The Proteins" (H. Neurath, ed.), Vol. I, pp. 54–189. Academic Press, New York, 1963.

R. B. Merrifield, New approaches to the chemical synthesis of peptides. *Recent Progr. Hormone Res.* **23,** 451–482 (1967).

CHAPTER **IV**

Proteins

1. Structure of Proteins

When a peptide chain is extended by more and more amino acids, until a chain length of between one hundred to about a thousand amino acid residues[1] is reached, it is classified as a protein.

The term *protein* (Greek: *proteuo*, I occupy first place) was first suggested by Berzelius and used by Mulder (about 1840) in his textbook. The name *protein* is well chosen; all the basic functions of life depend on specific proteins. Indeed, we know no form of life without proteins; they are present in every cell; they make up the contractile elements and the enzymes that catalyze the release of energy for maintenance of life; and they are present in blood, where they have a transport function.

The properties of proteins are extremely different. Consider two generally familiar proteins: egg white protein denatures on heating, dissolves easily in water, and is quite reactive, while keratin of nails and hoofs is wholly insoluble, tough, and chemically quite inert and resistant.

It is not easy to classify proteins. Here we distinguish simply between *scleroproteins*, which are insoluble and fibrous, and function as structural material, and *globular proteins*, which may be characterized for the present by the properties of such representatives as egg white or serum proteins. Proteins of this second group are soluble in water or salt solution and consist of spherical rather than threadlike molecules. (Globular proteins are further subdivided in Section 9.)

The Chemical Structure of Proteins. Comparatively simple in principle, numerous amino acids are joined to form peptide linkages, which, in turn form proteins. A

[1] This corresponds to molecular weights of from over 10,000 to more than 100,000. The distinction between protein and polypeptide (molecular weight below 10,000) is quite arbitrary.

segment of a peptide chain may be represented by the following structural formula:

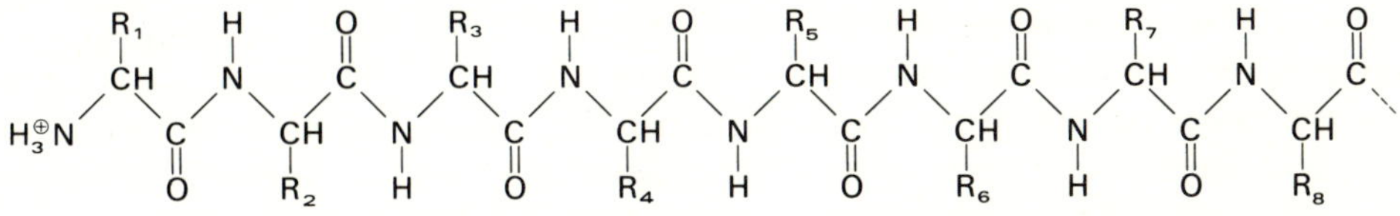

The problem of protein structure is primarily a question of the sequence of amino acids, i.e., what structures take the place of R_1 to R_x in the above formula.

The sequence of amino acids constitutes the *primary structure*, but this is not the only problem in protein chemistry. When several hundred amino acids join to form an extended chain molecule, it is easy to imagine that such a molecule may be arranged in space in various ways. For example, it may retain a straight linear form, or bunch up randomly in a ball, or arrange itself in a highly ordered helix (coil shape). The spatial arrangement of the chains is called *chain conformation.*

The term conformation[2] includes the two other concepts of secondary and tertiary structure which are used to distinguish the sequence of amino acids (called primary structure) from all other structural features. *Secondary* structure is defined as the three-dimensional arrangement of the peptide chain itself (regardless of the nature of the R groups in the above skeleton formula). *Tertiary* structure is the spatial position of all the atoms, including those of the side chain. The aggregation of several peptide chains making up a definite molecule is called *quaternary* structure.

The expression "secondary structure" is an unfortunate one. The above definition, eventually agreed upon, represents at best a highly arbitrary delineation against the concept of tertiary structure. On the other hand, the term secondary structure does encompass such important structural features of proteins as the pleated sheet and α-helix, both to be discussed below. The author generally prefers the concept of chain conformation, which includes both secondary and tertiary structures and which is applicable to other natural products as well. Chain conformation, i.e., the three-dimensional architecture of the macromolecule, cannot be studied by chemical analysis, but rather must be elucidated by physical or physicochemical methods.

2. The Sequence of Amino Acids

Determination of the Amino Acid Sequence of Proteins. The same methods can be used in principle as were described for peptides (Chapter II,2). However, the amino acid chains of proteins are usually so long that the successive removal of amino acids from one end to the other is most unpromising. The protein molecule must first be split up to a number of smaller fragments (e.g., by carefully controlled cleavage with trypsin, or by specific chemical cleavage at methionine residues with cyanogen bromide). The mixture of peptides obtained is then separated, and each individual peptide is analyzed for its own amino acid sequence.

The final task is to learn in what sequence the peptides, now successfully analyzed, occurred originally in the intact protein. The method of "overlapping cleavage" is

[2] Compare D. B. Wetlaufer, *Nature* (*London*) **190,** 1113 (1961).

used for this purpose. The protein is acted upon by another protease, e.g., pepsin instead of trypsin, and yields other peptide fragments from the same amino acid sequence. If now one of the new peptides contains, say, the last three amino acids of tryptic[3] peptide No. 2 and the first four amino acids of tryptic peptide No. 17, then one can conclude that these two tryptic peptides belong together. When all the peptide fragments have been assigned their position in the chain, the complete amino acid sequence of the protein is finally known.

The first extensive sequence of amino acids was determined on insulin (Sanger, 1953), a polypeptide consisting of 51 amino acids. The sequential analysis of true proteins was still beset with serious obstacles. Finally, in 1959, ribonuclease with 124 amino acid residues was analyzed completely (Hirs, Stein, and Moore; Anfinsen). In 1960–1961, there followed the publication of the complete primary structures of the protein of the tobacco mosaic virus with 158 amino acid residues (Schramm and co-workers), of hemoglobin with 574 acid residues (Braunitzer and co-workers), and of cytochrome c (Tuppy; Margoliash and Smith). Presently the complete sequences of well over 200 proteins are known.

Regularities in Sequences. Earlier it had been supposed that the arrangement of amino acids in the chains would be subject to certain periodicities. None of these hypotheses hold, however. If any rules can be derived from the available data, it may be that acidic amino acids tend to group themselves together, and the basic and aromatic acids do the same. Often the same amino acid appears several times in a row.

The question may be turned around: Are all possibilities of the amino acid sequence realized? A simple consideration will show that this is utterly impossible. From 20 different amino acids and with a chain length of 100 amino acid residues $20^{100} = 10^{130}$ different sequences can be generated—more than can be accommodated in the entire universe, even if there were only a single molecule of each kind.

Homologous Proteins. The sequence of amino acids is determined genetically; this will be discussed in much greater detail in Chapter VII,5. Occasionally striking similarities of sequence appear among different proteins, which are then called homologous proteins. These proteins often possess a similar function, such as trypsin and chymotrypsin (cf. Chapter VIII,2) or the α- and β-chains of hemoglobin (see below). Homology of various peptide chains may have arisen when at one time proteins evolved from one another (cf. Chapter VII,8).

The Sequence of Hemoglobin. As an example of the amino acid sequence of a protein we can use that of *hemoglobin*. Normal human hemoglobin is composed of four polypeptide chains (and four heme groups; cf. Chapter IX,2), or two identical pairs, where each pair consists of an α- and a β-chain. The α-chain is made up of 141 amino acid residues and contains relatively more acidic groups than the β-chain (with 146 residues); the two chains are held together by ionic and hydrogen bonds. The exact sequence of the two chains is reproduced in the Appendix (the reverse side of the foldout chart). The presentation chosen emphasizes certain homologous sequences between each other and in comparison with *myoglobin*. Homology is

[3] Peptides arising from cleavage with trypsin are called "tryptic peptides." They are numbered in the order in which they are eluted from the chromatographic separation.

achieved by leaving certain spaces blank. In this way a large number of positions contain the same amino acid residues in the different chains. This observation suggests that the two chains have both evolved from one common ancestral hemoglobin (cf. Chapter VII,8).

Instances of exchanges of amino acids have, indeed, been discovered. In man, one finds a fetal hemoglobin, a normal hemoglobin of adults, and a number of variants some of which cause pathological conditions. Sickle-cell anemia (a blood disease predominantly of the Negro race, characterized by morphological changes of erythrocytes in oxygen-poor venous blood) results from an altered hemoglobin and is inherited. Careful analysis has revealed that in sickle-cell hemoglobin glutamate in position 6 of the β-chain is replaced by valine:

Normal Hemoglobin:	Val—His—Leu—Thr—Pro—*Glu*—Glu—Lys—
Sickle-cell Hb S:	Val—His—Leu—Thr—Pro—*Val*—Glu—Lys—
Hemoglobin C:	Val—His—Leu—Thr—Pro—*Lys*—Glu—Lys—

In the case of hemoglobin C, another pathological hemoglobin, lysine occupies that position.

More than 100 hemoglobins with defined amino acid exchanges are currently recognized, all manifestations of mutations. The clinical symptoms caused by these "pathological hemoglobins" frequently are negligible.

3. Principles of Chain Conformation

Protein molecules have a definite three-dimensional shape. One speaks of the peptide chain being "folded" in a certain way. Further, the sequence of amino acids itself probably determines the particular kind of folding. In other words, folding is not a random process, but is the result of binding forces between different segments of the

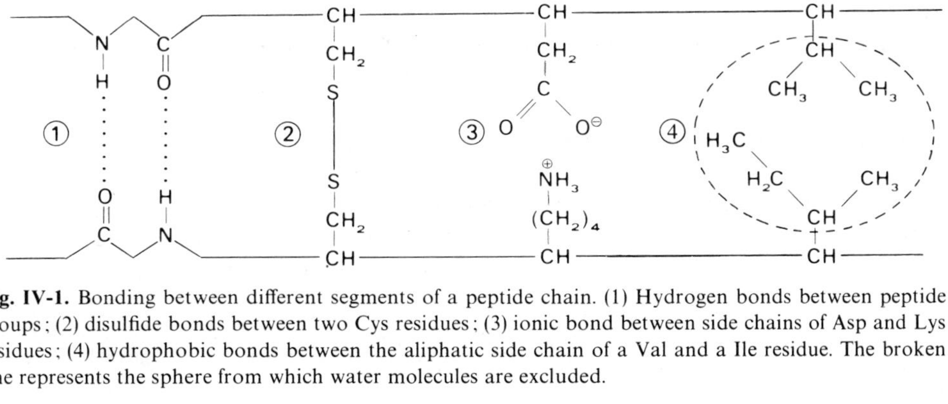

Fig. IV-1. Bonding between different segments of a peptide chain. (1) Hydrogen bonds between peptide groups; (2) disulfide bonds between two Cys residues; (3) ionic bond between side chains of Asp and Lys residues; (4) hydrophobic bonds between the aliphatic side chain of a Val and a Ile residue. The broken line represents the sphere from which water molecules are excluded.

peptide chain. The extent of folding exclusively dependent on the chain itself, i.e., the atoms of the peptide bonds, is designated as *secondary structure*. In globular proteins, however, it is the functional groups of the side chains that effect the particular kind of folding. These situations, represented diagrammatically in Fig. IV-1, will be discussed briefly here.

Hydrogen Bonds. The most important secondary valence forces that maintain the conformation of the protein molecule are hydrogen bonds. An interaction sets in between a C=O group and the proton of an NH or OH group if these groups come within a distance of about 2.8 Å.[4] This secondary valence bond is symbolized by a dotted line, CO···H—N. The binding energy of a hydrogen bond amounts to only about one-tenth of that of a primary valence; but, since a protein molecule has a large number of hydrogen bonds, the aggregate is quite significant.

Disulfide Bonds. The most important primary valence bond between side chains of amino acids is the disulfide bond formed between two cysteine SH groups by dehydrogenation. An example of this has already been seen in insulin (Chapter III,3). In every protein known so far, only certain specific cysteine residues are linked together.

Other primary valence bonds between side chains are, of course, possible in principle. The carboxyl group of glutamate or aspartate could form a peptide bond, e.g., with the ε—NH_2 group of lysine, or it could form an ester bond with the hydroxyl group of serine or threonine. Examples of this have, however, not yet been observed.

Ionic Bonds. The occurrence of bonding between positively charged (side chains of lysine and arginine; to a smaller extent, histidine) and negatively charged groups ($COO^{\ominus}$ group of glutamate and aspartate side chains) is called ionic bonding.

Hydrophobic Bonds. Another very important type of secondary valence bond arises from the close approach of hydrocarbon side chains primarily of the amino acids valine, leucine, isoleucine, and phenylalanine. The same forces that hold the molecules together in crystals also operate here. In addition, water molecules are displaced out of this region of interaction. A crude analogy to this situation is the coalescence of two oil droplets to form a larger one. Hydrophobic bonding appears to predominate in the interior of protein molecules.

Stereochemistry of Peptide Chains. All proteins consist of amino acids exclusively of the L-configuration. This fixes the steric arrangement at the α-carbon atom. The dimensions of a peptide chain are also known exactly (see Fig. IV-2). Another requirement is that the atoms of the peptide bond all lie in a plane, since the resonance between the limit structures shown here favors the planar configuration.

---

[4] Å is the abbreviation of the Angstrom, an atomic-physical unit of length equivalent to 10^{-10} m or 0.1 nm (nanometer).

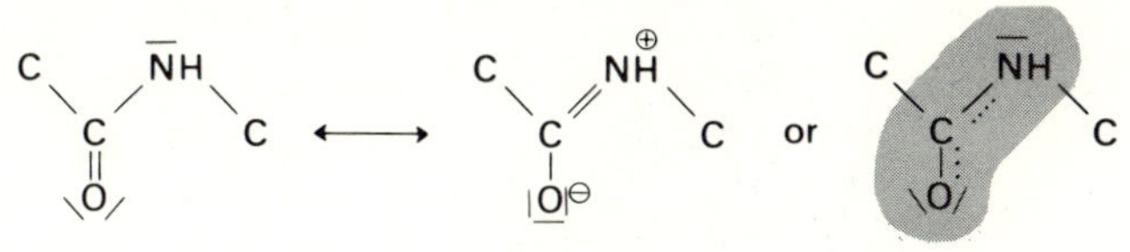

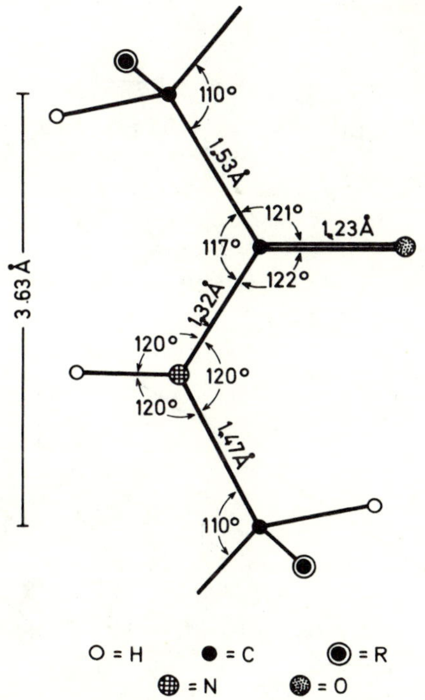

Fig. IV-2. Dimensions within the peptide chain.

Building Model Structures. By extending the structure drawn in Fig. IV-2 for some distance, a straight polypeptide chain, a threadlike molecule, is obtained. Hydrogen bonds, C=O···H—N, could form if a second chain were to lie alongside of it. However, such a model does not provide sufficient space for the side chains of amino acids. Pauling made a small correction, therefore, by having the planar "peptide grid" (Fig. IV-3) folded in the fashion of an accordion. As a result the side chains emerge nearly perpendicularly up or down, as can be demonstrated by a model (cf. Fig. IV-3). This model is called the *pleated-sheet structure*; it occurs in several fibrous protein molecules.

The pleated-sheet model accommodates the observed discrepancy that the identity period is actually shorter by 5 to 10% than calculated for the straight chain. In addition, it provides sufficient space for the side chains, whereas in the straight model the side chains come in contact. Pleating also permits hydrogen bonding between peptide chains running in the same direction. Consequently, we must accept the existence of two pleated-sheet structures (one with parallel, unidirectional chains and another with antiparallel chains in opposing directions).

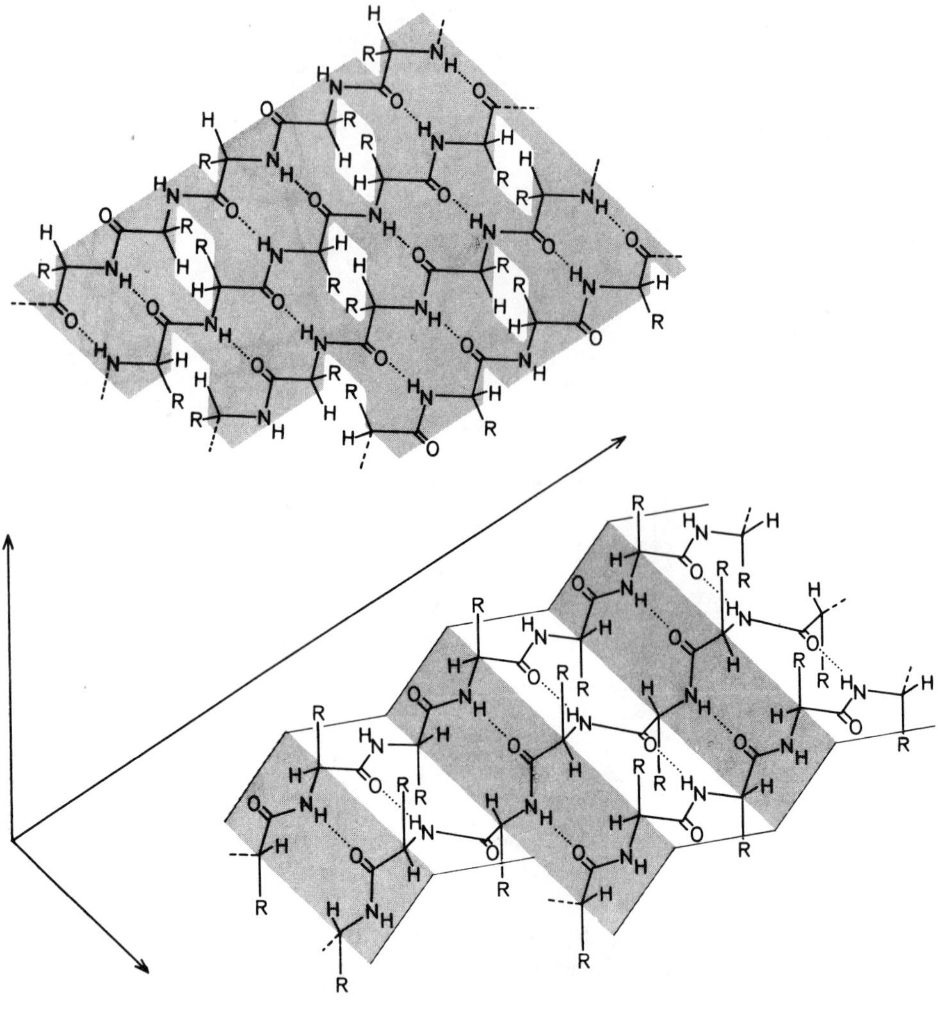

Fig. IV-3. Planar peptide grid (above) and pleated-sheet structure (below).

The α-Helix (α-Coil). In pleated-sheet structures or in "peptide grids," hydrogen bonds link the individual chains, but a structure in which the hydrogen bonds are all satisfied within a single chain should be favored. Such a preferred structure is achieved by winding the peptide chain around an imaginary cylinder in a way that from one turn to the next, CO and HN face one another at a suitable distance. Several models of this concept are possible. The one that is found widely distributed in nature is the α-helix with 3.7 amino acid residues per turn and with an identity period of about 5.44 Å (Pauling and Corey).

During the construction of an α-helix, care must be taken to keep the peptide groups planar. Free rotation is possible only at the α-C atom. These peptide planes form angles to each other of 80°, so that when viewed from the top the planes are arranged as indicated in Fig. IV-4c in which eight amino acid residues are shown. At the points of intersection, the CO and HN groups approach each other with a distance of one hydrogen-bond length, i.e., 2.8 Å, as seen from the side (Fig. IV-4b). Each amino acid residue advances the chain in the direction of the cylinder axis by 1.47 Å;

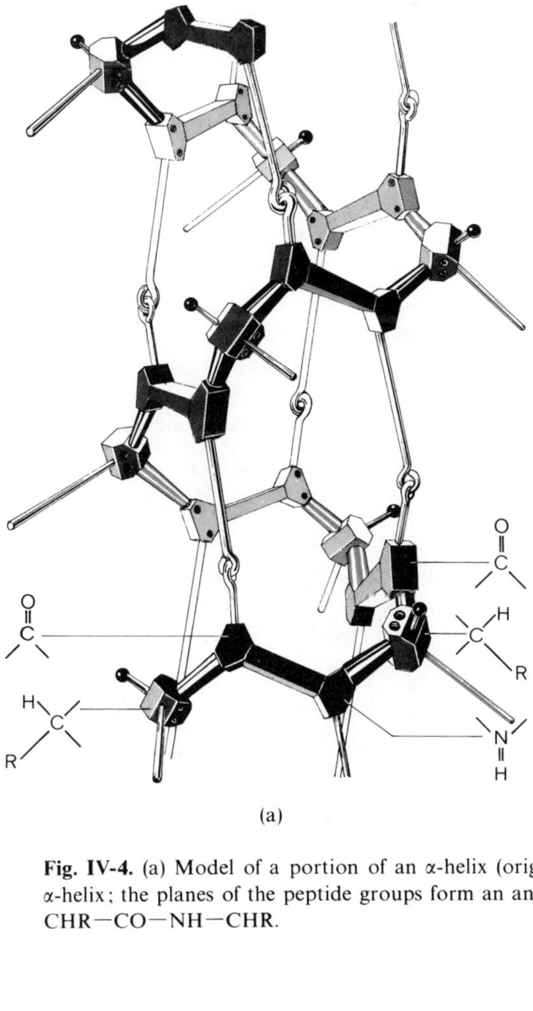

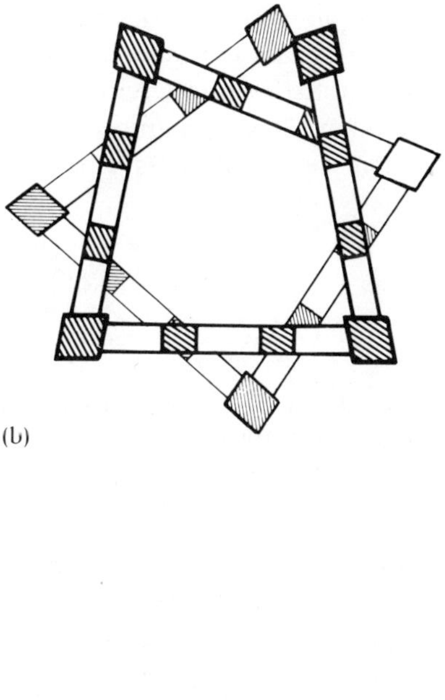

Fig. IV-4. (a) Model of a portion of an α-helix (original by the author). (b) View on top of the α-helix; the planes of the peptide groups form an angle of 80°; each bar represents the grouping CHR—CO—NH—CHR.

with 3.7 amino acid residues per turn, the identity period becomes 5.4 Å. As shown, hydrogen bonds connect consecutive turns; this feature grants unusual stability to the α-helix. In this model, which may be built up either as a right- or left-handed coil, side chains stick out radially from the body of the coil proper. They can interact among themselves or with the solvent. The secondary amino acid proline does not fit into the helix, and wherever it occurs in the amino acid sequence there is a deviation from the regular structure.

Both structures just discussed, the pleated-sheet and the α-helix, are stabilized by hydrogen bonds between atoms belonging to the peptide bonds and consequently are "secondary structures" of proteins, as defined earlier.

4. Conformation of Scleroproteins

Scleroproteins comprise the insoluble support and structural proteins, e.g., keratin of hair, nails, and feathers, and collagen. They are fibrous, i.e., the molecules are ordered in a certain way. This property makes them amenable to structural analysis by X-ray diffraction. Much has been learned about the conformation of proteins by this method. Regularities in molecular structure—regularly recurring groups—with sufficient order in the molecules are recognized as periods of identity.

From X-ray interference patterns, the identity periods are calculated and expressed in Angstrom units (1 Å = 0.1 nm). Scleroproteins may be classified according to their identity periods (and other properties) into three divisions:

1. Silk fibroin-β-keratin group:	Identity period 6.5–7.0 Å
2. α-Keratin-myosin-fibrinogen group:	Identity period 5.1–5.4 Å
3. Collagen group:	Identity period 2.8–2.9 Å

X-ray data alone do not permit rigorous calculation of spatial arrangement, except where a series of analogous crystals can be measured thoroughly. (Crystals are far more ordered than fibers.) As a result, it is necessary to conceive of models and to test these models empirically. It is possible to calculate precisely what type of X-ray diffraction will be shown by any given model.

β-Keratin. This protein has the pleated-sheet structure. The structure of silk fibroin, a representative of this group, is known in considerable detail: the chains are antiparallel (alternating direction of the CO → NH bond). Stretched hair, according to X-ray data also of the β-keratin type, probably has the pleated-sheet structure with parallel, unidirectional chains.

α-Keratin. The basic structure of all proteins of the α-keratin type is the α-helix. Sheep wool has been studied very thoroughly and, like other natural fibers, was found to be not simply a bundle of α-helices. Their structure is more complicated. Two or three helices are twisted together to form a "cable." Eleven such "cables" make up a protofibril, in which two central cables are surrounded by nine others. These protofibrils with a diameter of about 70 Å are embedded in a matrix substance. When wet hair is stretched to about twice its length, it changes from the α-keratin to the β-keratin structure.

Configuration of Collagens. This important group of scleroproteins is the chief component of supportive and connective tissue, primarily skin and the organic part of the bone. The cells of connective tissue at first form a *tropocollagen*, with a molecular weight of 300,000 and consisting of three chains, helical themselves (three amino acids per turn and an identity period of 8.6 Å), intertwined like the strands of a cable, and held together by hydrogen bonds. The whole molecule is 2800 Å long, but only 14 Å thick. Of three consecutive amino acid residues, two may be proline or hydroxyproline—these amino acids predominate in collagen—the third invariably is glycine (Fig. IV-5a). The fibers arise when many tropocollagen molecules line up alongside each other, shifted by one-fourth of the length along its neighbor molecule, and become cross-linked by primary valence bonds. This cross-linking increases with age. Basic and acidic amino acids occur in much greater proportions at certain places of the fiber and are responsible for the transverse stripes observed with the electron microscope after staining with metal salts (Fig. IV-5e).

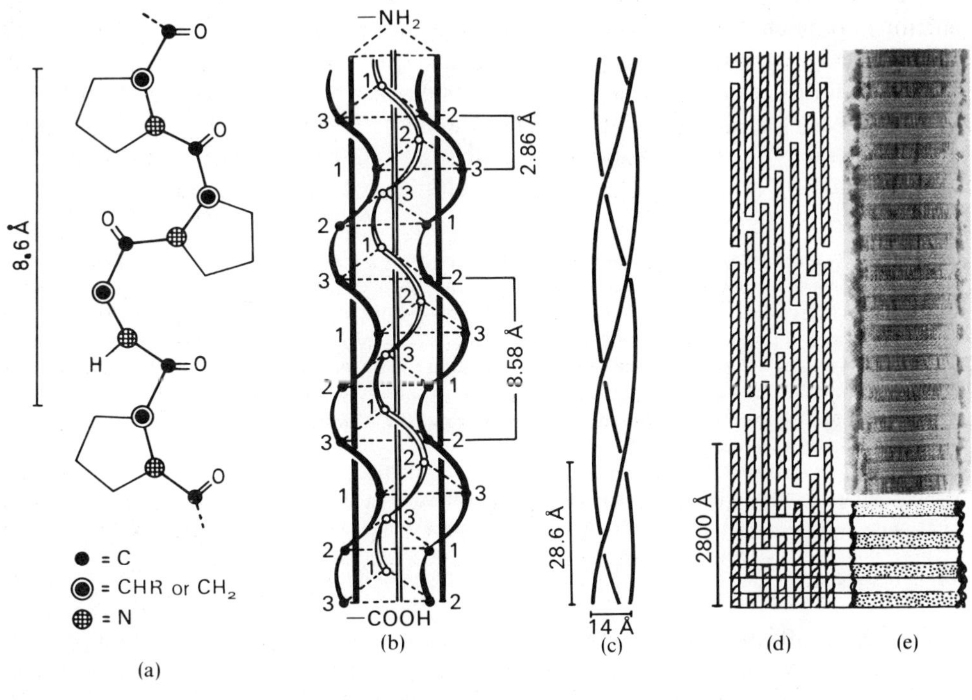

Fig. IV-5. Structure of collagen. (a) Shows the peptide chain with two proline or hydroxyproline and one glycine residue in regular sequence. This peptide chain forms a steep helix rising 2.86 Å per amino acid residue (b). Three of these helices are held together in a small bunch by hydrogen bonds. The axes of the steep helices, represented in (b) as vertical straight lines, are in reality twisted around one another as in a cable (c). The individual cablelike tropocollagen molecules, each 2800 Å long, now aggregate, but are staggered by one quarter of the length and leave a little space between themselves where they join lengthwise (d). The spacing and staggering causes the light and dark transverse stripes seen in (e). The same stripes or bands show up in the electron microscope. The negative staining used here lets the light stripes appear dark, and the darker ones, densely packed with material, come out light.

Collagen can be dissolved partially by the treatment with warm alkali and becomes gelatin. Solutions of gelatin become stiff gels upon cooling (see also Section 7).

Elastin. Another protein of connective tissue is elastin. It is completely insoluble in dilute sodium hydroxide, cannot be denatured by heat, and has elastic properties. The molecular structure of elastin is not yet known. Proteases do not attack it, but it can be digested by a specific enzyme called elastase.

5. Conformation of Globular Proteins

The protein fibers discussed above are aggregates of many molecules orderly lying side by side. Globular proteins in dilute solution, in contrast, exist as individual independent molecules. A structure of this type can arise only if the protein chain is "folded" in a certain way. This definite three-dimensional order evidently depends on portions of the chain being held in place by certain forces, primarily disulfide bonds (if present) and secondary valence bonds. The pleated sheet and the α-helix predominate in this regard as structural types. The contribution of the helix to the total structures varies greatly with individual proteins: 75% in the case of myoglobin, 30% with lysozyme, 17% with ribonuclease, and 10% with cytochrome *c*. The helix often is interrupted by other irregular structures after only two or three turns.

No methods are available as yet that would allow an estimation of the conformation of proteins in solution. In some cases, X-ray diffraction of protein crystals has been used to learn the exact structure of the molecules as they exist in the crystal. This was done with greater exactness than had been possible with fibrous proteins. For physical reasons, the method requires a series of derivatives crystallizing isomorphously and containing some heavy atoms (e.g., heavy metal atoms). With the availability of such derivatives, the method unequivocally produces three-dimensional representations of the structure. Resolution depends on the number of inflections measured and on the accuracy of the measurements. In the case of myoglobin, resolution has been achieved to 1.5 Å.

Beyond X-ray analysis, methods for assessing the contribution of α-helix to the total structure have become important, primarily the measurement of optical rotation. As a highly ordered system, the α-helix contributes to the overall optical rotation to a certain extent, which disappears during denaturation (= unfolding). Furthermore, the exchange of H atoms for deuterium is clearly diminished in the helical region of the chain.

The representation of a protein molecule as a sphere or a rotational ellipsoid is but a crude approximation, which at best serves as a model for calculating flow resistance. The surface of a protein molecule is in reality highly irregular due to the numerous amino acid side chains. Polar and charged groups are on the surface surrounded, in an aqueous medium, by a hydrate envelope that clings tightly to the molecule.

Quaternary Structure and Allosterism. Many proteins in solution, especially enzyme proteins, have molecular weights between 70,000 and several hundred thousand. These proteins commonly consist of several peptide chains (called *subunits*) occasionally linked by disulfide bonds, but usually only by secondary valence bonds. The manner in which subunits form a complete molecule is described as *quaternary structure.*

Extreme pH's often dissociate proteins to their subunits; also high concentrations of urea, or SH compounds (cysteine, thioglycolate, and others) when there are disulfide bonds, will do this. Most dissociations are reversible.

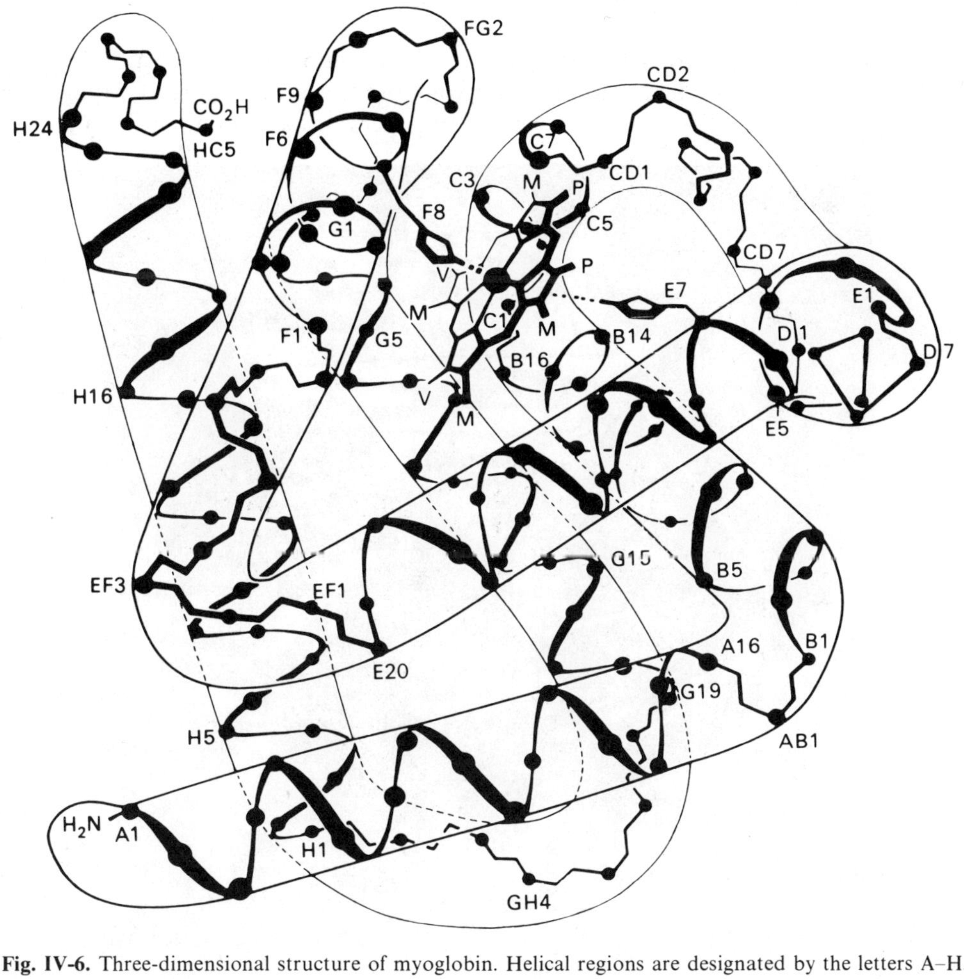

Fig. IV-6. Three-dimensional structure of myoglobin. Helical regions are designated by the letters A–H. B5 stands for the fifth amino acid in region B (cf. the reverse side of the foldout chart in the back of this book). The N-terminal amino acid residue with the free amino group is in the lower left corner; the C-terminal residue with the free carboxyl group is seen in the upper left corner. From R. E. Dickerson, X-Ray Analysis and Protein Structure. *In* "The Proteins" (H. Neurath, ed.), Vol. II. Academic Press, New York, 1964.

The conformation of the overall structure of proteins consisting of subunits, notably enzymes, frequently is changed by low-molecular weight substances. This has been termed the *allosteric effect*. A much-studied example is the change of conformation of hemoglobin induced by the loading of O_2 (see below and Chapter IX,4). Allosteric effects are highly significant in the case of enzymes (cf. Chapter V,8).

In some respects, allosteric change comes close to the old concept of "folding isomerism," according to which one and the same peptide chain could exist in several spatial conformations. These different foldings were postulated to be the basis for different biological properties. Now it appears that different conformations can only be stabilized by additionally bound molecules.

Structure of Myoglobin and Hemoglobin. The proteins studied most thoroughly by X-ray diffraction are hemoglobin and myoglobin (Perutz; Kendrew). Myoglobin consists of only one chain (with one heme). In general, it is made up of long helical

Fig. IV-7. Model of the hemoglobin molecule (according to Perutz). One "white" chain is removed for the sake of clarity. The lines are drawn in to show the course of the peptide chain (usually in the form of the α-helix). The light disk in the "black" chain at right represents the porphyrin ring. M. F. Perutz, M. G. Rossman, A. F. Cullis, H. Muirhead, and G. Will [Structure of hemoglobin. *Nature* (*London*) **185**, 416 (1960)].

portions and shorter nonhelical regions that constitute the turning points of the loops (the helix itself is a rigid structure which cannot be bent). The proportion of helical structure is rather high in this molecule with a content of about 75% (see Fig. IV-6). The shape of this molecule is determined by secondary valences formed between chain sections; there are no disulfide bridges.

Hemoglobin consists of four chains (two α- and two β-chains; see Section 2), each folded in a fashion similar to myoglobin. Certain minor deviations stem from certain differences in the amino acid sequence. In Fig. IV-7 the α-chains are shown white; the β-chains, black. For a clearer view, the frontward α-chain has been left out. It can be seen that the heme groups (shown as flat disks) are embedded in pocketlike cavities, but are still accessible from the outside. This is where oxygen attaches.

Recently the reduced (oxygen-free) form of hemoglobin has also been analyzed. One interesting difference observed is that the distance between β-chains has increased (37 Å instead of 30 Å from SH group to SH group.) With respect to reduced hemoglobin, the attachment of oxygen causes a shrinking of the molecule. To what extent the orderly arrangement changes within each chain is not yet known.

Structure of Lysozyme. Lysozyme consists of 129 amino acid residues and has a considerably more complex structure. Four α-helical segments constitute only 20% of the total molecule. Portions of pleated-sheet structure (twice 5 amino acids) are linked with additional secondary valence bonds. There are also four disulfide bonds (positions 6–127, 30–115, 64–80, and 76–94). The foldout chart in the back of this book contains a schematic drawing of the whole peptide chain. The molecule has a deep groove that accommodates the substrate (cf. Chapter V,9), a groove similar to the one found in the enzyme ribonuclease. This structural feature is essential for the catalytic action of lysozyme (Chapter V,9).

General Principles of Three-Dimensional Structure. Several general conclusions can be drawn from the observations made on hemoglobin. The validity of these conclusions has been tested on other proteins. The amino acid sequence and the disulfide bridges (if present) very probably determine precisely the spatial folding or arrangement of the chains. Some segments exist in the α-helical form, while other segments are held in an arrangement fixed by secondary valences (hydrogen bonds and hydrophobic bonds; see Section 3). In the proteins studies made thus far, all the charged groups were found to be on the surface of the molecule, while the hydrophobic side chains were found in the interior.

The three-dimensional arrangement is determined as precisely as the amino acid sequence itself. All molecules of a protein are identical in their conformation. The image of conformation, however, must not be too rigid. Some groups protruding from the surface of the molecule are mobile, and the spatial arrangement changes somewhat with the attachment of low molecular weight substances (e.g., substrates of enzymes) because certain groups of the protein tend to approach the attached substance.

Denaturation. Almost all soluble proteins coagulate upon heating, particularly in slightly acidic solution. This is a common, but not a necessary, sign of denaturation. The term *denaturation* is applied to structural changes of protein with concomitant loss of biological properties such as enzymatic or hormonal action, with a large drop in solubility, and with general changes of chemical and physical properties. Under special conditions denaturation may be reversible; usually it is irreversible.

There is an abundance of substances with denaturing action, namely, acids, bases, organic solvents, concentrated solutions of urea or guanidine, aromatic acids, like salicylic acid, and detergents, such as dodecyl sulfate. Proteins are not all equally susceptible to these substances or even to physical treatment like heat or radiation. Susceptibility itself depends to different degrees on pH, salt content, etc.

Denaturation is really the transition from a highly ordered to a less ordered state, a *random coil*. In this state, the chain can assume different arrangements which are statistically distributed. This amounts to a disordered structure, as evidenced by the large gains in entropy.[5] Denaturation is a strongly endothermic reaction; yet, it becomes exergonic above a certain critical temperature. There is more than a superficial analogy to the process of melting. As melting destroys the lattice structure of a crystal, so denaturation breaks secondary valence bonds that stabilize the globular protein in its native conformation. The peptide chain unfolds (denatured proteins give an X-ray pattern resembling one of β-keratin, i.e., one with a nearly straight peptide chain). New secondary valence bonds are formed purely randomly (statistically) among the chains, which makes the protein insoluble. At the same time, several other groups (e.g., SH groups of tryosyl residues), that had been hidden and "masked" in the interior of the molecule become exposed and more reactive. Proteases can attack denatured proteins much more readily than native ones.

6. Molecular Weights of Proteins

In discussing the classification of peptides we have described proteins as high-molecular weight substances. The actual size of protein molecules, however, was measured first with the ultracentrifuge developed by T. Svedberg (1925–1930).

The measurement of *sedimentation* requires that a protein solution be exposed to gravitational forces of over 100,000 times gravity. Such fields of gravity are obtained in very rapidly spinning centrifuges (up to 60,000 rpm). Protein molecules, which are denser than water, slowly sink to the bottom of the tube, i.e., they sink to the outside. The rate of sedimentation is recorded by optical methods that do not interfere with the operation of the centrifuge.

The rate of sedimentation is defined as the distance traveled by the molecule in the direction of centrifugal force per unit time (second), dx/dt, and is proportional to the centrifugal acceleration:

$$dx/dt = s \cdot \omega^2 x$$

[5] Entropy is a measure of the probability of a certain state; unordered states are always more probable and therefore of higher entropy content (see also Chapter V,2).

where ω = angular velocity, x = distance from center of rotation (radius). The factor of proportionality, s, is called the *sedimentation constant* and is a physical constant of the macromolecule. It is expressed in Svedberg units (S) (1 Svedberg unit = 10^{-13} second).

The molecular weight in turn is calculated from the measured sedimentation constant s, according to the formula:

$$M = \frac{R \times T \times s}{D(1 - \rho_s/\rho_{\text{prot}})}$$

where R = gas constant, T = absolute temperature.

It is evident that the diffusion constant D must be known (usually it is obtained from the same solution); further, the density of the dissolved protein ρ_{prot} and the density of the solvent ρ_s. A large sedimentation constant indicates a high molecular weight. Because of the hydrate layer enveloping the protein molecules, the measured values exceed those calculated from the amino acid sequence by about 5%. The latter are, of course, most accurate.

The molecular weight can also be calculated from light scattering instead of sedimentation measurements. Protein solutions opalesce and show the Tyndall effect. A measure of the intensity of light scattered rectangularly permits calculation of molecular weight. Finally, X-ray measurements on crystalline proteins yield very exact data for molecular weights.

Still another method of molecular weight determination is chromatography with "molecular sieves." As explained in more detail in Section 8, molecular sieves can fractionate according to molecular size. Chromatographic columns of this material are calibrated with proteins of known molecular weight, and the location of the peak of an unknown protein in the elution diagram can then be related to molecular weight. Since the method does not require any complicated apparatus it has rapidly been adopted in very many laboratories. A very critical interpretation of results, however, is essential.

The molecular weights of several proteins are listed in Table IV-1. Data for plasma proteins are in Table IV-2 (Section 10). The very high-molecular weight proteins listed there consist of subunits. Judging from the amino acid sequences determined so far, there seem to be few chains with more than 300–600 amino acid residues in uninterrupted series.

TABLE IV-1

Molecular Weights of a Few Proteins[a]

Protein	Molecular weight	Number of subunits
Cytochrome *c*	11,600	1
Ribonuclease	13,500	1
Trypsin	24,000	1
β-Lactoglobulin	35,000	2
Hemoglobin	64,500	4
Hexokinase	96,000	4
Lactate dehydrogenase	150,000	4
Apoferritin	480,000	20
Urease	483,000	6
Myosin	620,000	3

[a] See also Table IV-2, Section 10.

Since the mean molecular weight of a single amino acid residue is 110–120, such chain lengths correspond to molecular weights between 33,000 and 70,000. The higher molecular weights often seen with enzymes then must arise from an aggregation of such chains, as discussed in the example of hemoglobin. For further information, see the compilation of data on molecular weights and subunits of a large number of proteins by I. M. Klotz [*Science* **155,** 697–698 (1967)]. A few examples have been excerpted in Table IV-1 above.

7. Colloidal Nature of Proteins

The application of knowledge of colloid chemistry to biological phenomena and to proteins caused considerable confusion during the first half of the twentieth century. It has been overlooked for a long time that by the mere use of the statement, "the cytoplasm is a colloidal system," no knowledge was gained at all; on the contrary, deeper insight was actually foregone.

Colloid chemistry distinguishes molecular (or micromolecular) solutions, colloidal (or macromolecular) solutions, and coarse suspensions (e.g., mud). *Molecular solutions* are solutions of salts or any simple organic or inorganic molecules which exist individually in the solution. Such a solution often is converted into a *colloidal solution* when large numbers of molecules aggregate so that the individual particles no longer are single molecules, but rather consist of whole collections of molecules. Colloidal particles may be large or small, i.e., they may consist of many or of a few individual molecules. Such colloidal solutions are contrasted with "true solutions," i.e., molecular solutions.

With this distinction in mind, we find that most protein solutions are molecular solutions since the individual molecules are free in solution; there are no molecular aggregates.[6] The protein is truly dissolved. A finer subdivision is impossible without destroying the molecules.

Protein molecules possess high molecular weights, between 10,000 and several millions and molecular diameters of between 50 and 500 Å. For comparison, the molecular weight of glucose is 180 and its molecular diameter is about 5 Å. Protein macromolecules belong to the same size range as colloidal molecular aggregates, and it is for this reason that in the beginning of colloid chemistry, inorganic colloids and protein molecules were thought to be essentially of the same nature.

Modern colloid chemistry makes an important distinction between *monodisperse* and *polydisperse* systems. In monodisperse solutions all particles of the dissolved substances are of the same size and same shape, and have the same properties; in polydisperse solutions we find the most diverse particle sizes and varying charges—in short, different properties.

Protein solutions are monodisperse solutions whose particles, namely the protein molecules, are of the same size as the composite particles of colloidal solutions. It is to be expected, therefore, that protein solutions will manifest certain properties in common with colloidal polydisperse solutions of organic or inorganic materials, as well as certain different properties.

[6] A structure such as hemoglobin is taken as one molecule, even though it consists of four subunits, because these units do not normally dissociate in solution.

We are forced to make one more minor restriction. Some proteins tend to aggregate; two, three, four, or more molecules may associate. The resulting particles have a range of definite sizes, which differ by whole-number factors of two, three, four, or more. Such systems are called "paucidisperse."

Sol and Gel. Globular proteins in general give solutions of normal viscosity often called sols. Gelatinous solutions of high viscosity called gels result from fibrous or threadlike macromolecules that are loosely cross-linked; interstitial spaces are filled with water. A well-known example of this is gelatin, a degraded collagen. Carbohydrates (pectins, mucopolysaccharides) also can form gels.

Dialysis. Molecular size of proteins precludes their diffusion through membranes whose pores measure less than a few nanometers, for example, collodium, parchment, and cellophane, while small molecules and ions can pass through. This phenomenon is applied in dialysis, which separates salts from proteins. Proteins do not dialyze. The modern method of desalting with dextran gels also depends on the size differences of protein molecules and ions (see Section 8).

When equilibrium is attained between the inside and outside of a membrane, the solution inside is under greater pressure—in principle the same situation exists as in an ordinary osmotic cell where no solute at all can pass through the membrane. This pressure difference is called the colloid osmotic pressure. Its magnitude depends on the molarity of the protein solution, and is generally rather low due to the high molecular weight. A serum albumin solution of 48 g protein per liter (as for example in blood) is only 0.7 millimolar (m*M*); its osmotic pressure is 17.8×10^{-3} atm (at 37°). Nevertheless, colloid osmotic effects do play a role in physiology.

8. Purification and Check of Purity of Proteins

To purify proteins is often very difficult. In the first place, body fluids and tissue extracts are comprised of very complex mixtures; second, proteins are very unstable, thus precluding the use of many of the methods of organic chemistry; and third, the check for purity—necessary for the control of each purification step—presents a formidable problem.

Ultracentrifugation, discussed before, may be used for the separation of proteins with different particle weights and is very valuable as a criterion for homogeneity. It is one of the standard methods of protein research.

Precipitation of Proteins. Proteins can be precipitated from solution in various ways. Irreversible denaturation is of interest only for the total analysis of precipitable protein (trichloroacetic acid or uranyl acetate being used) or for the removal of interfering proteins. A reversible process is attempted whenever a certain component has to be isolated in pure form from a mixture of proteins. The most common examples are the isolation of a peptide hormone or an enzyme from tissue extracts.

One of the most important methods is salting-out with a neutral salt, such as ammonium sulfate, sodium sulfate, or magnesium sulfate, since highly concentrated solutions of these salts can be prepared. The salt content of a protein solution is increased stepwise, and after each addition the precipitated protein is centrifuged off. The pH may be varied to allow multiple fractionation (around their isoelectric points proteins become least soluble). Another method, advanced by E. Cohn, is precipitation with organic solvents (alcohol, acetone), carried out under refrigeration.

Chromatography. Separation of proteins is now being achieved by chromatographic methods. The absorbing material is usually a chemically modified cellulose to which charged groups had been added. In principle, this is ion-exchange chromatography (Chapter II,3). Chromatographical homogeneity is another criterion of purity.

Another principle operates in "exclusion" chromatography with dextran gels. Separation is based on the size of the molecules. The gels are hollow, porous beads, filled with water. Ions and low-molecular weight substances easily reach the interior of the beads, whereas larger molecules are excluded. Since the smaller molecules readily penetrate all the meshes of the gels, they are retarded in their passage through a column of this material. The high-molecular weight substances consequently are eluted first. Protein solutions can be desalted conveniently in this way. The use of gels with larger pore size also permits separation of proteins according to their molecular size. These gels often are called "molecular sieves"; one of the commercially available products used is Sephadex.

Electrophoresis. As discussed in Chapter II,1, acidic and basic amino acids are largely ionized in aqueous solution. The charges of the carboxyl group and the α-amino group disappear, of course, during formation of peptide bonds. Nevertheless, proteins are multivalent ions: acidic groups in the side chains of glutamyl and aspartyl residues are dissociated at physiological pH and bear negative charges, and the basic groups of lysine and arginine (and, to a lesser degree, histidine) bear positive charges. Insulin, for example, carries six acidic and six basic groups (including the end groups); with all of them ionized, the insulin molecule carries six positive and six negative charges.

The degree of ionization depends on the pH of the solution. An excess of $H^{\oplus}$ ions (low pH) retards dissociation of the acidic groups; at high pH, the basic groups remain uncharged. At a certain pH, the number of positive and negative charges will be exactly equal. This particular pH is called the *isoelectric point* (pI). At the isoelectric point, proteins are found to be least soluble and can be precipitated most easily.

If an electric field is applied to a solution of proteins, the multivalent protein ions will migrate at rates depending on the voltage applied. The velocity of migration depends also on the number of charges (which determines the effective electric force) and on size and shape of the molecule (which determines the extent of fluid friction; cf. Fig. IV-8). At the isoelectric point no net electric force exists, since the effect of positive and negative charges cancel each other. On the acid side of this point, the protein migrates toward the cathode. If the pH exceeds the pI, then the molecules are negatively charged and migrate toward the anode.

Electrophoresis is a very effective method for separating proteins and for determining impurities in a protein preparation. It is used primarily as an analytical tool, but has also been applied to the preparative separation of various proteins.

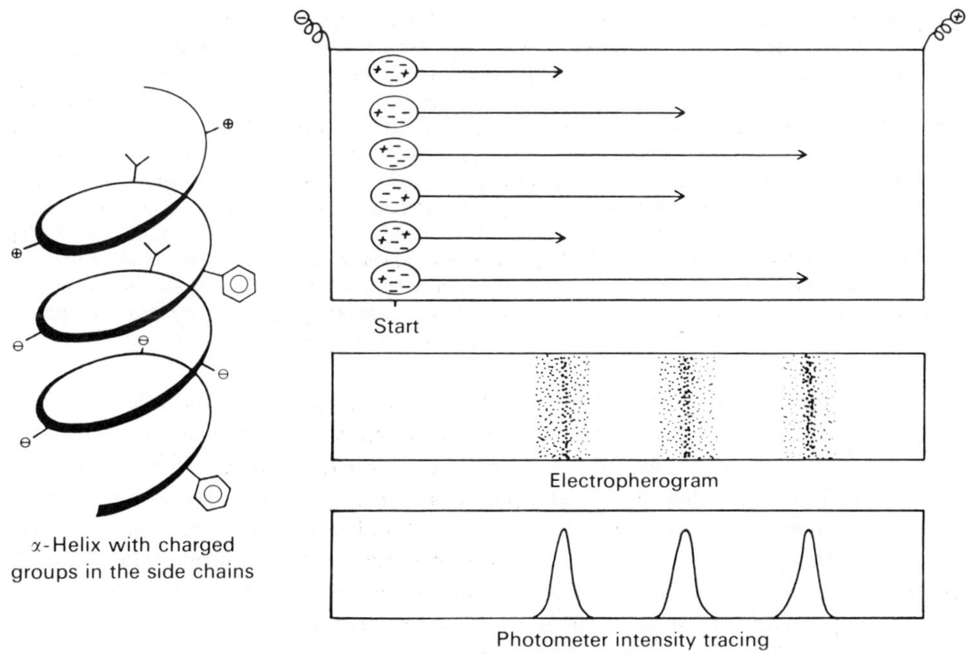

Fig. IV-8. Schematic representation of electrophoresis. At left, a segment of an α-helix which shows the location of certain charged groups. At right, three proteins with different charges are applied to the supporting medium; these proteins migrate during a given time (e.g., 15 hours) as far as the arrows indicate. The "electropherogram" is made visible by staining and evaluated quantitatively by photometry.

Electrophoresis of serum proteins has become an outstanding diagnostic tool in medicine. It is usually carried out on some inert support (paper or starch gel or acrylamide gel), and the proteins are made visible by staining. The resulting "electropherogram" can be evaluated quantitatively. Normally a curve is obtained as shown in Fig. IV-9. For further description of the fractions see Table IV-2.

Even better separation is achieved by "disc electrophoresis," in which technique two acrylamide gels with different pH are layered on each other. This results in a concentration of proteins before their separation and, consequently, in very sharp bands (Fig. IV-9b).

Immunological Differentiation of Proteins. Immunology provides the basis for one of the most sensitive methods for differentiation and identification of proteins. It entails the reaction of a protein solution with certain serum proteins of an experimental animal.

When proteins or carbohydrates foreign to an organism called *antigens*, or also more complex structures like bacteria or viruses, have penetrated into the organism (or have been injected), the organism responds by forming *antibodies*. Antibodies are proteins that combine in a specific manner with the antigen, i.e., with a protein or polysaccharide foreign to the body or with determining groups of bacterial cell walls. In the case of bacterial toxins or viruses, this combination amounts to inactivation of the antigen. The formation of antibody is the basis for immunity.

Under certain conditions, the antigen–antibody complexes are insoluble; upon mixing a solution of antigen with an antibody-containing serum (called simply "antiserum"), the insoluble complex precipitates. This highly specific reaction is used often for the characterization or determination of certain proteins. Two ways of carrying this out in practice are gel diffusion and immunoelectrophoresis. In both techniques, antigen and antibody are allowed to diffuse toward each other in a carrier gel (starch gel or agar). At the point where the two meet, precipitation occurs, recognizable by a white cloudiness in the gel.

Antibodies constitute the γ-globulins of blood plasma. Their chemical structure is discussed in Section 10. They are formed in the *plasma cells* found primarily in the spleen, lymph nodes, thymus, and bone marrow. According to Burnet's theory, there is a large number of different kinds of plasma cells, each determined genetically to produce one specific antibody protein. An antigen penetrating plasma cells, evidently by attachment to antibodylike structures on the cell surface and subsequent phagocytosis, stimulates only the cells that produce the corresponding antibody to divide and to multiply rapidly. The many daughter cells of this type of plasma cell then produce and secrete the specific antibody in quantity.

Each antibody molecule has two specific sites of attachment; the middle portion evidently lacks specificity. So-called antiserum usually contains different antibodies which react with different groups or areas on the causative protein, the antigen. This reaction is highly specific, particularly when carried out quantitatively; it permits the most sensitive distinction between closely related proteins. Nevertheless, cross-reactions can occur; an antiserum to protein A reacts with the closely related protein B, but not as strongly. This is taken as a sign of close kinship and may possibly be due to the presence of some identical areas on the protein surfaces.

Antibodies can also be generated against specific chemical groups (e.g., the benzenesulfonyl or iodotyrosyl residue) by attaching such groups to proteins and then using the modified proteins as antigens. The determining groups are called *haptens*. Low-molecular weight substances alone (not bound to protein) cannot elicit the formation of antibodies. They are nevertheless able to saturate the binding sites of the antibody and thus can prevent precipitation. These relationships have been employed in recent years in the development of highly sensitive and specific assay methods, the so-called *radioimmunoassays*. A great variety of substances in very small amounts can be measured quantitatively by these new methods.

9. Classification of Globular Proteins

In the past, globular proteins were classified according to their solubilities and some of their chemical characteristics. This classification is no longer satisfactory, especially because the "classes" cannot be defined precisely enough. We want to distinguish[7] only among (1) histones, (2) albumins, and (3) globulins.

We have relegated the strongly basic *protamines*, which in the past were considered members of a special class of proteins, to the polypeptides because of their relatively low molecular weight (around 4000).

[7] Often additional classes are listed for glutelins, gliadins, and prolamines, which are small groups of vegetable proteins occurring in grain kernels. There is no reason to grant them special significance by the establishment of exclusive classes.

Histones. Basic proteins, histones are found in the cell nucleus, where they are bound to nucleic acids. It is suspected that histones participate in the regulation of gene activity. When the amino acid sequence of several histones was elucidated, it was noted that histones from pea cotyledons and thymus glands differ only in a few amino acids. Histone molecules can, however, be modified quite specifically by enzymatic methylation, acetylation, and phosphorylation. The significance of the latter reactions is not yet clear.

Albumins. These are proteins which dissolve in water and precipitate from solution only at high ammonium sulfate concentrations (70–100 % of saturation).

Globulins. Slightly soluble in water, globulins dissolve easily in dilute solutions of neutral salts, and precipitate again when the ammonium sulfate concentration reaches half-saturation.

In many individual cases, globulins and albumins can not easily be classified. Furthermore, such a distinction provides very little insight; its value is historical only.

Complex Proteins. These compounds consist of a protein part and an additional, nonprotein, "prosthetic" group. It is rather difficult to maintain the distinction between proteins which adsorb metals and carbohydrates or incorporate them in small quantities, and proteins with a definite metal or carbohydrate component. The usual division of complex proteins comprises the following: (1) metalloproteins; (2) phosphoproteins; (3) lipoproteins; (4) nucleoproteins; (5) glycoproteins; (6) chromoproteins.

(Instead of lipoproteins, glycoproteins, etc., the terms lipoproteid, glycoproteid, etc., are sometimes used.)

Such a heterogeneous combination of components is not always of biological significance. For example, it is uncertain whether the protein in nucleoproteins possesses any specific functions. On the other hand, we are well aware of the role of both the protein component and the prosthetic group in many enzymes (cf. Chapters V and VII).

In some modern textbooks and handbooks proteins are classified according to their origins: blood proteins, milk proteins, plantseed proteins, etc.; or alternatively, they are classified by biological properties: enzymatic proteins, hormonal proteins, etc. Following these principles we shall discuss in some detail blood proteins and use them as examples for other proteins. Then enzyme proteins will be treated mainly with regard to their catalytic properties, but without overlooking their protein nature.

10. Plasma Proteins

Blood plasma is the clear liquid which remains as the supernatant after removal of the erythrocytes by centrifugation. It contains 7–8 % protein, actually a mixture of different proteins, including the clotting factors. If plasma (or whole blood) is allowed to clot, serum is obtained after centrifugation. It differs from plasma essentially in the absence of fibrinogen (cf. this section below). The red blood pigment, hemoglobin, is localized inside erythrocytes; its function is discussed in Chapter IX,4.

Separation of Plasma Proteins. In the past, separation of plasma proteins was carried out by fractional salting-out either with ammonium or sodium sulfate. The analytical characterization of individual plasmas or sera is now accomplished by

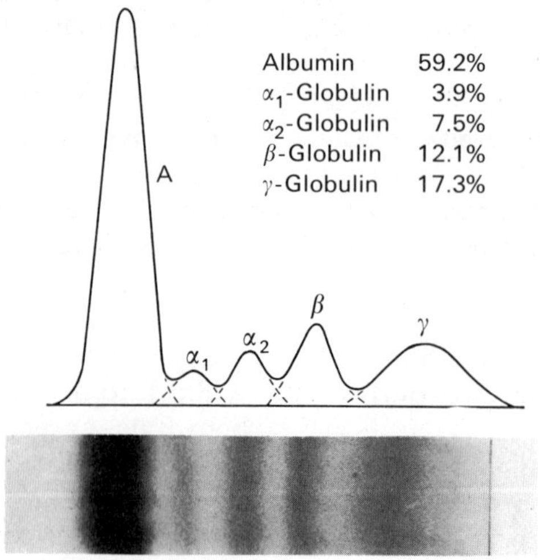

Fig. IV-9a. Electropherogram of a human serum: below, the stained paper strip; above, the photometric curve. (By Dr. Knedel.)

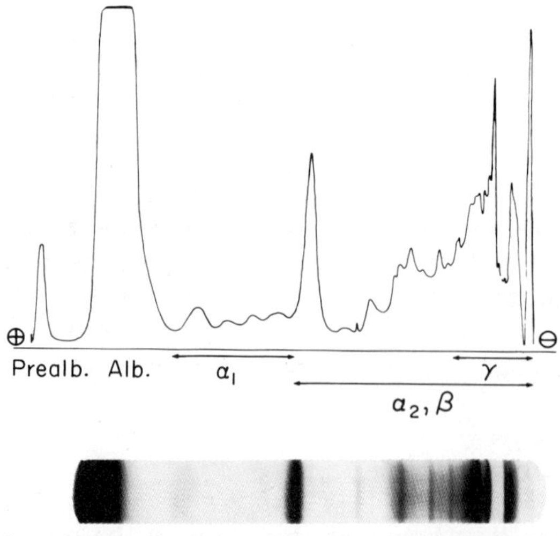

Fig. IV-9b. Separation of components of a serum by disc gel electrophoresis. Below, photograph of the gel after separation. Above, the photometric curve. The globulin fractions clearly can be resolved further. (By Dr. R. Baier.)

electrophoresis, usually paper electrophoresis (Turba; Grassmann and Hannig). In a normal electropherogram, run at pH 8.6, all proteins migrate toward the anode (cf. Fig. IV-8). Prealbumin and albumin travel fastest, followed by the α_1-, α_2-, β-, and γ-globulins. The relative amounts of these fractions permit important clinical conclusions. The electropherogram of a normal serum with its evaluation diagram is presented in Fig. IV-9a. Pathological sera deviate considerably from the normal pattern; "para-proteins," absent in normal sera, occasionally appear.

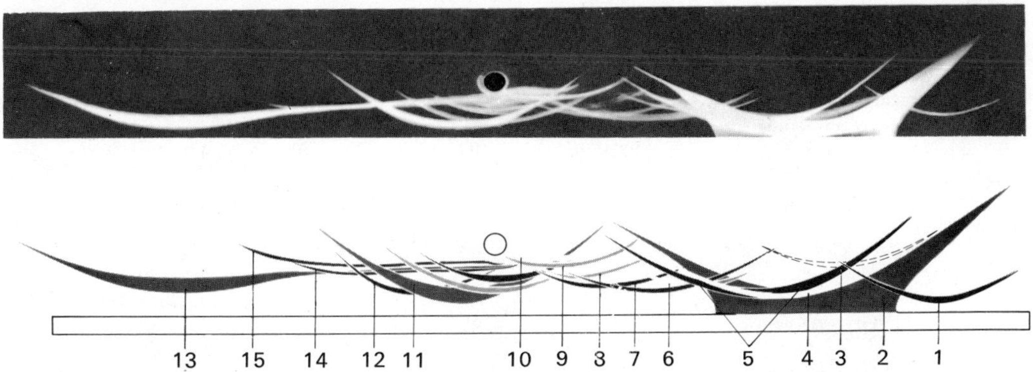

Fig. IV-10. Immunoelectrophoresis of a human serum. The serum sample was applied where the circle is. The trough (the elongated rectangle along the lower edge) was filled with anti-human serum. The place where a serum protein and antibody meet by diffusion (after electrophoresis of the serum) is marked by a precipitation band. The numbers correspond to the serum proteins listed in Table IV-2.

A further separation of serum proteins is possible with immunoelectrophoresis. Electrophoretically separated proteins are allowed to diffuse toward an antiserum; wherever the protein meets an antibody a precipitation ring develops (cf. Fig. IV-10). Apart from electrophoretic mobility, the rate of diffusion and the concentration in plasma govern the position of these precipitation lines. More than thirty serum components have been distinguished this way; the main ones are listed in Table IV-2.

Albumin. With 52–62% of the total protein in plasma, albumin constitutes the largest fraction. The chief functions of albumin are the regulation of osmotic relationships in blood and the provision of a reserve of protein for the organism. In addition, albumin easily and reversibly binds various substances, so that it plays the role of a carrier; in other words, it also has a transport function.

Lipoproteins. Because of their high lipid content (50–90%), lipoproteins are low in density and during centrifugation in appropriate salt solutions rise to the surface. The following categories of density are recognized: very low-density lipoproteins (VLDL, below 1.00), containing about 90% lipid; low-density lipoproteins (LDL, 1.00–1.065), that contain about 20% protein and 80% lipid, have the mobility of β-globulins, and have a high molecular weight of several million; and finally

TABLE IV-2

The Chief Proteins of Blood Plasma

No.[a]	Designation	Molecular weight	Carbohydrate (%)	Concentration in plasma (mg/100 ml)	Biological function
	Albumin fraction				
1	Prealbumin	61,000	0.4	10–40	Binding of thyroxine
2	Albumin	69,000	0.0	3900–5500	Osmotic regulation; transport
	α_1-Globulin fraction				
3	Acidic α_1-glycoprotein (orosomucoid)	44,000	41.4	70–110	—
4	α_1-Lipoprotein	200,000	1.4	290–770	Transport of fats and lipids
5	α_1-Antitrypsin	54,000	12.4	190–500	Trypsin inhibitor
	α_1-Antichymotrypsin	68,000	22.7	14–35	Inhibitor for chymotrypsin
6	α_{1x}-Glycoprotein	—	13.0	10	—
	α_2-Globulin fraction				
7	Ceruloplasmin	160,000	8.0	10–40	Transport of copper
8	α_2-Macroglobulin	820,000	8.4	130–380	Inhibitor of plasmin and trypsin
9	Haptoglobin	10,000	19.3	10–220	Binding of free hemoglobin
	β-Globulin fraction				
10	β-Lipoprotein	$3\text{–}20 \times 10^6$	1.8	290–950	Transport of fats and lipids
11	Transferrin	90,000	5.8	200–450	Binding and transport of iron in plasma
12	β_{1c}-Globulin	—	3.0	80–140	Complement factor C′3[b]
	Fibrinogen	341,000	5.0	200–600	Clotting of blood (fibrin precursor)
	γ-Globulin fraction				
13	Immunoglobulin G (IgG) or γG globulin (γG)	160,000	2.9	800–1800	Antibody against foreign proteins and bacterial antigens
14	Immunoglobulin A (IgA) or γA globulin (γA)	150,000	8.1	120–400	Antibody against foreign proteins and bacterial antigens
15	Immunoglobulin M (IgM) or γM globulin (γM)	960,000	10.9	80–170	Specific antibody reaction, e.g., against isoagglutinins

[a] See immunoelectrophoresis diagram, Fig. IV-10 above.

[b] Complement factors are a group of proteins involved in immune reactions and in immunopathology.

high-density lipoproteins (HDL, above 1.065) with about 50% protein and the electrophoretic mobility of the α_1-globulin fraction, overlapping partially with albumin.

The various lipoproteins probably are aggregates of three main classes of apoproteins that bind lipids (triglycerides, cholesterol, cholesteryl esters, and phospholipids) in varying proportions. Lipoproteins are important because they transport the water-insoluble fats and lipoid materials in the aqueous medium blood by attachment to a hydrophilic protein layer.

Metalloproteins. *Ceruloplasmin* contains copper. It is an oxidase catalyzing the oxidation of $Fe^{2\oplus}$ ions to $Fe^{3\oplus}$ ions. In addition, it acts as a laccase (see Chapter X,6), although the latter appears to be a nonphysiological attribute. *Transferrin* (iron siderophilin) is important for the transport of iron and especially for the new formation of erythrocytes (cf. Chapter XXI,7).

Glycoproteins. Most proteins of the globulin fraction are glycoproteins. Acidic α_1-glycoprotein, formerly called *orosomucoid*, has a particularly high content of carbohydrate. Several glycoproteins have specific binding functions, e.g., for thyroxine or cortisol. The physiological role of other glycoproteins is not known well enough.

Immunoglobulins. These constitute the fraction of the γ-globulins. Three types are distinguished. The *immunoglobulins G* (IgG or γG) with molecular weights around 160,000 and a 3% carbohydrate content comprise the largest portion of the γ-globulin fraction of human serum.

Figure IV-11 schematically shows the structure of a γG-globulin. It consists of four chains, two identical H-chains of molecular weight around 50,000 and two identical L-chains of molecular weight around 23,000. Various main types and subtypes are

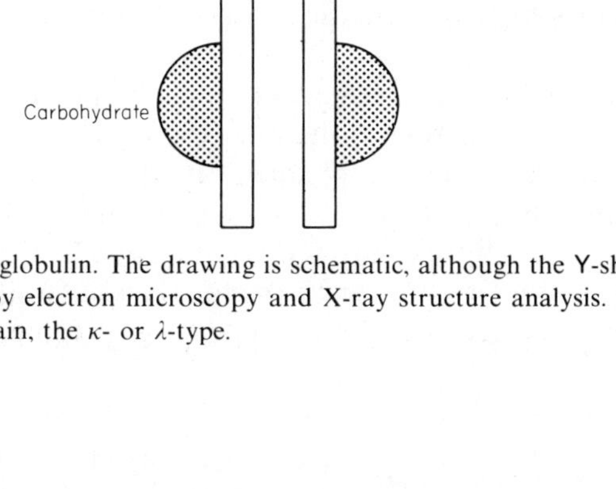

Fig. IV-11. Structure of γ-globulin. The drawing is schematic, although the Y-shape of the molecule has been substantiated both by electron microscopy and X-ray structure analysis. The H-chain can be the γ-, α-, or μ-type; the L-chain, the κ- or λ-type.

known of both the H- and the L-chains. The chains are tied together by disulfide bridges and secondary valences. L- and H-chains share in the composition of the binding site specific for the antigen. Binding between antigen and immunoglobulin is effected by secondary valences. The three-dimensional pattern of reactive groups on the surface of the protein very likely is responsible for binding.

The determination of the amino acid sequence of several L-chains has revealed that the 220 residues are grouped in two halves. The C-terminal half is constant, but the N-terminal half is highly variable, i.e., L-chains of different origin differ in numerous positions. The H-chains also have a variable portion. With this variability, it is possible in principle to explain the specificity. Each specific antibody possesses the amino acid sequence required for optimal secondary valence bonds with the antigen. It is still obscure how these many different sequences are coded genetically.

Besides the immunoglobulins G, there are the *immunoglobulins A* (IgA or γA) with the same molecular weight but a carbohydrate content of 7.5%; and the *immunoglobulins M* (γM, also called macroglobulins) with molecular weights between 900,000 and 1,000,000 and a 10% carbohydrate content. The latter are hexamers of a subunit of similar structure as IgG.

Isoagglutinins are also macroglobulins which have an antibodylike specificity against foreign blood corpuscles, but which are present normally (see Chapter XVII,8).

In patients with tumors of the spleen and bone marrow, plasma cells are far more abundant. As a result anomalous proteins, the *paraproteins*, appear in blood plasma and often even in urine. Paraproteins are closely related to the immunoglobulins. The so-called *Bence Jones proteins* have been studied intensively since they pass the renal filter because of their small molecular weights and since they can readily be isolated from the urine of patients. They are monomers or dimers of L-chains, of either the κ or λ-type. Most of their amino acid sequence is known.

Enzymes. Normal plasma contains only a few esterases in addition to ceruloplasmin and the blood clotting enzymes. However, in certain pathological conditions enzymes from tissue can enter the blood plasma, for example, glutamate-oxaloacetate transaminase (with cardiac infarction) or lactate dehydrogenase (with several liver diseases). Hence, enzyme assays in plasma have acquired diagnostic value.

Fibrinogen, Fibrin, and the Clotting of Blood. The clotting of blood when it leaves the vessels is an important protective feature of the body. It prevents excessive loss of blood. Numerous factors participate in the process of clotting, as shown in Table IV-3 and in Fig. IV-12.

The blood-clotting mechanism is triggered either by factors of the tissue, the *extrinsic* system, or by factors existing in the blood, the *intrinsic* system; the latter are connected with the thrombocytes or platelets. In the extrinsic system, trauma of tissue releases tissue *thromboplastin* (factor III), which converts factor X into its active form Xa, presumably involving the activation of factor VII. Factor Xa is an enzyme that catalyzes the conversion of *prothrombin* (factor II) circulating in the blood into the active form *thrombin*. The last conversion is really a limited proteolysis (as are

TABLE IV-3

Blood Clotting Factors

International designation	Name
I	Fibrinogen
II	Prothrombin
IIa	Thrombin
III	Tissue thromboplastin
IV	Calcium ions
V and VI	Proaccelerin, accelerator globulin
VII	Proconvertin
VIII	Antihemophilic globulin (AHG)
IX	Christmas factor, plasma thromboplastin component (PTC)
X	Stuart-Prower factor, autoprothrombin III
XI	Plasma thromboplastin antecedent (PTA)
XII	Hagemann factor
XIII	Fibrin-stabilizing factor

presumably all activations discussed here), analogous to the conversion chymotrypsinogen → chymotrypsin (cf. Chapter VIII,2). Prothrombin and chymotrypsinogen are indeed homologous proteins.

Thrombin acts on prothrombin in a type of autocatalysis (cf. Chapter VIII,2), but it acts primarily on *fibrinogen* (factor I). Two small peptides are split off the very long, soluble molecule of fibrinogen (molecular weight, 341,000). The monomer of *fibrin* arising from this cleavage polymerizes both end-on-end and side-by-side and thus produces a clot. Last, the clot is stabilized by factor XIII (*fibrin-stabilizing factor*). Unstabilized fibrin is much more susceptible to endogenous fibrinolysis than the stabilized form. The resulting fibrin clot, together with cellular elements of the blood, closes the wound.

In the intrinsic system, the clotting process begins with the activation of a neuraminic acid-containing glycoprotein, the *Hagemann factor* (factor XII). This activation occurs *in vitro* merely by contact with unphysiological surfaces. What releases it *in vivo* is not yet known. The activated Hagemann factor (XIIa) mobilizes a cascade of successive activations (see diagram), which again culminates in the activation of factor X and, subsequently, in the conversion prothrombin → thrombin. Many of these reactions require as cofactors $Ca^{2\oplus}$ ions and phospholipid proteins released by the blood platelets (*thrombocyte factor 3*). The cascade mechanism achieves a tremendous amplification of the primary effect, by a factor of 10^6 or more. However, some of the active factors are very unstable and rapidly become inactivated. This can interrupt the chain of reactions, critical for the maintenance of homeostatic equilibrium.

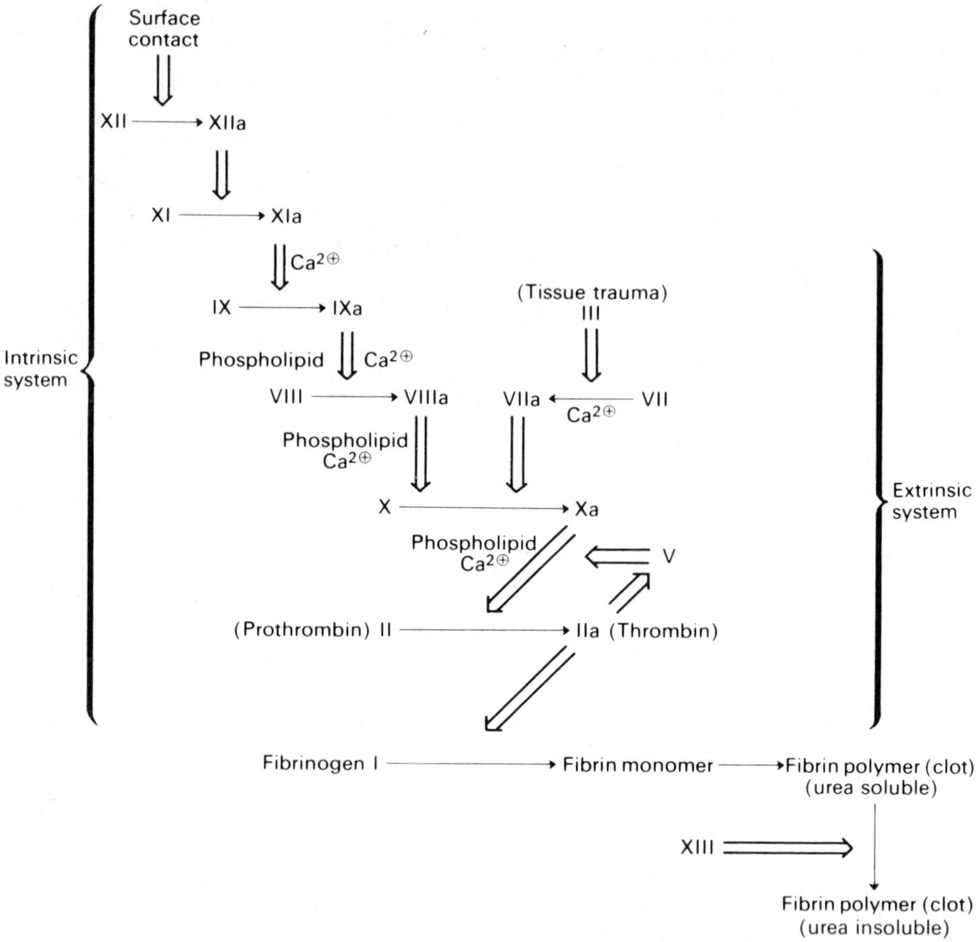

Fig. IV-12. Diagram of blood clotting.

Some details of the scheme presented here are still hypothetical. Many factors have not yet been isolated as homogeneous proteins and are characterized only by their activities. It is also debated whether factors VII, IX, and X are really independent protein components of plasma, or whether they are constituent parts of prothrombin—as suggested by Seegers. According to this suggestion, prothrombin would be a complex made up of the true thrombin precursor and the still inactive factors VII, IX, and X. Many of the clotting factors are very similar chemically. In fact, the suspicion arises that they are all homologous proteins (cf. Section 2) closely related to trypsin.

Clotting may be inhibited or disturbed in several ways. The well-known bleeding condition hemophilia is caused by the absence (or inactivity) of some plasma factor. In classic hemophilia A it is factor VIII (*antihemophilic globulin*); in hemophilia B, factor IX (*Christmas factor*). Because of a deficiency in these factors the transition prothrombin → thrombin is delayed greatly or even prevented.

One physiological inhibitor of clotting is *heparin* which is formed in mast cells. Heparin acts at different steps in the complicated process: First, it inhibits the formation of factor Xa; second, the activation of prothrombin; third, the enzymatic activity of thrombin. Many animal poisons (e.g., the blood poison hirudin extracted from leeches) also inhibit blood clotting. *In vitro*, clotting is prevented commonly by binding $Ca^{2\oplus}$ ions (precipitation with fluoride or oxalate; complex formation with citrate, etc.).

Prothrombin is produced in the liver by a process which requires vitamin K. The drug dicumarol blocks the action of vitamin K; consequently little prothrombin (factors VII, IX, and X) is made available and the clotting time is prolonged.

Fibrinolysis. Physiologically it is of paramount importance that any fibrin clots appearing in the blood stream can be dissolved again. This is accomplished by the fibrinolytic system. As was the case with clot formation, we can distinguish three phases. In the prephase, an activator is released either from the tissue or from the blood itself. Little is known about this phase. The activator then acts in the first phase of fibrinolysis on *plasminogen* (also called *profibrinolysin*) and converts it to *plasmin* (*fibrinolysin*). The latter is a protease with very high affinity for fibrin and capable of splitting fibrin into soluble peptides. The fibrinopeptides act back on the clotting system. They inhibit thrombin action and the polymerization of the fibrin monomers. Fibrinolysis thus also contributes to the maintenance of this important homeostatic equilibrium.

BIBLIOGRAPHY

G. J. Brewer (editor), "Hemoglobin and Red Cell Structure and Function." Plenum, New York, 1972.

E. S. Lennox and M. Cohn, Immunoglobulins. *Annu. Rev. Biochem.* **36,** 365–406 (1967); a concise review in a confusing field.

R. G. Macfarlane, An enzyme cascade in the blood clotting mechanism, and its function as a biochemical amplifier. *Nature (London)* **202,** 498–499 (1964).

H. Muirhead and M. F. Perutz, Structure of hemoglobin. *Nature (London)* **199,** 633–638 (1963); M. F. Perutz *et al.*, Structure of hemoglobin. *Nature (London)* **203**, 687–690 (1964).

H. Neurath (editor), "The Proteins," 2nd ed., 5 vols. Academic Press, New York, 1963–1970.

H. E. Schultze and J. F. Heremans, "Molecular Biology of Human Proteins," 2 vols. Elsevier, Amsterdam, 1966.

S. N. Timasheff and M. J. Gorbunoff, Conformation of proteins. *Annu. Rev. Biochem.* **36,** 13–54 (1967).

CHAPTER V

Enzymes and Biocatalysis

1. Chemical Nature of Enzymes

The enzymes are a very important group of proteins. All the chemical reactions in the organism (i.e., metabolism) are made possible only through the actions of the catalysts that we call "*enzymes*." The substance transformed by an enzyme is termed "*substrate*."

Although the German literature still uses both the terms "ferment" and "enzyme," only the latter is used in the English language, and "fermentation" is restricted to describing bacterial actions. The use of the word "enzyme," proposed by Kühne in 1878, for soluble "ferments" avoids the historical controversy concerning "formed ferments" (yeast and other microorganisms, i.e., intact organisms) and "unformed ferments" (pepsin, trypsin, saccharase). After Buchner's epoch-making discovery that alcoholic fermentation is indeed possible outside the living cell, the concept of formed ferment was dropped, and the designations "ferment" and "enzyme" became synonymous.

Chemically every enzyme known so far is a protein. About one hundred enzymes have been prepared in pure and crystalline form by the methods of protein chemistry; the first of these was urease (Sumner, 1926). Ribonuclease, trypsin, chymotrypsin, and lysozyme are a few of the enzymes whose structure, i.e., amino acid sequence, have been analyzed completely; the sequences of other enzymes are only partially known. The assumption is that a certain sequence of amino acids at and around the active site of the enzyme is responsible for the catalytic effect. This theory is supported by the observation that part of the molecule may be split off some enzymes without loss of activity. Denaturation abolishes catalytic activity, of course, without changing the sequence of amino acids.

Many enzymes are complex proteins; they consist of a protein component and a "prosthetic group." Some enzymes in their active form bind the prosthetic group in a reversible manner. In that case the protein portion is called "*apoenzyme*,"[1] and the prosthetic group, "*coenzyme*":

$$\text{Coenzyme} + \text{Apoenzyme} \rightleftharpoons \text{(Holo-) enzyme}$$

The protein component decides the *substrate specificity*; it determines which substances react and which do not. In many cases this same protein component also determines the direction of the reaction (*reaction specificity*)—in other words, the reaction out of the numerous possible ones undergone by the substrate. This point becomes especially clear in cases where the same coenzyme, i.e., the same prosthetic group, catalyzes different reactions, as does, for instance, pyridoxal phosphate (see Chapter VIII,4), or heme (see Chapter IX,3).

The frequently heard statement "substrate specificity resides in the apoenzyme; reaction specificity resides in the coenzyme," needs therefore to be corrected in its second part, but in any case, coenzymes are involved essentially in enzyme activity as will be discussed in Chapter VI.

Enzymes have been called catalysts or, loosely speaking, accelerators of reactions. In order to understand their action we must first ask why chemical reactions really take place.

2. Chemical Equilibria and Chemical Energetics

Equilibria of Chemical Reactions. A large number of chemical and biochemical reactions attain a measurable equilibrium among their reactants. Two well-known examples are given here:

$$CH_3COOH \rightleftharpoons CH_3COO^{\ominus} + H^{\oplus} \quad (1)$$

$$CH_3COOH + C_2H_5OH \rightleftharpoons CH_3COOC_2H_5 + HOH \quad (2)$$

General chemistry teaches that the law of mass action applies to equilibria. For reaction (1) it may be formulated as follows:

$$\frac{C_{CH_3COO^{\ominus}} \times C_{H^{\oplus}}}{C_{CH_3COOH}} = K$$

[1] The apoenzyme, the protein itself, has also been called a "colloidal carrier." This terminology is based largely on Willstätter's idea that "the molecule of an enzyme consists of a colloidal carrier and an active group with purely chemical activity." In Willstätter's time (around 1920) this statement nevertheless represented scientific progress, because it identified the action of enzymes with chemical properties. Today the concept of a colloidal carrier must be rejected, because it implies that the protein component is inactive, and we now know that it is not. The "group with purely chemical activity" is now designated the *active site* of the enzyme protein.

Here, C_{CH_3COOH}, $C_{CH_3COO^\ominus}$, and $C_{H^\oplus}$ stand for the concentrations (in moles per liter) of the reactants as they are found when equilibrium is established. In the case of the dissociation of acetic acid the equilibrium is established nearly instantaneously; the ester formation of reaction (2), for which we can write an analogous concentration equation, takes more time to reach equilibrium. Finally, the state of equilibrium can be attained from either side of the reaction, i.e., from a mixture of acetic acid and alcohol as well as from ethyl acetate and water.

Every chemical reaction is, in theory, reversible; the equilibrium of some reactions, however, lies too far on one side for the reverse reaction to be detectable.

Chemical Energetics. Let us look at a general equation:

$$A + B \rightleftharpoons C + D; \qquad \frac{C_C \times C_D}{C_A \times C_B} = K \tag{3}$$

It is easy to appreciate that the more vigorously the reaction between substance A and B proceeds, the farther the equilibrium lies to the right, i.e., to the side of C and D, or in other words, the greater is the equilibrium constant K.[2] We can say then that the reaction mixture possesses a high potential energy and during the course of the reaction it drops to a lower potential. This energy change is related quantitatively to the equilibrium constant by the following expression:

$$\Delta G^0 = -RT \ln K \tag{4}$$

In this expression, R stands for the gas constant 1,987 cal/mole · degree, T for absolute temperature, ln for the natural logarithm, and K for the equilibrium constant at temperature T.

The superscript (^{0}) indicates that the quantity ΔG^0 refers to the "standard state," where the reactants are dissolved in a pure solvent at a concentration of 1 mole/liter. It is further supposed that exactly 1 mole of material is reacted. ΔG^0 corresponds to that amount of energy which can be released under these prescribed conditions and is called "free energy," or more precisely, " the change of free energy of the system" during the reaction.

Starting with any arbitrary concentration other than the "standard state," the value for the free energy change is calculated from the standard value by the relation:

$$\Delta G = \Delta G^0 + RT \ln \left(\frac{C_C \times C_D}{C_A \times C_B} \right) \tag{5}$$

Here, C_A, C_B, etc., are again the concentration of the reactants. The expression within the parentheses is the same as is found in the law of mass action, Eq. (3). At equilibrium, i.e., at the concentrations of equilibrium, $\Delta G = 0$.

A reaction which releases energy to its surroundings is called "*exergonic*."[3] In this case, the free energy ΔG bears a negative sign, because, by convention, all quantities of energy which are lost by a system (in the form of heat or work) are negative.

---

[2] This presupposes an uninhibited reaction. To *initiate* a reaction, the energy of activation is of decisive importance (cf. Section 3).

[3] In contrast, reactions in which energy is taken up by the system are termed *endergonic*.

A comparison may be drawn from a business accounting system: expenditures (released energy) are entered with a negative sign; income (added heat or work) is given a positive sign.

Every spontaneous reaction is exergonic. In other words, every reaction proceeds as long as free energy is still decreasing. If no further decrease is possible, then the state of equilibrium is reached, and $\Delta G = 0$. This concept is shown in the diagram below (Fig. V-1).

Again we use ester formation, reaction (2), as an example. The potential energy depends on the concentrations. At the left is the mixture of acid and alcohol with its own potential; at right, ester and water. The equilibrium mixture has the lowest potential. Hence, either of the two reaction mixtures will approach the equilibrium mixture with a release of corresponding amounts of energy in an exergonic reaction.

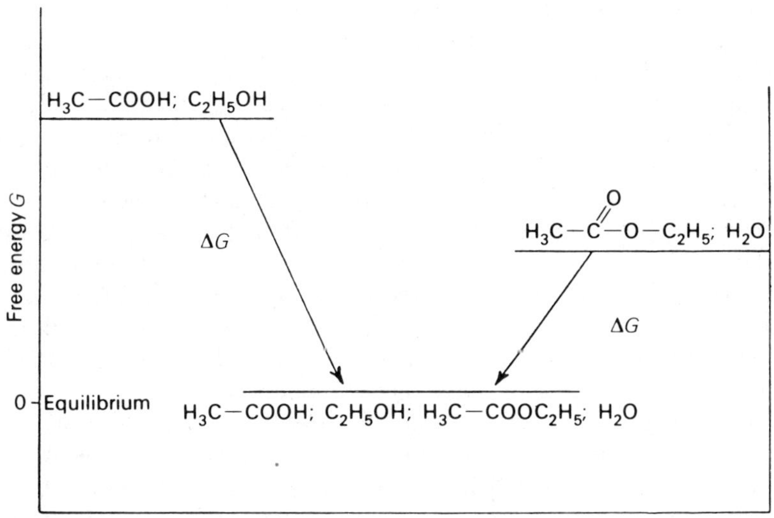

Fig. V-1. Energy diagram of an equilibrium reaction.

The direction of the reaction, therefore, depends on the concentrations, and in many cases it is important for us to be able to calculate from the *standard change* of free energy ΔG^0 by Eq. (5) the actual free energy change ΔG, which depends on the concentrations actually at hand. If this ΔG is negative, then the reaction runs from left to right, as usually written; if ΔG is positive, then it proceeds in the reverse direction.

Physical chemistry teaches that the change of free energy during a chemical reaction is defined by two other thermodynamic quantities of state:

$$\Delta G = \Delta H - T\Delta S \tag{6}$$

ΔH means change of heat content[4] (enthalpy; basically the change of "internal energy"), while ΔS means change of entropy content.[5]

A more detailed explanation of the thermodynamic concepts cannot be included here,[6] save to mention that in several processes, such as protein denaturation, the increase in entropy, the rise of molecular disorder, is the driving force of the reaction.

3. Catalysts and Enzymes

If the equilibrium were always reached quickly there could be neither organic chemistry nor any life on our planet. The equilibrium of most organic compounds in the presence of atmospheric oxygen greatly favors the oxidation products CO_2 and H_2O. If the reactivity of such substances is raised by heating them, they burn. At room temperature they are metastable; though they are not at equilibrium, they do not change. Only when a certain amount of energy, the energy of activation, has been applied will they react with atmospheric oxygen. The same holds true for other reactions. We must then correct the above diagram of energy levels as shown in Fig. V-2a. At first some energy of activation must be absorbed (usually from thermal motion), before the reactants—hydrogen and oxygen in this case—can react to attain equilibrium and form water. The energy of activation can frequently be calculated from empirical data. The greater this energy, the less reactive the substances.

There is one way to lower the energy of activation: the addition of a catalyst. An example may help to explain this. A mixture of oxygen and hydrogen at room temperature will remain unchanged for a long time because the energy of activation is quite high. Local heating, as through an electric spark, can initiate the reaction, which is violent because of the large release of free energy. If instead we introduce a catalyst, say a platinum sponge, then the reaction starts at room temperature and will finish quickly, with the release of heat. The catalyst's job here is to lower the energy of activation, or in other words, to make the molecules more reactive.

The explanation given, using the formation of water as an example, is also applicable to biochemical reactions. The vast majority of these reactions become possible solely through the aid of catalysts, which we call enzymes. Thermodynamically, catalysts lower the necessary energy of activation of the reaction and thus facilitate reaching equilibrium. Enzymes achieve this according to the principle of "intermediary catalysis": initially an enzyme-substrate complex, ES, is formed which, in the reaction proper, becomes a complex of enzyme and product, EP. The latter complex then dissociates into enzyme + product, whereby the enzyme is regenerated and free to

[4] The change in enthalpy differs from change of heat content (developed or absorbed quantity of heat) only by its sign: $\Delta H = -Q_p$.

[5] Entropy is a measure of the lack of molecular order. According to Boltzmann, entropy is a measure of the probability of a state: $S = k \cdot \ln W$. Disordered states always are more probable than ordered ones. Entropy is expressed in cal/°K or "entropy units;" the dimension of $T \cdot \Delta S$ thus becomes cal, or the dimension of energy.

[6] For an introduction to biochemical thermodynamics, see I. M. Klotz, "Energy Changes in Biochemical Reactions." Academic Press, New York, 1967.

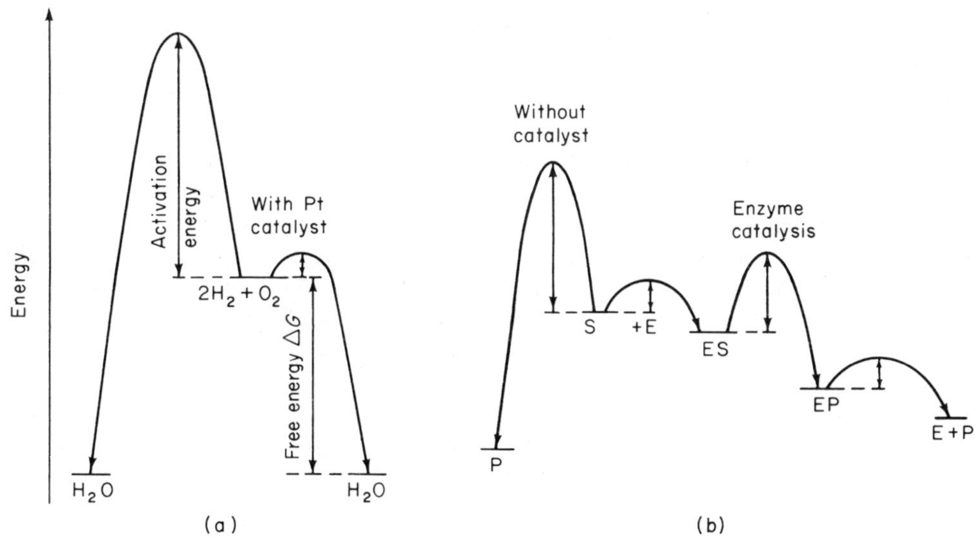

Fig. V-2. Energy of activation and the function of catalysts.

associate with another substrate molecule. The energy of activation of each of these steps is considerably smaller than it is for the overall noncatalyzed reaction (cf. Fig. V-2b). The catalyzed reaction therefore proceeds much faster.

The net amount of free energy (ΔG) remains unchanged in the process, and as a result the equilibrium position also remains the same. Catalysis "beyond the state of equilibrium," i.e., a shift of equilibrium, is not possible, however. Every reaction, though catalyzed by an enzyme, proceeds only until equilibrium is reached. This same equilibrium would have been reached in the presence of some inorganic catalyst, or even without the aid of a catalyst; the equilibrium is defined alone by the equilibrium constant K.[7] This fundamental law of the action of enzyme must never be forgotten.

There is only one possible way to maintain a continuous reaction: two enzymatic reactions in tandem, in which the second reaction continually uses up a substance—often present in low concentration—that is being produced by the first reaction. Such a situation is also known in noenzymatic reactions, e.g., in inorganic chemistry. Loosely speaking, the equilibrium is "shifted" by the removal of the reaction product. More correctly we say that equilibrium concentrations are never established, simply because one of the reaction products reacts further and continuously disappears and has to be resupplied by the first reaction.

In this way even an endergonic (energy-requiring) reaction can proceed in continous fashion. The subsequent reaction, however, must be exergonic so that the overall

[7] Were it different, i.e., were an enzyme able to influence the position of equilibrium, one could construct a *perpetuum mobile*; according to the laws of thermodynamics, however, this is not possible.

ΔG for both reactions has a negative value, i.e., the combination of reactions must be exergonic.

4. Dynamic Equilibria and the Steady State

The above-mentioned familiar example of the formation of water also points out the limits of the equilibrium theory. In biochemistry, this theory is very useful in explaining energies and the nature of biocatalysis, but it would be a gross mistake to assume that the organism is anywhere near chemical equilibrium ($\Delta G = 0$). L. von Bertalanffy put is succinctly:

> "A closed system[8] at equilibrium neither needs energy for its maintenance, nor can energy be obtained from it. The chemical equilibrium for this reason, is unable to produce work. In order for a system to perform work, it must *not* be at equilibrium, but rather it must *tend* toward equilibrium. And in order for the system to be able to persist in its tendency, it must be kept in a steady state. Such is the situation with the organism, whose constant capacity for work is insured by the fact that it is an open system."

In a dynamic equilibrium or steady state, substances flow in continuously and reaction products escape. It is part of the nature of dynamic equilibria to have transport phenomena at the boundaries of the system. Along with other parameters, the transport phenomena determine the steady state concentrations of individual substances within a system. Evidently the organism fulfills these conditions: food and oxygen are taken up; CO_2 and waste products are eliminated. The organism is truly an open system and can develop and maintain itself only as an open system, i.e., a system in exchange with its environment. Different laws apply for an open system at dynamic equilibrium than for a closed system at true equilibrium. The theory behind this is very complicated and has not yet been developed completely; we will not discuss it further save for one point: in a closed system a catalyst, or an enzyme, does not disturb the position of equilibrium. In an open system, in contrast, an enzyme is very well able to regulate the steady state concentration of a reacting substance. For example, the substrate concentration may drop because more substrate is turned over than is supplied from the environment whenever there is more enzyme present. Furthermore, the system is capable of developing forces to counteract disturbances of the steady state.

A very good example from biochemistry is the respiratory chain, which in a limited way represents a system in dynamic equilibrium. It depends on steady state concentrations of the various redox systems. It will be discussed further in Chapter X,4.

In summary, it is impossible to comprehend the action of enzymes and to experiment with biochemical systems and their energetics without a firm foundation in chemical equilibria. In many experimental situations, the enzyme governs the arrival

---

[8] In physical chemistry, the expression "system" is used for that segment of the universe which is being studied at the time. A *closed system* exchanges with its environment neither matter nor energy; material transport and exchange of energy is the mark of an *open system*.

at equilibrium, but not its position. In contrast, the organism, like its own individual cells, must be considered as an open system which normally maintains a dynamic equilibrium. In a dynamic equilibrium, steady state concentrations will always be established differently from those dictated by a true chemical equilibrium, which is governed by thermodynamics. Hence, reactions tending toward equilibrium take place continuously, and it is these reactions which provide the energy necessary for the organism's vital processes.

5. Energetic Coupling and Energy-Rich Compounds

For thermodynamic reasons endergonic reactions proceed neither spontaneously nor through enzymatic catalysis. An endergonic reaction has to be coupled with another reaction which is sufficiently exergonic to let the sum of ΔG for both be equal to zero or be negative (cf. this section). The most important mechanism of coupling is the formation of an energy-rich (hence, very reactive) intermediate compound. The same principle is employed in organic chemistry when acid chlorides are prepared for the synthesis of amides and esters (cf. Chapters I,2 and II,1). In biochemistry, this energy-rich "activated" component is very often *adenosine triphosphate* abbreviated as ATP:

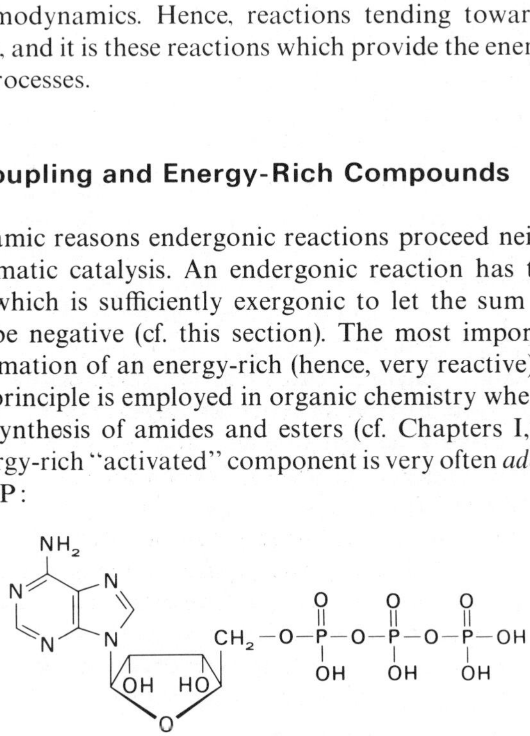

ATP

Let us consider for example the formation of glucose 6-phosphate, i.e., an ester of phosphoric acid.[9] In the reaction between H_3PO_4 and the alcohol group of glucose $\Delta G^0 = +3$ kcal/mole, and at equilibrium one would find only minute amounts of glucose 6-phosphate. If phosphoric acid is, however, first raised to a high energy level by being attached to *adenosine diphosphate* (ADP) to give the triphosphate (ATP), then the transfer of phosphate from ATP to glucose becomes exergonic ($\Delta G^0 = -4$ kcal/mole) and runs nearly to completion. Here phosphoric acid had to be bound first in an energy-rich way, which we can formalize[10]:

$$\text{ADP} + \text{P} = \text{ATP} + H_2O \qquad \Delta G = +7 \text{ kcal}$$

The reaction requires an energy expenditure of 7 kcal/mole, which has to be provided by some other exergonic reaction. This may come, e.g., from the transport of electrons in the respiratory chain (Chapter X,5). In the following diagram the necessary energy

[9] This reaction is discussed in more detail in Chapter XV,7.

[10] The phosphate radical, symbolized in the text by P, will appear as Ⓟ in the drawings.

is provided by the oxidation of glyceraldehyde to glyceric acid:

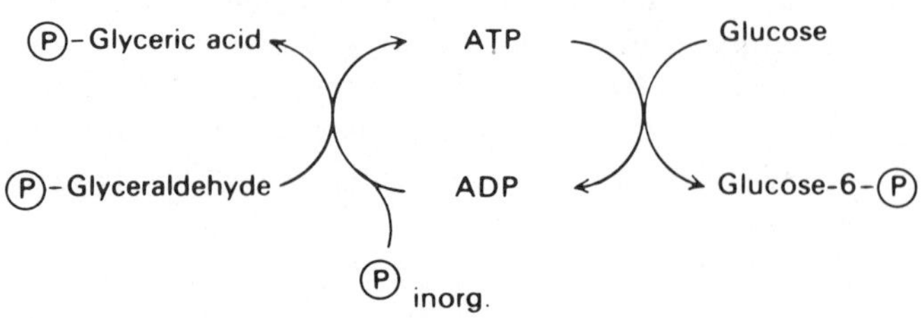

Here the adenylic acid system is subjected to a cyclic process: ATP is consumed, but regenerated through a coupled reaction. The decisive feature is that the coupling mechanism is a chemical reaction. In our example, the phosphate group is transferred with the help of an enzyme from ATP directly onto glucose. The regeneration of ATP similarly proceeds over a corresponding intermediate reaction. We frequently encounter this type of interrelationship in the study of metabolism; the living cell is comparable to a chemical machine which converts chemical energy in a variety of ways and tries to maintain it in the form of chemical energy as long as possible. For this purpose several versatile "energy carriers," such as ATP, acyl-coenzyme A, and others, are of particular importance, since they can exchange energy among various reactions (cf. also Chapter VI).

At this point we must warn of a possible misunderstanding. Although chemical energy (free energy of chemical reactions) is measured and expressed in kcal units, it is nevertheless not identical with heat energy. The organism is incapable of utilizing heat energy applied to it; to maintain the operations of life the organism depends on chemical energy in the form of food.[11] Chemical energy is easily converted to heat, but not the reverse. One ought to be aware of the confusion arising from the use of the same units (kcal) for different forms of energy.

Chemical energy which is released from "combustion" processes of the cell is largely trapped in the form of these energy-rich bonds and then used in a great variety of tasks. The number of ATP molecules produced is sometimes taken as a measure of the useful energy from biochemical exergonic processes.

Group Transfer Potential and the "Energy-Rich Bond." Bonds such as ester, acid anhydride, and similar ones are described as "energy-rich" because their hydrolysis releases relatively large quantities of energy (more than about 5 kcal/mole). The choice of the term "energy-rich bond" is unfortunate because it leads to confusion with the physiochemical concept of "bond energy." The latter refers to the free energy of breakdown of a compound into its component atoms or radicals. Such *molecular cleavages* are strongly *endergonic*; the cleavage of a primary valence bond requires 50–100 kcal/mole; that of a secondary valence bond (e.g., a hydrogen bond),

---

[11] Plants are able to utilize the radiation energy of sunlight for biochemical reactions. This process, of course, makes life on this planet possible. No other forms of energy described in physics are used profitably.

about 5–8 kcal/mole. Furthermore, the strongly *exergonic hydrolysis* of "energy-rich" compounds is categorically avoided in the cell. Instead, parts of molecules are transferred—the phosphate group in the above example. It is thus more appropriate to employ the term *group transfer potential*, and in formulas to mark those groups that are bound in such a manner with the squiggle sign instead of the single bond. This symbol will be used regularly in this text to designate groups that are transferred easily (i.e. in exergonic reaction). For the sake of comparing the group transfer potentials of metabolites such as ATP, acyl-CoA, and other "energy-rich" compounds, it has become conventional to refer to one common reaction partner, namely water. In effect, the free energies of hydrolysis are compared. The values for several important compounds have been compiled in Table V-1.

TABLE V-1

Standard Free Energy of Hydrolysis of Some Energy-Rich Compounds

Compound	ΔG^0 (kcal/mole)
Adenosine triphosphate, ATP (→ADP + orthophosphate)	7.0[a]
Adenosine triphosphate, ATP (→AMP + pyrophosphate)	8.6
Pyrophosphate (→2 orthophosphate)	6.7
Creatine phosphate	10.2
Phosphoenolpyruvate	12.7
3-Phosphoglyceryl 1-phosphate (→3-phosphoglycerate + orthophosphate)	13.6
Acetyl-coenzyme A (Chapter VI,7)	8.2
Aminoacyl-AMP (Chapter VII,6)	7.0

[a] It can be estimated that in the cell the cleavage of ATP releases not 7 but 9–10 kcal per mole of available chemical energy. (See above discussion.)

These "standard energies" are merely theoretical values; the energy actually available to the cell depends fundamentally on the ratio of concentrations of reactants. For the cleavage of ATP, for example, under physiological conditions the energy amounts to 9 to 12 kcal/mole.[12]

The estimation of this value involves the following consideration: Applying equation (5) in Section 2 above to the hydrolysis of ATP, we get

$$\Delta G = \Delta G^0 + RT \ln \frac{[ADP^{2\ominus}][HPO_4{}^{2\oplus}]}{[ATP^{4\ominus}]}$$

Physiological concentrations are of the order of magnitude of $10^{-3}M$ for ADP and ATP, and $10^{-2}M$ for phosphate.[13] In the special case where [ADP] = [ATP] these two values cancel, and with

[12] See R. C. Phillips, P. George, and R. J. Rutman, "Thermodynamic data for the hydrolysis of adenosine triphosphate as a function of pH, $Mg^{2\oplus}$ ion concentration, and ionic strength. *J. Biol. Chem.* **244,** 3330–3342 (1969).

[13] The abbreviation "*M*" stands for the unit of concentration "molarity," i.e., moles per liter. See also Chapter XXI.

$[HPO_4^{2\ominus}] = 10^{-2} M$ we obtain

$$\Delta G = -7000 + 1.98 \times 310 \times (2.3 \times \log 10^{-2}) = -9820 \text{ cal/mole}$$

If the ratio ADP/ATP changes to 1 : 10, then ΔG shifts even more, by 1.4 kcal/mole in this case. If the pH deviates from 7.0, then that too leads to a correction, because it affects the ionic relationships of the phosphates. For many biochemical reactions, the pH value is a separate term in the equation.

6. Specificity of Enzyme Catalysis

Reaction Specificity. The foregoing discussion of the role of enzymes has emphasized that enzymes cannot shift the equilibrium; rather they accelerate the establishment of equilibrium or, in cases of very sluggish reactions, they make the reactions possible. In this way the enzyme is also capable of selecting a single reaction from a number of possible reactions and catalyzing that one alone. The energy of activation is depressed enough to restrict the attainment of equilibrium exclusively to this particular reaction. This phenomenon is called "*reaction specificity*" of the enzyme. The enzyme catalyzes only one out of numerous thermodynamically possible conversions of a substance. (The same applies, incidentally, for many inorganic catalysts employed by the chemical industry.) Another enzyme with another reaction specificity initiates another reaction. Here is an example:

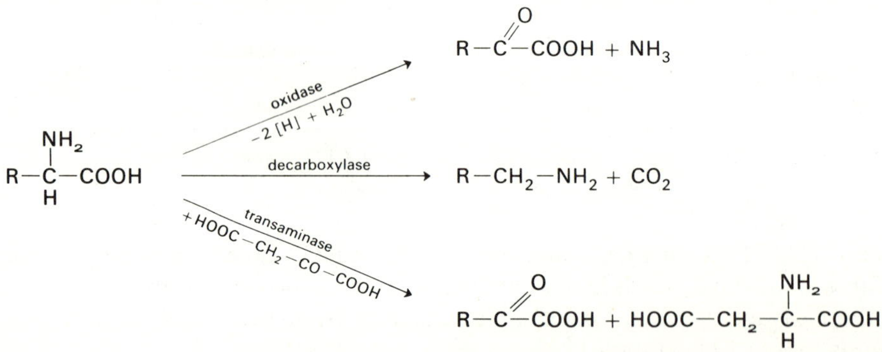

Amino acid oxidase oxidatively deaminates the amino acid at the left (one of many); the enzyme specifically catalyzes this particular reaction as described extensively in Chapter VIII,7. Another possible reaction, decarboxylation, does not occur. The catalysis of CO_2 loss requires a different enzyme, and a third reaction, transamination, requires a third enzyme assisting in the exchange of functional groups (the keto group of oxaloacetic acid with the amino group). Obviously each of the three enzymes possesses a characteristic reaction specificity; this is true for all enzymes.

The last two enzymes happen to use the same coenzyme (pyridoxal phosphate, Chapter VIII,4). The reaction specificity therefore depends on the protein component, not on the coenzyme (cf. Chapter VI,1). This selectivity of reactions by the enzymes is of prime importance for life processes in general.

The following mechanism furnishes a likely (but still incomplete) explanation of reaction specificity: Enzyme and substrate form a weak bond from which the reaction on the substrate may then proceed. The kind of bond and/or steric arrangement between enzyme and substrate prepares for the specific reaction. In bimolecular reactions, two substrates must be bound simultaneously to the enzyme surface to allow the substrates to react with each other. Frequently, one of the reaction partners is a coenzyme (see Chapter VI).

Substrate Specificity. The principle of selection operates furthermore in the binding of the substrate to the enzyme. Certainly not every substance which might be able to undergo a particular reaction is bound. In the example of the amino acid decarboxylase, a few amino acids are bound tightly, some loosely, and others not at all. The latter do not react, i.e., their decarboxylation is not catalyzed. This kind of selectivity is called "*substrate specificity.*" It is especially characteristic for optical antipodes (mirror-image isomers). Usually only one of the two isomers reacts, or at least one antipode reacts much faster.

The extent of substrate specificity varies from enzyme to enzyme. A few hydrolases are relatively nonspecific, others require substrates containing certain groups (group specificity; e.g., β-galactosidase and α-glucosidase, which cleave all β-galactosides and α-glucosides, respectively; cf., Chapter XVII,3). Other enzymes react very specifically with one substrate only; closely related substances may be bound by the enzyme but are not brought to reaction and thus behave as inhibitors (for example, *p*-aminobenzoate and *p*-aminosulfonate or -sulfonamide in the biosynthesis of folic acid).

The metabolism of fluoroacetate, $CH_2F{-}COO^{\ominus}$, illustrates this point nicely. Fluoroacetate is very similar to acetate, $CH_3{-}COO^{\ominus}$, and is converted just like the latter to the CoA derivative, in which form it condenses with oxaloacetate to fluorocitrate (cf. Chapter XI,1). Both the activating enzyme and the condensing enzyme, citrate synthase, treat fluoroacetate like acetate itself. Aconitase, on the other hand, the next enzyme in the citric cycle, is very strongly inhibited by fluorocitrate. Fluorocitrate is bound by the enzyme surface (thus replacing citrate) but cannot undergo the reaction (hence competitive inhibition, Section 8). Aconitase appears to be an enzyme of higher specificity than citrate synthase. The great toxicity of fluoroacetate, therefore, is explained by its blocking of the citrate cycle.

Isoenzymes. These are very similar chemically and physically and catalyze the same reaction, but they possess different, genetically determined primary structures. They can be separated from each other by ordinary means of protein chemistry, such as electrophoresis. Often they differ in properties that are important for the regulation of activity; the biological function of the multiple forms may be found in this observation.

This phenomenon has been studied thoroughly especially in the case of the five forms of *lactate dehydrogenase*, which can be separated by electrophoresis. These five isoenzymes can be reduced to two genetically determined subunits A and B. The enzyme molecule invariably consists of four subunits. Thus the five combinations A_4, A_3B, A_2B_2, AB_3, and B_4 are possible. If the isoenzymes A_4 and B_4 are isolated and broken down to their subunits and if the mixture is allowed to reaggregate, then a mixture of all five isoenzymes is obtained, as demonstrated by electrophoresis. The relative proportions of isoenzymes of lactate dehydrogenase vary from organ to organ. It is not yet known what determines the kind of aggregation.

7. Enzyme Kinetics

Chemical kinetics deals with velocities of reactions. As discussed above, enzymes (or catalysts in general) alter the velocities of reactions. Thus we can take the reaction velocity achieved as a measure of the amount and activity of an enzyme. The rate (= velocity) of a reaction is defined as the *conversion of a substance per unit of time*; the conventional unit for enzyme-catalyzed reactions is micromoles per minute. The amount of enzyme that converts 1 micromole of substrate in 1 minute under optimal conditions is called *one unit of enzyme*.[14]

Enzyme units are determined even in impure enzyme preparations. Indeed, they are important control parameters during the isolation and purification of enzymes. The purer a preparation of an enzyme, the more units per milligram of protein will be found.

The simplest case of a chemical reaction is the irreversible change of substance A to substance B. An example of this is the radioactive decay of a nuclide. The following relationships apply for the reaction rate dc/dt and the amount present at time t:

$$-\frac{dc}{dt} = k \cdot c; \qquad c = c_0 \cdot e^{-kt}$$

(c, concentration; c_0, concentration at $t = 0$; k, rate constant.)

It can be seen that the rate is always proportional to the amount present. Another helpful concept is half-life, the time required for one-half of a given amount to decay; it is a constant for each radioactive isotope. By analogy, the biological half-life stands for that time during which one-half of a certain substance is broken down or eliminated. Many biological reactions follow this simple exponential law.

In a bimolecular reaction $A + B \rightarrow AB$, the rate of reaction is a function of the concentrations of both reaction partners. The equation in this case becomes

$$-\frac{dc}{dt} = k \cdot c_A \cdot c_B$$

Most reactions are reversible and tend toward attainment of equilibrium. In the state of equilibrium, the forward and backward reactions proceed equally fast (dynamic equilibrium). Applying these boundary conditions to a simple case, e.g., a dissociation

$$AB \underset{k_{-1}}{\overset{k_1}{\rightleftharpoons}} A + B$$

then we obtain the familiar law of mass action:

$$\frac{c_A \cdot c_B}{c_{AB}} = \frac{k_1}{k_{-1}} = K$$

[14] In a new proposal of nomenclature the amount of an enzyme is to be expressed in "catal" units. One *catal* effects the conversion of 1 mole of substrate per second. In practice, microcatals would be used (i.e., the turnover of 1 μmole/second); 1 μcat is equivalent to 60 enzyme units.

Michaelis' Theory. If indeed our assumption holds true that enzyme and its substrate form a complex,

$$\text{Enzyme} + \text{substrate} \rightleftharpoons \text{enzyme-substrate complex, or}$$

$$E + S \underset{k_{-1}}{\overset{k_1}{\rightleftharpoons}} ES \tag{7}$$

then we can again formulate the relationship of the law of mass action,

$$\frac{[E] \cdot [S]}{[ES]} = K_s \tag{8}$$

According to the fundamental precepts of "catalysis of intermediates," the enzymatic reaction itself can proceed only *via* the complex. The products are formed this way:

$$ES \xrightarrow{k_2} P_1 + P_2 + E \tag{9}$$

This reaction is assumed to run practically to completion (presupposing the likely situation that the reaction is exergonic and the equilibrium lies far to the right). At the same time, the catalyst is being regenerated. Michaelis and Menten then supposed that this last reaction (9) is rate limiting, i.e., that it runs much slower than the formation of the enzyme-substrate complex. The speed of the overall reaction, which is directly measurable, is therefore proportional to the concentration of enzyme-substrate complex [ES], and conversely, the speed of the reaction can be taken as a measure of the ES concentration.

Equation (8) can be interpreted as follows: Starting with a given amount of enzyme, if the substrate concentration is raised gradually, more and more enzyme will be converted into the complex ES and the rate of reaction will increase until finally virtually all the enzyme is in the form of ES. The enzyme is then saturated; at this point the reaction rate is maximal. This relationship is represented graphically in Fig. V-3.

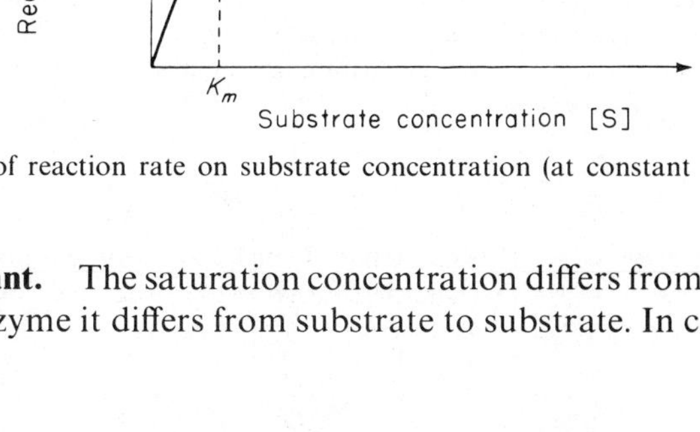

Fig. V-3. Dependency of reaction rate on substrate concentration (at constant enzyme concentration).

Michaelis Constant. The saturation concentration differs from enzyme to enzyme, and for any one enzyme it differs from substrate to substrate. In contrast to maximal velocity, it cannot be read easily from the graph. But the situation is better defined by choosing as a point of reference the half-maximal velocity. According to our hypothesis, at this point half of the entire enzyme is in form of the ES complex and the other half as free E, since the reaction rate should be proportional to the ES concentration. In Eq. (8) the terms E and ES cancel, leaving

$$[\mathrm{S}]_{\text{at half-maximal velocity}} = K_m$$

To express this in words: That substrate concentration at which half-maximal reaction velocity is reached equals the dissociation constant of the enzyme-substrate complex. This constant is named the "Michaelis constant," K_m, after the originator of the theory. Its dimensions are those of substrate concentration (moles/liter); when $[\mathrm{S}] = K_m$ the enzyme is only half saturated.

A large Michaelis constant means that a high substrate concentration is necessary to attain half saturation; the enzyme possesses a low affinity for the substrate in question. It will preferentially bind another substrate whose Michaelis constant is smaller and probably bring that substrate to reaction, too. Michaelis constants usually range between 10^{-2} and 10^{-5} moles/liter.

It is relatively easy to determine the Michaelis constant, K_m, experimentally (i.e., to find the concentration of substrate at which half-maximal reaction velocity is attained). We are assuming that the reaction velocity v is proportional to the concentration of ES:

$$v = k_2 \cdot [\mathrm{ES}] \tag{10}$$

Maximal velocity, $V_{\max}$, is reached when the enzyme is saturated with substrate—in other words, is present virtually entirely in the form of the ES complex. Designating the total amount of enzyme as E_t, then we obtain

$$V_{\max} = k_2[\mathrm{E}_t]$$

Rearrangement of Eq. (8) results in

$$\frac{([\mathrm{E}_t] - [\mathrm{ES}]) \cdot [\mathrm{S}]}{[\mathrm{ES}]} = K_m; \qquad [\mathrm{ES}] = \frac{[\mathrm{E}_t] \cdot [\mathrm{S}]}{K_m + [\mathrm{S}]} \tag{11}$$

and the substitution of Eq. (10) yields

$$v = \frac{k_2[\mathrm{E}_t] \cdot [\mathrm{S}]}{K_m + [\mathrm{S}]} = \frac{V_{\max} \cdot [\mathrm{S}]}{K_m + [\mathrm{S}]} \tag{12}$$

This expression contains, with the exception of K_m, only easily measurable variables, the reaction rate v at a given substrate concentration and the rate $V_{\max}$ achieved at saturation of the enzyme. These measurements then permit the calculation of K_m. The evaluation is simplified greatly if the reciprocal of Eq. (12) is used instead (procedure of Lineweaver and Burk):

$$\frac{1}{v} = \frac{K_m + [\mathrm{S}]}{V_{\max} \cdot [\mathrm{S}]}; \qquad \frac{1}{v} = \frac{K_m}{V_{\max}} \cdot \frac{1}{[\mathrm{S}]} + \frac{1}{V_{\max}} \tag{13}$$

Since $1/v$ and $1/[\mathrm{S}]$ are the variables, the equation to the right represents the function of a straight line, $y = ax + b$. By plotting $1/v$ against $1/[\mathrm{S}]$, $V_{\max}$ and K_m are found easily from the coordinate intercepts (see Fig. V-4).

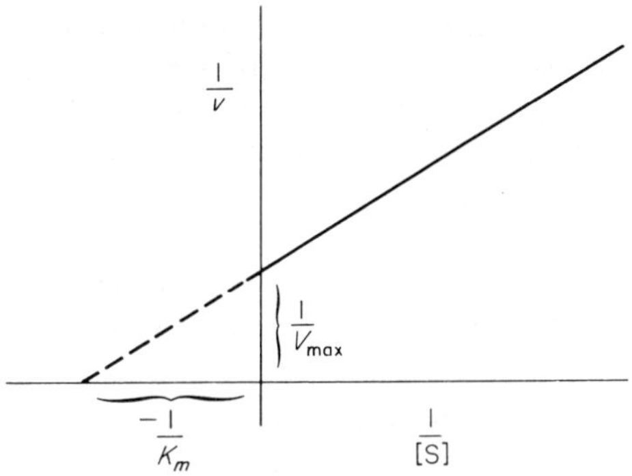

Fig. V-4. Determination of V_{max} and K_m from the Lineweaver–Burk plot of enzyme kinetics found experimentally.

The Michaelis constant determined in this way (= [S] at $v = 1/2\ V_{max}$) is a dynamic value. More rigorous calculation shows that the rate constant k_2 must be included:

$$K_m = \frac{k_{-1} + k_2}{k_1}; \qquad K_s = \frac{k_{-1}}{k_1}$$

For this reason a distinction is made between Michaelis constant K_m and the substrate constant K_s, which is defined by Eq. (8) or kinetically by the expression immediately above. If k_2 is very small, i.e., the reaction according to Eq. (9) is rate determining (the essential premise of Michaelis and Menten), then the term k_2 becomes insignificant compared to k_{-1} and K_S and K_m become numerically identical. This is true for many enzyme reactions.

Molecular Activity. Many enzymes have thus far been isolated in pure and crystallized form. In these cases, the *molecular activity* can be determined; it is defined as the number of molecules of substrate transformed per minute per molecule of enzyme (or the number of μmoles of substrate per μmole of enzyme, i.e., the number of enzyme units per μmole of enzyme). The term "*turnover number*" has also been applied to this definition (and similar ones). For the calculation of molecular activity, the activity of the pure enzyme and its molecular weight must be known. A large molecular activity indicates a rapid reaction. Very large molecular activities have been found in the cases of catalase (5×10^6) and acetylcholinesterase (3×10^6). The usual values range from several thousand to ten thousand molecules of substrate per enzyme molecule per minute—still rather rapid turnover rates.

8. Conditions for Enzyme Activity

Enzymes manifest their highest activity only under optimal conditions. Hence this is one of the requirements for the calculation of molecular activity. Saturation with substrate and coenzyme, when applicable, is not enough. The enzyme must also be in an optimal milieu of ionic strength and often in the presence of activators.

Temperature is just as important for the speed of enzymatic reactions as it is for uncatalyzed reactions; a temperature rise of 10° accelerates the rate of reaction two- to fourfold, but the phenomenon quickly reaches a limit because of the heat lability of the proteins; some enzymes are damaged irreversibly even at 40 to 50°, and only a few continue to be active above 60°.

pH Optimum. Since enzymes are proteins, their properties are greatly affected by the pH of the solution. The majority of enzymes exhibit their specific catalytic action only at one particular range of pH values. The most favorable $H^{\oplus}$ concentration is called the "*pH optimum.*" This is related to the state of dissociation of the acidic and basic groups at the active site of the enzyme (on occasion also of the substrate). Charged groups often are involved in binding the substrate or in an acid–base type of catalysis. In more acidic or basic pH regions, these groups loose their charge and consequently their participation in catalytic action ceases.

The dependence of activity on pH may be more or less striking and is usually expressed in the form of a curve, by plotting enzyme activity (e.g., micromoles converted per unit time) versus pH (Fig. V-5). For most enzymes the optimal pH is in the neutral or slightly acidic region. Extreme pH optima occur with several digestive enzymes, e.g., pepsin (1.5–2.5), trypsin (7.5–10), and pancreatic lipase (8.0).

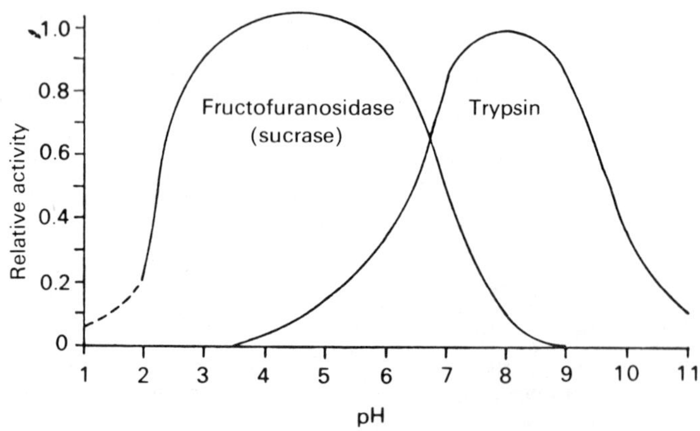

Fig. V-5. Dependency of enzyme activity on pH.

Activators. Many enzymes require certain ions for their activity. Amylases need $Cl^{\ominus}$ ions; enzymes turning over ATP nearly always require $Mg^{2\oplus}$. Numerous peptidases are activated by $Mn^{2\oplus}$, $Zn^{2\oplus}$, or $Co^{2\oplus}$. Occasionally, divalent ions can replace one another. The mechanism by which ions exert their influence, however, is known only in a few cases.

"Poisoning" of Enzymes. Many enzymes can be poisoned more or less specifically by different substances, i.e., they are *completely* and *irreversibly* inactivated. A good example is the inhibition of cytochrome oxidase with cyanide ($CN^{\ominus}$); several other heavy metal-containing enzymes are also poisoned by $CN^{\ominus}$. Irreversible inactivation of so-called SH enzymes is accomplished by the reaction with iodoacetamide or *N*-ethylmaleimide. Diisopropylfluorophosphate reacts with enzymes that have a serine residue at the active site. Such a serine residue is blocked through the formation of a phosphate triester with the diisopropylphosphate group. This specific marking of the active site has greatly aided the elucidation of the mechanism of action of enzymes. Similarly, the specific poisoning of certain enzymes in a natural cluster of enzymes has helped significantly in the investigation of reaction sequences.

Reversible Inhibition and Types of Inhibition. The reversible inhibition of enzymes is an important aspect of the regulation of metabolism. We can distinguish different types of inhibition, formally characterized by deviations from the Lineweaver-Burk diagram of enzyme kinetics. In "competitive inhibition" (Fig. V-6a) the ordinate intercept $1/V_{max}$ remains unchanged, while the slope of the plot increases. In "non-competitive inhibition" (Fig. V-6b) the abscissa intercept $1/K_m$ is constant, but slope and $1/V_{max}$ become greater (i.e., V_{max} decreases). Finally, "substrate inhibition" (Fig. V-6c) results in a nonlinear Lineweaver–Burk diagram.

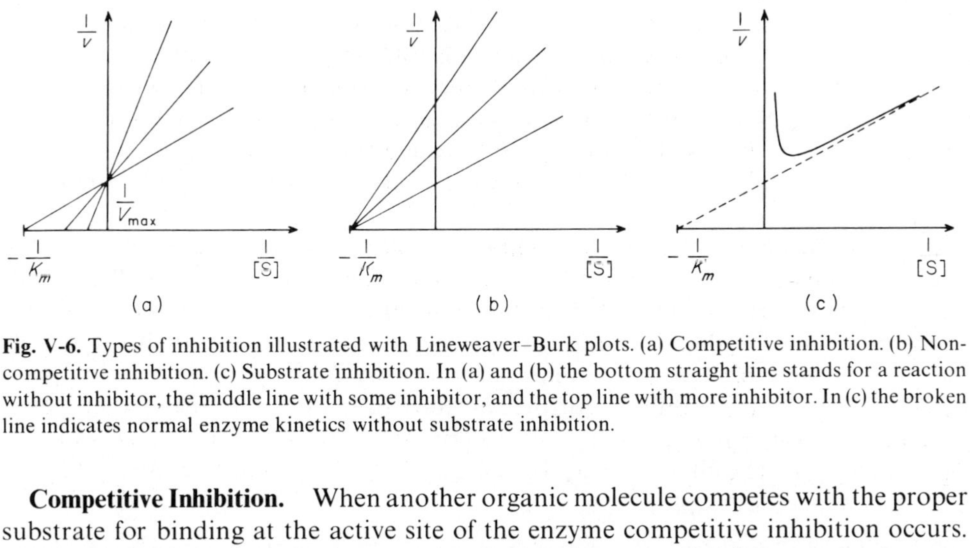

Fig. V-6. Types of inhibition illustrated with Lineweaver–Burk plots. (a) Competitive inhibition. (b) Non-competitive inhibition. (c) Substrate inhibition. In (a) and (b) the bottom straight line stands for a reaction without inhibitor, the middle line with some inhibitor, and the top line with more inhibitor. In (c) the broken line indicates normal enzyme kinetics without substrate inhibition.

Competitive Inhibition. When another organic molecule competes with the proper substrate for binding at the active site of the enzyme competitive inhibition occurs. Generally this presupposes a structural similarity between inhibitor and substrate. If the inhibitor is present in sufficiently high concentration it can displace the substrate entirely and thus block the reaction. An increase of substrate concentration conversely displaces the inhibitor. This competition also follows the law of mass action. The extent of inhibitor binding may be expressed by the inhibitor constant

$$K_i = \frac{[E] \cdot [I]}{[EI]}$$

(in analogy to the substrate constant K_s). In the Lineweaver–Burk plot (Fig. V-6a) one recognizes a shift in the apparent Michaelis constant; the abscissa intercept is

$$-\frac{1}{K_m} \cdot \left(1 + \frac{[I]}{K_i}\right)$$

Noncompetitive Inhibition. When an inhibitor affects the active site in such a way that the substrate can still be bound but the rate of reaction is slowed down noncompetitive inhibition occurs. One example is the reversible binding of a reactive SH group by copper or mercury compounds. This situation is of little significance for the regulation of metabolism.

Substrate and Product Inhibition. The dependency of activity on substrate concentration represented in Fig. V-2 does not apply in all cases. Some enzymes react more slowly with very high substrate concentrations, i.e., the rate decreases again. This phenomenon is called substrate inhibition. The explanation seems to be that more than one molecule of substrate attaches to the protein, thereby interfering with the reaction. In similar fashion, the product of the reaction in higher concentrations can inhibit as well, for the product is the substrate for the reverse reaction. With high concentrations of product the reverse reaction is favored; but, even when the reaction cannot reverse, for thermodynamic reasons (unfavorable equilibrium position), the product often is bound to the enzyme and thus prevents access of the substrate. Product inhibition therefore kinetically resembles competitive inhibition.

Allosteric Inhibition. Here the inhibitor molecule (or, more generally, the effector substance) is structurally different from the substrate molecule and is bound to the enzyme protein at a site distant from the substrate. Consequently, the inhibition cannot be reversed by high substrate concentrations. As a mechanism, it is thought that the reversible binding of the inhibitor alters the conformation of the protein, thereby, making it more difficult for the substrate to bind and, in the extreme case, preventing it from binding (increase of K_S), or actually decelerating catalysis itself (decrease of V_{max}). The allosteric effector alters both the $1/V_{max}$ intercept and the slope in the Lineweaver–Burk plot.

Enzymes that are subject to allosteric inhibition frequently consist of several subunits which can interact with each other. This "cooperative effect" results in a sigmoid (S-shaped) curve of substrate binding even in absence of an inhibitor. The sigmoidicity can be characterized quantitatively by the so-called Hill coefficient n in the equation

$$\frac{v}{V_{max} - v} = \frac{[S]n}{K_m}$$

An allosteric inhibitor can shift the entire curve toward greater substrate concentrations and reinforce the sigmoid effect. One example of such cooperative effects is the oxygen binding curve of hemoglobin (see Fig. IX-2).

Allosteric inhibition is of great physiological significance in the *feedback inhibition* found in long metabolic reaction sequences. The accumulation of the end product slows down the whole reaction sequence by inhibiting an earlier step and finally brings it to a halt. As the end product is consumed, synthesis resumes because allosteric inhibition is reversible (see also Chapter XIX,6).

9. The Mechanism of Enzyme Catalysis

In modern organic chemistry, reaction mechanisms are explained on the basis of the electron theory of the chemical bond. In principle, this explanation should also apply to enzymatic transformations. The difference is that in the reactions catalyzed by proteins reactive groups, especially those at the "active site," participate in the reactions.

The study of the mechanisms of action of enzymes calls for a combination of various methods. The chemical blocking of certain groups (cysteine, serine, and others) often yields information about the nature of the active site. The effect of pH on the kinetics of catalysis can be related to dissociable groups. The geometry of the active site can be probed with chemical analogs of the substrate. The most useful information, however, is obtained from the total X-ray structure analysis, i.e., the complete resolution of the chain conformation of the enzyme molecule and the enzyme–substrate complex.

Very often enzymatic reactions are assisted by coenzymes or prosthetic groups. The chemical structure of these cofactors is discussed in the following chapter. They are low-molecular weight substances whose reactivity can be explained by modern electron theory (cf. Chapter VIII,4).

Mechanism of Action of Lysozyme. Lysozyme is a hydrolase that cleaves chains of carbohydrates ("mureins," cf. Chapter XVII,7). It is the first enzyme whose three-dimensional structure has been elucidated completely (Phillips *et al.*). From the structure of lysozyme and that of the enzyme–substrate complex (a noncleavable substrate analog substitutes for the real substrate) the essential characteristics of catalysis can be gleaned.

As illustrated on the reverse of the foldout chart (in back of this book), the protein has a deep groove ready to accept the substrate. When the substrate molecule is thus in place, part of the peptide chain shifts (e.g., Trp-62 by 0.75 Å) and forms a bond with it. This corresponds exactly to Koshland's "*induced fit*" postulate. All together a large number of secondary valence bonds between substrate and the reactive groups of the protein have been detected; atoms of the peptide bonds also participate.

In the key step one carbohydrate ring is distorted from its chair form to a strained, largely planar conformation. The energy for this distortion is supplied by neighboring secondary valence bonds. The change of conformation of the substrate energetically favors the cleavage of the C—O—C linkage. At this point, a $H^{\oplus}$ from a —COOH group (Glu-35) approaches the oxygen atom of the glycosidic linkage and the bond is broken. The remaining carbonium ion (at C-1 of the strained ring) is stabilized by the $—COO^{\ominus}$ group of Asp-52. Following the dissociation of the first fragment, a water molecule attacks the carbonium ion. The addition of $OH^{\ominus}$ completes the

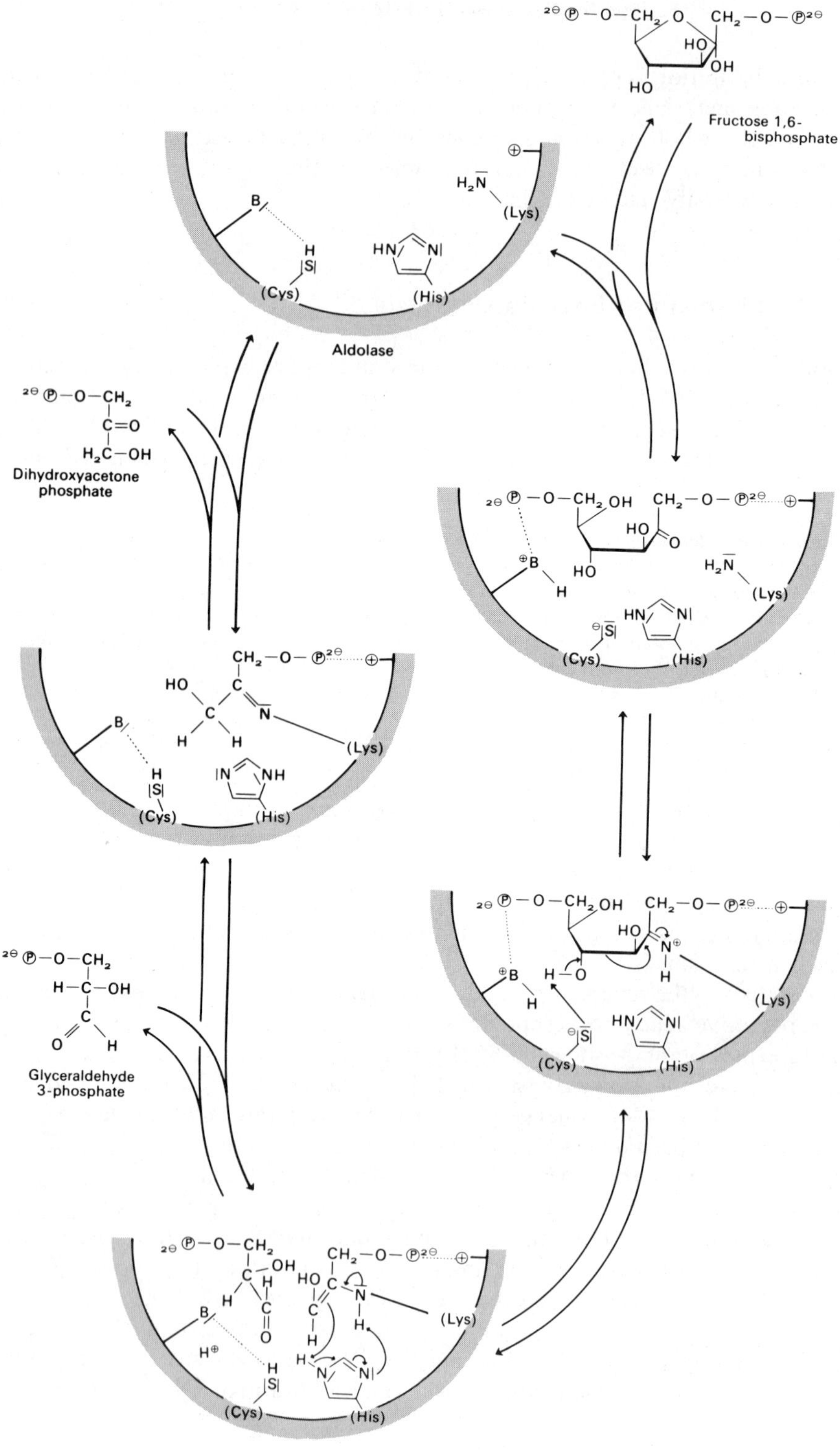
Fructose 1,6-bisphosphate
Aldolase
Dihydroxyacetone phosphate
Glyceraldehyde 3-phosphate
(Lys)
(Cys)
(His)

hydrolysis, and the second fragment dissociates from the enzyme, which is now free to accept another substrate molecule.

The reaction sequence described here for lysozyme probably includes a few generally applicable principles of enzyme action. A "groove" for the substrate is also a feature of the structures of ribonuclease and chymotrypsin. Substrate molecules are bound through secondary valences. Charged groups and proton donors and acceptors commonly cooperate in catalysis. The spatial arrangement is crucial in reactions between two or more substrate molecules. This is particularly striking in instances where first a coenzyme (= "auxiliary substrate") is bound and then the substrate.

Two-Step Mechanism. The reaction on the enzyme protein may also proceed in two steps. A reactant (or part of it) at first is bound to the protein by primary valence bonds, and only in a second step does the formation of the product proceed from the intermediate complex. The reaction with aldolase follows this type of mechanism, as shown schematically in Fig. V-7.

The diagram can be read in both directions, since the reaction is reversible. We can start with the cleavage of fructose 1,6-bisphosphate and proceed clockwise. The substrate is bound to the protein by two positively charged groups, presumably in a groove or pocket. The keto group then reacts with the NH_2 group of a lysine residue to form a Schiff's base. Aldol cleavage proper is initiated by a basic group (histidine in the diagram) which accepts the proton of the OH group at C-4. A subsequent electron shift results in the scission of the bond with the formation of a carbanion of dihydroxyacetone phosphate. The histidyl residue provides the proton to neutralize the carbanion, while simultaneously the first fragment (glyceraldehyde 3-phosphate) is liberated. The Schiff's base is then cleaved hydrolytically, dihydroxyacetone phosphate migrates away, and the enzyme is ready to begin the cycle anew. The reaction sequence reverses if we start with a mixture of the triosephosphates. (For the equilibrium position, see Chapter XV,7.)

The substrate is bound to reactive groups of the polypeptide by primary valence bonds in but a few enzymatic reactions. In some instances, this task is handled by prosthetic groups or coenzymes, but in most cases, primary valence bonds are not formed at all. The reaction between substrates occurs as a many-center reaction, and the role of the enzyme is to hold the reaction partners in a certain fixed orientation to one another. The enzyme often acts through its dissociable functional groups (e.g., the histidyl residue) as an acid–base catalyst.

Multienzyme Complexes. Aggregates of several enzymes, namely, multienzyme complexes, catalyze a sequence of reactions of many steps. The key feature of these complexes seems to be that they pass intermediate products along from enzyme to enzyme by transfer reactions. One well-studied example is the synthesis of fatty acids described in Chapter XII,6. The enzymes of oxidative decarboxylation of α-keto

Fig. V-7. The mechanism of action of aldolase. The semicircle represents the groove in the enzyme molecule. The amino acid side chains shown, which together constitute the active site, do not occur in that sequence in the primary structure of the enzyme.

TABLE V-2

Classification of Enzymes

Main class and subclasses	Example	Chapter
1. Oxidoreductases (Enzymes of biological oxidation and reduction)		X
1.1 Acting on the >CH—OH group of donors		
1.1.1 With NAD or NADP as acceptor	Alcohol dehydrogenase, lactate dehydrogenase	XV,7
1.1.3 With O_2 as acceptor	Glucose oxidase	
1.2 Acting on the aldehyde or keto-group of donors		
1.2.1 With NAD or NADP as acceptor	Glyceraldehyde-3-phosphate dehydrogenase	XV,7
1.2.3 With O_2 acceptor	Xanthine oxidase	VII,2
1.3 Acting on the >CH—CH< group of donors		
1.3.1 With NAD or NADP as acceptor	Dihydrouracil dehydrogenase	VII,2
1.3.2 With a cytochrome as acceptor	Acyl-CoA dehydrogenases	XII,3
1.4 Acting on the >CH—NH_2 group of donors		
1.4.3 With O_2 as acceptor	Amino acid oxidases	VIII,7
2. Transferases (Group-transferring enzymes)		
2.1 Transferring C_1-groups		
2.1.1 Methyltransferases	Guanidinoacetate methyltransferase	VI,6
2.1.2 Hydroxymethyl- and formyltransferases	Serine hydroxymethyltransferase	VIII,12
2.1.3 Carboxyl- and carbamoyltransferases	Ornithine carbamoyltransferase	VIII,8
2.3 Acyltransferases	Choline acetyltransferase	XXIII,7
2.4 Glycosyltransferases		
2.6 Transferring N-containing groups		
2.6.1 Aminotransferases	Transaminases	VIII,6
etc.		

3. Hydrolases (Enzymes catalyzing hydrolytic cleavage)		
3.1 Cleaving ester linkages		
3.1.1 Carboxylic ester hydrolases	Esterases, Lipases	XII,2
3.1.3 Phosphoric monoester hydrolases	Phosphatases	VII,10
3.1.4 Phosphoric diester hydrolases		XIII,3
etc.		
3.2 Cleaving glycosides		XVII,3
3.2.1 *O*-Glycosidases	Amylase	XVII,6
	β-Glycosidase, etc.	XVII,3
3.2.2 *N*-Glycosidases	Nucleosidases	VII,10
3.4 Cleaving peptide linkages		VIII,1
3.4.11 α-Aminoacyl-peptide hydrolases	Aminopeptidase	VIII,3
3.4.12 Peptidyl amino-acid hydrolases	Carboxypeptidases	VIII,3
3.4.14–15 Peptidylpeptide hydrolases, etc.		VIII,2
4. Lyases (Enzymes cleaving C—C, C—O, C—N bonds)		
4.1 C—C lyases		
4.1.1 Carboxy-lyases	Pyruvate decarboxylase	XV,7
4.1.2 Aldehyde-lyases	Aldolases	XV,5
4.2 C—O lyases		
4.2.1 Hydro-lyases	Fumarate hydratase	XI,2
4.3 C—N lyases		
4.3.1 Ammonia-lyases	Histidine ammonia-lyase (= histidase)	VIII,13
5. Isomerases (Enzymes catalyzing geometric changes within one molecule)		
5.1 Racemases and epimerases		
5.1.3 Acting on carbohydrates	Ribulose-5-phosphate epimerase	XV,5
5.2 *Cis-trans* isomerases	Maleylacetoacetate isomerase	VIII,11
5.3 Intramolecular oxidoreductases		
5.3.1 Interconverting aldoses and ketoses	Glucose-phosphate isomerase	XV,7
5.4 Intramolecular transferases	Methylmalonyl-CoA mutase	XII,4

TABLE V-2—continued

Main class and subclasses	Example	Chapter
6. Ligases (Enzymes that join together two molecules. Also called synthetases.)		
6.1 Forming C—O bonds		
6.1.1 Amino acid-RNA ligases	Amino acid-activating enzyme	VII,6
6.3 Forming C—N bonds		
6.3.1 Acid-ammonia ligases	Glutamine synthetase	VIII,8
6.4 Forming C—C bonds		
6.4.1 Carboxylases	Acetyl-CoA carboxylase	XII,6

acids (such as pyruvate and α-ketoglutarate) also form multienzyme complexes, which can be made visible in the electron microscope.

10. Classification and Nomenclature of Enzymes

Enzymes known for a long time have trivial names that are still in use (e.g., *trypsin*, *pepsin*, and *emulsin*). More recently, names have been devised more systematically with the suffix "*-ase*." To designate a group of enzymes with a specific property, the suffix is attached to the name of the reaction catalyzed. Thus, dehydrogenating enzymes are called dehydrogenases, enzymes that transfer groups are called transferases, etc. The suffix *-ase* attached to the name of a substrate indicates hydrolytic enzymes: glycosidases are glycoside-cleaving enzymes; phosphatases those that hydrolyze phosphate esters. The complete designation of individual enzymes includes the full chemical name of the substrate and the reaction. Examples are lactate dehydrogenase, cytochrome oxidase, and glucose-6-phosphatase.

Until very recently the naming of the individual enzymes has been entrusted largely to the discoverers. This resulted often in such descriptive names as "zwischenferment" or "pH 5 enzyme." An international commission meanwhile has drafted specific rules for the classification and nomenclature of enzymes.[15] The commission has established six main classes, which are further subdivided into subclasses and sub-subclasses, according to the nature of the reaction catalyzed and to the type of bond formed or severed. In Table V-2, several examples of each main class are listed to illustrate the system.

Obviously only a small selection of the more than 1500 known enzymes has been incorporated into the table; the mere enumeration of all subgroups would have occupied too much space. The first group is comprised of *dehydrogenases*, *reductases*, *oxidases*, and *oxygenases* (see Chapter X). Group 2 includes group-transferring enzymes, called *transferases*. Group 3 includes *hydrolases*, whose name is generally derived by attaching the suffix *-ase* to the name of the substrate. Group 4, the *lyases*, encompasses all enzymes that catalyze the cleavage of C—C, C—O, C—N, and other bonds by elimination reactions, i.e., nonhydrolytic reactions in which the reaction products end up with a double bond. In the reverse direction, the addition of a reactant to a double bond can be taken as the reversal of a lyase reaction and thus would also be catalyzed by a lyase. When in a lyase-catalyzed reaction the addition is the predominant reaction, then the enzyme can also be called a *synthase*. Group 5 includes *isomerases*, *epimerases*, etc. Group 6, the *ligases*, contains enzymes that catalyze the linkage of two substrates in a reaction coupled with the hydrolysis of a pyrophosphate group of ATP or another energy-rich triphosphate. Compounds synthesized in this manner generally are those with a high potential for group transfer. These enzymes frequently are called *synthetases*.

[15] A discussion and an outline of the original Report is given by R. H. S. Thompson, *Nature* (*London*) **193**, 1277 (1962). The current report is "Enzyme Nomenclature," Recommendations (1972). M. Florkin and E. H. Stotz (editors), Comprehensive Biochemistry, vol. 13, 3rd edition, Elsevier, Amsterdam, 1973.

BIBLIOGRAPHY

L. von Bertalanffy, "Problems of Life." Wiley, New York, 1952.

P. D. Boyer (editor), "The Enzymes," 3rd ed., Vol. I: "Enzyme Structure-Control," Vol. II: "Kinetics-Mechanism," Academic Press, New York, 1970. For specific enzymes see Vols. III–X, 1971–1974.

M. Dixon and E. C. Webb, "Enzymes," 2nd ed. Academic Press, New York, 1964.

"Enzyme Nomenclature," Recommendations (1972) of the Commission on Biochemical Nomenclature on the Nomenclature and Classification of Enzymes together with their Units and the Symbols of Enzyme Kinetics; published as M. Florkin and E. H. Stotz (editors), "Comprehensive Biochemistry," 3rd edition, vol. 13, Elsevier, Amsterdam, 1973.

U. Henning, Multienzyme complexes. *Angew. Chem. Intern. Ed. Engl.* **5,** 785–790 (1966).

I. M. Klotz, "Energy Changes in Biochemical Reactions," Academic Press, New York, 1967.

I. M. Klotz, "Introduction to Chemical Thermodynamics," 2nd ed. Benjamin, New York, 1972.

E. R. Stadtman, Allosteric regulation of enzyme activity. *Advan. Enzymol.* **28,** 41–154 (1966); contains 339 references.

CHAPTER **VI**

Coenzymes

1. Coenzymes and Prosthetic Groups

We have already indicated (Chapter V,1) that all enzymes are proteins and that many have in addition a "prosthetic group," the coenzyme. These groups often dissociate:

$$\text{Coenzyme} + \text{Apoenzyme} \rightleftharpoons \text{Holoenzyme}$$

The terms "*coenzyme*" and "*prosthetic group*" are frequently used synonymously.[1] Lately, the tendency has been to call the tightly bound groups that cannot be removed (for example by dialysis) prosthetic groups, and those which dissociate easily, coenzymes. Naturally it is often difficult to decide what is easily dissociable and what is not, for there is a complete range of possibilities.

The following considerations may lead to a sensible distinction between coenzyme and prosthetic group. Both types of groups participate chemically in the catalytic process; they are thereby altered and their original condition is restored only in a second reaction, usually through the mediation of a second enzyme. In this sense they do not conform to the older definition of catalyst, because they do not emerge from the reaction unaltered. This process of restoration is not the same for coenzyme and prosthetic group.

Typical dissociable coenzymes might better be called "*cosubstrates.*" They assume the role of a hydrogen or group donor; e.g., ATP contributes a phosphate radical in the hexokinase reaction (Chapter V,5, diagram), and nicotinamide-adenine dinucleotide (formerly coenzyme I or DPN) accepts hydrogen from the substrate. Each substance functions therefore as a second substrate that reacts with the real

[1] "Prosthetic group" is a generic term which comprises all groups bound to proteins including even those without any catalytic effect (cf. Chapter IV,9). In this chapter we always refer to prosthetic groups of *enzymes*.

substrate strictly stoichiometrically; i.e., mole per mole and *not* catalytically. In a second reaction, catalyzed by another enzyme protein, the hydrogenated NAD can then relinquish its hydrogen (or ADP can again pick up phosphate); even at this point the coenzymes are "cosubstrates," reacting stoichiometrically. The catalytic nature of a coenzyme begins to appear only when it is coupled to two enzymes in an enzyme system. This process, illustrated schematically here, is realized in anaerobic glycolysis.

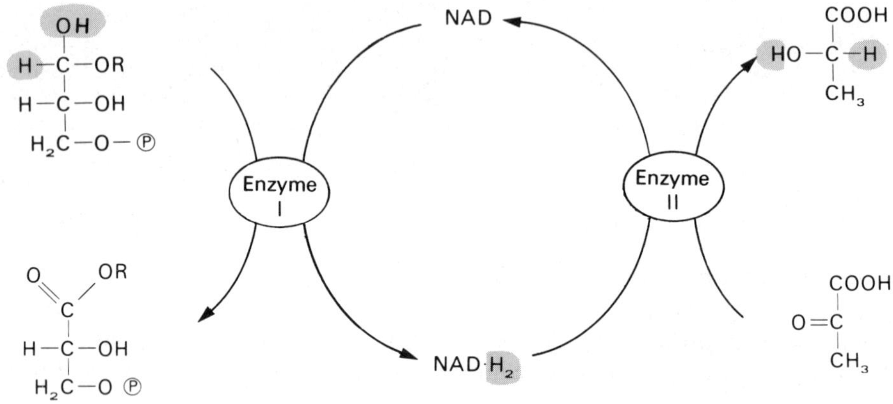

It is precisely because coenzymes couple different enzymes that they are important in metabolism: They are the links by means of which the large-scale exchange of material—be it hydrogen, phosphoric acid, or some organic group—becomes possible. Bücher for this reason very aptly calls coenzymes "transport metabolites."

In other enzymes systems, in contrast, the "active group" is tightly bound to the enzyme protein. In this case catalysis occurs when the holoenzyme reacts with two different substrates in rapid succession. This is illustrated in the following diagram.

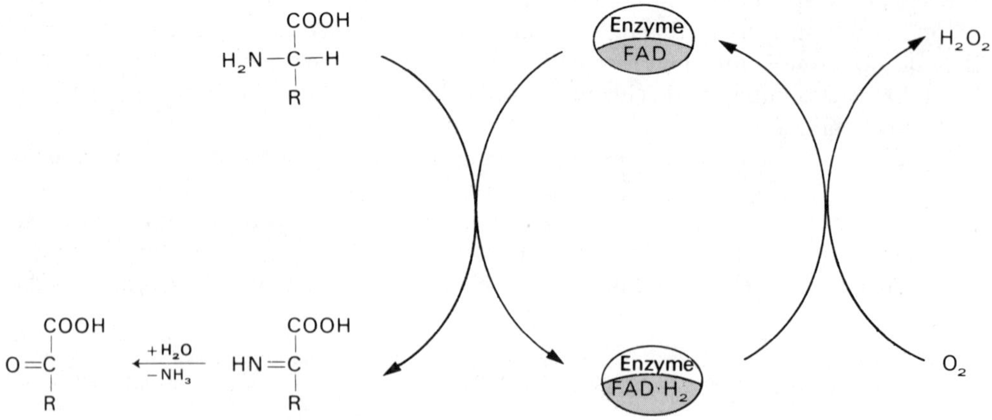

The amino acid is dehydrogenated by the enzyme; the hydrogen is accepted by the prosthetic group and transferred to a molecule of oxygen in a second reaction (with the formation of H_2O_2). The active group reacts stoichiometrically even here, but is returned to the original state on the same enzyme by reaction with the second substrate. Further examples of this are found in transamination (Chapter VIII,6) and in the respiratory chain (Chapter X,4). In these cases we would speak of the "prosthetic group" of the enzyme.

The distinction drawn here between coenzyme and prosthetic group has not yet been adopted generally; on the other hand, the concept of coenzymes as cosubstrates has been accepted universally. It is true even with our terminology that coenzymes play a decisive role in the activity of the enzyme. The individual specificity of the enzyme—including the reaction specificity—does reside, however, with the protein component. Many cases are known in which the same coenzyme is involved in completely different reactions, depending on the protein to which it is bound. Not a single case is known, however, where one protein in conjunction with various coenzymes shows various activities, although this is theoretically possible.

The chemical structure of coenzymes has been known for a much longer time than that of the enzymes themselves. The explanation is that all enzymes are proteins and, as indicated before, the methods for their structural determination (sequential analysis) have been developed only very recently. Coenzymes, on the other hand, are low-molecular weight compounds which can be studied by ordinary methods of organic chemistry.

2. Coenzymes and Vitamins

Many coenzymes are closely related to vitamins. Vitamins are active compounds which must be ingested in food. They are necessary for the proper functioning of life processes and cannot be replaced by other substances. The organism is itself incapable of synthesizing them, except possibly from direct precursors, the *provitamins*. The amounts of vitamins which must be supplied daily are relatively minute (cf. Table XXII-1, Chapter XXII,3). This means that vitamins are not "foodstuffs" in the ordinary sense, but rather function catalytically. The biocatalytic function is known for many vitamins; generally, a vitamin is the main (or sole) component of a coenzyme (see Table VI-1). An insufficient supply of one of the vitamins results in arrested growth and deficiency diseases; avitaminoses were the starting point of vitamin research. The connection between the biochemical role of the coenzyme and the clinically observable manifestations of vitamin deficiency is still unknown. It is remarkable that many deficiency diseases have skin and mucosal symptoms. In animal experiments it is occasionally possible, however, to observe deviations from normal metabolism, based on vitamin deficiency, before the characteristic symptoms of the avitaminosis become discernible.

In this chapter the chemical structures of those vitamins which are components of presently known coenzymes will be discussed; their nutritional significance will be treated in Chapter XXII.

3. Structure and Classification of Coenzymes

Almost all coenzymes contain phosphoric acid as an essential component, often in a type of bond called "nucleoside phosphate," or better "*nucleotide*." Nucleotides consist of a base, a monosaccharide (nearly always ribose), and phosphoric acid. The base and sugar are linked through an *N*-glycoside bond (cf. Chapter VII,1); the phosphoric acid is esterified with one hydroxyl group of the ribose. Nucleotides were originally discovered as components of nucleic acids and received their name in that way (cf. Chapter VII,1).

It seems convenient to classify coenzymes according to the reactions in whose catalysis they are instrumental. Such a classification (Table VI-1) parallels that of the enzymes. Accordingly, the hydrogen-transferring coenzymes are listed first. A second and particularly comprehensive class is made up by the group-transferring coenzymes. The hydrolases frequently require metal ions for activity but have no true coenzymes. Group-transferring coenzymes also participate often in reactions catalyzed by the isomerases, lyases, and ligases; a few special cases will be discussed in Section 9, at the end of this chapter.

Understanding of coenzymes necessarily involves knowledge of the reactions in which they participate. These are treated thoroughly later in this book, and this chapter will be referred to in many places. It is hoped that any initial difficulties in understanding this topic will be gradually overcome as the reader proceeds.

4. Coenzymes of the Oxidoreductases

The Nicotinamide Nucleotides. The hydrogen-transferring enzymes of fermentation, of glycolysis, and of many other reactions utilize as their coenzymes dinucleotides one of whose bases is the pyridine derivative nicotinamide. These dinucleotides are therefore pyridine nucleotides.

Pyridine nucleotides are the coenzymes known the longest. In 1906, Harden and Young noticed that Buchner's "zymase" could be separated by ultrafiltration into a high-molecular weight component and a low-molecular weight component, or coenzyme. These results were later extended and served as a basis for theories of enzyme structure. This "coenzyme" (then called "*codehydrogenase I*," later "*coenzyme I*," "DPN," and now NAD) was isolated by H. von Euler and co-workers in 1931; shortly afterward, Warburg and Christian discovered the closely related "*coenzyme II*" ("TPN," now NADP).

Pyridine nucleotides are intimately associated with the B vitamin *nicotinic acid* (formerly also called niacin). It is a pyridine derivative, pyridine-3-carboxylic acid. The amide *nicotinamide* (formerly niacinamide) also is active as vitamin and appears in the coenzyme in this form:

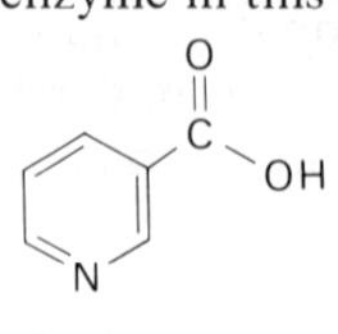

Nicotinic acid
(Niacin)

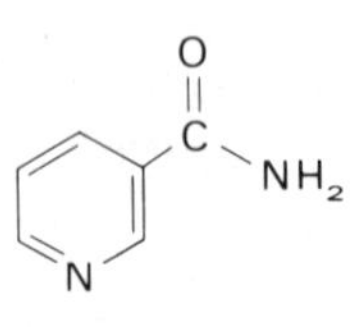

Nicotinamide
(Niacinamide)

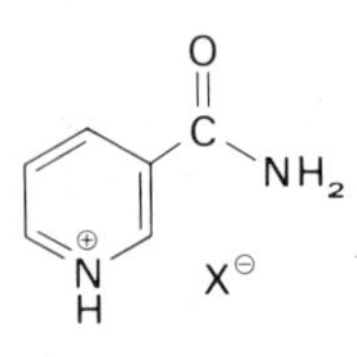

Nicotinamide
pyridinium salt

TABLE VI-1

Coenzymes and Prosthetic Groups

Coenzymes	Usual abbreviation	Group transferred	Corresponding vitamin	See Chapter
1. Hydrogen-transferring coenzymes:				
Nicotinamide-adenine dinucleotide	NAD (formerly DPN)	Hydrogen	Nicotinamide	X,4
Nicotinamide-adenine dinucleotide phosphate	NADP (formerly TPN)	Hydrogen	Nicotinamide	X,4
Flavin mononucleotide	FMN	Hydrogen	Riboflavin	X,4
Flavin-adenine dinucleotide	FAD	Hydrogen	Riboflavin	X,4
Ubiquinone (coenzyme Q)	Q	Hydrogen	—	X,4
Cell hemins	—	Electrons	—	X,4
Lipoate	Lip(S_2)	Hydrogen and acyl groups	—	VIII,10
2. Group-transferring coenzymes				
Adenosine triphosphate	ATP	Phosphate group (and AMP residue)	—	VI,5
Phosphoadenosyl sulfate	PAPS	Sulfate group	—	VI,5
Pyridoxal phosphate	PALP	Amino group	Pyridoxine	VIII,6
Cytidine diphosphate	CDP	Phosphoryl choline and related groups	—	XIII,3
Uridine diphosphate	UDP	Sugar, uronic acid	—	XVII,4
Coenzymes for C_1-transfer				
Adenosylmethionine	—	Methyl group	Methionine	VI,6
Tetrahydrofolate	CoF	Formyl group	Folate	VI,6
Biotin	—	Carboxyl group (CO_2)	Biotin	XII,6
Coenzymes for C_2-transfer				
Coenzyme A	CoA	Acyl group	Pantothenate	XII,6
Thiamine pyrophosphate	TPP	C_2-aldehyde groups	Thiamine	XV,5
3. Coenzymes of isomerases and lyases				
Uridine diphosphate	UDP	Sugar isomerization	—	XVII,4
Pyridoxal phosphate	PALP	Decarboxylation	Pyridoxine	VIII,6
Thiamine pyrophosphate	TPP	Decarboxylation	Thiamine	VIII,10
B_{12} coenzyme	—	Carboxyl displacement	Cobalamin	XII,4

The pyridine ring in the coenzyme is attached in *N*-glycosidic linkage (cf. Chapter VII,1) to ribose. Such a linkage evidently is possible only with the pyridinium cation, which bears one hydrogen atom on the nitrogen. Pyrophosphoric acid provides the link between nicotinamide riboside and adenosine. The complete formula for nicotinamide-adenine dinucleotide is given here:

Nicotinamide

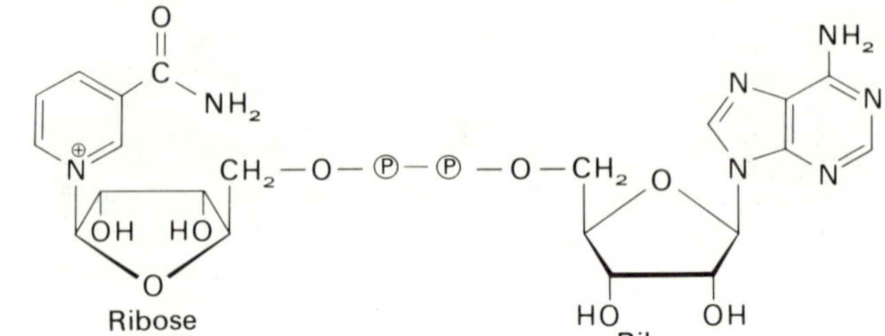

Adenine

Ribose

Ribose

Pyrophosphoric acid

Adenosine

Nicotinamide-adenine dinucleotide = Diphosphopyridine nucleotide
NAD or NAD⊕ DPN⊕

The common designation "*diphosphopyridine nucleotide*" is incorrect and confusing. We are not at all dealing with the nucleotide of diphosphopyridine. We are concerned, however, with a *dinucleotide*, one of whose base components is a pyridine derivative, nicotinamide. The Commission on Enzymes of the International Union of Biochemistry, therefore, has chosen the name "*nicotinamide-adenine dinucleotide*," abbreviated NAD, for the so-called diphosphopyridine nucleotide. "*Nicotinamide-adenine dinucleotide phosphate*," abbreviated NADP, is the new name for the coenzyme triphosphopyridine nucleotide.[2] We will adopt this improved and internationally agreed upon nomenclature throughout this book (see also footnote 3, Chapter XIX,1).

In nicotinamide-adenine dinucleotide phosphate the adenosine moiety carries the additional phosphate group in the 2′-position:

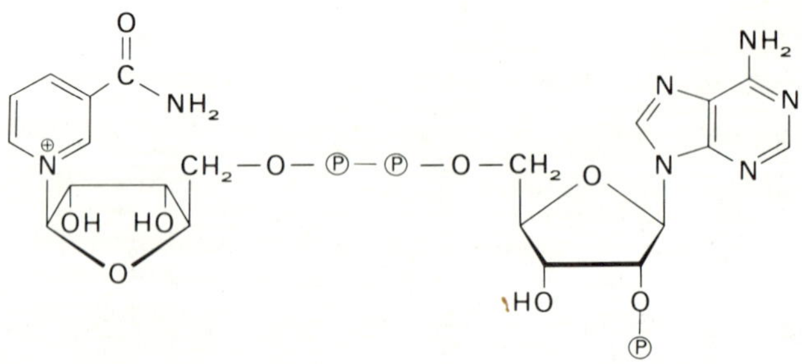

Nicotinamide-adenine dinucleotide phosphate = Triphosphopyridine nucleotide
NADP or NADP⊕ TPN⊕

[2] See M. Dixon, *Science* **132,** 1548 (1960); see also R. H. S. Thompson, *Nature* (*London*) **193,** 1227 (1962).

Because of the positive charge in the pyridine ring these coenzymes are also abbreviated NAD$^{\oplus}$ and NADP$^{\oplus}$. Their function is the reversible uptake of hydrogen. The pyridine ring becomes reduced, retaining only two double bonds, while the nitrogen loses its positive charge:

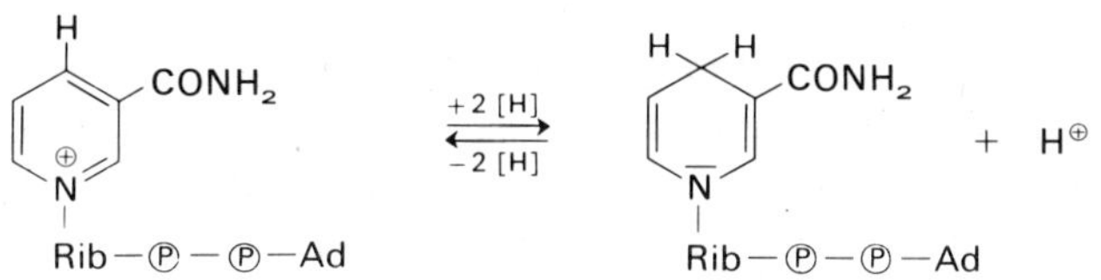

$$NAD^{\oplus} + 2[H] \rightleftharpoons NADH + H^{\oplus}$$

Obviously, the aromatic nature of the pyridine ring is lost, too. This changes the light absorption very characteristically. The dihydropyridine system has a broad absorption maximum at 340 nm, whereas the pyridine system does not absorb at that wavelength (cf. Fig. VI-1). If NADH is produced during a reaction, the absorption at 340 nm rises. This rise in light absorption can be measured quite easily and thus the transition NAD$^{\oplus}$ $\rightleftarrows$ NADH can be followed optically. The increase of absorption per unit time is proportional to the enzyme concentration (cf. Fig. VI-2). This optical test for enzyme activity, or enzyme assay, is of extreme practical importance in the laboratory.

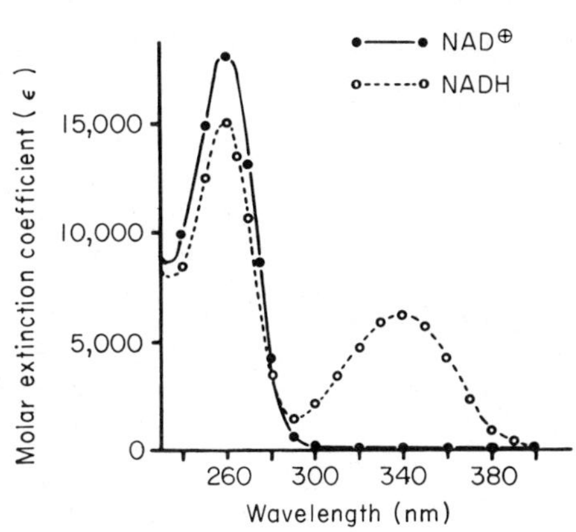

Fig. VI-1. Ultraviolet absorption of pyridine nucleotides. There is a maximum at 340 nm for the reduced form.

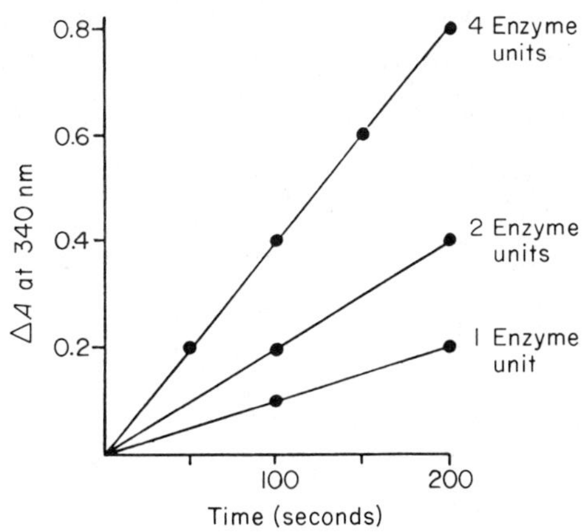

Fig. VI-2. Optical assay of a dehydrogenase. The change in absorption is plotted against time. The reaction runs faster when more enzyme is used. *A* is the absorbance of light of indicated wavelength.

Since the reduction of pyridine nucleotides occurs very frequently in biochemical equations, the structural formulas are usually not written out and the reaction is symbolized in the following manner:

$$NAD^{\oplus} + CH_3{-}CH_2OH \rightleftharpoons NADH + H^{\oplus} + CH_3{-}CHO$$

The pyridine nucleotides NAD and NADP are coenzymes (more correctly, co-substrates) of a large number of dehydrogenases, especially for the dehydrogenation of primary and secondary alcohol groups. The reactions generally are reversible. The particular significance of these coenzymes consists in the reversible transfer of hydrogen according to the scheme in Section 1. Bücher very aptly has called them "transport metabolites" because they take care of the hydrogen transport within the cell. The two coenzymes have different functions: NADPH usually supplies the required hydrogen, or generally the reduction equivalent for biosynthesis (cf. Chapters X,4, XII,6, and XIV,1), while NADH usually releases its hydrogen to enzymes of the respiratory chain, in which the reaction with O_2 to form H_2O is used for the synthesis of ATP (cf. Chapter X).

Hydrogen bound to NADPH thus follows another metabolic pathway than that bound to NADH. However, certain enzymes, called *transhydrogenases* can transfer the hydrogen of NADPH to $NAD^{\oplus}$. Since the system $NADP^{\oplus}$/NADPH is reduced to a very high degree, the reaction is exergonic and under certain circumstances can even be coupled to the synthesis of ATP. In the metabolism as a whole, transhydrogenation is, however, of subordinate importance.

Careful analysis of the reaction mechanism for the addition of hydrogen to the pyridine ring, resulted in two noteworthy observations. First, the hydrogen that is added to the ring is transferred *with its electron pair*, i.e., as *hydride ion*. $NAD^{\oplus}$ reacts through one of its resonance structures with the charge in position 4. An example (with alcohol dehydrogenase) will help to explain this (cf. also Chapter X,2):

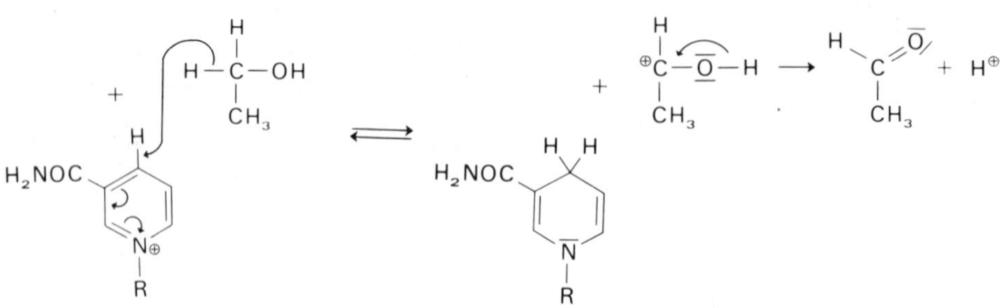

Second, the reaction is stereospecific, as was shown by deuterium labeling. It is easy to imagine that the deuterium ion can approach the ring only from the top when

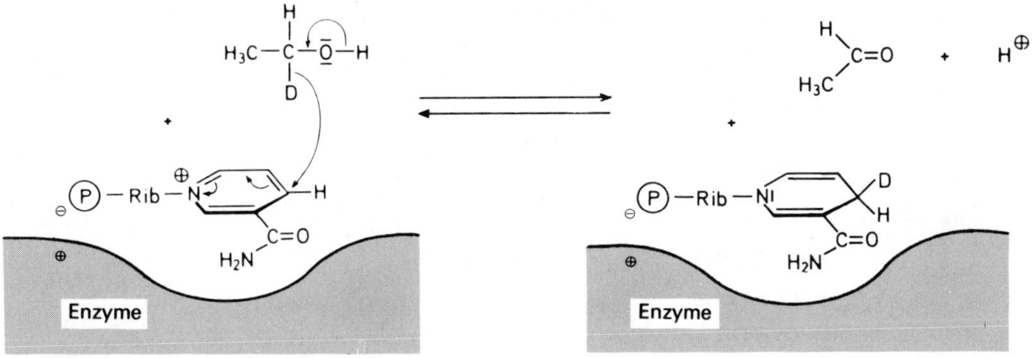

Fig. VI-3. Stereospecificity of the catalysis of pyridine nucleotides. Deuterium is transferred to the "A side" of the coenzyme.

the ring is assumed to rest on the protein surface (Fig. VI-3). The points of attachment on the enzyme surface for rib-Ⓟ and for $-CONH_2$ will then determine exactly which side of the pyridine ring is available for reduction.

By convention, that side which accepts a hydrogen during the dehydrogenation of ethyl alcohol by alcohol dehydrogenase and of lactate by lactate dehydrogenase is designated as the A side; the other side, the B side, is attacked by glucose dehydrogenase and triose phosphate dehydrogenase. This type of specificity is but a special case of common stereospecificity: During the reduction of pyruvate, for instance, hydrogen is also attached on one particular side, yielding only L-(+)-lactate, or D-(−)-lactate, depending upon the particular enzyme used.

Two enzymes which approach the nicotinamide-adenine dinucleotide from opposite sides are able to cooperate especially closely. At first, hydrogen is transferred to the A side, then the B oriented H is removed and a new H added on the A side. (In the dehydrogenated, aromatic form of $NAD^{\oplus}$, hydrogen is in the plane of the ring, and therefore between the A and B position.) There is evidence that in addition to freely dissolved enzymes there are also ordered enzyme complexes which work in just this manner.

Many apoenzymes react with NAD analogs, too. These are compounds in which the amide group of the pyridine part has been replaced by some other group. Thus nicotinic acid, 3-acetylpyridine (with a $-COCH_3$ group), and thionicotinamide (with $-CS-NH_2$) can be coupled to give pyridine nucleotides, all of which are reduced enzymatically. These compounds have achieved importance in several studies of reaction mechanism and in distinguishing different apoenzymes.

Flavin Nucleotides. Vitamin B_2 or *riboflavin* is a component of the prosthetic group of flavoproteins. The isolation of the vitamin from biological material was accomplished by György, R. Kuhn, and T. Wagner-Jauregg in 1933. Almost simultaneously, the first "yellow enzyme" was obtained in pure form and the close relationship between these active substances was recognized.

Riboflavin is an isoalloxazine derivative, i.e., a pteridine ring with a benzene ring fused on to it. The side chain is a C_5-polyhydroxy group. Note that this is not an *N*-glycoside of ribose, but rather a derivative of ribitol, a pentahydroxy compound.

The systematic name is 6,7-dimethyl-9-(ribityl)isoalloxazine:

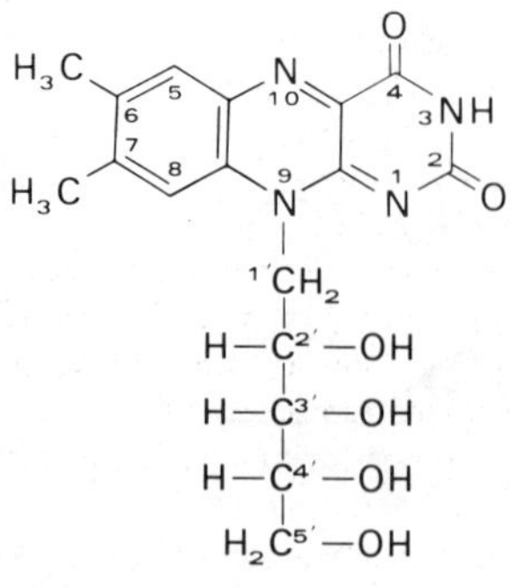

Riboflavin

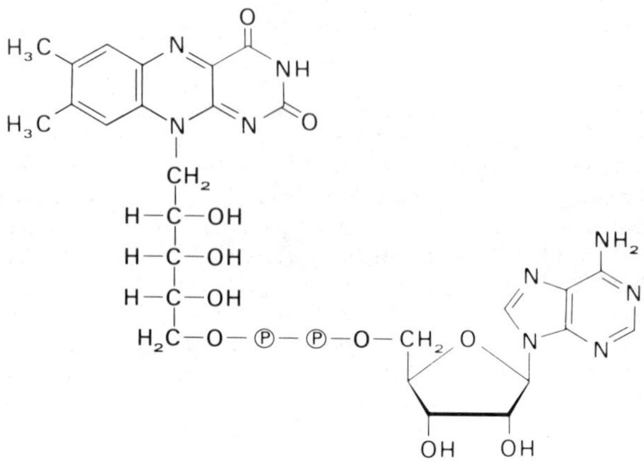

Flavin-adenine dinucleotide (FAD)

The active component of the "old yellow enzyme" is riboflavin 5′-phosphate, usually called *flavin mononucleotide* and abbreviated FMN. This nomenclature is somewhat incorrect again, since we do not have a nucleotide, i.e., an *N*-glycoside of a ribose phosphate. In spite of this, the name and abbreviation have been adopted, and actually the close similarity to real nucleotides must be conceded.

Most flavoproteins do not contain the mononucleotide, but rather flavin-adenine dinucleotide, abbreviated FAD. As in the pyridine nucleotides, adenosine monophosphate and riboflavin phosphate are joined by a pyrophosphate bond; the formula is shown above.

The isoalloxazine ring acts as a reversible redox system when functioning as a prosthetic group of dehydrogenases. Hydrogen is added at N-1 and N-10 as shown:

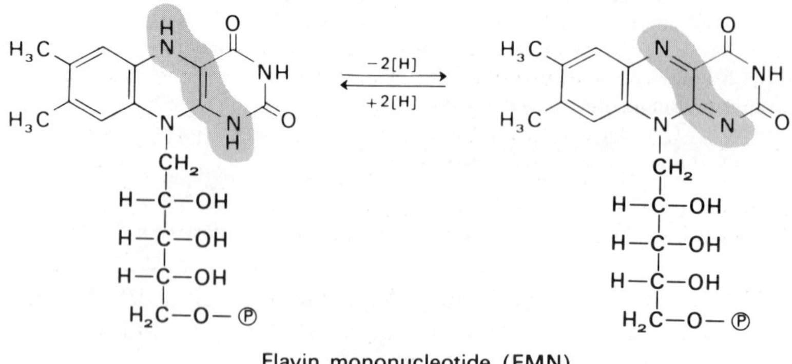

Flavin mononucleotide (FMN)
(oxidized and reduced form)

Hydrogen is transferred to the prosthetic group through the action of the enzyme. For the enzyme to retain its catalytic property the flavin system must be reoxidized.

This is usually accomplished by another enzyme system (Chapter X,4). A few yellow enzymes are able to transfer their hydrogen onto oxygen directly, yielding H_2O_2. Such enzymes are "aerobic dehydrogenases" (cf. Chapter X,7).

Flavin, as the name reveals, is yellow; the dihydro compound, however, is colorless (the situation is similar with other dyes, e.g., methylene blue, indigo). This property is used to follow the reduction of flavin by optical methods, as previously discussed in connection with the pyridine coenzyme.

There is evidence that in the reduction of several flavoproteins only *one* hydrogen atom is taken up. The result is a semiquinone with radical properties (unpaired electron). The semiquinone is very reactive and can easily donate the accepted hydrogen atom, for example, to $Fe^{3\oplus}$ ions or to oxidized cytochrome.

Some flavin enzymes can be separated into the apoenzyme and coenzyme components by a shift of pH and dialysis, or protein precipitation procedures. The old yellow enzyme was the first example of this, in that the active enzyme protein could be regenerated from coenzyme and apoenzyme (Theorell, 1934). However, the bonds often are much tighter, and the flavin groups generally do not dissociate from the protein after reduction, but are reoxidized while still attached to the same enzyme protein. According to our definition, they are prosthetic groups. Some flavoproteins also contain tightly bound metal ions which probably participate in catalysis (cf. Chapter X,4).

Ubiquinone. Yet another redox system of the respiratory chain is ubiquinone or coenzyme Q (cf. Chapter X,4). *Plastoquinone* has a similar function in higher plants (cf. Chapter XVI,2). The structure of these quinones is discussed in Chapter XIV,9.

Hemin as a Prosthetic Group. Cytochromes often are associated with the flavoproteins. Cytochromes contain a complex porphyrin system which is either identical with or closely related to the heme of hemoglobin. They are oxidoreductases because they transfer electrons. Heme-bound iron is involved in the electron transport by a reversible change of its valency between $Fe^{2\oplus}$ and $Fe^{3\oplus}$. Because of their importance, a separate chapter (Chapter IX) has been devoted to hemoproteins. The cooperation between pyridine nucleotides and the flavoproteins in the respiratory chain is discussed in Chapter X,4.

Lipoic Acid. Another hydrogen-transferring cofactor participates in the rather complex reaction of "oxidative decarboxylation."

Lipoic or *thioctic acid* was discovered around 1950 as a growth factor for certain microorganisms; it was then isolated, and its structure finally determined. It is a cyclic disulfide and in its side chain contains a carboxyl group, which is usually bound to the enzyme protein in amidelike linkage. Lipoic acid is another hydrogen-transferring cofactor. Its prime role is in oxidative decarboxylation, a very complicated reaction.

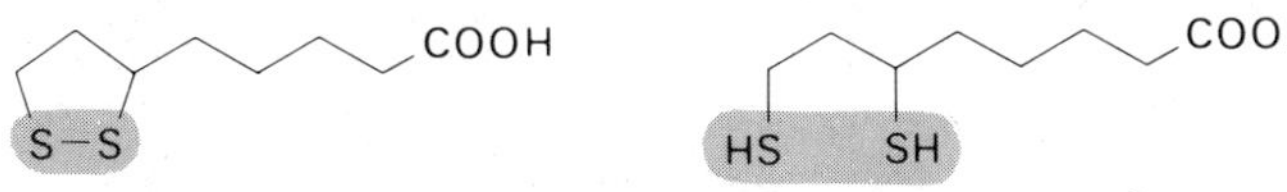

Lipoic acid   Dihydrolipoic acid

The reaction sequence of oxidative decarboxylation and the role of lipoic acid therein are discussed in detail in Chapters VIII,10 and XI,2. Here it suffices to mention that it cooperates in the conversion of the primary decarboxylation product, an active aldehyde, to the activated acid. Its job, therefore, is both dehydrogenation and group transfer. When acetate arises from active acetaldehyde, chosen here as the example, dihydrolipoic acid with an opened ring and two sulfhydryl groups remains. The cofactor must now be returned to the oxidized form by a dehydrogenase in order to be able to repeat its part in the reaction; this explains its classification as a hydrogen-transferring coenzyme.

The dehydrogenase has been identified and studied. It transfers hydrogen from dihydrolipoic acid to nicotinamide-adenine dinucleotide. Peculiarly, it is a flavoprotein, but its redox potential (cf. Chapter X,3) is much farther on the negative side than that of the familiar flavoproteins, so that hydrogen is able to switch over to NAD.

5. Adenosine Triphosphate as Coenzyme

Adenosine triphosphate, commonly abbreviated as ATP, is composed of the base adenine, the sugar ribose, and phosphoric acid. The nucleotide with one phosphate esterifying the OH group in position 5 of ribose is called *adenosine monophosphate*, abbreviated AMP (cf. also Chapter VII,1). With one further phosphoric acid attached in anhydride linkage we have *adenosine diphosphate*, ADP; with three phosphoric acid residues we come to the triphosphate ATP.

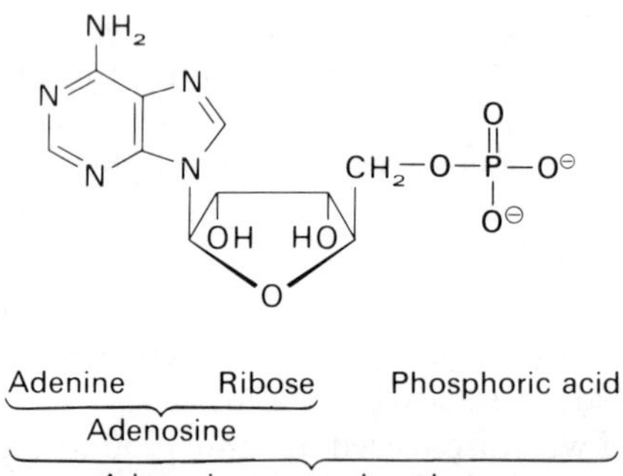

Adenosine monophosphate
AMP

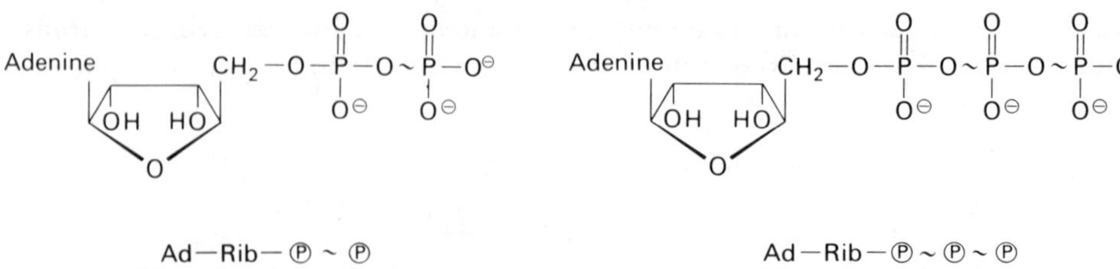

Ad—Rib—Ⓟ ~ Ⓟ
Adenosine diphosphate
ADP

Ad—Rib—Ⓟ ~ Ⓟ ~ Ⓟ
Adenosine triphosphate
ATP

Our symbol for the phosphoric acid residue is P, or as used in the above legends Ⓟ.

ATP contains two energy-rich bonds; it has a high potential for group transfers. Different types of groups can be transferred. Depending on what bond reacts the result is:

(a) Transfer of the orthophosphate group—release of ADP;
(b) Transfer of the diphosphate group—release of AMP;
(c) Transfer of adenosyl monophosphate group—release of pyrophosphate;
(d) Transfer of the adenosyl group—release of both orthophosphate and pyrophosphate.

These reactions are illustrated in Fig. VI-4.

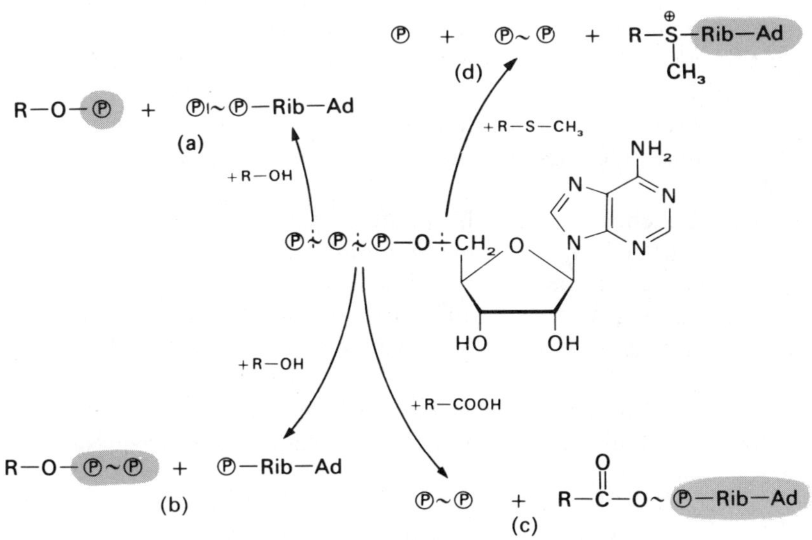

Fig. VI-4. Reactions of adenosine triphosphate.

Reaction (a) is the most common. If the orthophosphate residue is transferred onto water, HOH, we have hydrolysis. Enzymes catalyzing this reaction are called adenosine triphosphatases or ATPases. The free energy of hydrolysis is strongly negative, $\Delta G^0 = -7.0$ kcal/mole. *In vivo* this cleavage is almost invariably tied in with some particular function of the cell, e.g., muscle contraction, or active transport of compounds across membranes.

Kinases. The highly exergonic nature of the phosphate transfer onto water explains the generally great transfer potential of the phosphate groups of ATP. ATP can transfer phosphate onto alcoholic hydroxyl groups, onto acyl groups, or amide groups. All these transfer reactions are catalyzed by specific enzymes, which collectively are called "*kinases.*" The hexokinase and pyruvate kinase reactions have already been mentioned under the section on coupled reactions (cf. also Chapter V,5 and XV,7). The reaction is reversible whenever the compound formed is energy rich as, e.g., creatine phosphate (formula, Chapter XXIII,6; cf. Table V-1, Chapter V,5). Otherwise, the equilibrium would be too unfavorable to let the reverse reaction proceed to more than a minute extent. The phosphate donor is ATP. The resynthesis of ATP must take place through another enzymatic reaction.

Another important function of the kinases is the *synthesis of ATP* from ADP and energy-rich bound phosphate in "substrate-linked phosphorylation" (cf. Chapter XV,7) and in "respiratory-chain phosphorylation" (Chapter X,6). ATP hence represents a "pool" for energy-rich phosphates—and, in a sense, for chemical energy in general (Chapter XIX,5).

ATP is not the only reactive triphosphate. Other purine or pyrimidine bases (guanine, cytosine, hypoxanthine, uracil; cf. Chapter VII,1) may take the place of adenine in the molecule; the corresponding triphosphates replace ATP in several metabolic reactions.

Reaction (b) a transfer of diphosphate, occurs comparatively rarely. One example is the conversion of ribose 5-phosphate to 5-phosphoribose 1-pyrophosphate (cf. Chapter VII,2).

The transfer of adenosine monophosphate to an acceptor with removal of pyrophosphate, reaction (c), is again quite common. A compound with a high potential for group transfer, i.e., an "activated compound," is usually generated. Such compounds include, among others, the intermediates of fatty acid activation (Chapter XII,6) and of amino acid activation in protein synthesis (Chapter VII,6).

Cyclic AMP. Intramolecular transfer onto the 3′-hydroxyl with liberation of pyrophosphate results in the formation of "cyclic adenosine-3′,5′-monophosphate," shown here:

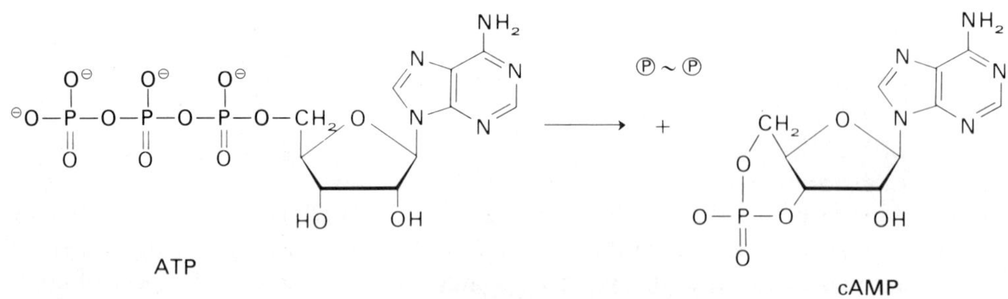

The requisite enzyme, *adenylate cyclase*, is localized in cell membranes. Cyclic AMP (also abbreviated cAMP) plays a cardinal role in the regulation of metabolism, primarily in the mechanism of action of many hormones (see Chapters XVII,6 and XX,1).

The reaction mentioned under (d) above, the transfer of the adenosyl residue, plays a part in the formation of the "active methyl groups" (Section 6, below).

The equilibrium of reaction (c) is occasionally unfavorable to the formation of the new activated compound R ~ P—Rib—Ad. In that case, the free energy of hydrolysis of the P ~ bond (−5 kcal) can be drawn upon through the action of pyrophosphatase, which splits pyrophosphate and forces the reaction to proceed to completion by removing one of the reaction products from the equilibrium mixture. A fine example of a coupled energetic reaction of this type is provided by the formation of active sulfate, according to Lipmann's scheme:

(1)	$\text{Ad—Rib—P} \sim \text{P} \sim \text{P} + H_2SO_4 \rightleftharpoons \text{Ad—Rib—P} \sim SO_3H + \text{P} \sim \text{P}$	$\Delta G^o = +11$ kcal
(2)	$\text{P} \sim \text{P} \rightleftharpoons \text{P} + \text{P}$	$\Delta G^o = -6$ kcal
(3)	$\text{Ad—Rib—P} \sim SO_3H + \text{ATP} \rightleftharpoons \text{Ad—}\underset{\text{P}}{\text{Rib}}\text{—P} \sim SO_3H + \text{ADP}$	$\Delta G^o = -5$ kcal
		$\Delta G^o = 0$ kcal

Adenosine phosphorylsulfate from reaction (1) can scarcely be detected in the equilibrium mixture. Only in the presence of pyrophosphatase are appreciable amounts produced, because then the reverse of reaction (1) (thermodynamically enormously favored) is no longer feasible. Even so, the energy is balanced only when the activated sulfate reacts with a further ATP and takes up a phosphate at the 3′-hydroxyl group of ribose. (The values for the standard free energies are approximate.)

Active sulfate therefore has the formula:

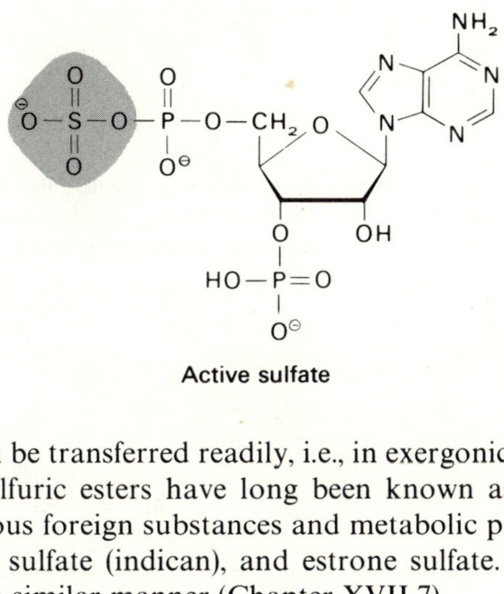

Active sulfate

The sulfate group can be transferred readily, i.e., in exergonic reactions, to phenols and alcohols. These sulfuric esters have long been known as "detoxication" and excretion forms of various foreign substances and metabolic products. They include phenol sulfate, indoxyl sulfate (indican), and estrone sulfate. The sulfate esters of carbohydrates arise in a similar manner (Chapter XVII,7).

6. Coenzymes of C_1 Metabolism

Various one-carbon fragments are involved in metabolism. They may be derived in the following fashion:

(a) the methyl group, $—CH_3$, from methanol, $HO—CH_3$;

(b) the hydroxymethyl group, $—CH_2OH$, from formaldehyde, H_2CO, or rather its hydrate $HO—CH_2—OH$;

(c) the formyl group, —CHO, from formic acid HO—CHO;
(d) the carboxyl group, —COOH, from carbonic acid HO—COOH.

The following coenzymes or prosthetic groups carry out the transfer of these C_1 fragments.

Adenosylmethionine, "Active Methyl." Methionine, one of the sulfur-containing amino acids, is the chief donor of methyl groups, which are transferred to various substances, but especially to amino groups (cf. Chapter VIII,12). The methyl group bound in thioether linkage does not possess a high potential for group transfer; it is activated by reaction with ATP, yielding, according to Fig. VI-4d, a very reactive sulfonium compound:

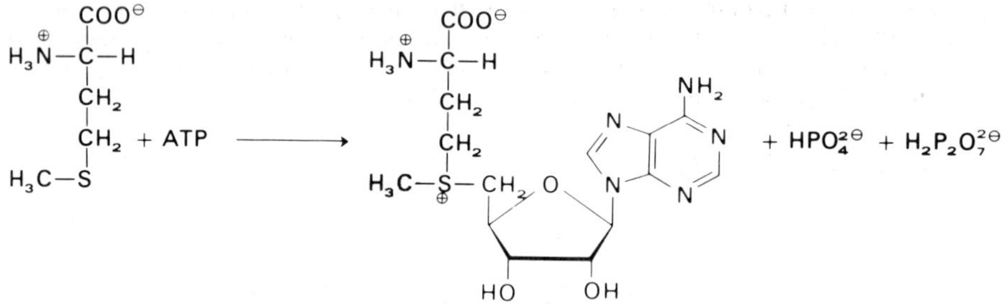

Methionine — Adenosylmethionine, "active methyl"

The group bound to sulfur is transferred as $^{\oplus}CH_3$ to an atom with a free electron pair (e.g., N); thus the sulfonium structure is lost and an adenosine thioether remains. The following example shows the conversion of guanidinoacetate to creatine, which then can pick up phosphate in an energy-rich bond (cf. Chapters V,5, V,9, and XXIII,6).

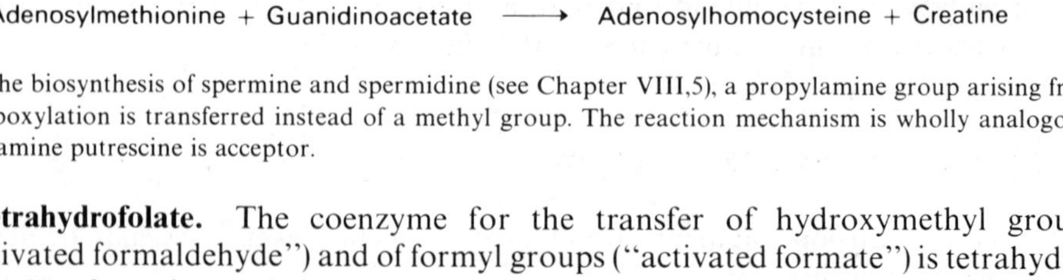

Adenosylmethionine + Guanidinoacetate ⟶ Adenosylhomocysteine + Creatine

In the biosynthesis of spermine and spermidine (see Chapter VIII,5), a propylamine group arising from decarboxylation is transferred instead of a methyl group. The reaction mechanism is wholly analogous; the diamine putrescine is acceptor.

Tetrahydrofolate. The coenzyme for the transfer of hydroxymethyl groups ("activated formaldehyde") and of formyl groups ("activated formate") is tetrahydrofolate. For formulas see below under "active formate."

The parent substance, *folic acid*, was originally discovered and isolated as a vitamin independently by several different laboratories. The compound contains a pteridine ring and a heterocyclic ring system with one pyrimidine ring fused onto one pyrazine ring. Pterins are colored substances occurring in butterfly wings, e.g., *xanthopterin* in the brimstone butterfly, and in wasps. The work of H. Wieland, C. Schopf, and co-workers on these insect pigments gained particular interest with the discovery of folic acid.

Biopterin is widely distributed in very minute amounts; it is found in urine and in the food juice of queen bee larvae (royal jelly) (Rembold). Tetrahydrobiopterin appears to donate hydrogen in the hydroxylation of phenylalanine (see Chapter X,7).

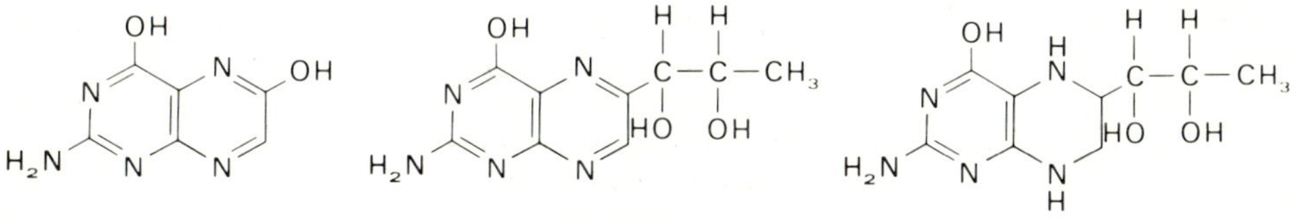

Xanthopterin Biopterin Tetrahydrobiopterin

Folic Acid. In addition to the substituted pteridine ring, folic acid contains *p*-aminobenzoic acid, which is a growth factor for a number of microorganisms. The growth effect can be repressed by sulfonamides, and the mode of attack of the sulfonamides, which are common chemotherapeutic agents, involves interference with folic acid formation in bacteria.

The structure of folic acid also contains glutamic acid. The complete formula, therefore, looks like this:

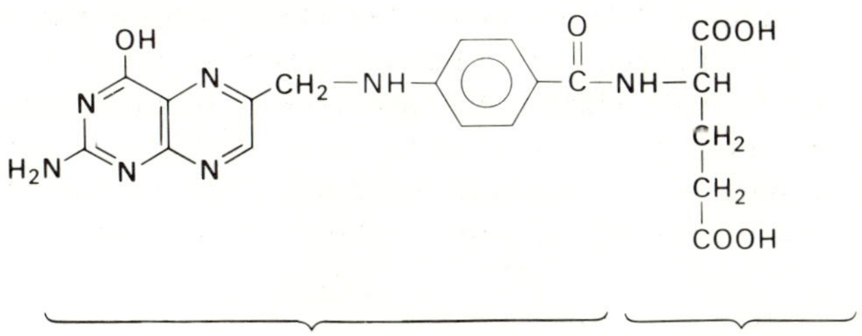

Pteroic acid Glutamic acid

Pteroylglutamic acid = Folic acid

In addition to the folic acid with but one glutamate residue, various polyglutamate conjugates are known, which stimulate growth of certain bacterial strains and which cannot be replaced by folic acid. Pteroylheptaglutamate, in particular, has drawn attention; it has seven glutamate residues in peptidic linkage (through the γ-carboxyl groups). It is really an oligopeptide with one folic acid residue.

Leucovorin. A folic acid derivative closely related to the true coenzyme. Leucovorin is a growth factor for the bacterium *Leuconostoc citrovorum*. It contains a reduced ring and bears a formyl group on the 5 nitrogen atom (N-5), but has only a low potential for group transfer.

Active Formaldehyde. This probably is bound initially to N-10 as the hydroxymethyl group. A ring can be closed easily by forming a bridge to N-5:

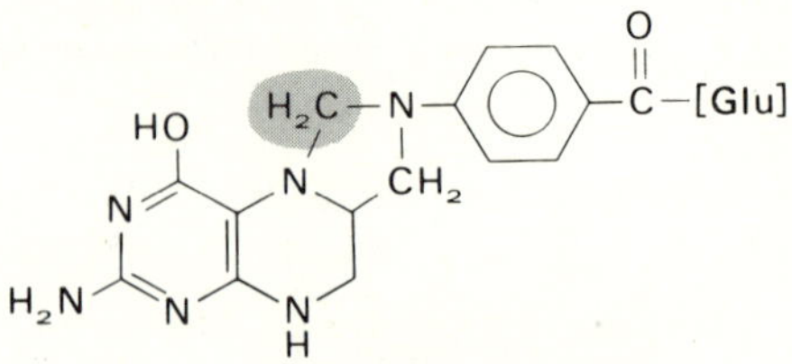

N-5,*N*-10-Methylenetetrahydrofolate
(= Active formaldehyde)

One important donor of the hydroxymethyl group is serine, which in reversible reaction can deliver its β-carbon atom to tetrahydrofolate (cf. Chapter VIII,12). Dehydrogenation with NADP converts active formaldehyde to active formate.

Another form of active formaldehyde recently discovered by Holzer is *hydroxymethyl thiaminepyrophosphate.* It arises from the decarboxylation of glyoxalate and has a similar structure as active glyoxal (Chapter XV,5).

Active Formate (*N*-10-Formyltetrahydrofolate). Another immediately reactive intermediate with an even greater energy content is the anhydro derivative *N*-5,*N*-10-methenyltetrahydrofolate. It can revert to the *N*-10-formyltetrahydrofolate in slightly alkaline media by the addition of $OH^{\ominus}$ (Jaenicke). The cooperation of ATP is required to convert free formate to the *N*-10-formyl derivative or its anhydro form. Ordinarily, however, it is not free formate which is activated; the C_1 fragment arises rather from metabolism and is bound immediately to tetrahydrofolate (cf. Chapter VIII, 12). Donors of active formate are, for example, histidine and tryptophan. Active formate is consumed, among other things, in purine synthesis (Chapter VII,2).

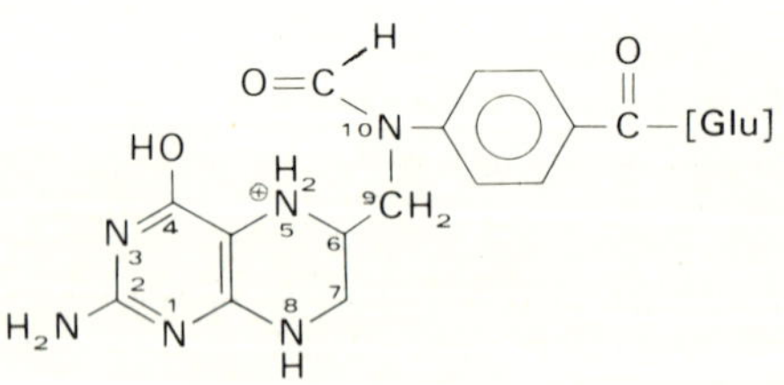

N-10-Formyltetrahydroformate
(= Active formate)

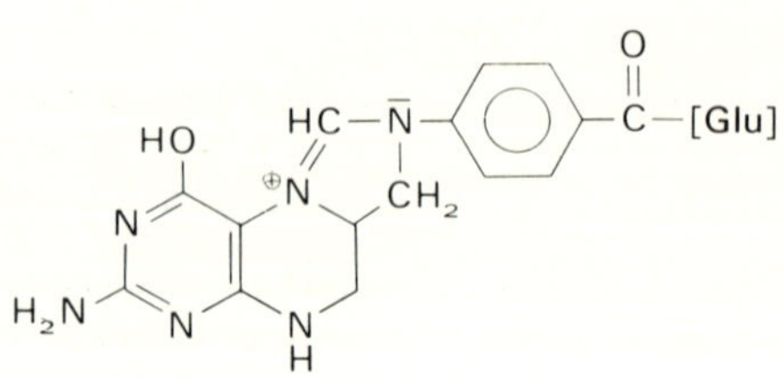

N-5,*N*-10-Methenyltetrahydrofolate

Biotin. The function of biotin is in the transfer of carboxyl groups. It has been isolated as "vitamin H" from liver extracts (DuVigneaud and co-workers) and as a yeast growth factor from egg yolk (Kogl and co-workers). It is a cyclic derivative of urea with a thiophane ring attached. Biotin adheres very tightly to the specific protein of egg white *avidin*, and consequently is inactivated by it. Large doses of raw egg white fed to animals can evoke biotin deficiency. Being a protein, avidin is denatured and inactivated by heat.

Biotin is linked to the enzyme protein by a peptide bond to the ε-amino group of a lysyl residue; hence, it represents a prosthetic group. *In vivo* this linkage is mediated by ATP by forming the intermediate biotyl-AMP.

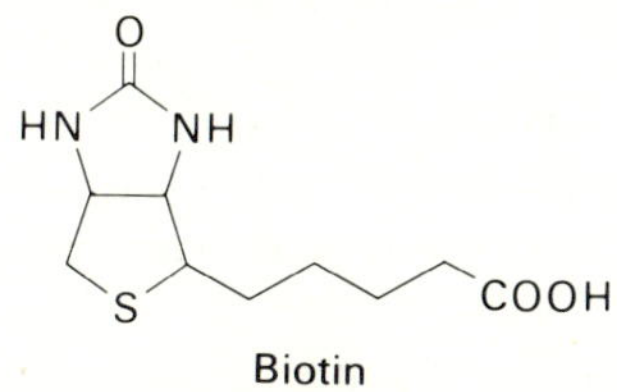

Biotin

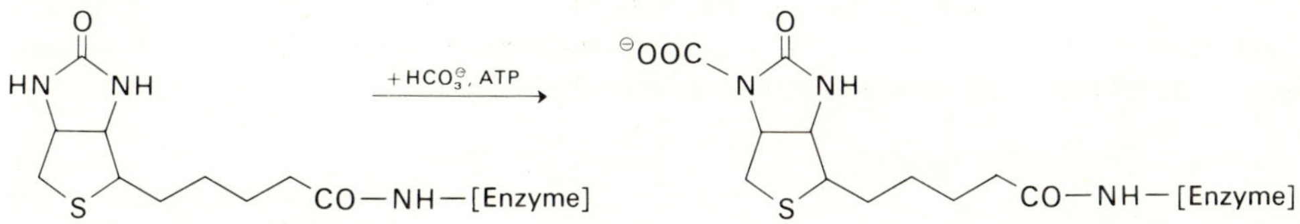

Biotin enzyme

Carboxybiotin enzyme
(= Active carbon dioxide)

Charging the biotin enzyme with carbon dioxide is an endergonic process requiring the assistance of ATP. The CO_2 attached to the nitrogen of biotin is the *active form of carbon dioxide* which participates in numerous carboxylation reactions (e.g., acetyl-CoA to malonyl-CoA, Chapter XII,6, and related reactions). The very labile carboxybiotin has been isolated as a methyl ester (Lynen and co-workers).

7. Coenzymes of C_2 Metabolism

Three C_2 fragments have major significance in metabolism: acetaldehyde, glyoxal (=hydroxyacetaldehyde), and acetate.

Thiamine Pyrophosphate. The coenzyme responsible for transferring "active acetaldehyde" and "active glyoxal" is thiamine pyrophosphate. In the decarboxylation of pyruvate it simultaneously acts as "codecarboxylase,"[3] i.e., as the coenzyme for a lyase. The characteristic component is *thiamine* (vitamin B_1), one of the vitamins whose biological role has been known longest. The name thiamine refers to its sulfur content. Thiamine has two heterocyclic rings that are not condensed, namely one

[3] The older term *cocarboxylase* is still used; it stems from *carboxylase*, the older name of the decarboxylating yeast enzyme. Now the term "carboxylase" is reserved for those enzymes that introduce a carboxyl group into the substrate; they contain biotin as prosthetic group.

pyrimidine and one thiazole ring. They are connected at the quaternary nitrogen of the thiazole ring. Thiamine, therefore, always carries a charge.

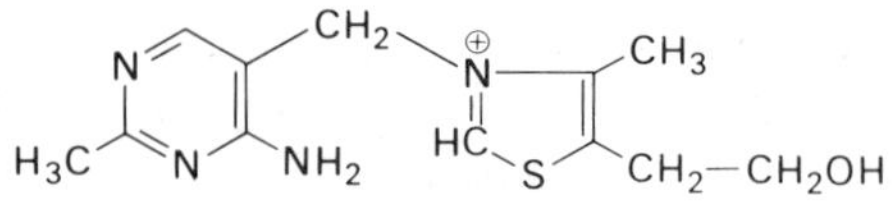

Thiamine

The active form of thiamine is its *pyrophosphate*. Aside from pyridine nucleotide coenzymes, thiamine pyrophosphate was one of the earliest in the group of factors to be recognized as being catalytically active. By our definition it is really a prosthetic group, since it remains linked to the enzyme protein.

C-2 of the thiazene ring readily becomes a carbanion, which can react with the polarized keto group of an α-keto acid and thus can catalyze its decarboxylation. "Active acetaldehyde" bound to C-2 is then accepted by the amino group (Schellenberger) and finally released or transferred to other acceptors:

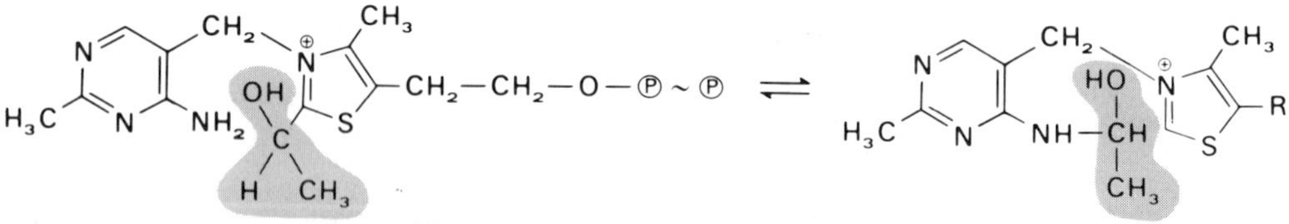

Active acetaldehyde

The most important reaction in which thiamine pyrophosphate collaborates is the oxidative decarboxylation of α-keto acids (cf. formulas in Chapters VIII,10 and XI,2). CO_2 is split off and the aldehyde residue is transferred by thiamine pyrophosphate to *lipoic acid*, which functions as an oxidizing agent (cf. end of Section 4, above). However, in the transketolase reaction thiamine pyrophosphate serves only as a group-transferring coenzyme; it transfers glyoxal (cf. Chapter XV,5).

Coenzyme A. This coenzyme transfers the acyl residue of acetic and other carboxylic acids. Acyl groups bound to CoA (the usual abbreviation for coenzyme A, the "A" standing for acylation) have a high potential for group transfer. Hydrolysis of acyl-CoA compounds is exergonic to the extent of about 8 kcal/mole.

The chemical structure of coenzyme A is appreciably complex. It is helpful to divide it mentally into adenosine 3′,5′-diphosphate and pantetheine phosphate. Pantetheine is a growth factor for several microorganisms, e.g., *Lactobacillus bulgaricus*. It, in turn, consists of pantoic acid, β-alanine, and mercaptoethylamine, the decarboxylation product of cysteine, which bears the SH group to which acyl residues are linked with high-energy bonds.

The combination of pantoic acid (α,γ-dihydroxy-β,β-dimethylbutyric acid) and β-alanine is also called *pantothenic acid*; this name is derived from its nearly universal occurrence. Pantothenic acid is listed as one of the B vitamins (see Chapter XII,5). The biochemical significance of the vitamin resides in its being a necessary precursor in the biosynthesis of CoA. Depending on the capabilities of the organism, pantoic acid (or β-alanine), or pantothenic acid, or pantetheine must be supplied.

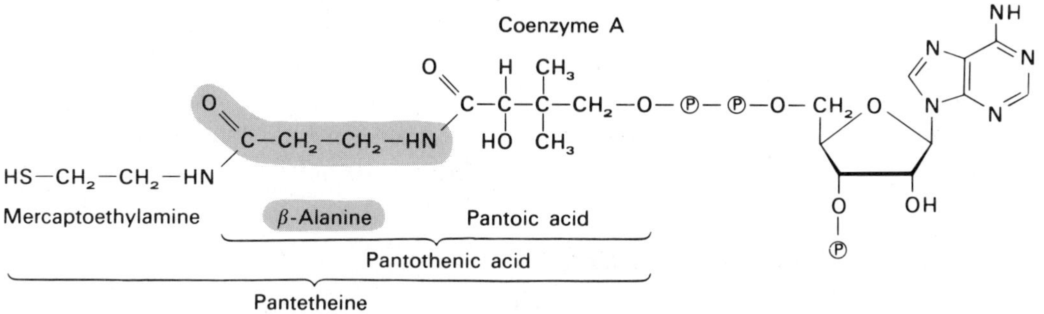

The most important CoA compound undoubtedly is acetyl-CoA, *activated acetate*. Here the acetyl residue $CH_3CO—$ is bound to the free SH groups.[4] Chemically this constitutes a thioester, and thioesters are known to be very reactive.

To bring acetic acid (or any other carboxylic acid) into this compound, with its high potential for group transfer, it is, of course, necessary to expend energy. This energy can be derived either from cleavage of ATP—whereby an intermediate arises, an anhydride with adenylic acid according to reaction (c) in Fig. VI-4—or from a strongly exergonic reaction (oxidative decarboxylation; cf. Chapter VIII,10).

The reactions of acetyl-CoA can only be touched upon here. There are two groups: (1) reactions of the carboxyl group, and (2) reactions of the methyl group or, in the case of higher homologs, of the α-methylene group.

Reactions of the Carboxyl Group. Ester and amide formation may be mentioned first; one H atom of an alcohol or an amine is replaced by the acyl group, but C-acylation ("ester condensation" in organic chemistry) is also known, along with the reverse reaction, thioclastic cleavage:

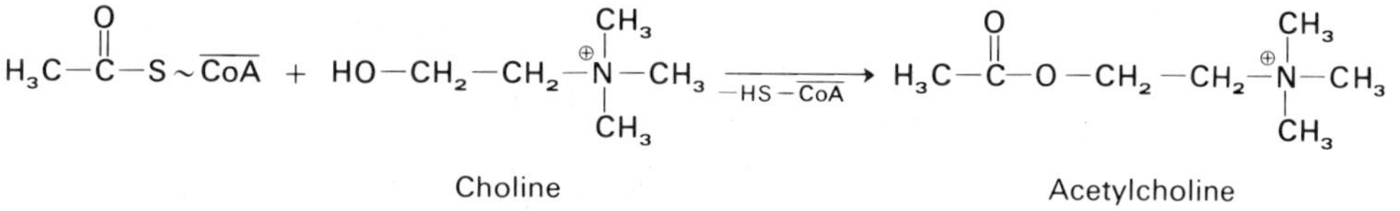

Reactions of the Methyl Group. These reactions usually occur with carbonyl groups and can be understood more easily from the polarized form of the acetyl-CoA in which the α-C atom receives a partial negative charge from the coenzyme (i.e.,

[4] If it is desired to emphasize the reactive group itself, then $HS—\overline{CoA}$ (or $H_3C—CO—S—\overline{CoA}$) is written, where $\overline{CoA}$ stands for "coenzyme A less —SH."

through the thioester bond). The polarized α-C atom then joins with the partially positive C atoms of the carbonyl group:

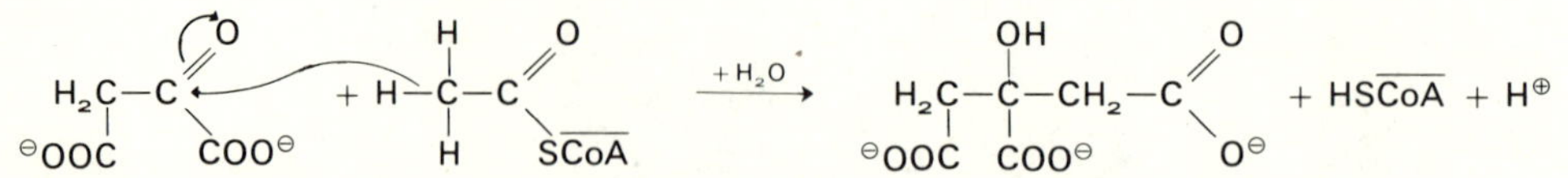

The formation of citrate from oxaloacetate shown here is quantitatively most important. These reactions are treated more thoroughly in Chapter XI.

8. Further Group-Transferring Coenzymes

Uridine Diphosphate is the carrier of "active glucose." Uridine triphosphate, entirely analogous to ATP, can exchange the third phosphate group for a sugar molecule and hold it in an energy-rich bond:

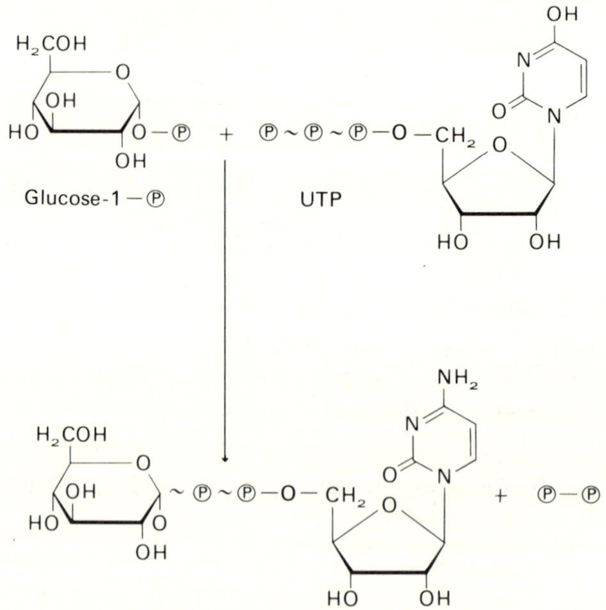

Uridine diphosphate glucose (UDP-Glc)

The sugar molecule—glucose in the above example—can be transferred to other molecules with OH groups. Glycosides and disaccharides are synthesized in this manner. These reactions will be discussed more extensively in Chapter XVII,4.

Cytidine Diphosphate. The coenzyme of phosphatide biosynthesis cytidine diphosphate generates phosphodiester bonds. In the following example, cytidine triphosphate and choline phosphate give rise to choline phosphate and in high-energy linkage to cytidine phosphate (cytidine diphosphate-choline, CDP-choline).

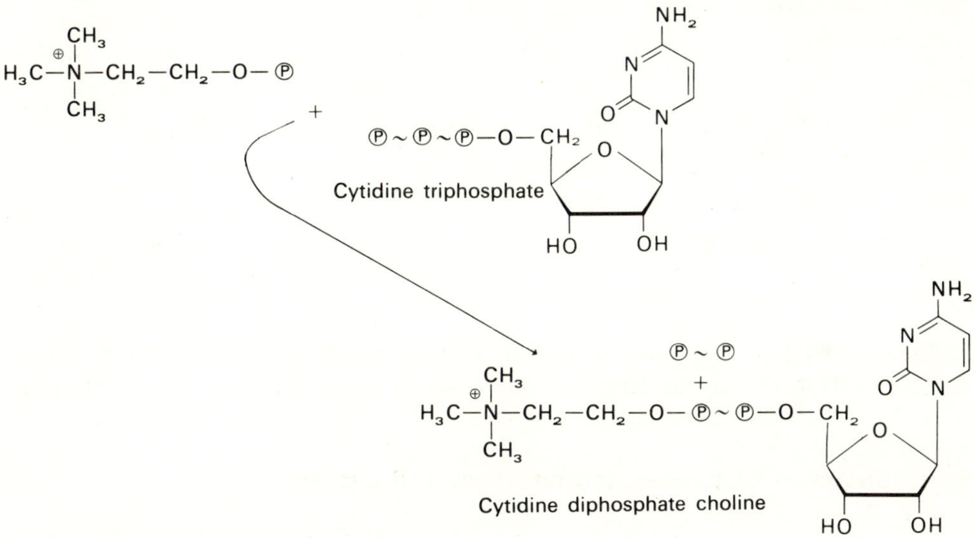

The choline moiety can then be transferred to some acceptor molecule, e.g., a diglyceride, forming a phosphodiester bond while cytidine monophosphate is split off (cf. Chapter XII,3). In contrast to the reaction with uridine diphosphate glucose, one phosphate group enters the product. Ethanolamine, instead of choline, can also be activated in this manner. Furthermore, both serine *N*-acetylsphingosine and glycerophosphate can function as acceptors in place of diglycerides. Thus a great number of phosphatides are synthesized (cf. Chapter XIII,3).

Pyridoxal Phosphate. The coenzyme of amino acid metabolism, pyridoxal phosphate is closely related to pyridoxine, a vitamin of the B group, which is also called vitamin B_6.[5]

Pyridoxine, *pyridoxal*, and *pyridoxamine*, as the names suggest, are all pyridine derivatives; the various substitutions are seen in the following formulas:

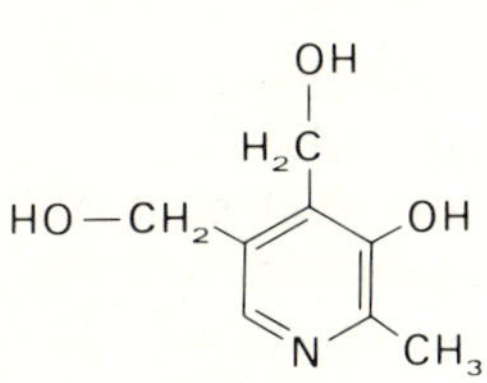
Pyridoxine

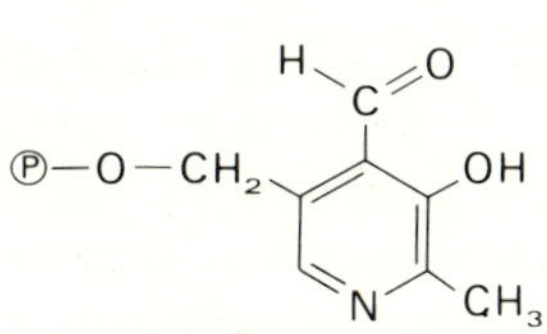
Pyridoxal phosphate

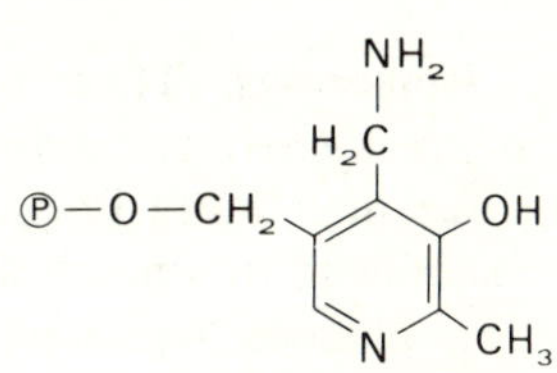
Pyridoxamine phosphate

[5] *Vitamin B_6* is still in use as a generic term for pyridoxine, pyridoxal, and pyridoxamine.

Pyridoxal phosphate is an excellent example of the fact that a single coenzyme is able to catalyze completely different reactions. It is the active group not only for the aminotransferases, but also for the decarboxylases and various lyases and synthases as well. A hypothesis by Snell endeavors to derive all types of reactions from one intermediate product, a type of *Schiff's base*:

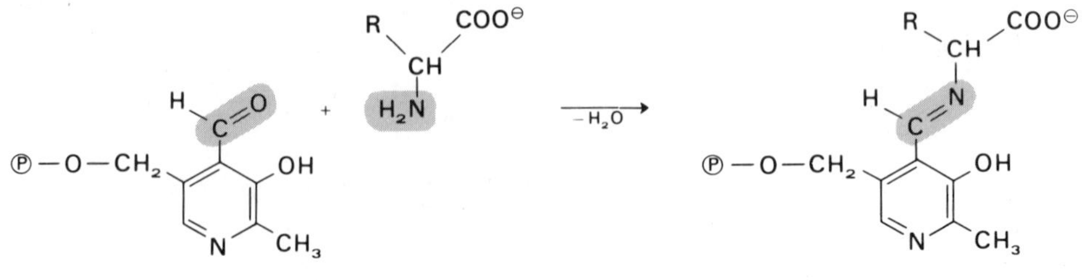

Pyridoxal phosphate — Schiff's base

The intermediate can react further in various ways, as discussed extensively in Chapter VIII,4. The apoenzyme determines which reaction is to be catalyzed.

9. Coenzymes of Lyases, Isomerases, and Ligases

Lyases and Synthases. These are enzymes which catalyze the cleavage of a compound into two fragments and, in reverse, which catalyze the joining of two substances to form a third, respectively. The latter reaction is frequently equivalent to a group transfer. While it is still possible to classify the enzymes, the coenzymes cannot be classified in this way. Numerous groups activated by coenzymes participate in the reactions of synthases, such as acetyl-CoA, carboxybiotin, and thiamine-bound active aldehyde.

One important subgroup of the lyases are the *decarboxylases*. The decarboxylation of amino acids is assisted by pyridoxal phosphate as prosthetic group, whereas in the decarboxylation of pyruvate to acetaldehyde, thiamine pyrophosphate (TPP) plays that role. Last, the oxidative decarboxylation of α-keto acids depends on the cooperation of no fewer than five cofactors: thiamine pyrophosphate, lipoic acid, coenzyme A, flavin-adenine dinucleotide, and nicotinamide-adenine dinucleotide (cf. Chapters VIII,10 and XI,2).

Isomerases. These usually do not require cofactors. Isomerization reactions of sugars sometimes involve uridine disphosphate and firmly bound NAD.

An interesting phenomenon is presented by the *mutases* which shift phosphate residues in the metabolism of carbohydrates. They catalyze, for example, the transition glucose 1-phosphate $\rightleftarrows$ glucose 6-phosphate. In this case, the intermediate glucose 1,6-diphosphate is at the same time "coenzyme" (i.e., phosphate donor), and the stoichiometrically reacting "coenzyme" is regenerated from the substrate (cf. Chapters XV,7 and XVII,6).

In some reactions of carboxyl shifts, yet another coenzyme, a derivative of vitamin B_{12}, is involved.

Vitamin B_{12}. Called cobalamin (because of its content of tightly bound cobalt), Vitamin B_{12} protects against pernicious anemia. Its structure is rather involved; there is a certain kinship to the hemin system. There is a whole series of compounds of this type with vitamin properties; they are distinguished by the various base components in nucleotidelike linkage and by the cobalt ligands ($CN^{\ominus}$, $OH^{\ominus}$, $Cl^{\ominus}$, etc.).

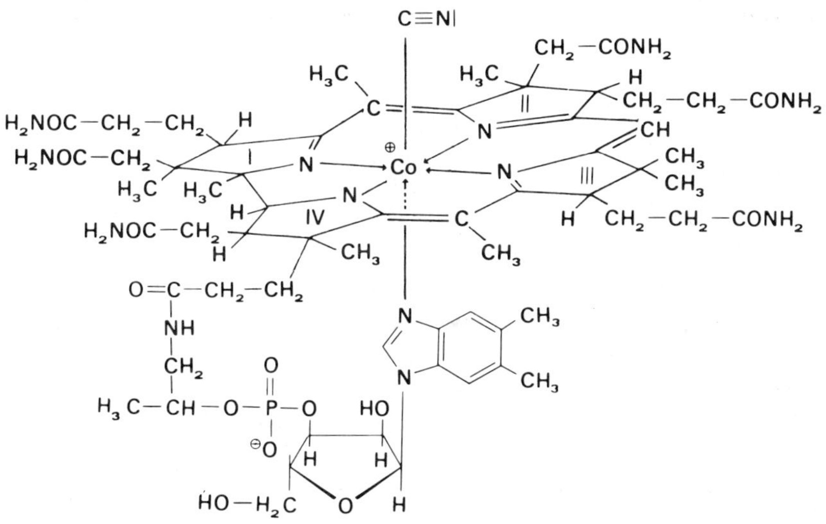

Vitamin B_{12}

How this vitamin enters the process of blood formation is still entirely unknown. More can be said, however, about its biochemical action: It is a component of the coenzyme to certain carboxylation reactions. The coenzyme B_{12} in which the cyanide group is replaced by 5′-deoxyadenosine (with direct C—Co linkage) is instrumental in a number of reactions, such as carboxyl shifts within molecules. The most important of these is the conversion of methylamalonyl-CoA to succinyl-CoA in which the —CO ~ SCoA moiety migrates (see Chapter XII,4). In addition, coenzyme B_{12} can transfer C_1 fragments from the tetrahydrofolate system to homoserine and thus synthesize methionine. Finally, it is the coenzyme of ribonucleotide reductase, which reduces ribonucleotides to deoxyribonucleotides.

Ligases. These are defined by the new system of nomenclature as enzymes which aid the formation of new bonds coupled with the elimination of a pyrophosphate bond. Consequently either ATP or a similar triphosphate must participate; occasionally biotin is involved (in carboxylation reactions).

The amino acid-activating enzymes are also ligases. "Transfer ribonucleic acids" may well be called coenzymes, although they are not low-molecular weight substances (molecular weight is around 25,000). The structure and function of nucleic acids will be discussed in the following chapter.

BIBLIOGRAPHY

S. Chaykin, Nicotinamide coenzymes. *Annu. Rev. Biochem.* **36,** 149–170 (1967) and references 4 and 5 listed therein.

P. Fasella, Pyridoxal phosphate. *Annu. Rev. Biochem.* **36,** 185–210 (1967).

L. O. Krampitz, Catalytic functions of thiamine diphosphate. *Annu. Rev. Biochem.* **38,** 213–240 (1969).

F. Wagner, Vitamin B_{12} and related compounds. *Annu. Rev. Biochem.* **35,** 405–434 (1966).

CHAPTER VII

Nucleic Acids and Protein Biosynthesis

Nucleic acids were discovered by Miescher in 1869; he found them in pus corpuscles, in fish sperm, and other biological material. They are high-molecular weight polymeric substances which are hydrolyzable into heterocyclic organic bases, carbohydrate (pentose), and phosphoric acid.

The chemistry of the carbohydrates will be treated extensively in Chapters XV and XVII. Here we mention that they are polyhydroxyaldehydes or ketones, which exist in the form of cyclic hemiacetals. The OH group arising from the hemiacetal structure is particularly reactive and can react with amines with loss of water (cf. formulas in Section 1 below). Compounds resulting from this reaction are called *N*-glycosides (cf. Chapters VI,3 and XVII,1).

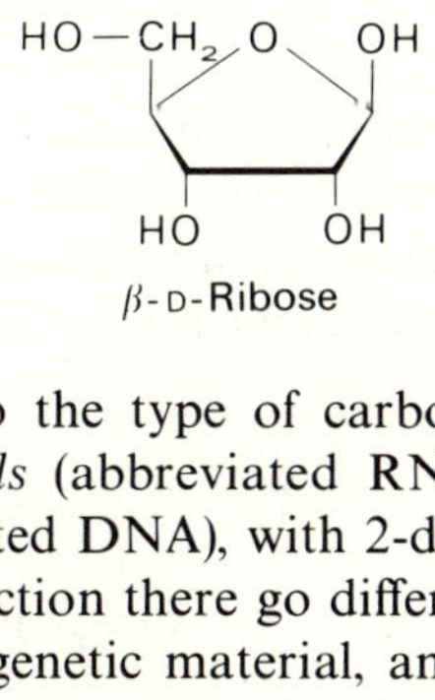

β-D-Ribose

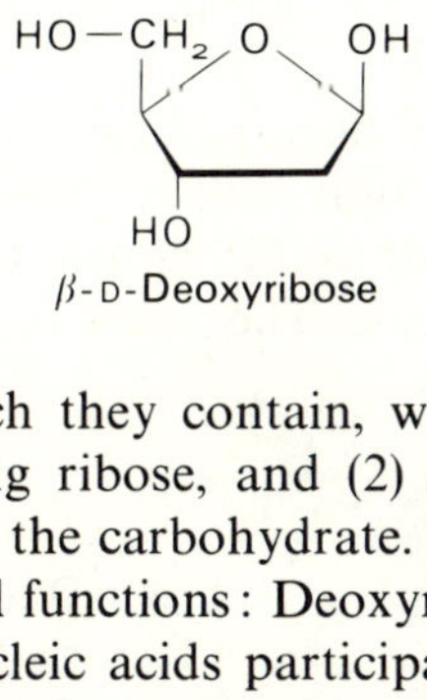

β-D-Deoxyribose

According to the type of carbohydrate which they contain, we distinguish (1) *ribonucleic acids* (abbreviated RNA), containing ribose, and (2) *deoxyribonucleic acids* (abbreviated DNA), with 2-deoxyribose as the carbohydrate. With this purely chemical distinction there go different biological functions: Deoxyribonucleic acids constitute the genetic material, and the ribonucleic acids participate intimately in all phases of protein synthesis. First, however, the cleavage products of nucleic acids will be discussed: the bases, the nucleosides, and the nucleotides.

1. Bases, Nucleosides, and Nucleotides

Pyrimidines. The pyrimidine ring is a six-membered ring with two nitrogen atoms; the pyrimidine bases of nucleotides bear an amino or hydroxyl group in position 4 (for numbering see formula below) and always have an oxygen function in position 2. This situation allows for tautomerism: the H atom can belong either to the oxygen or to the ring nitrogen, as shown below for uracil. Among these tautomeric forms, the most interesting is the one that has a hydrogen on N-1 (in nucleosides, ribose has replaced this hydrogen; the ribose is bound in *N*-glycosidic linkage).[1]

The main pyrimidine bases are *cytosine*, *uracil*, and *thymine*. Uracil is found virtually only in RNA; thymine almost exclusively in DNA. The formulas of all the bases and corresponding nucleosides are presented in Table VII-1.

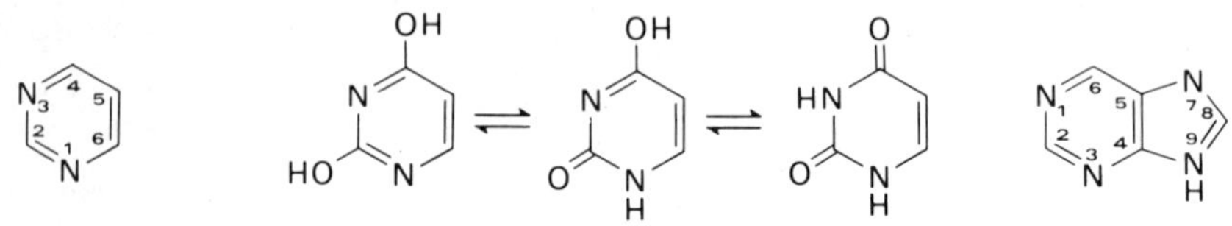

Pyrimidine Uracil (tautomeric forms) Purine

Purines. Emil Fischer named the ring system that is the skeleton of uric acid "purine" (from Latin *purum uricum*). The positions on the ring are numbered as shown.

Adenine (6-aminopurine) is a component of ATP and other low-molecular weight nucleotides (nucleotide coenzymes). The replacement of the amino group with a hydroxyl gives *hypoxanthine*.

Another common component of nucleic acid is *guanine* (2-amino-6-hydroxypurine). Its name stems from its original discovery in guano. Guanine has the guanidine grouping [$NH{=}C(NH_2)_2$] in its molecules, and indeed, guanidine (imino urea) was first obtained by the degradation of guanine.

Nucleosides. The structural features of nucleosides have already been discussed briefly in connection with coenzymes. Base and sugar are linked through a C—N bond (*N*-glycoside bond; cf., also Chapter XVII,1), which formally derives from loss of a water molecule. The arrow indicates that the equilibrium favors the side of the components:

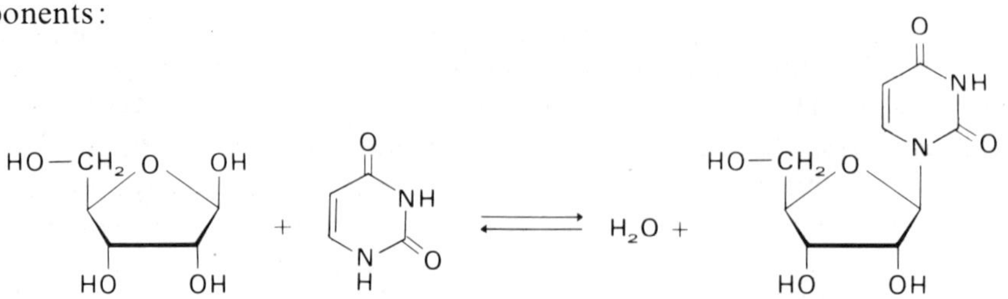

Ribose Uracil Uridine

Nucleosides have trivial names, derived from the bases; derivatives of pyrimidine end in "*idine*" and those of purine end in "*-osine*" (see structures in Table VII-1).

[1] In the old numbering system this nitrogen was N-3, and the other N-1.

TABLE VII-1

Purine derivatives:

Bases:

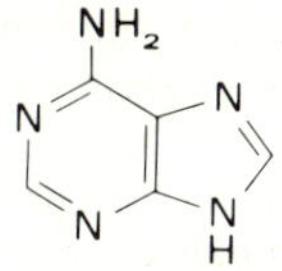

Adenine
(Ade)

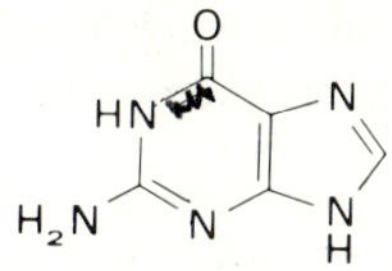

Guanine
(Gua)

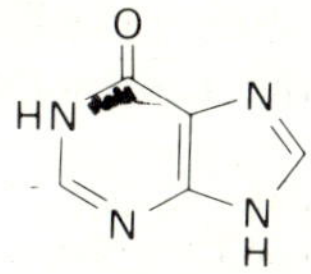

Hypoxanthine
(Hyp)

Nucleosides:

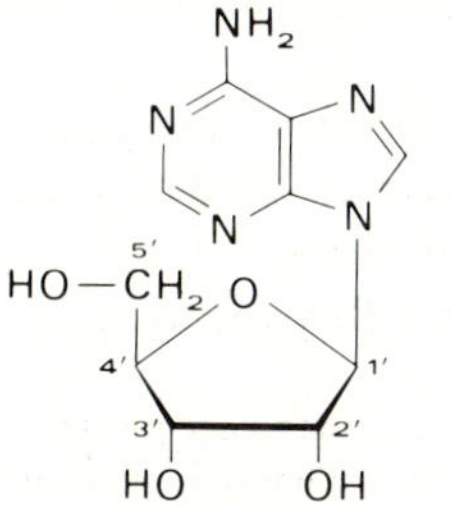

Adenosine (A)

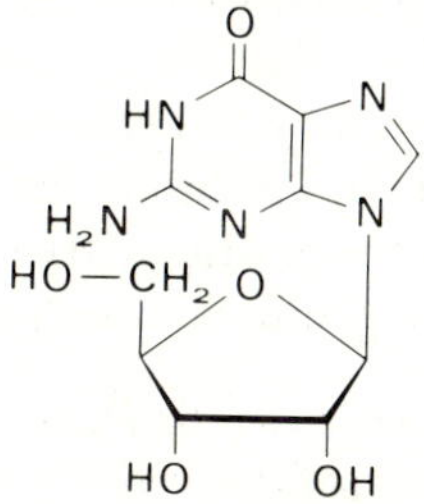

Guanosine (G)

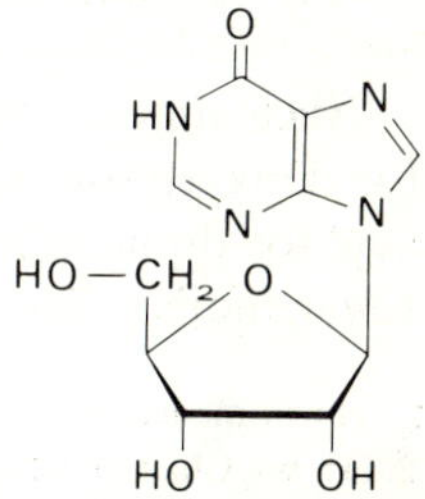

Inosine (I)

Pyrimidine derivatives:

Bases:

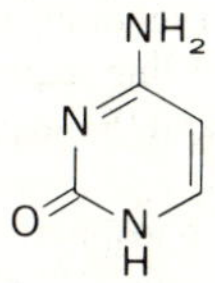

Cytosine
(Cyt)

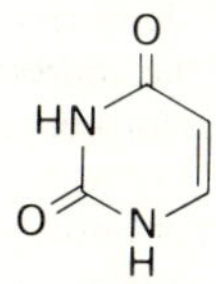

Uracil
(Ura)

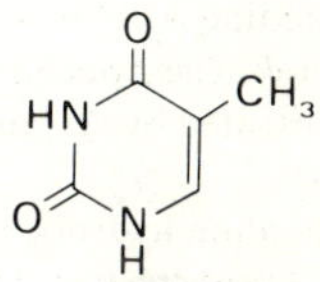

Thymine
(Thy)

Nucleosides:

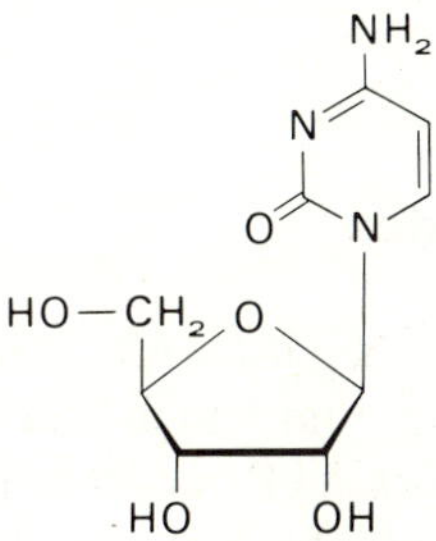

Cytidine (C)

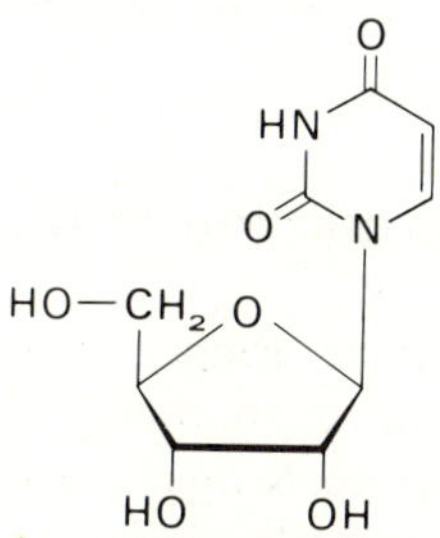

Uridine (U)

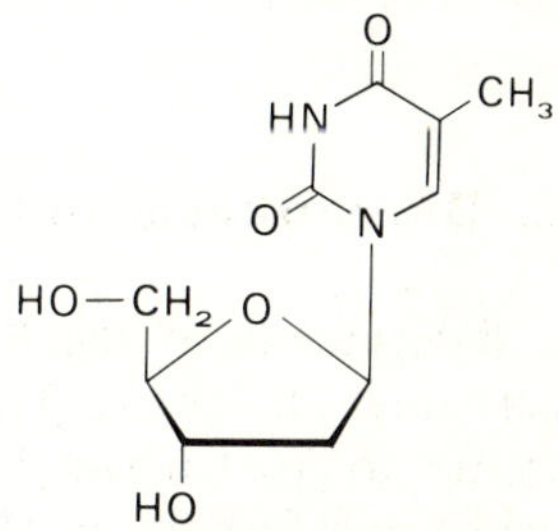

Thymidine (dT)

Rare bases:

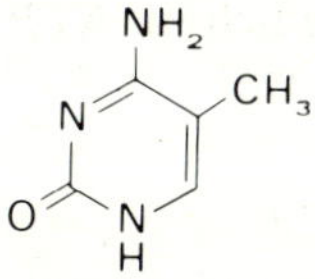

5-Methylcytosine
(mC)

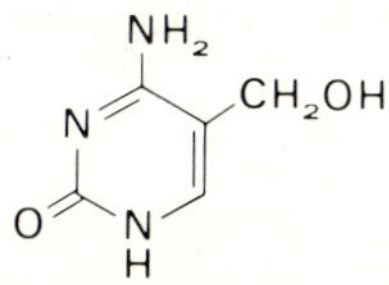

5-Hydroxymethyl-
cytosine (hmC)

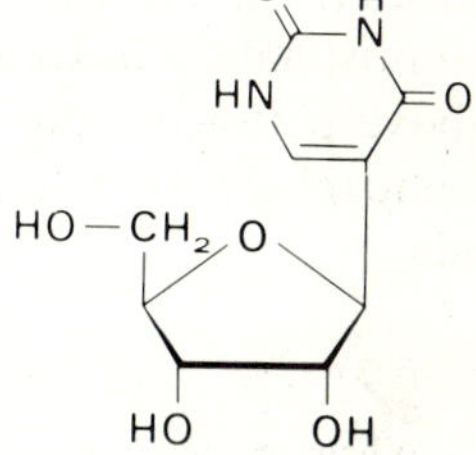

Pseudouridine
(ψ)

In distinction to the ring numbering, the C atoms of the sugar components are numbered 1′ to 5′ (see formula of adenosine in Table VII-1).

Beyond the well-known base components, several other bases and nucleosides occurring only in special nucleic acids have been discovered. Some bacterial viruses (bacteriophages) contain in their nucleic acids the bases 5-*hydroxymethylcytosine* and 5-*hydroxymethyluracil*. Transfer ribonucleic acid (see Section 6) contains several so-called rare nucleosides, e.g., N^6-isopentenyl adenosine, 5-ribosyluracil (or pseudouridine), 5,6-dihydrouridine, 5-methylcytidine, and other methyl derivatives of the main nucleosides.

Nucleotides. Nucleotides carry phosphoric acid in ester linkage with the ribose (or deoxyribose) component. In the biologically important 5′-esters, the phosphoric acid (or diphosphoric or triphosphoric acid) is bound to the $-CH_2OH$ group of the sugar moiety. The phosphate group is often represented by the symbol Ⓟ.

The nomenclature of nucleotides has already been seen in the case of adenosine phosphates (AMP, ADP, ATP; see Chapter VI,5). The remaining, similarly built nucleoside phosphates are named and abbreviated analogously. The nucleoside is usually abbreviated just with its first letter; G for guanosine, C for cytidine, etc. Deoxyribonucleotides, the components of DNA, are designated by the prefixed letter "d"; dAMP for deoxyadenosine monophosphate, dCTP for deoxycytidine triphosphate, for example. To be consistent thymidine is abbreviated as dT since it is a deoxyribonucleoside. Pseudouridine is indicated by the Greek letter ψ. The Appendix contains the rest of the abbreviations. In writing sequences, the phosphate group is represented by "p" or just a hyphen (see the formula in Section 3 and the sequences on the back of foldout chart).

Alkaline hydrolysis of RNA, however, does not release the 5′-phosphates, but rather a mixture of 2′- and 3′-phosphates. This mixture arises because an intermediary cyclic 2′,3′-phosphodiester[2] forms during hydrolysis (see formulas in Section 10). The cyclic phosphate ester can then open up either as the 2′-monoester or the 3′-monoester. Deoxyribonucleotides, of course, are prevented from forming such cyclic 2′,3′-phosphates. DNA consequently is stable with alkali.

2. Biosynthesis and Degradation of Nucleotides

Biosynthesis of the Pyrimidine Ring. Biosynthesis begins with aspartate and carbamyl phosphate. The latter is an energy-rich compound which reacts with the former to give carbamyl aspartate. Ring closure consumes ATP and in principle is an acid amide formation. The intermediate *dihydroorotate* is dehydrogenated to orotate, probably by action of a flavoprotein. *Orotate* is the key precursor of pyrimidine nucleotides. It reacts with phosphoribosyl diphosphate. The removal of pyrophosphate yields the nucleotide of orotate, whose enzymatic decarboxylation produces uridine 5-phosphate. Phosphorylation with ATP yields UDP and finally, UTP. Besides the above pathways, there is the further possibility of converting free uracil and ribose 1-phosphate to the nucleoside and subsequently with the help of ATP to the nucleotide.

---

[2] Whenever two hydroxyl groups of phosphoric acid have been engaged in ester bonds, one speaks of phosphoric diesters or simply phosphodiesters.

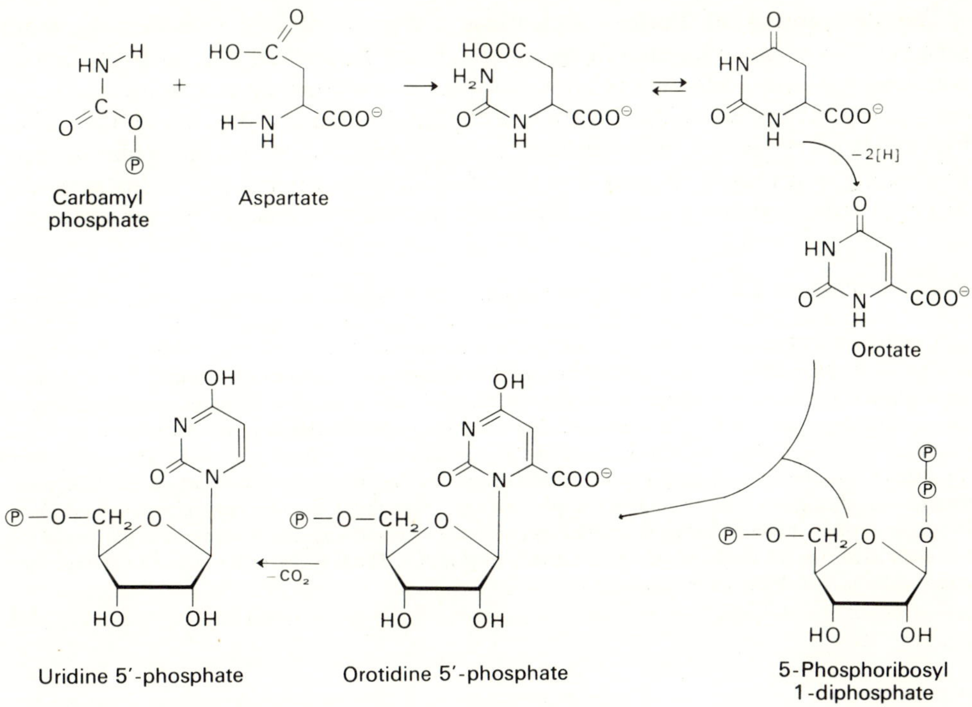

The other pyrimidines arise from uridine derivatives. One enzyme converts the C-4 hydroxyl group of uridine triphosphate with the aid of ammonia or glutamine to an amino group. Uridine triphosphate thus becomes cytidine triphosphate. Thymidine derivatives are synthetized by methylation of uridine or deoxyuridine monophosphate with the aid of *methylenetetrahydrofolate* (cf. Chapter VI,6).

The reductive step yielding deoxyribosides, the important building blocks of DNA, in similar fashion utilizes the diphosphates of pyrimidine and purine nucleotides; it involves the reductive removal of the 2′-hydroxyl group. The hydrogen donor for the reduction is *thioredoxin*, a protein consisting of 106 amino acid residues. It contains two reactive HS groups instrumental in the reduction of the sugar. The exact mechanism of the reduction is not yet known. Oxidized thioredoxin (the —S—S— form) is reduced again by NADPH. In some bacteria there exists, in addition, another biosynthetic pathway in which a cobalamin coenzyme participates in the reduction of triphosphates to deoxynucleoside triphosphates. Thioredoxin is also possibly involved.

The Degradation of Uracil. This process is the reverse of its biosynthesis; in other words, the ring is hydrogenated partially and cleaved hydrolytically between N-3 and C-4, but since CO_2 was split off in the conversion of orotate→ uridine, the degradation product is β-alanine instead of aspartate.

The Biosynthesis of Purine Nucleotides. This process is considerably more involved. Two principles should be noted. In the first place, the sugar-phosphate component of the nucleotide is prepared first, and subsequent reactions take place on it (every chemist would first synthesize the ring system and then attach the carbohydrate moiety); second, this synthesis uses very small units and molecular groups. Everything is arranged around one glycine molecule. Figure VII-1 illustrates the origin of the individual groups of atoms and provides the basis of the biosynthesis.

In detail, the synthesis as studied by Buchanan and Greenberg takes the following route: 5-phosphoribosylamine (stemming from phosphoribosyl diphosphate and glutamine, as mentioned under pyrimidines) with the aid of ATP, condenses with glycine to form the amide; the central piece of the purine ring is established. The five-membered ring is then prepared by attaching a C atom which is transferred by formyltetrahydrofolate. After the addition of another N group—later to become N-3—the imidazole ring is closed, whereby 1 mole of ATP is consumed. Carbon dioxide (as bicarbonate) next enters the molecule. Remarkably, no activation by ATP is necessary here; the reaction is freely reversible. The acid is then converted to the amide; the nitrogen (later N-1 of the purine ring) is supplied by aspartate which becomes fumarate (in analogy to urea formation, Chapter VIII,8). Throughout all these enzymatic reactions phosphoribose remains attached to the molecule. Last, the six-membered ring is closed, just as in the case of the five-membered ring, by the insertion of an active formyl group. The first purine derivative to emerge then is inosinic acid, which occurs free in muscle.

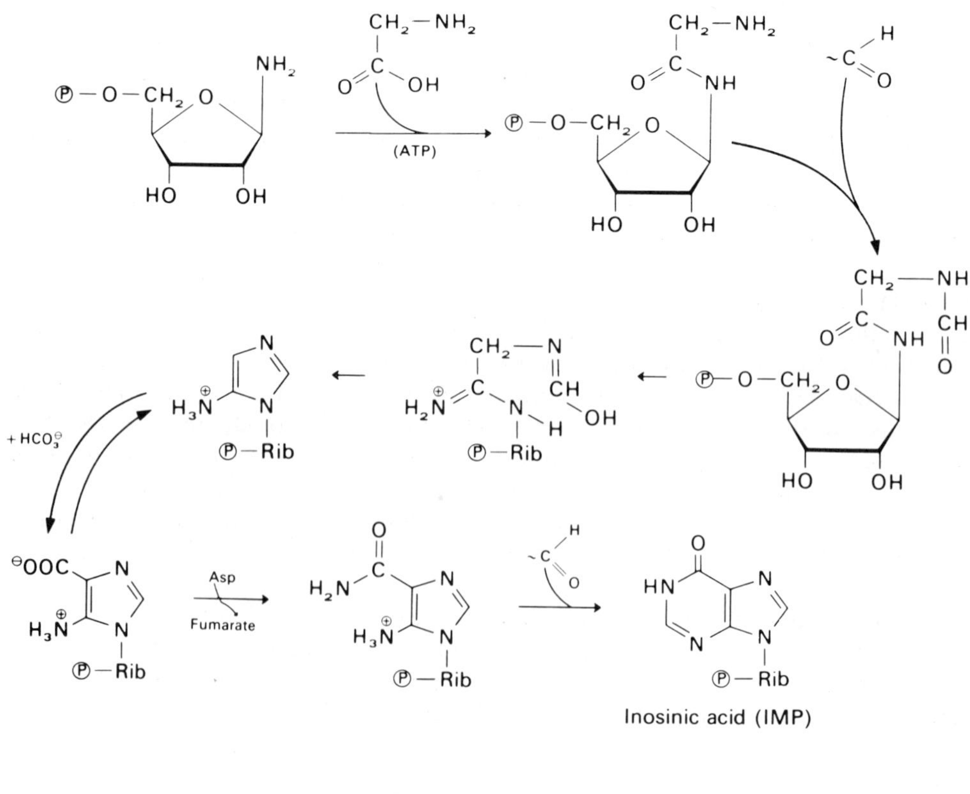

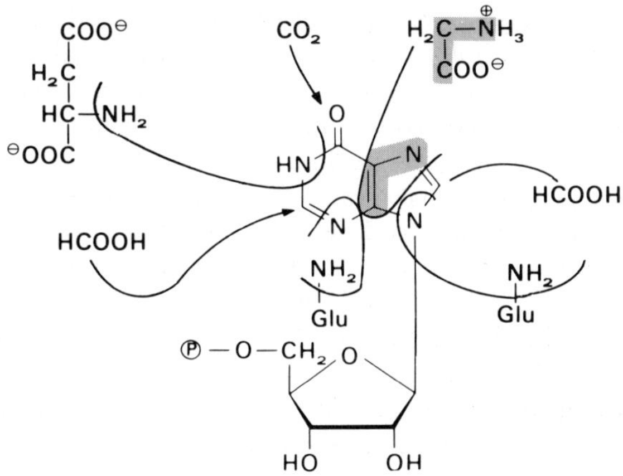

Fig. VII-1. Components of the biosynthesis of purine nucleotides.

Inosinic acid is the parent substance of the two purine nucleotides adenosine 5′-phosphate (*adenylic acid*) and guanosine 5′-phosphate (*guanylic acid*), both components of nucleic acids. It is also the parent compound of uric acid, which is the final excretion form of nitrogen in birds and reptiles.

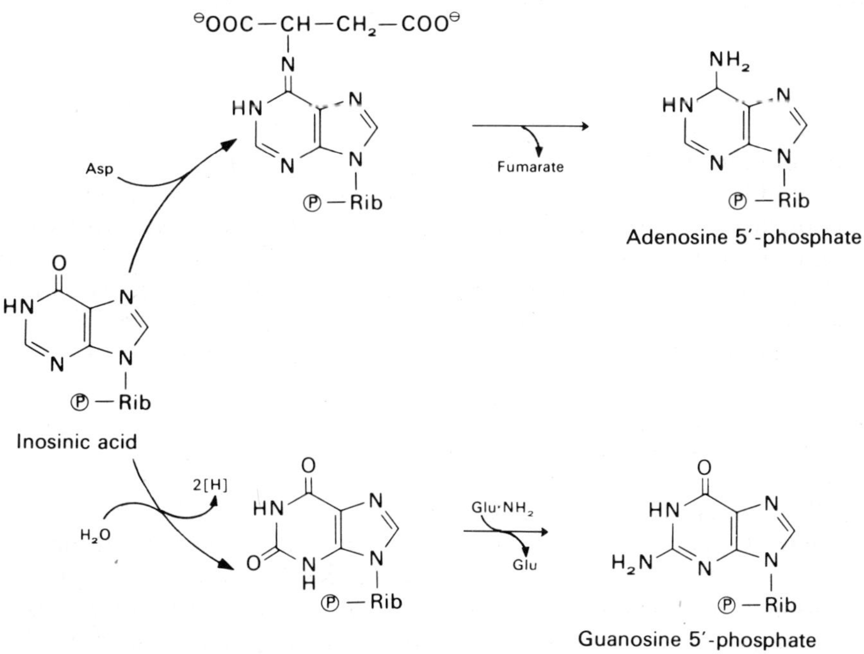

Degradation of Purine Bases. A portion of the purines freed by the hydrolysis of nucleosides is reutilized in metabolism; reaction with 5-phosphoribosyl 1-diphosphate reconverts free purines to mononucleotides. The excess of purine bases is converted to uric acid. Ribose is first removed from guanosine; the remaining guanine is then deaminated to *xanthine*. Adenosine is deaminated as a nucleoside. Phosphorolytic cleavage of the ribose produces *hypoxanthine*, which is oxidized to *uric acid* via xanthine by the action of xanthine oxidase, a flavoprotein with very low substrate specificity. In most mammals uric acid is further degraded to *allantoin*; in man and other higher primates, however, uric acid is excreted as such.

Methylated xanthines have pharmacological applications (*theophylline*, 1,3-dimethylxanthine; *caffeine*, 1,3,7-trimethylxanthine). Certain analogs of the purines (8-azaguanine, 6-mercaptopurine) have antibiotic and cytostatic properties; they have been used as chemotherapeutic agents against tumors.

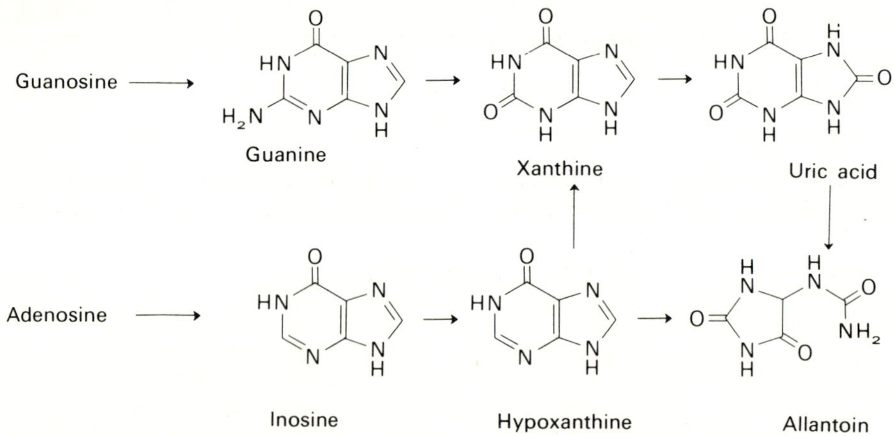

Errors in Purine Metabolism. With the pathological condition of *gout*, the level of uric acid in blood is elevated, which results in the deposition of uric acid crystals in joints or in the formation of kidney stones. The disease is often caused by an increased biosynthesis of purines. *Xanthinuria* is an inborn metabolic defect, characterized by the absence of xanthine oxidase, in which hypoxanthine is excreted in place of uric acid. In the *Lesch-Nyhan* syndrome the reutilization of purines is interfered with; the specific enzyme defect is at the phosphoribosyl transferase.

3. Structure of Nucleic Acids

The Chain Structure of Deoxyribonucleic Acids (DNA). Nucleic acids are high-molecular weight polynucleotides; nucleosides are linked together with phosphoric acid in diester bonds. Because in deoxyribonucleosides, which we shall consider first, position 1′ of 2′-deoxyribose is occupied by the base and position 4′ is in the furanose ring, phosphoric acid has only the hydroxyls of 3′ and 5′ available for linkage.

By general agreement the chain is written with its 5′—OH end to the left and its 3′—OH to the right. A section of a hypothetical chain is represented below.

The molecular weight of native DNA is very high. Carefully isolated preparations have measured more than 10^9 daltons; with the lower values frequently found in the literature one suspects that fragments may have arisen during preparation.

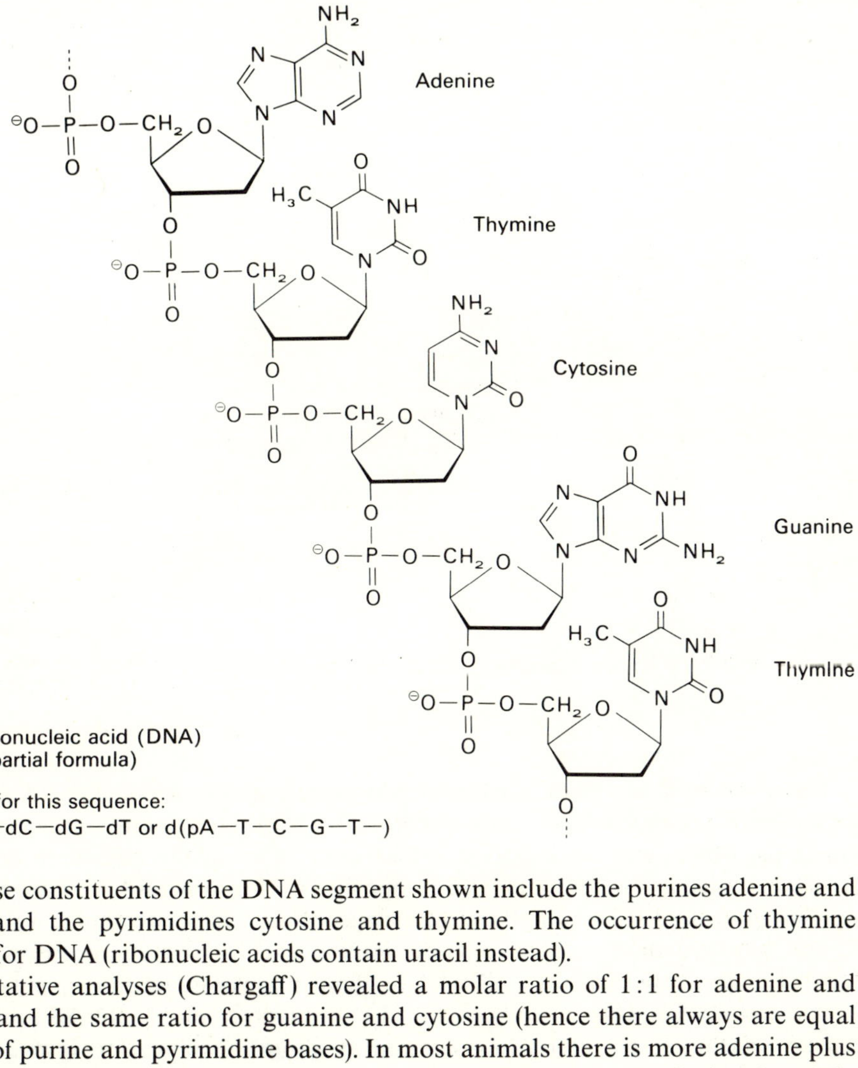

Deoxyribonucleic acid (DNA)
(partial formula)

Shorthand for this sequence:
—dA—dT—dC—dG—dT or d(pA—T—C—G—T—)

The base constituents of the DNA segment shown include the purines adenine and guanine and the pyrimidines cytosine and thymine. The occurrence of thymine is typical for DNA (ribonucleic acids contain uracil instead).

Quantitative analyses (Chargaff) revealed a molar ratio of 1:1 for adenine and thymine, and the same ratio for guanine and cytosine (hence there always are equal portions of purine and pyrimidine bases). In most animals there is more adenine plus thymine than guanine plus cytosine (molar ratios of the pairs range from 1.3 to 1.5); in bacteria the opposite may be found.

The order of occurrence of bases in the molecule is almost entirely unknown since methods for sequential analysis are still lacking. The sequence is of great significance as we shall see below.

Ribonucleic Acids (RNA). RNA is constructed very similarly to deoxyribonucleic acid. It consists of numerous nucleosides held together by phosphoric acid. The diester of phosphoric acid constitutes the link between the individual ribose units. The 3′-5′-type of linkage is found in RNA. A segment of such a chain is represented by the following formula:

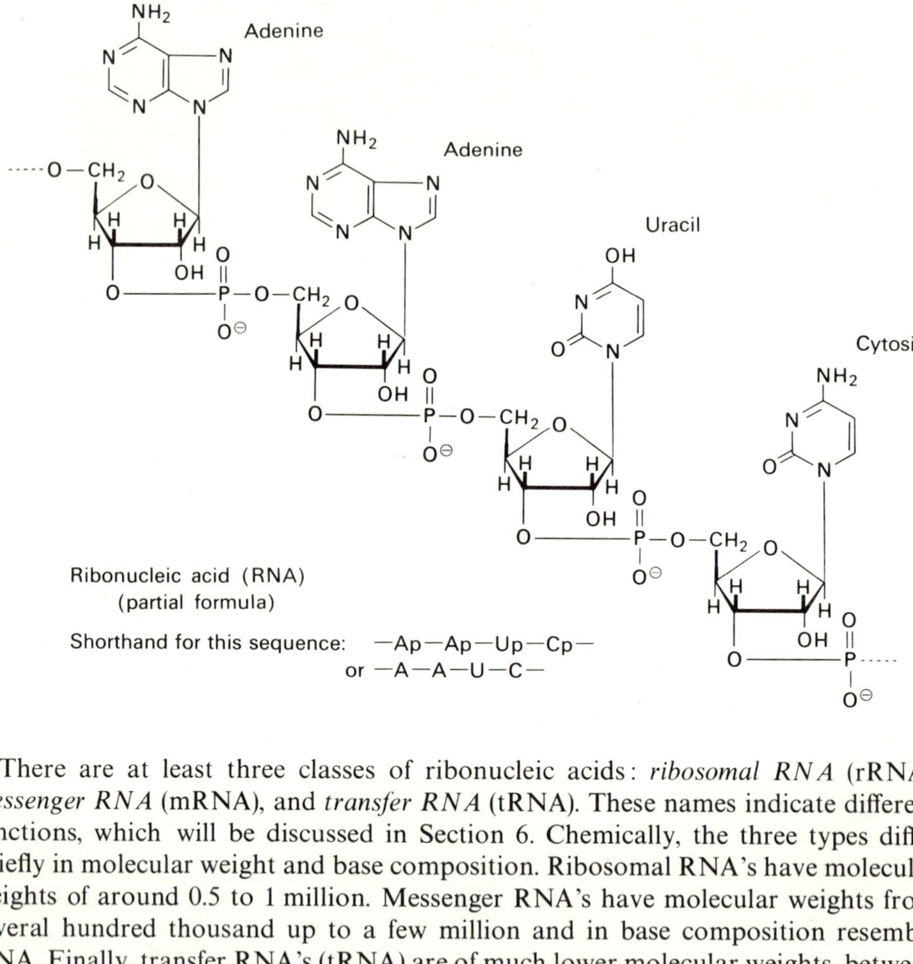

Ribonucleic acid (RNA)
(partial formula)

Shorthand for this sequence: —Ap—Ap—Up—Cp—
or —A—A—U—C—

There are at least three classes of ribonucleic acids: *ribosomal RNA* (rRNA), *messenger RNA* (mRNA), and *transfer RNA* (tRNA). These names indicate different functions, which will be discussed in Section 6. Chemically, the three types differ chiefly in molecular weight and base composition. Ribosomal RNA's have molecular weights of around 0.5 to 1 million. Messenger RNA's have molecular weights from several hundred thousand up to a few million and in base composition resemble DNA. Finally, transfer RNA's (tRNA) are of much lower molecular weights, between 25,000 and 30,000. They have been called collectively "soluble RNA" because they remain unsedimented after centrifugation at about 100,000 g for 1 to 2 hours, in contrast to RNA bound to ribosomal structures.

Base Sequence of Transfer RNA. It has been possible to isolate from the natural mixture of transfer RNA's individual representatives in pure form. More recently the base sequences of many ribonucleic acids have been established. tRNA specific for alanine consists of 77 nucleotides (Holley and co-workers). Zachau and collaborators have analyzed two different serine-transferring RNA's and found them to differ by only three bases (U in place of C at position 50, and G for A at positions 66 and 68). Several nucleotide sequences have been reproduced on the back of the foldout chart of this book. One structural feature represented by the cloverleaf pattern is common to all. The "stem" of the cloverleaf bears the aminoacyl residue in a high-energy bond (see Section 6); the "leaf" opposite the stem incorporates the anticodon (also Section 6). The other two leaves contain ribothymidine or dihydrouridine. In general, disproportionately more of the rare nucleosides such as pseudouridine, methylguanosine, isopentenyladenosine, inosine, among others, are found in these sequences.

Three-Dimensional Structure of Deoxyribonucleic Acid. Watson and Crick have developed a model on the basis of X-ray data obtained by Wilkins and co-workers which was subsequently substantiated by other analyses.

The basis for the structural model is the assumption that pairs of bases always interact through hydrogen bonds (cf. Chapter IV,3); this is very favorable for the pair adenine-thymine and for the pair guanine-cytosine. The analytical data referred to firmly support the assumption of base pairing.

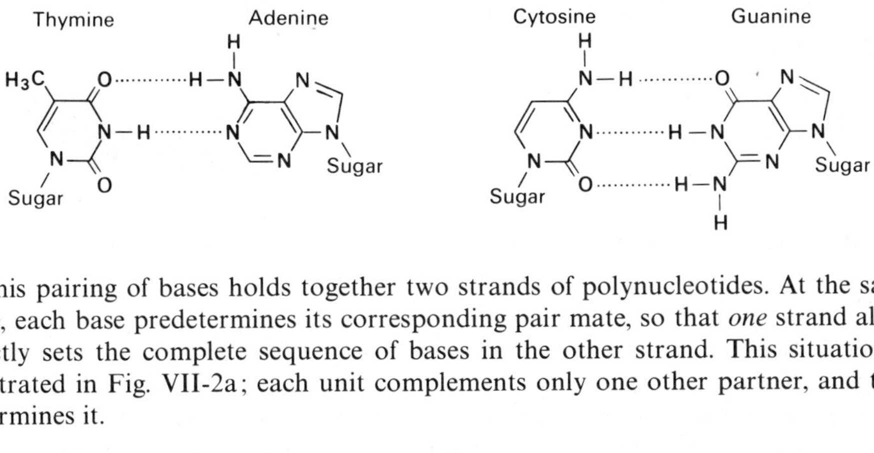

This pairing of bases holds together two strands of polynucleotides. At the same time, each base predetermines its corresponding pair mate, so that *one* strand alone exactly sets the complete sequence of bases in the other strand. This situation is illustrated in Fig. VII-2a; each unit complements only one other partner, and thus determines it.

The Helical Nature of Nucleic Acids. If the two-stranded tape is imagined as twisted, then one gets a fair picture of Watson and Crick's model for DNA: two molecules (or two halves of one molecule) form a double strand and turn around one another like threads of a screw. Figure VII-2b is a schematic drawing of such a helix model. The base pairs lie horizontally; the five-membered sugar rings are perpendicular to the base pairs. Each revolution of the helix requires ten base pairs, and the total diameter of the double helix amounts to about 20 Å. The sugar phosphate backbone forms the outer edge of the helix. The two strands are of opposite "polarity" (i.e., direction of the p $\rightarrow$ 5′ Rib 3′ $\rightarrow$ p $\rightarrow$ 5′ Rib 3′ $\rightarrow$ linkages).

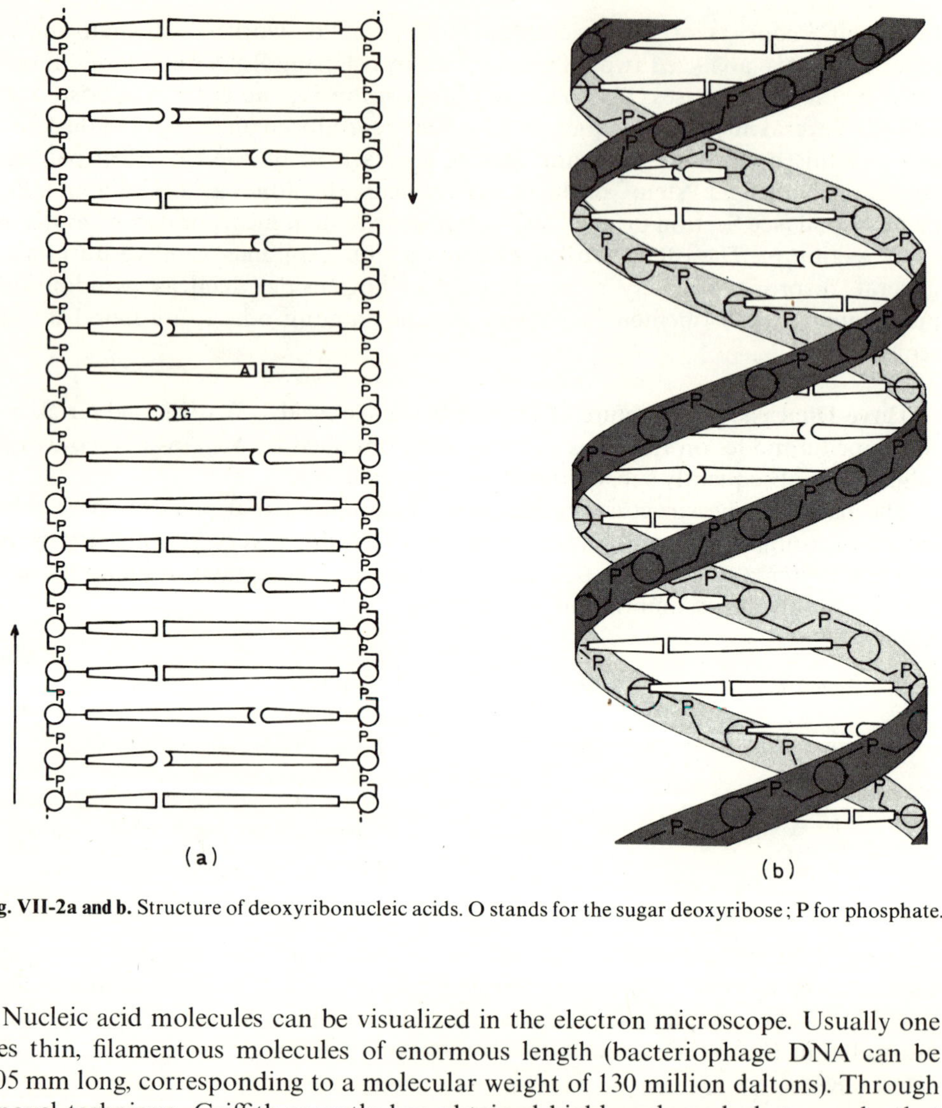

Fig. VII-2a and b. Structure of deoxyribonucleic acids. O stands for the sugar deoxyribose; P for phosphate.

Nucleic acid molecules can be visualized in the electron microscope. Usually one sees thin, filamentous molecules of enormous length (bacteriophage DNA can be 0.05 mm long, corresponding to a molecular weight of 130 million daltons). Through a novel technique, Griffith recently has obtained highly enlarged photographs, that show some structural details of the molecule. A helical structure can clearly be recognized on some pictures (see Fig. VII-3). The possibility of artifacts has, however, not yet been eliminated in this case.

The double helix does not withstand the disruptive forces of heating (denaturation) any more than does the conformation of proteins (cf. Chapter IV,5). When a salt solution of DNA is heated to 70°–80°, the structure separated into single strands and certain physical properties change (viscosity, light absorption, and optical rotation). By cooling the solution very slowly, the molecules are given the opportunity to partially order themselves back to the double strand, whereas with rapid cooling the single

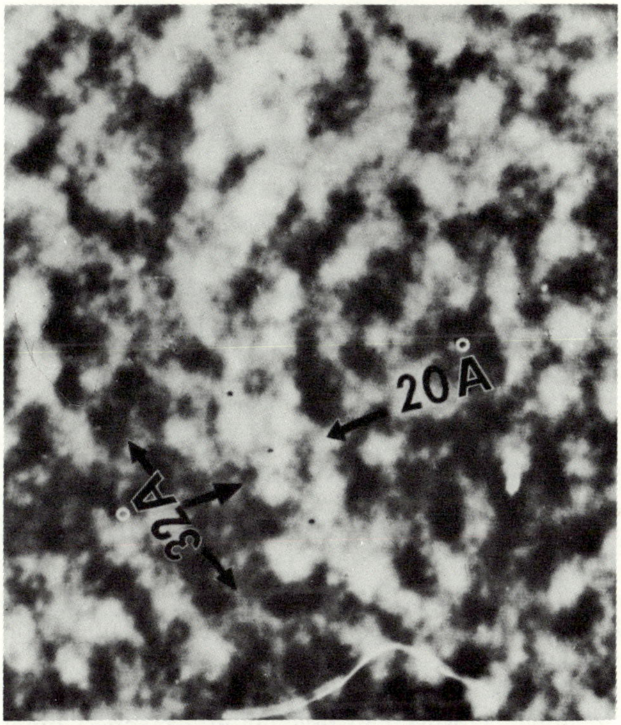

Fig. VII-3. Electron micrograph of DNA at very great enlargement (about 7 million fold). One can distinguish a portion of a molecule—stretching from the right lower corner to the upper margin of the picture—that clearly appears to be helical. (Original photograph by Jack Griffith.)

strands remain largely separated. Since this breaking-up of the helical structure resembles the melting process, in which the crystal lattice disintegrates, one talks about the "melting point" of DNA.

Three-Dimensional Structure of Ribonucleic Acid. Less is known about the spatial arrangement of the RNA chain compared to that of DNA. Base pairing and possibly double strandedness likely play a role here too. The recently determined base sequences of transfer RNA's reveal that base pairing is possible over short stretches only. The loops of the cloverleaf structure are left unpaired in this model. According to information from X-ray photography it seems possible that the loops crowd close together and that the entire molecule rather has the shape of a short rod with a bulge in the middle. The anticodon would be at one end; the activated amino acid at the other.

Hybridization. Very closely related nucleic acids, i.e., those possessing complementary base sequences over greater portions of the molecules, can form double helices consisting of one strand of each of the two nucleic acids. *Hybridization*, as this phenomenon is called, can also occur between a molecule of DNA and a complementary RNA molecule when the dissolved components are heated above their melting points and then allowed to cool slowly.

Hybridization is achieved only, of course, if the two nucleic acid strands possess complementary structure. Only then can hydrogen bonds form between the individual bases along the whole length of the molecule and hold the strands together. Moreover, the observation of hybridization affords proof that different nucleic acids possess complementary structure. This observation is particularly significant in the analysis of RNA fractions.

4. Deoxyribonucleic Acid as Carrier of Genetic Information

The principle of base pairing, developed by Watson and Crick has been important in elucidating the structural problems of DNA. Far exceeding this role, the same principle has become immensely fruitful in explaining the chief *functions* of DNA as carrier of hereditary information and the transmission and utilization of this information.

Genetic information is contained in DNA as the sequence of bases. The four bases adenine, guanine, cytosine, and thymine in a manner comprise an alphabet with the four letters AGCT. In accord with the rule of base pairing, each of these bases unambiguously determines the complementary base. Information can thus be transferred and transmitted.

Definition of the Gene. Hereditary factors or genes initially were defined as biological entities with the *ability to induce characteristics*, to undergo *identical reproduction*, and to *mutate*.

In genetic experiments, certain inherited characteristics (e.g., skin color, morphological peculiarities, presence or absence of certain substances or metabolic functions) are studied. The genes are localized on the chromosomes and are inherited according to the well-known Mendelian laws. It is from the appearance of a certain characteristic (taking dominance and recessivity into consideration) that the presence of a gene is inferred.

In all these experiments we can recognize a characteristic to be truly hereditary only when we also find individuals that do not possess this characteristic. Today we have evidence for the notion that every important property and development potential is determined by a gene; i.e., the information for it is carried by a gene.

The fact that genes can be handed over unchanged from generation to generation, often to many thousand offspring and to millions of generations of cells, compels us to ascribe to genes the capacity for *identical replication.* Only in this way can daughter nuclei contain the same stock of genes after division.

Very rarely one can observe the phenomenon of sudden changes in hereditary characteristics. The altered gene is then handed on in the same way as the original one. Such *mutation* is part of the evolutionary mechanism accounting for the diversity of species.

Deoxyribonucleic Acids as the Genetic Material. The substance that makes up genes is deoxyribonucleic acid. In general, each individual gene has a chemical equivalent, an appropriate segment of DNA characterized by a definite base sequence. The *information* carried by the gene and finally expressed as a particular character consists in the sequence of the bases. The *base sequence* in turn determines the primary structure of proteins, i.e., the *sequence of amino acids.* There is a direct correlation: a certain group of three bases of the DNA stands for a certain amino acid in the final protein. The segment of DNA chain carrying all the information for one protein (one polypeptide chain) is called a *cistron.* The cistron is identical with the unit of function (the expression of a character), although mutations can arise at various points within a cistron and can combine by crossing-over.

The ability of DNA for identical replication becomes plausible with the aid of Watson and Crick's structural model. Through the pairing of bases each individual strand determines a second and complementary strand. The mechanism of the biosynthesis of DNA follows the same concept. More will be said about this in the following section.

DNA is localized predominantly in the cell nucleus, more specifically in the *chromosomes.* A small amount is also present in mitochondria (Chapter XVIII,3). This has been demonstrated by Feulgen's reaction (Feulgen's nuclear staining), which is specific for DNA and stains the chromosomes. Optical methods (measurements of ultraviolet absorption) have corroborated these observations.

The amount of DNA per cell is constant for a given organism; it is the same regardless of the type of cell. This is not surprising since the chromosomal set and gene content is the same for every cell (except in cases of polyploidy). Only gametes (germ cells and egg cells) have one-half as much DNA as soma cells. The gametes are haploid; they contain only a single set of chromosomes.

It was discovered, however, that the amount of nuclear DNA is considerably greater than required to carry all genetic information. A portion of the DNA moreover exists in form of multiple repetitive sequences. It is not clear yet what the function is of this *redundant DNA.*

Mutagenesis. Mutation means change of information. Chemically it is a change of the DNA, particularly of its base sequence. It has been known for some time that high energy radiation (such as X or γ rays of radioactive substances) can cause mutation. In recent years, mutagenic substances have been discovered. The latter can modify the bases of DNA by introducing alkyl groups and thus interfere with base pairing. As a result certain sections of the base sequence may not be replicated. 5-Bromouracil, a base analogous to thymine (the bromine is in place of the methyl group), is incorporated into DNA as thymine; but during replication it is mistaken occasionally for cytosine. This finally leads to an exchange of adenine for guanine and to an altered base sequence. Nitrous acid (HNO_2) also can change the base sequence; cytosine is converted to uracil:

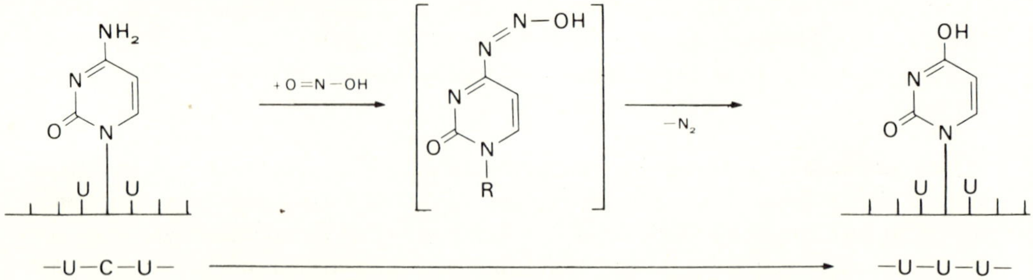

In the same manner, HNO_2 converts adenine to hypoxanthine. Uracil or thymine pair with adenine, while hypoxanthine pairs with cytosine and therefore takes the place of guanine. Systematic studies of controlled mutations of this kind have yielded valuable insights into the process of information transfer.

Transformation and Transduction. Transformation factors of bacteria have produced important proof for the genetic role of DNA. Of some bacteria, e.g., the pneumococci, different strains are known which produce different capsular material. The material (a polysaccharide; cf. Chapter XVII,7) must be regarded as one of the hereditary properties, just like hair coloration of mammals. It is possible to convert (to transform) bacteria of one type (e.g., type II) to another (e.g., type III) by treating them under certain conditions with an extract of type III. The active factor in the extract was prepared in pure form by Avery (1944); it is high molecular weight DNA. The explanation is that *transformation* is the transplantation of a gene. In the host bacterium, the transplanted nucleic acid, used in the process of transformation, realizes its capacity to induce characteristics—here the synthesis of type-specific capsular material. In the host it also manifests the capacity for identical replication, since the very same transformation factor can be isolated in quantity from a new harvest of transformed bacteria.

In other bacteria, genetic material is transferred by *transduction.* In this process part of a bacterial chromosome is transported (transduced) by a bacteriophage (virus; cf. Section 9) into the receptor bacterium. Bacterial genetics has become a very valuable tool for many fundamental questions in genetics.

5. The Transfer of Information: Biosynthesis of DNA and RNA

Replication of DNA. The passing on of genetic information requires, as we have seen, "identical replication" of DNA, which means *de novo* synthesis in strict conformity with an existing template. Watson and Crick's model makes identical replication intelligible. Each single strand of the helix becomes a template for the "stringing up" of the bases on the growing complementary strand. The assumption is that the primer DNA unwinds to single strands for a short stretch and that the nucleoside

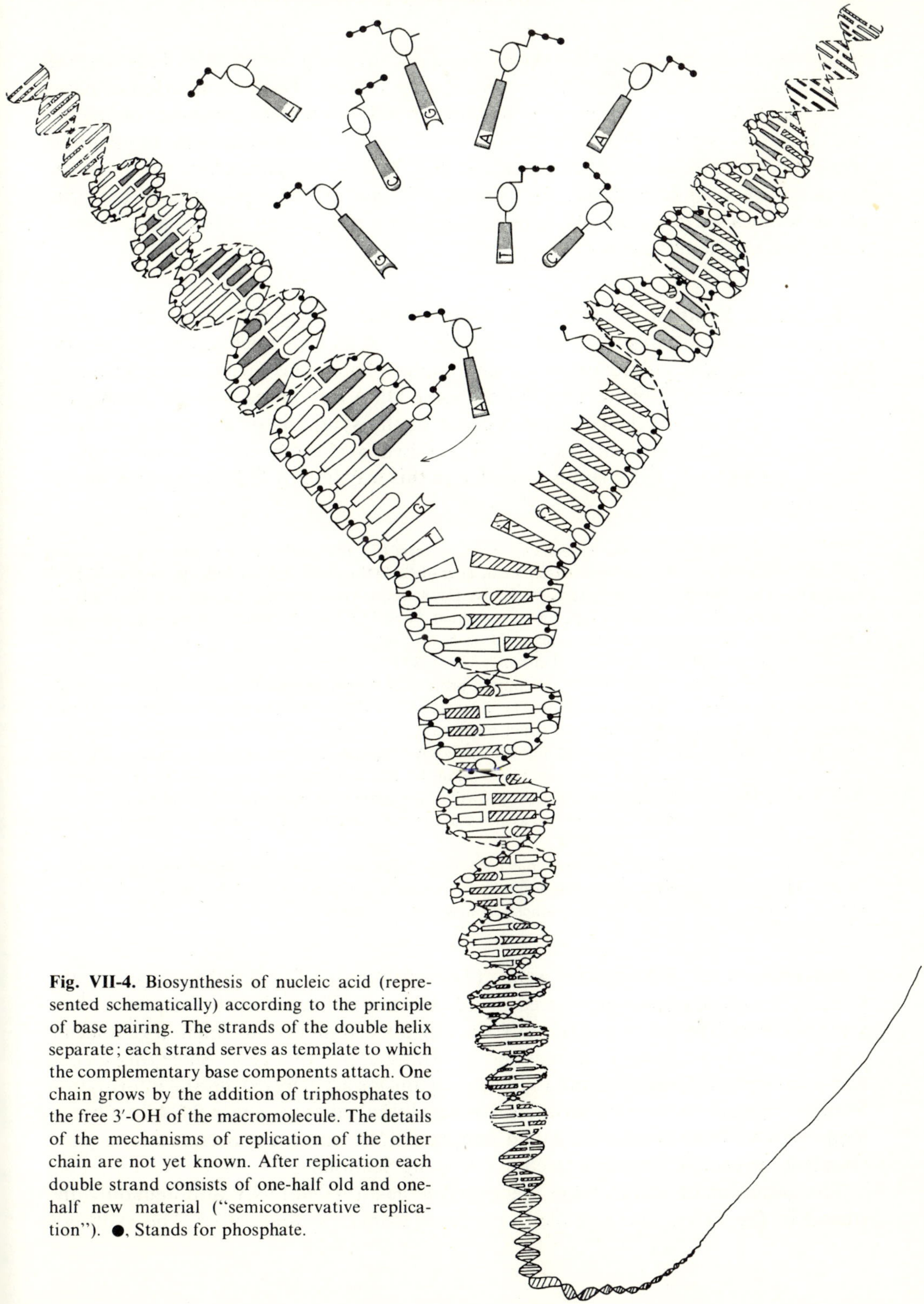

Fig. VII-4. Biosynthesis of nucleic acid (represented schematically) according to the principle of base pairing. The strands of the double helix separate; each strand serves as template to which the complementary base components attach. One chain grows by the addition of triphosphates to the free 3′-OH of the macromolecule. The details of the mechanisms of replication of the other chain are not yet known. After replication each double strand consists of one-half old and one-half new material ("semiconservative replication"). ●, Stands for phosphate.

triphosphates with the complementary bases line up along the single strands. The job of the enzyme is to establish the phosphodiester linkages (with elimination of pyrophosphate). At the conclusion of replication, each DNA molecule consists of one old and one new strand (*semiconservative replication*). This mechanism of identical replication is represented schematically in Fig. VII-4. Biological observations of the formation of viral nucleic acid in bacterial cells and of the replication of bacterial chromosomes have essentially substantiated this mechanism, although a host of individual problems remains to be solved.

Enzymatic Synthesis of DNA. DNA-synthesizing enzymes, called *DNA polymerases*, require as substrates the triphosphates of all four nucleosides, i.e., deoxyadenosine triphosphate, deoxyguanosine triphosphate, deoxycytidine triphosphate, and thymidine triphosphate. In addition, some DNA is required to serve as template for the reaction. The chemical mechanism of the synthesis of the macromolecule consists in a nucleophilic attack of the free 3′-hydroxyl group of deoxyribose on the triphosphate group. The chain is extended in this way link by link (in the direction of its polarity).

The enzyme "*DNA polymerase* I" isolated from bacteria (*E. coli*) and studied in detail by Kornberg can act simultaneously as an exonuclease. It can degrade a DNA chain link by link starting with the 5′ end. Its main function probably is to repair defective DNA sections by breaking down parts of the old DNA and replacing it with new material. The linkage of old and new DNA is achieved by special enzymes called "*DNA ligases.*" Some require NAD, others ATP as cosubstrates; they form the phosphodiester bonds at the expense of the diphosphate bonds of either NAD or ATP.

Mutants of *E. coli* have been found that contain practically no polymerase I and that still can multiply normally. This important finding shows that replication of DNA is not dependent on polymerase I. From these same mutants it was possible to extract membrane-bound enzymes (*DNA polymerase II* and *III*) that likewise require all four triphosphates and a DNA template. Polymerase II can bind more than 1000 units to DNA per second, a value that corresponds to the rate of synthesis expected for *in vivo* conditions. It is unclear what mechanism effects the growth of the chain in the direction of antipolarity. There is evidence that initially small segments are synthesized in the direction of polarity and joined together subsequently.

An "*RNA-dependent DNA polymerase*" has been found in tissue culture cells infected with carcinogenic viruses. This enzyme catalyzes the formation on an RNA template of a DNA–RNA double strand which, in turn, serves as template for the biosynthesis of new viral nucleic acid.

Biosynthesis of Ribonucleic Acids (Transcription). The information for the primary structure of proteins stored in DNA is not utilized directly for protein biosynthesis. First the pattern in the DNA is reproduced in a complementary molecule of RNA; it may be looked upon as the "working copy" of a gene. Since this special RNA transmits information and in higher organisms actually carries the information from the nucleus to the cytoplasm, it is generally called "messenger RNA" (abbreviated mRNA). It could also be called template RNA because it becomes in fact the template for protein synthesis (see the following Section).

The formation of mRNA takes place on double-stranded DNA, although only one strand of the DNA, the so-called *codogenic strand*, is transcribed. The synthesis commences at a particular DNA segment called the *promoter*. Total despiralization of DNA is not necessary; base pairing is broken only for a limited region.

Enzymes synthesizing RNA according to this mechanism have been extracted both from bacteria and mammalian tissue. The *E. coli* enzyme has been studied particularly thoroughly. It has a molecular weight of 500,000 and consists of four different subunits each with different functions. The σ subunit has aroused great interest since it is involved in the initiation process and appears to possess a specificity for certain DNA segments. Chain release appears to depend on the special factor ρ.

Mammalian cells contain two types of RNA polymerases, distinguished in that one type is inhibited by the amanita mushroom poison amanitine (see Chapter III,2), while the other is not. The amanitine-resistant enzyme occurs predominantly in the nucleolus, where, in higher organisms at least, ribosomal RNA is formed.

Another RNA-synthesizing enzyme has been isolated from bacteria by Ochoa and his collaborators. The enzyme polymerizes nucleoside *diphosphates* with the elimination of orthophosphates:

$$n(\text{nucleoside—P—P}) \rightleftharpoons \text{polynucleotide} + n\text{P}$$

The reaction is reversible. Cleavage with the uptake of orthophosphate is analogous to the phosphorolysis of glycogen (Chapter XVII,6), and probably the biological significance of the enzyme is due more to its lytic than to its synthetic activity.

6. Protein Biosynthesis

The problem of the biosynthesis of proteins has two parts: How is the peptide linkage made possible energetically, and how is the sequence of amino acids determined.

Activation of Amino Acids. Proteins cannot be formed from free amino acids by a reversal of proteolysis. We have emphasized that the equilibrium of this process lies very far to the side of hydrolysis; peptides decompose very readily to amino acids, but amino acids never recombine to give peptides. Consequently, the amino acid must first be activated, i.e., raised to a high potential for group transfer.

The reactions which result in the activation of amino acids have basically been elucidated in recent years. Chemical energy is supplied by adenosine triphosphate (ATP), which with elimination of pyrophosphate produces a mixed carboxylic-phosphoric anhydride with the acid group of the amino acid [see Fig. VI-4 in Chapter VI,5, reaction (c)]. As shown in the upper right corner of the diagram below, the reaction is reversible.

In a second reaction, the aminoacyl group is transferred onto a molecule of transfer ribonucleic acid, to the 3′-hydroxyl of ribose. An ester is evidently formed with the hydroxyl group of ribose. This ester has a sufficiently high potential for group transfer to form peptide bonds. The "activating" ribose is part of an adenosine residue which makes up the end group of a transfer RNA molecule. For every amino acid there is at least one specific activating enzyme and at least one specific transfer RNA.

Hence, "soluble RNA" of the cell consists of a mixture of different transfer ribonucleic acids. Several individual components of this mixture have been isolated in pure form. It was observed that some amino acids have several different transfer RNA's. They differ in their base sequence and, in part, in their particular "code region" or anticodon (see below). However, all transfer RNA's invariably end in the same CpCpA base sequence, which bears the aminoacyl group. These three end groups are easily split off and replaced in the course of general metabolism.

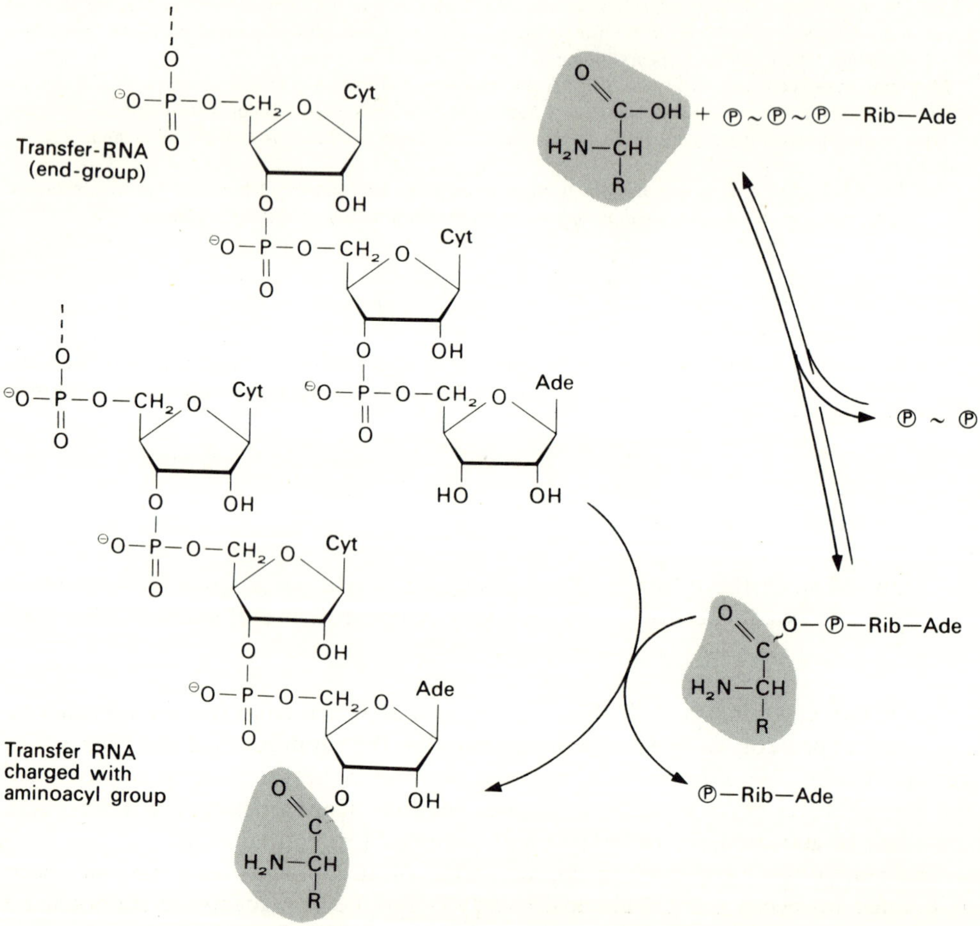

Determination of Amino Acid Sequence. We have learned that the sequence of amino acids in proteins is determined genetically and that the information for it resides in the base structure of the DNA. For protein synthesis a "working copy" of the gene is produced, but this is RNA (template or messenger RNA), not DNA. This particular step of information transfer is known as *transcription*. The formation of mRNA has been discussed in the preceding section. Figure VII-5 illustrates the manner in which mRNA is thought to attach to ribosomes and serve as template for protein synthesis. At this point the "language" of the bases (of the nucleic acids) is converted to that of amino acids (of the peptide formed). This step is called the *translation* of information.

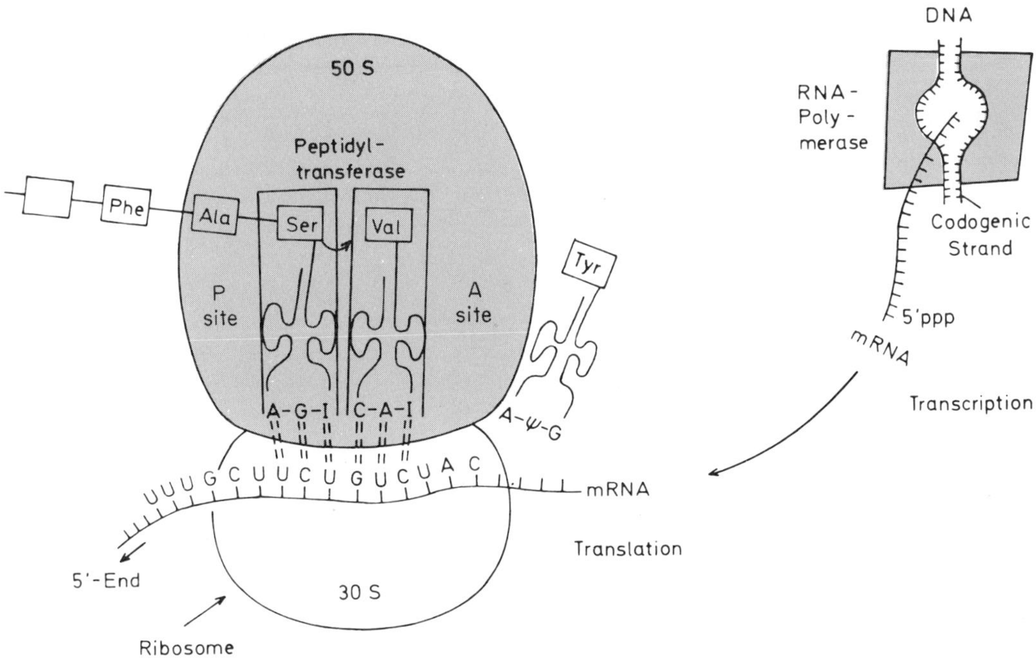

Fig. VII-5. Diagram of protein biosynthesis. Transcription is illustrated in the upper right corner. mRNA complementary to the codogenic strand of the DNA is formed through the action of RNA polymerase on the DNA template. The left part of the diagram shows the translation step on the ribosome. At the peptidyl-tRNA binding site (P site) serine-specific tRNA is bound with its partially completed peptide chain. At the A site (aminoacyl-tRNA binding site or acceptor site) valine-charged, valine-specific tRNA is attached. Shown to the right is tyrosine-charged $tRNA^{tyr}$, which is not bound yet, but is ready to interact with the next codon as soon as the mRNA is shifted to the left by one codon.

The Coding Problem. Now we may ask how the amino acid sequence is coded into the sequence of bases in the messenger RNA. RNA has only the four bases (adenine, guanine, cytosine, and uracil) with which to determine the twenty amino acids. In other words, there are only four letters, *A*, *G*, *C*, *U*, in the alphabet to make up the code words or *codons*. A codon consists of three adjacent bases, or a *triplet*.

If a code word consisted of only two letters, then $4^2 = 16$ combinations would result; evidently not enough. However, three symbols per combination brings it up to $4^3 = 64$ possibilities—more than enough to provide each of the twenty amino acids with one codon. There are, in fact, two or even three codons for each of several amino acids. This situation is called a *degenerate code*, which means that several combinations stand for the same thing.

A direct experimental approach to the code problem was afforded by the surprise discovery that in place of the regular messenger RNA relatively simple polynucleotides also can direct the incorporation of amino acids if they are incubated with ribosomes, activated amino acids, and enzymes (Matthaei and Nirenberg; Ochoa). With polyuridine phosphate in place of the natural messenger RNA, a polypeptide is formed which consists only of phenylalanine. The base sequence U—U—U therefore must be the code word for phenylalanine. If the polynucleotide also contains some cytidine, then leucine is included in the polypeptide as well. The codon for leucine apparently consists of the symbols C, U, and U.

TABLE VII-2[a]

Nucleotide Sequences of mRNA Codons in *Escherichia coli*

5'-OH terminal base	Middle base: U	C	A	G	3'-OH terminal base
U	Phe	Ser	Tyr	Cys	U
	Phe	Ser	Tyr	Cys	C
	Leu	Ser	"Stop"	"Stop"	A
	Leu	Ser	"Stop"	Trp	G
C	Leu	Pro	His	Arg	U
	Leu	Pro	His	Arg	C
	Leu	Pro	Gln	Arg	A
	Leu	Pro	Gln	Arg	G
A	Ile	Thr	Asn	Ser	U
	Ile	Thr	Asn	Ser	C
	Ile	Thr	Lys	Arg	A
	Met	Thr	Lys	Arg	G
G	Val	Ala	Asp	Gly	U
	Val	Ala	Asp	Gly	C
	Val	Ala	Glu	Gly	A
	Val	Ala	Glu	Gly	G

[a] From M. F. Singer and P. Leder, *Annu. Rev. Biochem.* **35,** 214 (1966), modified by filling in the blank spaces.

The entire genetic code has now been deciphered (Nirenberg; Khorana), in part by the use of chemically and enzymatically synthesized trinucleotides of known sequence, which under certain conditions can replace high-molecular weight messenger RNA, and in part by the use of polynucleotides of known sequence.

The complete dictionary, codons to amino acids, is presented in Table VII-2. It can be seen that the code is degenerated according to a definite rule: The only distinction made in the third position of the triplet (with two exceptions) is between "pyrimidine" and "purine," and in several cases no distinction is made between any of the four bases. The code is universal; i.e., it applies to all forms of life, from bacteria to man.

The Adaptor Hypothesis. The translation of the base code into amino acid sequence, according to a hypothesis of Crick, again depends on the principle of base pairing. The individual transfer nucleic acids bear in a highly prominent position (one of the loops) a base group—called *anticodon*—which is complementary to the codon on the messenger RNA (see Fig. VII-5) and which pairs with the codon. The transfer nucleic acids charged with amino acids arrange themselves along the template

(=mRNA) and according to this mechanism recognize the correct place along the mRNA which also is the correct position for the amino acid.

It appears to be of special significance that several anticodons contain rare nucleotides. Inosinic acid (I), for example, seems capable of pairing with U, C, and A. Thus several "degenerated" codons for an amino acid can be translated correctly by a single tRNA.

Ribosomes. The attachment of tRNA to the codon of the mRNA as well as the linking up of the activated amino acids take place on ribosomes, submicroscopic particles (150–200 Å diameter) that can be visualized only in the electron microscope (see Fig. XVIII-5 in Chapter XVIII,2). They consist up to 65% of ribonucleic acid and the rest is protein and some low-molecular weight material. In the cell, ribosomes are located in the "ergastoplasm"; in thin sections ribosomes can be seen lined up on membranes (see Fig. XVIII-4). The cytoplasm, in addition, contains *polysomes*, groups of ribosomes associated with mRNA but not attached to the endoplasmic reticulum.

Ribosomes of colon bacteria have been studied more thoroughly. Their particle weight is about 3 million (sedimentation constant 70 S) and they are aggregates of two different subunits sedimenting with 30 and 50 S. The two subunits aggregate reversibly requiring $Mg^{2\oplus}$ ions. The 30 S subunit contains 21 different proteins (one molecule of each), a 23 S RNA (molecular weight 1.1 million), and a molecule of a 5 S RNA (molecular weight 40,000). Ribosomes of eukaryotes have similar structure but are somewhat larger (sedimentation constant of 80 S with subunits of 40 and 60 S).

Initiation, Elongation, and Termination of Peptide Chains. Many details of the complex process of the start of protein synthesis are known in the case of *E. coli.* At first mRNA, which carries near its end the initiation codon AUG, attaches to a 30 S subunit. This step requires the participation of certain proteins, the initiation factors F_1, F_2, and F_3, as well as $Mg^{2\oplus}$ ions and GTP. Then tRNA corresponding to the AUG codon and bearing *N*-formylated methionine joins the complex. (The formyl group is derived from N^{10}-formyltetrahydrofolate; see Chapter VI,6). Last, a 50 S ribosomal subunit is bound, thereby making the complex functional. Protein synthesis starts when the next charged tRNA binds to the 50 S ribosomal subunit.

Elongation of the polypeptide chain is described in Fig. VII-5. The ribosome has two binding sites for tRNA, the peptidyl-tRNA binding site (P site) and the aminoacyl-tRNA binding site (A site). In the diagram both are occupied. The peptide bond is now established by action of the enzyme *peptidyltransferase*, one of the numerous ribosomal proteins. It transfers the peptide chain bound to tRNA to the amino group of the amino acid on tRNA at the A site. This action elongates the peptide chain by one aminoacyl residue. The chain is still attached to the tRNA at the A site, while the other tRNA leaves the P site. In the subsequent *translocation* step, the tRNA with attached peptide chain and with the mRNA is transferred from the A site to the P site. The protein factor G is required for translocation; the required energy is derived from GTP, which breaks down to GDP and P_i. The translocation step also brings the next codon to the A site; the corresponding tRNA binds, and the sequence starts anew.

The polypeptide chain is *terminated* as soon as one of the stop codons, e.g., UAA, is reached on the mRNA. The details of termination are less well known. With translation completed, ribosomes again dissociate to the 30 and 50 S subunits.

Inhibitors of Nucleic Acid and Protein Biosynthesis. Substances that inhibit the biosynthesis of protein have become important tools for a number of biochemical studies. Many inhibitors specific for certain reactions first became known as antibiotics.

The biosynthesis of nucleic acids can conceivably be inhibited at two places: at the template and at the enzyme protein. The antibiotics of the actinomycin, chromomycin, and anthracycline group attack the template by interposing themselves between base pairs of the helix; actinomycin (formula Chapter III,2) prefers the base guanine. They block transcription which must proceed by base pairing. Evidently RNA polymerase is much more susceptible to such interference than DNA polymerase. In *in vitro* experiments 10- to 100-fold lower concentrations can inhibit RNA polymerase. It is therefore possible to selectively inhibit the biosynthesis of RNA with the appropriate dosage of actinomycin.

Recently discovered rifamycin and some of its derivatives specifically bind to and thus inhibit the enzyme RNA polymerase. It is noteworthy that rifamycin binds only to bacterial polymerase; it does not inhibit the enzyme from animal sources. Amanitine, on the other hand, specifically inhibits animal polymerase (see end of Section 5).

Analogs of purine and pyrimidine bases can also inhibit the biosyntheses of nucleic acids *in vivo*. They inhibit the formation of purine and pyrimidine nucleotides with the result that the synthesis of nucleic acids comes to a halt for lack of substrate. Such analogs occasionally are incorporated erroneously into nucleic acids and give rise to mutations (see Section 4).

Protein synthesis on ribosomes is inhibited particularly by puromycin, chloromycetin, streptomycin, cyclohexamide, and others. Puromycin is of interest because it resembles the end group of transfer RNA. It also contains an aminoacyl group tied to a nucleoside, and can react with peptidyl-tRNA. The chain cannot be elongated, however, and is released from the ribosome instead as free peptidyl puromycin.

7. Mode of Action of Genes

In preceding sections we have developed the significance of DNA as the genetic material in protein synthesis. Again, let us remember that genes are defined primarily by the *expression of a character*. The process by which the gene determines the hereditary character in general is extraordinarily complex, especially with morphological characters and is as yet far beyond biochemical analysis. Some hereditary biochemical anomalies, or biochemically differentiated races, however, have been amenable to the study of how hereditary factors operate.

Biochemical Mutants. Mutants with such anomalies were first found and studied among insects (e.g., the flour moth *Ephestia* and the fruit fly *Drosophila*) (A. Kuhn, A. Butenandt). Later, biochemical mutants were also generated systematically from the easily grown bread mold *Neurospora crassa* (Beadle).

Often the result of a metabolic defect in a mutant is that some substance which was synthesized by the wild strain now must be provided as an essential growth factor. The mutant is said to be *auxotrophic* for that factor (e.g., for tryptophan or nicotinic acid); it is also called a deficiency mutant. The question then remains, what biochemical reaction has been interrupted.

In the above-mentioned insect mutants, a chain of syntheses forming the brown pigment of the eyes was interrupted; the insects became conspicuous because of their light eyes. Butenandt and co-workers showed that the pigment xanthommatin arises from tryptophan, and that in one group of mutants the step from tryptophan to kynurenine and in another mutant the step from kynurenine to 3-hydroxykynurenine was blocked as shown below (see formulas in Chapter VIII,11).

The same sequence was then discovered in *Neurospora*, but there 3-hydroxykynurenine yields nicotinic acid rather than the pigment (cf. Chapter VIII,11). The designations of the mutants are written near the arrows (letters refer to the mutants of *Drosophila*; numbers to those of *Neurospora*) wherever a particular reaction seems to be blocked. It is more correct to say that in the wild form the corresponding gene (the wild allele is designated by the addition of the + sign) permits the reaction while in the deficient mutant the altered gene no longer has that ability.

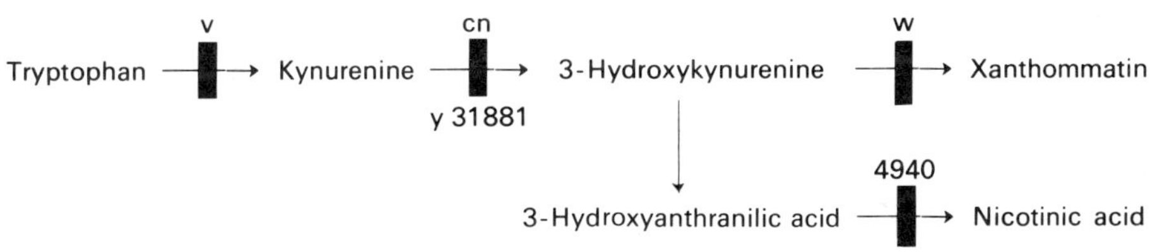

The simplest explanation is that the gene participates in the production of the enzyme that catalyzes the particular metabolic reaction. In the above diagram, the gene v^+ of *Drosophila* controls the production of the enzyme which effects the conversion of tryptophan to kynurenine; the gene cn^+ (*Drosophila*) or y 31881^+ (*Neurospora*) controls the biosynthesis of the enzyme kynurenine hydroxylase. In some cases it could be demonstrated that the mutant indeed lacked the enzyme. However, it is replaced occasionally by a very similar, but enzymatically inactive, protein.

Corresponding situations are found in man. The metabolism of phenylalanine and tryosine is especially subject to congenital defects (cf. diagram of formulas in Chapter VIII,11). In Fölling's imbecility (phenylketonuria) the step phenylalanine ⟶ tyrosine is blocked; in alcaptonuria, the breakdown of homogentisic acid. In albinism the side pathway leading to melanin is interrupted. These "inborn errors of metabolism" were recognized as effects of the genes by Garrod as early as 1909.

Generalizing, it can be stated that genes act by controlling the production of enzymes (or other proteins). As an example, sickle-cell hemoglobin, a congenital anomaly prevalent among Negroes (cf. Chapter IV,2), differs from "normal" hemoglobin by having valine in place of the usual glutamic acid. It has, therefore, one less negative charge and can be separated electrophoretically from normal hemoglobin. The difference in hereditary stock is manifested here in a changed protein molecule, through an exchange of amino acids. Numerous other examples of this kind have also been noted. This is direct proof for the assertion with which we began in Section 4, namely, that genes determine the amino acid sequence. We have come full circle in our exposition. Research on the biochemical action of genes, on the genetic code, and on protein biosynthesis has led to a uniform picture which is presented diagrammatically in outline form.

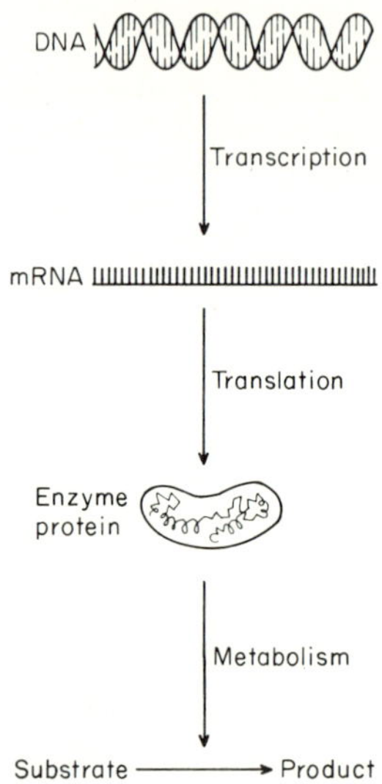

The Regulation of Gene Activity. This is an old but still unsolved problem of genetics and developmental physiology: all genes are, of course, present equally in every cell of the organism for its entire life-span. Yet genes exert their influence only in very specific organs or tissues and only during highly specific phases of development. How is this activity being directed?

Two kinds of regulatory mechanisms appear to operate in higher organisms: (a) long-term blockade, and (b) reversible on-and-off switching in accordance with particular metabolic needs.

A long-term or permanent blockade appears to be biologically appropriate for genes that determine certain developmental steps and that have already fulfilled their function. For this purpose Bonner has postulated the involvement of histones, the basic proteins of the cell nucleus. A functional blockade, on the other hand, can be explained better by the extensively studied model of bacterial enzyme induction.

Enzyme Induction. The formation of certain enzymes is enhanced dramatically by the substrate entering the cell. *Escherichia coli*, for example, produces β-galactosidase, a milk sugar cleaving enzyme, when the medium contains lactose. Formerly, this was looked upon as a mysterious enzymatic adaptation to the environment. Today the phenomenon is called *enzyme induction* and has been studied mainly in microorganisms. Not only the substrates, but related substances, even enzyme inhibitors can induce the formation of enzyme. Many investigations have borne out that the induced enzyme is identical with the enzyme protein normally present.

Studies of mutants having defects in enzyme induction have shown that control is applied at the level of gene activity or *transcription*, i.e., at the formation of mRNA. Furthermore, it is known that frequently the activity of a whole group of neighboring genes is regulated simultaneously. Jacob and Monod have called such a group an *operon*. At the start of the operon there is an *operator gene* possessing only regulatory functions; it is followed by *structural genes*, so named because they determine the structure of certain proteins (e.g., enzymes). Still another gene, the *regulator gene* (located elsewhere on the chromosome) is in charge of controlling the operator gene. The regulator gene accomplishes this by producing a so-called *repressor*, which has

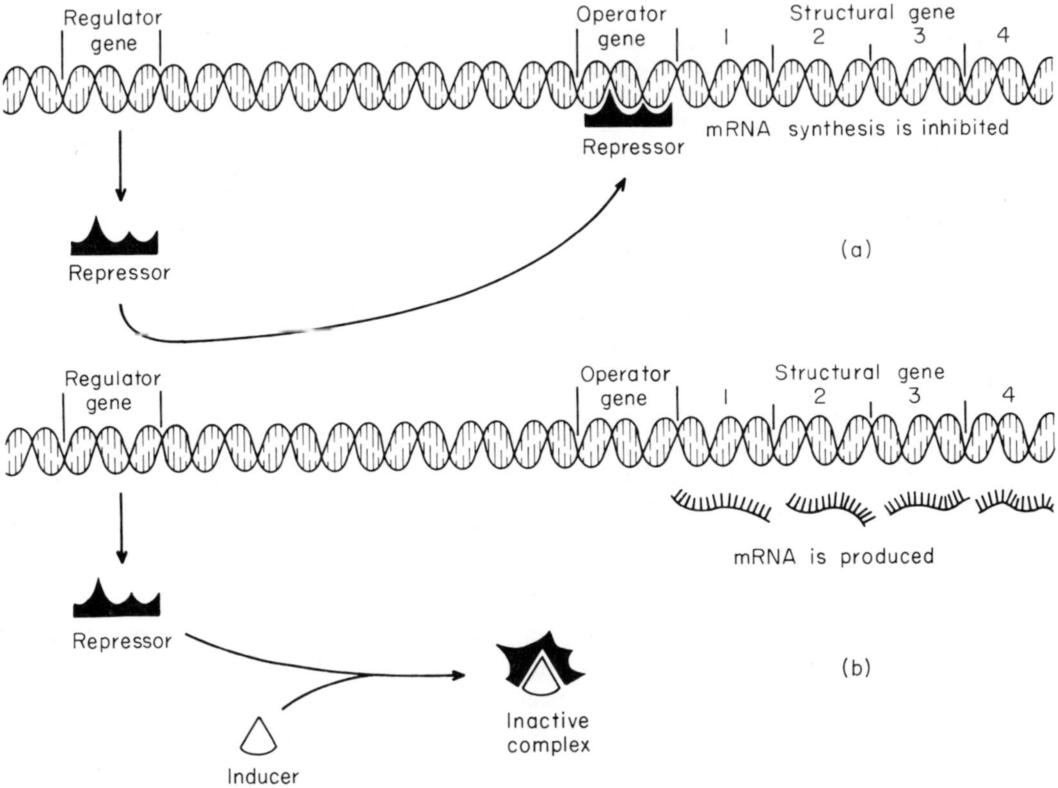

Fig. VII-6. Diagrammatic representation of the regulation of gene activity according to Jacob and Monod. In (a) the repressor molecule blocks transcription; in (b) the inducer inactivates the repressor resulting in unhindered transcription.

dual specificity. It is specific for its own operator gene, which it can block; this then prevents the subsequent structural genes from being read off. It is also specific for the enzyme *inducer*, i.e., the substrate or substrate analog. The inducer entering the cell combines with the repressor and inactivates it. The inducer–repressor complex has lost the ability to block the operator gene. Consequently, the synthesis of mRNA and with it the synthesis of the corresponding protein begins (since the unblocked operator gene permits the expression of the structural genes). With the disappearance of inducer, new repressor is supplied (through the continued activity of the regulator gene), and the operator gene is blocked again. The entire mechanism is shown in Fig. VII-6.

The repressor mechanism, initially deduced from genetic experiments, has been corroborated by the actual isolation of the *lac* repressor from bacteria. It is a protein with the approximate molecular weight of 150,000.

For chains of synthetic reactions there is an analogous mechanism of regulation. In this case, however, genes become active in the *absence* of the final product of the synthetic pathway and become repressed in the presence of final product. It is held that the complex of repressor and low-molecular weight products, termed "corepressors," blocks the operator gene, while the repressor molecule without its corepressor is inactive.

Jacob and Monod developed their theory on the *lac* operon of *E. coli*. Undoubtedly there are other mechanisms of regulation for protein biosynthesis. In particular, it is not yet clear to what extent the above hypotheses may be applied to higher organisms. Specifically, are several genes grouped together as operons and activated as units in the mammal? Though the phenomenon of enzyme induction has been observed in the mammalian organism, inducers frequently are hormones instead of substrates (see Fig. XX-3 in Chapter XX,1). Furthermore, there is evidence for regulation on the level of translation.

8. Biochemical Evolution

The evolution of organisms in the course of the development of this planet, according to current thinking, follows essentially the principles of mutation and selection. Whereas research into evolution at first traced hereditary lines exclusively on the basis of morphological characters, it now includes increasingly biochemical characters, especially the comparison of similar proteins (e.g., cytochrome *c* or hemoglobin) from different organisms. Such a comparison is only meaningful, of course, if the underlying principles of protein synthesis and particularly if the genetic code are the same in all organisms; and this seems to be true, indeed.

Origin of Life. Geologists tell us that life arose about three or four billion years ago, when earth was enveloped in a reducing atmosphere consisting largely of hydrogen, methane and other hydrocarbons, hydrogen cyanide, ammonia and hydrogen sulfide. There was virtually no free oxygen; it presumably is a product of photosynthesis. In such an atmosphere, there arose spontaneously, in part as a result of electric discharges, various organic molecules.

These substances most likely accumulated to a considerable extent. Urey has calculated that the total mass of carbon presently on the surface of the earth when distributed in the body of water presumably existing at that time (about one-tenth of the present volume) would have yielded a 10% solution! One has to remember too, that there were no bacteria to decompose organic substances.

This phase of evolution, the abiotic creation of organic molecules, can be reconstructed today by exposing an appropriate gas mixture to electric discharges. Monosaccharides (mainly pentoses), amino acids and peptides, and the bases of nucleic acids have been isolated as reaction products. Out of these building blocks, higher molecular weight compounds could conceivably have arisen spontaneously, for example, during the evaporation of aqueous solutions. The requisite chemical energy might have been contributed by inorganic polyphosphates or similar "energy-rich" compounds.

Although spontaneously formed polynucleotides are weakly catalytic, they do not have the capacity for autocatalytic multiplication, for self-instruction. Nucleic acids, however, have the latter properties due to the physicochemical interaction between bases bound complementarily. In principle, nucleic acids can propagate and multiply autocatalytically.

From known energies of binding, Eigen has calculated the probability of autocatalytic propagation of simple polynucleotides and has concluded that nucleic acids alone have essentially no chance to form larger molecules. This further development to more complex nucleic acids is possible, however, when coupled to simple, spontaneously arisen oligo- or polypeptides that provide the necessary catalytic action.

What is still lacking in the origination of life is the link between the potential "carriers of information," the nucleic acids, and the catalytically active polypeptides, which now, of course, are determined by the genetic code. We are not aware of any physicochemical principles that would explain the interaction between amino acids and polynucleotides. Eigen has postulated, however, that for further evolution it would suffice if a system consisting of polynucleotide and polypeptide would arise which initially autocatalytically produced identical polynucleotides but only "similar" polypeptides with comparable catalytic properties. Such a system not only has the capability for self-propagation, but also the potency for the selection of advantageous mutants. On this basis a strict correspondence between nucleic acids and polypeptides may gradually have evolved in the sense of the present-day code. This development may conceivably have passed through a phase utilizing only the two bases A and U and coding for only $2^3 = 8$ amino acids before arriving at the triplet code. The final fixation of the genetic code, however, must have occurred relatively early since the code is universal.

The battle for survival most likely was the key feature in the evolution of organisms. A strong weapon in the further development on the basis of mutation and selection is the increase in amount of nucleic acid per cell, because more "information" could be stored in this way. This also enables "experimentation" with various mutations. The expected rise in DNA content has indeed been found. The DNA of viruses comprises about 10^4 to 10^5 nucleotides; of microorganisms, 10^6 to 10^7; and in higher animals as much as 10^8 to 10^{10} nucleotides. This increase is probably based on the initial duplication of certain segments of the DNA (e.g., by unequal exchange), followed by mutation of these duplicated segments in the course of new generations.

First Structures. One essential feature of living things is their "milieu interne," the complete encapsulation against the environment. A model for this is found in coacervates, which are droplike aggregates that form spontaneously in aqueous solutions of two or three different colloids (=macromolecular substances). This process is often accompanied by considerable enrichment of certain substances in the interior of the droplet. Membranelike structures are formed on the surface, in some ways resembling membranes of living cells. These coacervate models were studied first by Bungenberg de Jong and later especially by Oparin.

We have of course, no certain knowledge of structure and composition of the precursors of life, the "eobiotas." At best we are making more or less educated guesses. Their metabolism must have been quite simple and probably involved conversion of organic compounds that had arisen spontaneously under these early conditions.

Evolution of Metabolism. How certain metabolic pathways have developed initially in the eobiotas is not known. A decisive role must have been played by the first organic catalysts, presumably precursors to our present enzymes. Another very early "invention" must have been the conservation of chemical energy through the synthesis of ATP. Many metabolic pathways, such as glycolysis and the citrate cycle, are identical in nearly all organisms known today. This would indicate that the pathways evolved very early and then did not change any further. Mutations in this area should not be viable.

Another milestone in the development of life surely was the harnessing of light energy, initially presumably through photosynthetic phosphorylation. Later there appeared photochemical synthesis utilizing organic and inorganic compounds as hydrogen donors. This process is still seen today in certain bacteria. The advent of photolysis of water finally was responsible for the production of molecular oxygen and the subsequent conversion of the atmosphere from a reducing to an oxidizing medium. Biological oxidation then appeared as a late consequence. The respiratory chain is the reverse of the photosynthetic chain (cf. Chapter XVI,2).

Evolution of Proteins. Different examples have shown that mutations—changes in DNA—engender corresponding changes in the sequence of amino acids. One can well imagine how several new, presently occurring proteins could have arisen in this manner from a single primitive protein. One good example is the evolution of the sequence of hemoglobin. The amino acid analysis of hemoglobins from a variety of species brought to light extensive agreement. Such agreements certainly have not evolved accidentally. The data strongly indicate that an "ancestral hemoglobin," a chain of about 160 amino acid residues, by stepwise mutation evolved into myoglobin and the hemoglobin chains known today, namely the α-, β-, γ-, and δ-chains. It is possible to devise a phylogenetic tree for these proteins (see Fig. VII-7) and to

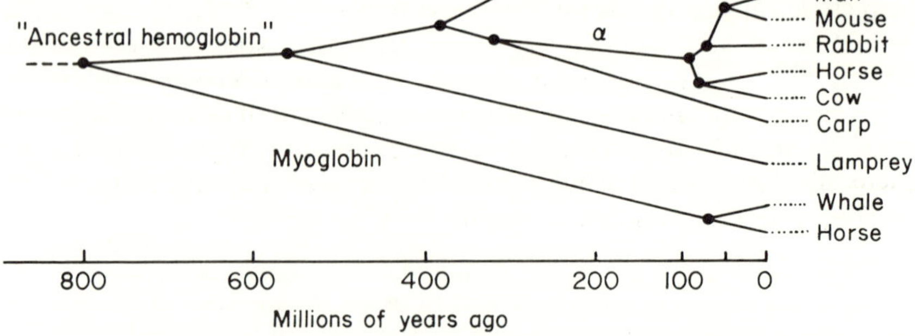

Fig. VII-7. Phylogenetic tree of myoglobin and hemoglobin.

estimate how many generations ago the evolutionary paths have parted. Such a tree agrees well with paleontological data.

The homologous enzymes trypsin, chymotrypsin, and thrombin in similar fashion arose from a primitive ancestral protein, presumably by gene duplication and subsequent mutation. Another group of homologous proteins are the immunoglobins of mammals.

9. Biochemistry of Viruses

Viruses can perhaps be defined as submicroscopic particles causing infectious diseases. They are the smallest units capable of identical replication and mutation. With that they have become valuable models for the study of genes; bacterial viruses in particular have provided many fundamental insights into molecular genetics. A discussion of viruses hence is included in this chapter.

Depending on the host organism, viruses have been categorized as bacterial viruses, also called bacteriophages, plant viruses, and animal viruses. Viruses have also been classified according to morphological and serological properties.

Viruses as Nucleoproteins. Simple viruses consist of nucleic acid and protein, and thus chemically are nucleoproteins; there are both RNA and DNA viruses. The nucleic acid alone is the infectious agent. It lends the virus the capacity for identical replication within the host cell and it bears the information for the virus protein as well as for certain enzymes, primarily RNA or DNA polymerases, that are needed for the multiplication of the virus.

The molecular weight of viral nucleic acid is between about 1 to 100 million. The classical case of viral DNA is a helical double strand which may or may not be closed to a loop. There are viruses with single-stranded DNA (e.g., the phage ϕX174), in which case replication nevertheless proceeds via a double-stranded step. RNA viruses always have single-stranded RNA as carrier of information.

Replication. Viral nucleic acid replication always takes place in a host cell of which there are two types: (a) Viral nucleic acid is incorporated into the genome of the host cell. Here the propagation of viral nucleic acid is subject to the same control as the DNA of the host cell. In such a "transforming infection" viral nucleic acid is always replicated in all daughter cells but without ever causing a discernible disease. (b) "Lytic infection," however, is more common. In this case the regulatory mechanisms of the host cell are disabled and the entire metabolic machinery is commandeered for the sole purpose of viral multiplication. Viral nucleic acid and virus-specific protein are formed in rapid succession and from these components complete, mature virus particles are assembled. The host cell frequently dies and lyses releasing the virus particles. The particles can, however, pass through the cell membrane and leave the cell without killing it.

Structure of Virus Particle. The nucleic acid of viruses is always packaged in a protein coat called *capsid*, which often consists of numerous, identical subunits,

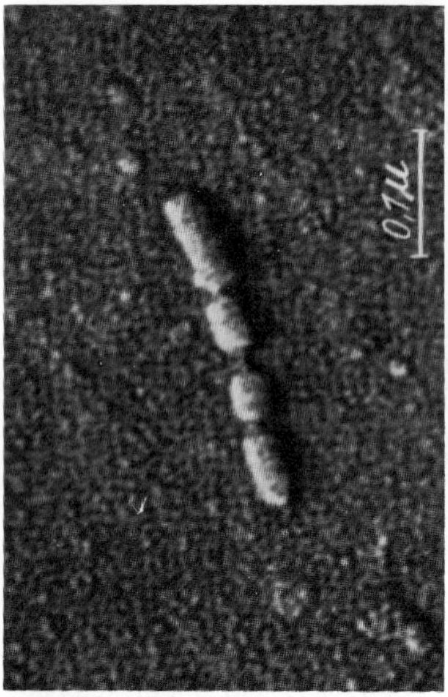

Fig. VII-8. Tobacco mosaic virus partially broken down. Careful treatment with a weakly alkaline salt solution removed the virus protein completely in some places, and thus exposed the thread of nucleic acid; 150,000-fold enlargement. Photograph by Professor Schramm.

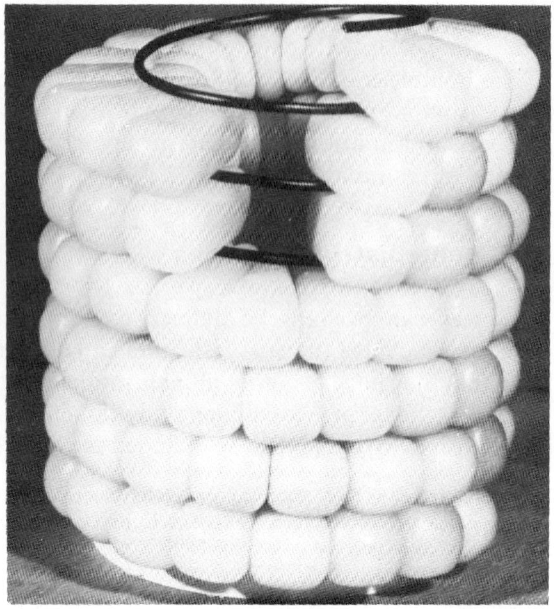

Fig. VII-9. Model of the tobacco mosaic virus (section). The black coil represents the nucleic acid molecule, the breadloaflike structures are the protein subunits of which there are altogether about 2300. The model is partly opened to show the internal structure of the virus. (Model and photograph by the author.)

the *capsomeres.* Higher viruses such as bacteriophages (see below) have more specialized structures. Animal viruses often contain lipids and carbohydrates originating with the cell membranes of the host and incorporated into the capsid during the expulsion of the viral particles.

An example of an RNA virus with a simpler structure is provided by the *tobacco mosaic virus* (TMV). It has a particle weight[3] of 40×10^6 daltons and is rod shaped, 2800 Å long and 150 Å thick (see Figs. VII-8 and VII-9).

Each tobacco mosaic virus particle consists of one long thread of nucleic acid embedded in protein. The RNA thread coils through the entire "supermolecule." About 95% of the material is protein; it consists of individual subunits with a molecular weight of 17,500, which exhibit a marked tendency to aggregate. At neutral or slightly acidic pH the protein molecules aggregate to little rods, very similar to the intact virus particles both in shape and size. The amino acid sequence is now known.

Many other viruses have their coat proteins arranged in the shape of a regular dodecahedron or icosahedron (20-sided) instead of the helical form. The nucleic acid molecule invariably is inside these hollow

[3] With viruses it is more meaningful to speak of particle weight than molecular weight, since the higher viruses especially are quite variable. Particle weight is also measured on the scale of atomic weights and expressed in daltons.

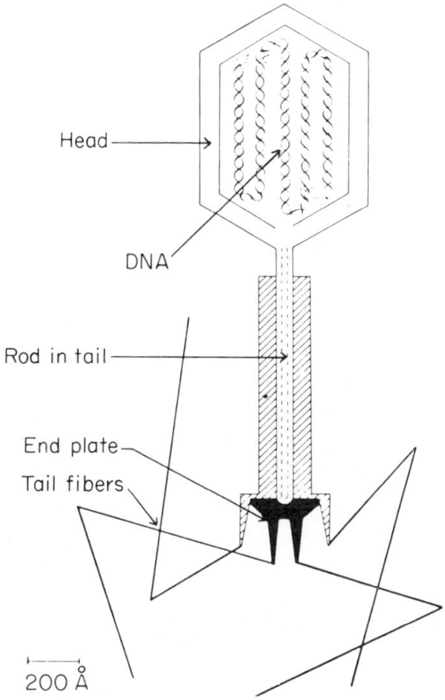

Fig. VII-10. Schematic representation of the T2 bacteriophage of *Escherichia coli*. The various structural features of the tail facilitate the penetration of deoxyribonucleic acid into the host cell. Compare with the accompanying photograph.

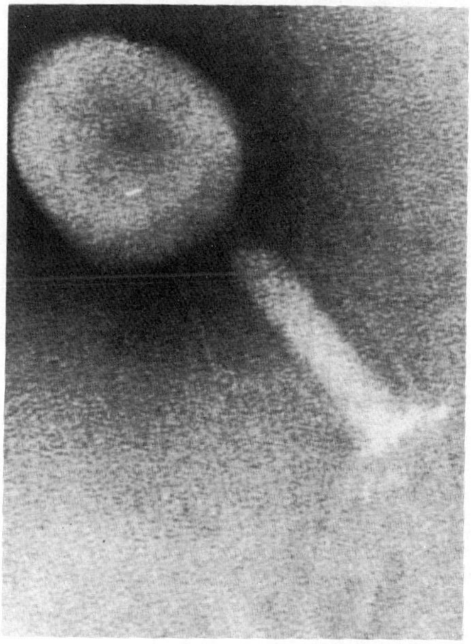

Fig. VII-11. Electron micrograph of the *E. coli* bacteriophage T2. Enlargement: 250,000 times. The technic of "negative staining" brought out the details of the end plate and the tail fibers. (Original photograph by Dr. Hofschneider.)

bodies; the protein on the surface. Furthermore, morphologically and functionally differentiated structures can be discerned with several viruses. An excellent example is T2 phage of *Escherichia coli* depicted in Figs. VII-10 and VII-11.

The large hexagonal head is enveloped by a protein membrane. It contains the nucleic acid (of the DNA type). The tail has three different proteins. One is responsible for the attachment of the phage to the bacterial membrane. Another is an enzyme; it dissolves the membrane during infection. Finally, the DNA is injected through the tail portion into the bacterial cell and there begins the process of forming new phage material by using the host's metabolic setup.

When all components are formed and the finished phages assembled, the bacterial membrane bursts open (lysis of bacteria) and the phages stream out.

As an example of an animal virus, Fig. VII-12 shows a virus of influenza type A (also the virus of the classical fowl plague), particle weight of 350 million. Some interior structure is seen in one virus particle on the left; it is ribonucleoprotein (the virus is one of the RNA viruses). A particle to the right clearly shows the spikes on the coat containing a protein that readily sticks to erythrocytes ("hemagglutinin activity"). Also on the surface is the only enzyme of the virus, a neuraminidase, which presumably facilitates the penetration of the viral RNA into the cells.

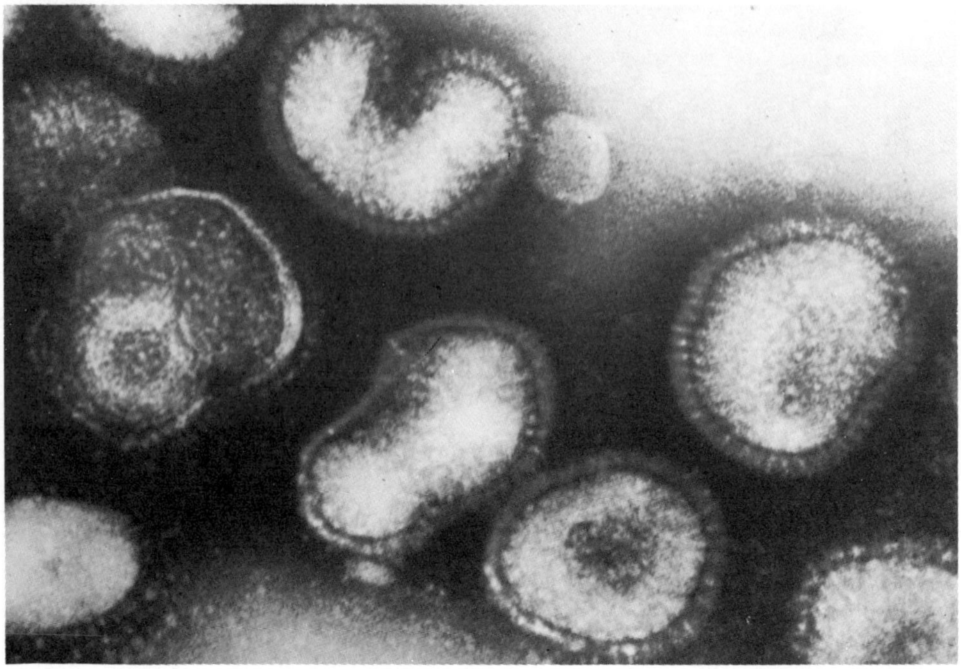

Fig. VII-12. A virus of influenza type A. At the left, internal structure (ribonucleoprotein) is discernible. Other particles show the structure of the coat. The kidneylike malformed objects probably are virus particles damaged during preparation. (Photograph by D. Peters.)

Among animal viruses the oncogenic (cancer-causing) viruses have recently gained prominence. Included are the Rous virus of chickens, SV40 (simian virus), various leukemia viruses, and others. The type of infection generally is transforming infection (see above). It has now been demonstrated that tissue of various human tumors also regularly contains virus particles.

Virus and Host. Viruses depend very generally on the metabolism of their host cells; hence they cannot be grown in artificial culture media, but can only be propagated in living cells, e.g., tissue cultures or incubated chicken eggs. They have neither energy metabolism nor enzymes for the biosynthesis of their own substance, the viral protein and the nucleic acid. They are, however, able to influence the metabolism of the host cell so that virus-specific substances are formed, usually at the cost of cellular material already present.

The question whether viruses are alive can only be answered on the basis of an exact definition of "alive." They are able to mutate and to reproduce identically (in the appropriate medium), but they are not able to metabolize and show no (or only very limited) reaction to stimuli. It can be stated that viruses acquire the attributes of life only in conjunction with the host cell, although they truly are at the threshold of life.

Defense Mechanisms. Viral disease very often leads to *acquired immunity*. Viral proteins are the antigens and the organism counteracts the infection with the formation of specific antibodies (see Chapter IV,8). In addition to this immune response there is a further defense mechanism, the specific formation of *interferon.* The penetrating viral nucleic acid stimulates the formation of a protein that can protect other cells of the organism against the infection of the virus. The mechanism of action of interferon is not yet known. The production of interferon under certain conditions can be stimulated by polynucleotides of relatively simple composition. Experiments are in progress to exploit interferon production in the battle against virus diseases.

10. Nucleic Acid-Cleaving Enzymes and Phosphatases

A number of different enzymes attack high-molecular weight nucleic acids. Systematically they are all classified as hydrolases, subgroup phosphodiesterases. A few monoesterases (cleaving only phosphomonoesters) and nucleosidases will also be mentioned in this section.

Deoxyribonucleases. The enzyme from the pancreas has been crystallized; its molecular weight is 60,000; the pH optimum is around 6.0 to 7.0. It splits the *3′-phosphoester bond*, freeing oligonucleotides. There remain also some nondialyzable, large fragments.

The spleen and other tissues contain an enzyme that splits the other bond, the *5′-phosphate bond.*

Ribonucleases. Pancreatic ribonuclease has already been mentioned in the chapter on proteins. Its molecular weight is not very high (13,500) and the sequence of its 124 amino acids has been unraveled. As an enzyme it has one remarkable property: it is relatively heat stable; solutions may be warmed briefly up to 80°.

Ribonuclease has a very pronounced specificity. It cleaves only the phosphodiester bond of a pyrimidine nucleoside 3′-phosphate. The first step of enzymatic catalysis is transphosphorylation (see page 162).

The ester bond going to C atom 5′ is transferred to the 2′-hydroxyl group, establishing first a cyclic diester (see also Section 1). This cyclic ester is then hydrolyzed specifically to the 3′-monoester. Purine nucleoside 3′-phosphodiester groups, strangely, are not changed to the cyclic form; they are resistant to the action of ribonuclease. The cleavage products therefore are, in addition to uridine 3′-phosphate and cytidine 3′-phosphate, various oligonucleotides with one pyrimidine nucleoside 3′-phosphate end group.

Besides pancreatic ribonuclease, an enzyme from a mold is known to split specifically at guanine (Egami). This is of practical importance for the sequence analysis of nucleic acids. In addition, there should be still other functional enzymes that specifically attack certain base sequences and dismantle long chains to a few well-defined fragments.

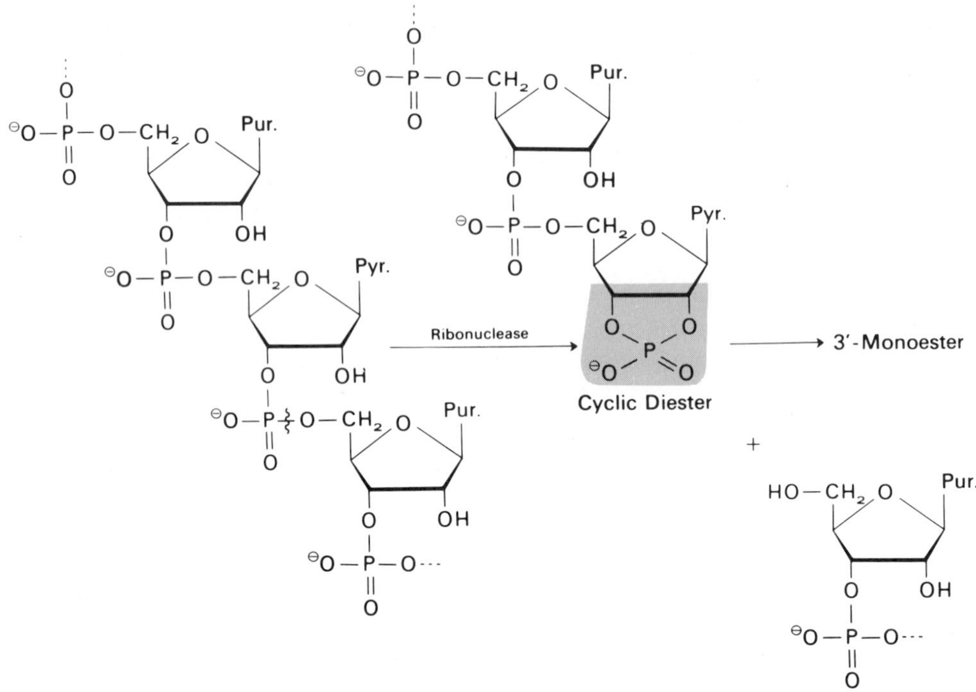

Nonspecific Phosphodiesterases. These enzymes cleave a variety of phosphoric acid derivatives, even synthetic substrates and lecithins (diesters of phosphoric acid with glycerol and choline). Such enzymes occur among other places in *snake venom*, and in the intestinal mucosa. Snake venom phosphatases specifically split *3′-phosphate bonds*, giving rise to 5′-monophosphates, whereas chemical hydrolysis yields a mixture of the 2′- and 3′-monophosphates. Free 3′-phosphate groups (the monoesters) are inhibitory in the enzymatic reaction.

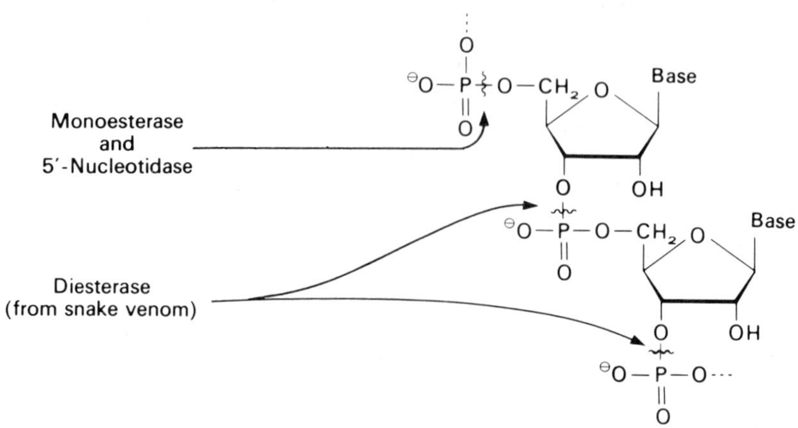

Phosphomonoesterases. Enzymes with this specificity are widely distributed. They are simply called "phosphatases;" we usually distinguish acid phosphatases with a pH optimum of about 5 and alkaline phosphatases with a pH optimum around 7 to 8. They split monoesters exclusively, the 3′-monophosphates as well as the 5′ monophosphates. One acid phosphatase occurs in remarkably high concentration in the human prostate gland. The enzyme active at the higher pH is found in the small intestine and in bones where it participates in the formation of bone tissue.

In addition to these enzymes, whose physiological substrates are probably not nucleotides, but more likely sugar phosphates, there exists specific 5′-nucleotidases. One of these enzymes was detected in muscle and in seminal fluid; it splits optimally in slightly alkaline media (pH 8.5).

Nucleosidases. Enzymes which cleave the bond between the sugar residue and the base are called nucleosidases. The cleavage of the nucleoside bond is not via hydrolysis, but preferentially through phosphorolysis. Usually orthophosphate is involved, but occasionally pyrophosphate in conjunction with an appropriate enzyme accomplishes the same result, which is equivalent to the reverse of nucleoside synthesis from base and phosphoribosyl diphosphate (cf. Section 2).

All the numerous cleavage possibilities listed here probably are more significant for research on nucleic acids than for *in vivo* processes. In the cell the decisive role is played by the nucleases that break high-molecular weight nucleic acids into fragments of oligonucleotide size. These enzymes are able swiftly to degrade nucleic acids and abolish their biological activity.

BIBLIOGRAPHY

Cold Spring Harbor Symposia of Quantitative Biology, A serial publication

Progress in Nucleic Acid Research and Molecular Biology, J. N. Davidson and W. E. Cohn (editors). A serial publication; Academic Press.

A. Butenandt, On the analysis of hereditary factors and their significance in biochemical problems. *Arbeitsgemeinschaft Forsch. Landes Nordrhein-Westfalen* **62,** (1960).

M. Eigen, Selforganization of matter and the evolution of biological macromolecules. *Naturwissenschaften* **58,** 465–523 (1971).

F. A. Eiserling and R. C. Dickson, Assembly of viruses. *Annu. Rev. Biochem.* **41,** 467–502 (1972).

B. A. Hamkalo and O. L. Miller, Electronmicroscopy of genetic activity. *Annu. Rev. Biochem.* **42,** 379–396 (1973).

R. Haselkorn and L. B. Rothman-Denes, Protein synthesis. *Annu. Rev. Biochem.* **42,** 397–438 (1973).

A. Klein and F. Bonhoeffer, DNA replication *Annu. Rev. Biochem.* **41,** 301–332 (1972).

J. Lucas-Lenard and F. Lipmann, Protein biosynthesis. *Annu. Rev. Biochem.* **40,** 409–448 (1971).

W. J. Rutter, R. L. Pictet, and P. W. Morris, Toward molecular mechanisms of developmental processes. *Annu. Rev. Biochem.* **42,** 601–646 (1973).

J. D. Watson, "Molecular Biology of the Gene." Benjamin, New York, 1965.

CHAPTER VIII

Metabolism of Proteins

In the organism, most of the proteins are constantly being built up and broken down again. In man, for example, the biological half-time of liver proteins amounts to 20–25 days; this is the span of time required to degrade half of the liver protein and replace it with new material. This turnover raises two problems, namely the mechanisms of degradation and of biosynthesis. The latter problem was discussed in Chapter VII in connection with the function of nucleic acids. The problem of degradation will be treated here, and at the same time the utilization of nutritional proteins will be discussed. Like many animals, man depends on a fairly regular diet of either animal or plant protein; its digestion and utilization will therefore be the focus of our discussion.

Proteolytic enzymes initiate protein degradation whereby amino acids are set free. The amino acids either are recycled to form new body protein or are broken down further: Nitrogen from the amino acids is transformed to urea, which is excreted as such; the carbon skeletons are oxidized through various intermediate steps to CO_2 and H_2O.

1. Proteolytic Enzymes

Proteases, along with amylases, probably are the most important hydrolytic enzymes. Both have been known for a long time. In systematic nomenclature the proteases would be called C—N hydrolases, since they catalyze the cleavage of peptide bonds, i.e., C—N bonds:

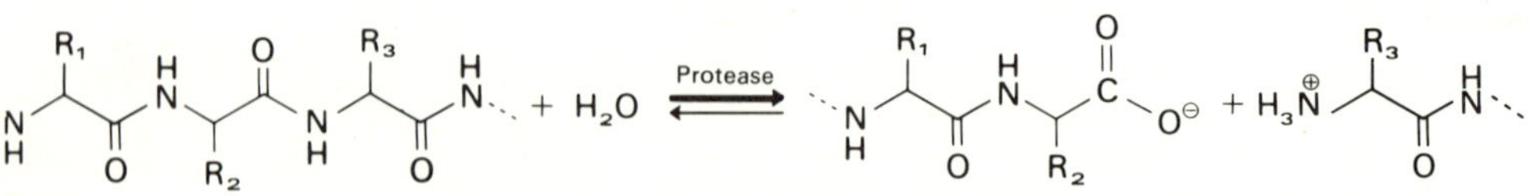

As mentioned in the chapter on peptides (Chapter III,1), the equilibrium greatly favors cleavage. Coupling to an energy-producing reaction is therefore not necessary. The organism does not appear to make use of the free chemical energy of hydrolysis; it is lost as heat.

Some proteases function as digestive enzymes. They usually occur extracellularly in the gastrointestinal tract, but some animals secrete proteases to the outside of their bodies. Intracellular proteases have been studied relatively little; they are called *cathepsins.*

Classification of Proteases. Peptide-cleaving enzymes customarily are classified as endopeptidases and exopeptidases, according to their specific mode of action. *Exopeptidases* are so called because they attack peptide chains only at the ends, in other words, they remove terminal amino acids only. A further distinction is made between *carboxypeptidases* acting on the carboxyl end and *aminopeptidases* acting on the amino end of the chains. The preferred substrates for many exopeptidases are smaller protein fragments, oligopeptides and polypeptides; hence they are also called simply "peptidases."

Endopeptidases split proteins at certain points along the chain and do not usually attack the ends. Consequently, they tend to act on proteins and larger polypeptides and are also called proteinases. The well-known digestive enzymes pepsin, trypsin, and chymotrypsin belong to this group. Several important proteases are listed in Table VIII-1.

A newly proposed nomenclature has proteinases categorized according to the mechanism of catalysis, i.e., the nature of the active center determines the following

TABLE VIII-1
Proteolytic Enzymes

Name	Active center	Occurrence	pH Optimum	Catalytic specificity
Chymotrypsin A	Ser . . . His	Small intestine	7.8	Tyr$\overset{\downarrow}{-}$, Trp$\overset{\downarrow}{-}$, Phe$\overset{\downarrow}{-}$, Leu$\overset{\downarrow}{-}$
Trypsin	Ser . . . His	Small intestine	7.5–8.5	Arg$\overset{\downarrow}{-}$, Lys$\overset{\downarrow}{-}$
Thrombin	Ser . . . His	Blood plasma	7.4	Arg$\overset{\downarrow}{-}$, (fibrinogen)
Cathepsin B_1	HS group	Intracellular	5–6	Arg$\overset{\downarrow}{-}$, Lys$\overset{\downarrow}{-}$, Leu$\overset{\downarrow}{-}$, Gly$\overset{\downarrow}{-}$
Papain	HS group	Papaya fruit	5	Arg$\overset{\downarrow}{-}$, Lys$\overset{\downarrow}{-}$, Leu$\overset{\downarrow}{-}$, Gly$\overset{\downarrow}{-}$
Pepsin A	COOH group	Stomach	1.3–3	($\overset{\downarrow}{-}$Tyr$\overset{\downarrow}{-}$, $\overset{\downarrow}{-}$Phe$\overset{\downarrow}{-}$)
Pepsin C (gastricsin)	COOH group	Stomach	3–3.5	($\overset{\downarrow}{-}$Tyr$\overset{\downarrow}{-}$, $\overset{\downarrow}{-}$Phe$\overset{\downarrow}{-}$)
Rennin	COOH group	Calf stomach		(—Caseinogen)
Cathepsin D	COOH group	Intracellular	3–4.5	Same as pepsin
Thermolysin	$Zn^{2\oplus}$	Bacteria	6–10	$\overset{\downarrow}{-}$Leu, $\overset{\downarrow}{-}$Phe
Clostridiopeptidase	$Ca^{2\oplus}$	*Clostridium* (Bacteria)	8.6	—Pro—X$\overset{\downarrow}{-}$Gly—Pro (Collagen)

subgroups:

1. Serine proteinases with serine and histidine in the active center (trypsin, chymotrypsin, etc.).
2. HS-proteinases bearing a cysteine residue in the active center, e.g., papain. They can be inhibited by sulfhydryl reagents.
3. Acidic proteinases, in which a carboxyl group is involved in catalysis, e.g., pepsin. They are active only at an acidic pH (less than 5).
4. Metalloproteinases which depend on a metal ion (often $Zn^{2\oplus}$, $Ca^{2\oplus}$, $Mn^{2\oplus}$) for catalysis and which can be inhibited by complexing or chelating agents such as ethylenediaminetetraacetate.
5. Enzymes with other or poorly understood reaction mechanisms. See Table VIII-1.

Specificity of Proteases. In contrast to most other enzymes, the proteases are not specific for individual substrates, i.e., for particular proteins, but are specific for certain structural features of peptide chains. Thus pepsin, trypsin, etc., attack all proteins (important for digestion). Denatured proteins are hydrolyzed more easily than native ones.

Within the peptide chain the proteases attack only at select positions determined by the kind of amino acid residue. In Table VIII-1, this is indicated by the column headed "catalytic specificity." The specificity of trypsin is particularly striking. It cleaves only lysyl and arginyl bonds in such a way that all resulting peptides have either Lys or Arg as their carboxy terminal amino acid (except for the peptide that carries the original terminal carboxyl group). Pepsin is somewhat less specific. Although bonds with aromatic or acidic amino acids are split preferentially, the neighboring amino acid residues seem to have an influence too.

Exopeptidases, as the name indicates, attack only the ends of a peptide chain: Carboxypeptidase attacks the carboxyl end, with a preference for some amino acids (see Section 3); aminopeptidase, the amino end. The first group of exopeptidases evidently requires, besides the peptide bonds, a negative charge; the second group requires, besides the peptide bonds, a positive charge. Last, there are also dipeptidases, which need both charges and cleave only dipeptides.

2. Endopeptidases

The endopeptidases of the gastrointestinal tract are synthesized in the form of inactive enzyme precursors called *zymogens*. The active digestive enzymes are released from the zymogens after certain (usually proteolytic) conversions have occurred.

Pepsin (recently renamed pepsin A). The most important protein-cleaving enzyme of the stomach is pepsin. The precursor, *pepsinogen*, is formed by the gastric mucosa. It is a protein with a molecular weight of 42,600 and has been obtained in crystalline form. In an acidic medium, or by the action of pepsin itself, the zymogen is converted to the active enzyme. Several peptides are released in the process. One of these peptides in a neutral medium inhibits pepsin, but the inhibitor–pepsin complex dissociates in an acidic medium and pepsin then digests the inhibitor. This conversion reaction is autocatalytic, because the catalyst, pepsin, appears as the reaction product.

Active pepsin was crystallized by Northrop. It is a protein with a molecular weight of 34,500, whose isoelectric point is unusually low (below 1). A phosphoric acid residue in the molecule, although dispensable for the catalytic action, is responsible for the very low pI. Pepsin has optimal activity at the acidity of the gastric juice (for pH optimum see Table VIII-1).

At the active site of pepsin, probably, the carboxyl group of a glutamyl residue is operating. It presumably undergoes transpeptidation with —NH— of the peptide group to be cleaved. Thus, the first cleavage product is a new carboxyl group; in the second step (regeneration of the enzyme) a new amino group is freed (cf. the reverse process in the action of trypsin, Chapter V,9). The product of hydrolysis is commonly a mixture of polypeptides, formerly called "peptones." The term has now been abandoned because the substances are ill-defined (the term persists, however, in describing one component of bacterial growth media). On the average, about 10% of the peptide bonds are broken. The peptides have molecular weights ranging between 600 and 3000, depending on the duration of digestion.

Rennin. This enzyme occurs in the calf stomach. Its main substrate is the caseinogen (casein) of milk, which is converted by the enzyme to insoluble casein (also called paracasein). This conversion is only a mild hydrolysis that can be catalyzed just as well by pepsin, trypsin, or other proteases. The counterpart in man to rennin from calf gastric juice is gastricsin (pepsin C).

Cathepsin. Cathepsin is the generic term for those proteases that are active at weakly acidic pH (5) and are usually found inside of cells. Intracellular proteases have not been studied as carefully as digestive enzymes because they exist only in low concentrations in the tissues. They are usually localized in the lysosomes (cf. Chapter XVIII,2).

Trypsin. In the intestine trypsin is formed from the product of the pancreatic cells, *trypsinogen.* This proenzyme or zymogen is converted to the active form by *enteropeptidase* (formerly called enterokinase) of the small intestine. Trypsin itself acts in the same way; the activation, hence, is autocatalytic.

$$\text{Trypsinogen} \xrightarrow{\text{Enteropeptidase or trypsin}} \text{Trypsin} + \text{Hexapeptide}$$

During activation, the hexapeptide Val-Asp-Asp-Asp-Asp-Lys is split off. This cleavage enables a conformational change to occur which brings the side chains participating in the active site closer together, similar to the situation occurring with chymotrypsin (see below).

Crystalline trypsin has a molecular weight of 24,000 and cleaves optimally at pH 7–9, i.e., in weakly alkaline media. As mentioned, it acts on lysyl and arginyl bonds of peptide chains. Trypsin attacks denatured proteins much more readily than native ones. The intact three-dimensional structure of a protein effectively protects against proteolysis.

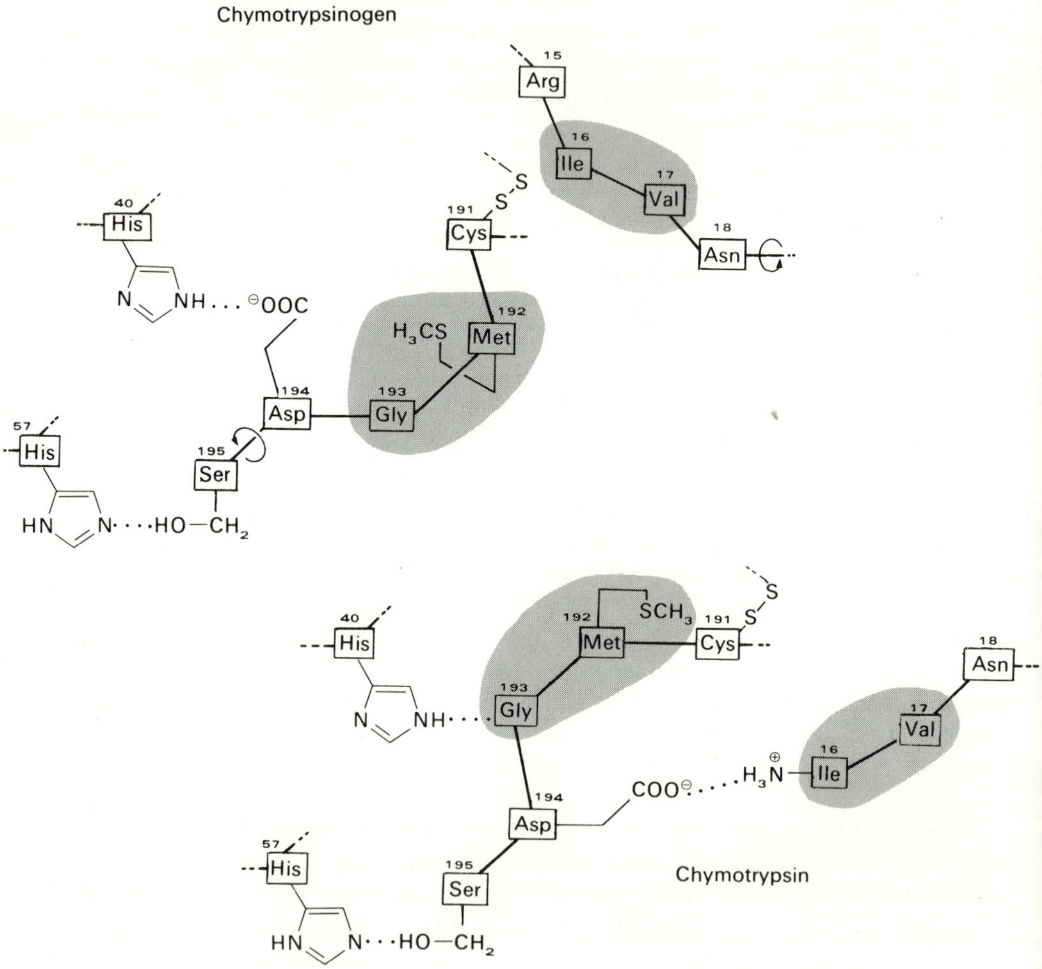

Fig. VIII-1. Conformational change during the transition of chymotrypsinogen to chymotrypsin, presented diagrammatically on the basis of X-ray structure analysis. Rotation of part of the peptide chain brings residues 192 and 193 pointing to the interior of the chymotrypsinogen molecule to the surface. The driving force for this rotation is the establishment of an ionic interaction between the negatively charged Asp-194 residue and the newly arisen amino end group of Ile-16.

In both animal and plant tissue there are proteins that inhibit trypsin. The inhibitors adhere to the enzyme, but are not cleaved themselves, and thus block the reaction with other substrates.

Trypsin consists of 223 amino acid residues whose sequence is known completely. At the active site of the enzyme we find a seryl residue whose hydroxyl group participates in the cleavage reaction. With diisopropyl fluorophosphate one can specifically block this hydroxyl group. The heightened reactivity of this seryl residue derives from the hydrogen bond to histidine, which in turn interacts with an aspartyl residue (cf. the partial formula of chymotrypsin, Fig. VIII-1). The cleavage mechanism is thought to involve a two-step reaction: The nucleophilic OH group of serine attacks the CO—NH bond and with the formation of an ester the first cleavage product with a new amino end group is released and dissociates off. Hydrolysis of the ester then frees the second cleavage product and regenerates the active enzyme.

Chymotrypsin. This enzyme is formed in the pancreas, just as trypsin is, in the form of inactive *chymotrypsinogen*, a protein with the molecular weight of 25,000. The sequence of its 246 amino acids is known. A total of five disulfide bonds are found in the molecule. Activation requires trypsin, which cleaves an Arg-Ile bond (see Fig. VIII-1).

The newly formed $H_3N^{\oplus}$ end group of isoleucine enters an ionic bond with the Asp residue 194, facilitated by the rotation of the chain segment 187–194 by 180°. The rotation brings Asp-194 and Met-192 to the surface of the protein molecule and generates the substrate binding site of the active center. The catalysis itself involves Ser-195 and His-57; even Asp-102 participates. A mechanism similar to that for trypsin (see above) has been advanced for the cleavage of the peptide bond of the substrate. Initially the active serine residue forms an ester which is then hydrolyzed.

The amino acid sequence of chymotrypsinogen is akin to that of trypsinogen. It is assumed that the two proteins are homologous in the sense discussed in Chapter IV,2. The different specificity of chymotrypsin (compared to that of trypsin), i.e., the predominating cleavage at aromatic groups, must be explained by the general environment surrounding the active site, or possibly also by a different chain conformation.

Papain. A plant protease, papain is obtained from the fruit of the papaya tree. The enzyme (molecular weight 21,000) has been prepared from the commercial product. In its specificity it resembles pepsin and chymotrypsin. At the active center there is an HS group which most likely in the course of catalysis is acylated and deacylated (analogous to the serine residue of trypsin and chymotrypsin). Papain can also cleave a number of esters and amides. It reacts optimally at neutral pH.

Bacterial Proteases. These enzymes characteristically have a broad substrate specificity toward a variety of amino acid residues. A protease from *Streptomyces griseus*, for example, cleaves over 80% of the peptide bonds of ovalbumin. Enzymes from different strains of *Bacillus subtilis* (subtilisin) have been analyzed in great detail. Although there is serine at the active site, these enzyme proteins are not homologous to trypsin and chymotrypsin, as was seen from the amino acid sequence. Bacterial proteases have been added to household detergents to digest proteinaceous spots (e.g., milk spots). Sensitive individuals may suffer allergic reactions when repeatedly exposed to such enzyme detergents.

3. Exopeptidases and Dipeptidases

Carboxypeptidases. Although found in the kidney and spleen, carboxypeptidases occur in particularly large amounts in the digestive juice. *Carboxypeptidase A* has been purified to the crystalline stage; it has a molecular weight of 34,000 and contains zinc. It is secreted by the pancreas in the form of an inactive precursor (procarboxypeptidase) with a molecular weight of 90,000.

The specificity of the enzyme reminds one of that of chymotrypsin: The end groups that are left exposed after the action of chymotrypsin are split off especially easily. *Carboxypeptidase B*, on the other hand, preferentially splits off carboxy-terminal basic amino acids. It is therefore capable of degrading peptides that have arisen from tryptic action. Carboxypeptidases can remove terminal amino acids from proteins as well. Thus they have found use in protein chemistry for end group analysis.

Aminopeptidases. These enzymes split off amino-terminal amino acids from peptide chains. They are utilized therefore for sequence determinations of peptides. The large number of aminopeptidases is roughly distinguishable by their specificity. Recently an enzyme was found that cleaves the last two amino acids as a dipeptide from the amino acid.

Dipeptidases. As implied by the name, dipeptidases hydrolyze only dipeptides. Here too, different enzymes must be distinguished. We should mention glycylglycine dipeptidase, prolinase (prolyl peptide cleaving), and prolidase (aminoacyl proline cleaving).

These peptidases contain bivalent metal ions or require addition of such ions for full activity. It is assumed that metal complexes (chelates) form and change the electron configuration and thus facilitate hydrolysis.

4. Outline of the Metabolism of Amino Acids

The cooperation of endopeptidases, which break the proteins at certain places into large fragments, and of exopeptidases, which hydrolyze these fragments further, results in a mixture of amino acids that is absorbed and allowed to enter the bloodstream. Part of the amino acid mixture is used to build new endogenous proteins. A large part, however, is degraded further. The following are the possible metabolic pathways: (a) Alteration of the side chains while maintaining the α-aminocarboxylic acid grouping; (b) decarboxylation; (c) transamination to α-keto acids; (d) oxidative deamination to α-keto acids.

In the reaction paths (a) to (c), the cofactor is usually pyridoxal phosphate. According to chemical electron theory all these reactions may be derived from one common intermediate, a Schiff's base.

The illustration below shows first how a Schiff's base arises from the amino acid and pyridoxal phosphate. The electron-attracting effect of the positively charged N in the pyridine ring favors one of the

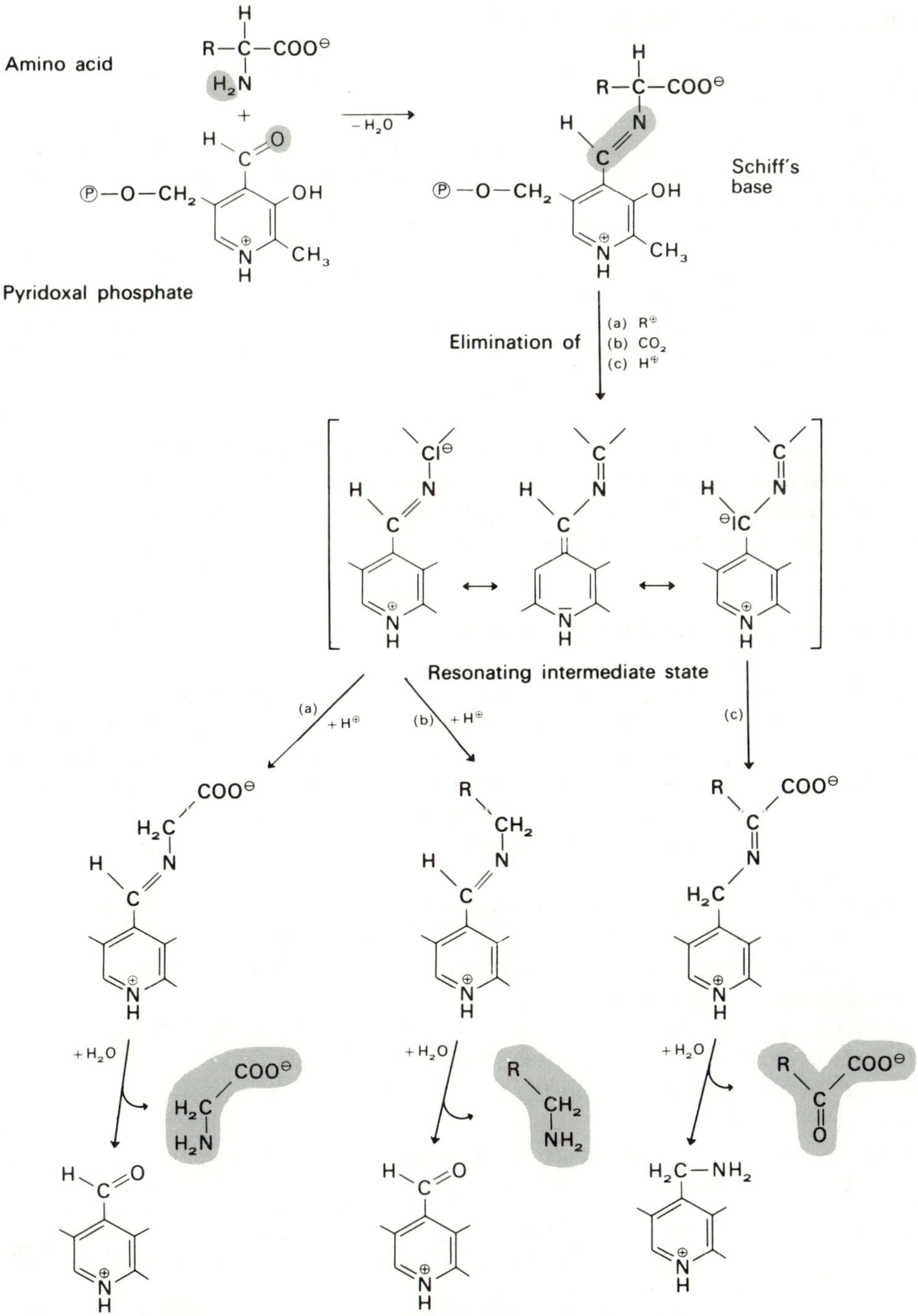
Amino acid
Pyridoxal phosphate
$-H_2O$
Schiff's base
Elimination of
(a) $R^\oplus$
(b) CO_2
(c) $H^\oplus$
Resonating intermediate state
(a) $+H^\oplus$
(b) $+H^\oplus$
(c)
$+H_2O$
$+H_2O$
$+H_2O$

contributing resonance structures, but this structure is realized only when one substituent is eliminated as a cation from the α-C atom of the amino acid. That could be the group R [reaction path (a) taken only in special cases, e.g., with serine; cf. Section 12] or the C atom of the carboxyl group [reaction path (b); decarboxylation]. In the second case, the resonating intermediate state is stabilized further by the addition of a proton to the α-C atom and concomitant hydrolysis of the Schiff's base to form a primary amine. Third—and usually—the eliminated substituent can be the hydrogen on the α-C atom; the loss of the proton aids the shift of the double bond to the α-C atom, and the resulting intermediate is hydrolyzed to the α-keto acid and pyridoxamine phosphate. This is reaction (c), a transfer of the amino group to the coenzyme with a simultaneous exchange of levels of oxidation (univalent for bivalent).

After the loss of a $H^{\oplus}$ from the α-C atom, stabilization can also be attained by eliminating $X^{\ominus}$ from the β-C atom and forming a double bond. In nonenzymatic model experiments *serine* and *cysteine* react according to this scheme of "α,β-elimination" to produce aminoacrylic acid and eventually pyruvic acid and NH_3. In bacterial metabolism, *tryptophan* is degraded in this fashion to indole + pyruvic acid + NH_3.

In higher plants we find metabolic reactions of much greater complexity; for example, amino acids enter the synthesis of alkaloids.

Reaction (d), which takes place without the participation of pyridoxal phosphate, also leads to α-keto acids. Both reactions (c) and (d) are most important for the break-down of amino acids and the metabolism of nitrogen. But before considering this reaction, we shall briefly discuss decarboxylation.

5. Decarboxylation of Amino Acids

The mechanism of this enzymatic reaction, catalyzed by pyridoxal phosphate, has been discussed above as pathway (b) (Section 4). The net result is the production of CO_2 and a primary amine, whose formula can easily be derived from that of the amino acid which was decarboxylated. Amines of this type are called *biogenic amines* (Guggenheim); many of them possess strong pharmacological effects, and others are important as precursors of hormones and as components of coenzymes and other active substances.

The occurrence and significance of some biogenic amines are listed in Table VIII-2. The biogenic amines derived from aromatic amino acids are of both pharmacological and physiological interest. They will be discussed in Chapter XX, because of their close relationship to hormones.

Amino acid decarboxylases are especially common in bacteria; their biological significance there is unclear. The enzymes are usually specific for one amino acid, and more particularly for the L-form.

Some of the "bacterial" biogenic amines recently have enjoyed renewed interest: *Cadaverine* (from lysine) and *putrescine* (from ornithine) have been found in ribosomes (cf. Chapter XVIII,2). In seminal fluid as well as in ribosomes and some viruses there is *spermine*, $H_2N-(CH_2)_3-NH-(CH_2)_4-NH_2$, and *spermidine*, $H_2N-(CH_2)_3-NH-(CH_2)_4-NH-(CH_2)_3-NH_2$, derived from putrescine by alkylation with aminopropyl groups, which in turn arise from *S*-adenosylmethionine by decarboxylation (cf. Chapter VI,6).

Decarboxylation of serine gives rise to *ethanolamine* (or aminoethanol), a regular component of phosphatides and the parent substance of choline. The analogous compound containing sulfur, called *β-mercapto-*

TABLE VIII-2
Biogenic Amines

Amino acid	Decarboxylation product	Occurrence and significance
Lysine	Cadaverine	Ribosomes, bacteria
Ornithine	Putrescine	Ribosomes, bacteria
Methionine	(→spermidine, spermine)	Ribosomes; sperm
Arginine	Agmatine	Bacteria (intestinal flora)
Serine	Ethanolamine	Phosphatides
Threonine	Propanolamine	Vitamin B_{12}
Cysteine	β-Mercaptoethylamine	Coenzyme A
Aspartic acid	β-Alanine	Coenzyme A, pantothenic acid
Glutamic acid	γ-Aminobutyric acid	Brain (ganglia inhibitor)
Histidine	Histamine	Effective on blood pressure
Tyrosine	Tyramine	Uterus contracting
3,4-Dihydroxyphenylalanine	Dopamine (→epinephrine)	Tissue hormone (→hormone)
Tryptophan	Tryptamine	Hormone(?)
5-Hydroxytryptophan	Serotonin (→melatonin)	Tissue hormone (hormone)

ethylamine (or cysteamine), occurs in coenzyme A, where it is the carrier of the reactive SH group. In the pantothenic acid part of coenzyme A there is *β-alanine* (the decarboxylation product of aspartic acid). Vitamin B_{12} contains *propanolamine* (or aminopropanol from threonine). It is possible that *γ-aminobutyrate*, derived from glutamate, plays a role in brain function (cf. Chapter XX,13).

Amine Oxidases. The inactivation of biogenic amines, which often are biologically highly active, requires the action of amine oxidases. They are flavoproteins; they dehydrogenate the amines to imines and transfer the hydrogen to molecular oxygen, whereby H_2O_2 arises. The C=N bond of the imines is easily hydrolyzed and yields an aldehyde and ammonia (compare oxidative deamination, Section 7). Different substrate specificities distinguish *monoamine oxidase* (abbreviated MAO) and *diamine oxidase* (DAO), which requires two basic groups in the substrate. The most important substrate of the latter is histamine (Chapter XX,13).

6. Transamination

The amino groups of many, if not all amino acids, react identically. They are transferred by transaminases to the prosthetic group *pyridoxal phosphate.* The diagram shows the reaction with alanine as an example (for the mechanism of the reaction see Section 4).

By exchanging the levels of oxidation one obtains pyridoxamine (which remains bound to the enzyme) and an α-keto acid with the same carbon skeleton as the amino acid (upper half of diagram). Further degradation of the keto acid leads eventually to a common metabolic pool. These catabolic steps, differing for each individual amino acid, will be discussed in Section 9 ff.

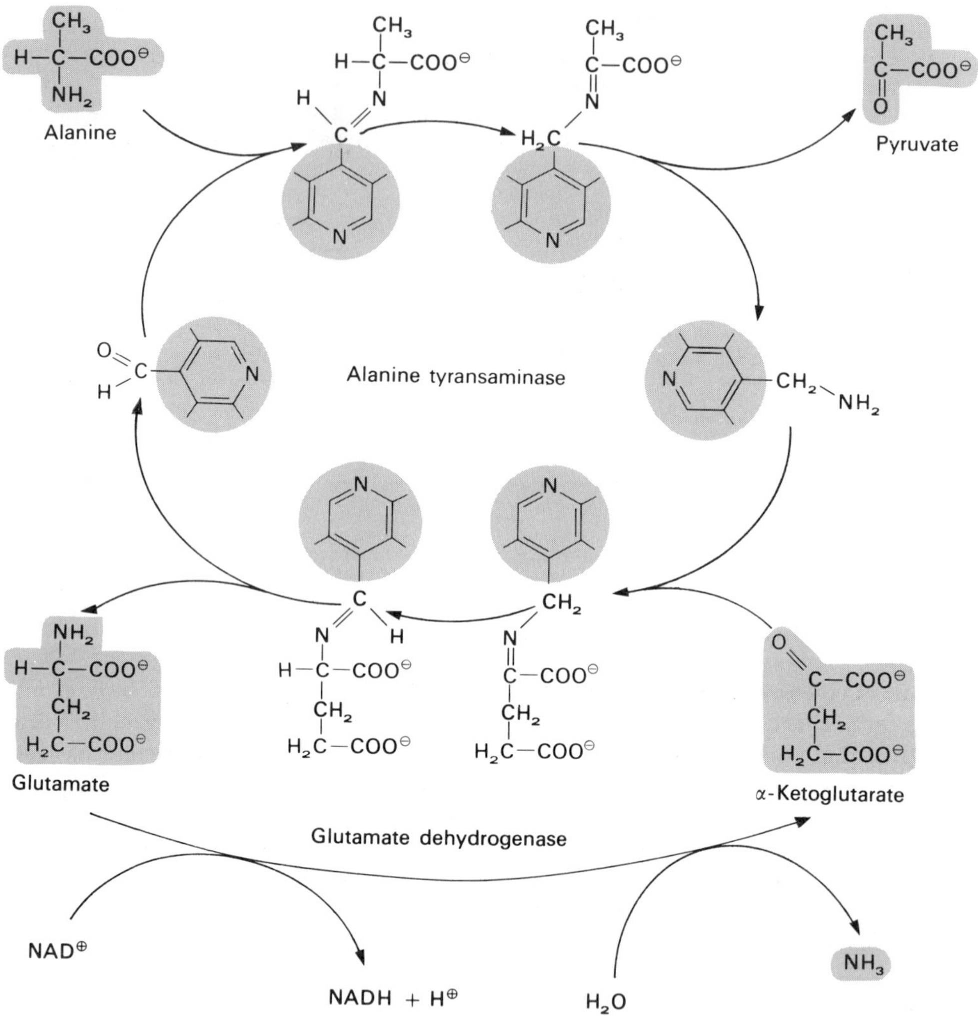

Pyridoxal phosphate is regenerated from pyridoxamine phosphate in a reaction with a keto acid; α-ketoglutarate and oxaloacetate are very effective NH_2 acceptors (lower half of diagram) and thus yield glutamate or aspartate. The significance of transamination reactions is primarily that nitrogen is passed on from glutamate or aspartate to the final excretion product, urea (cf. Section 8). Glutamate dehydrogenase has an important role in this step (cf. Section 7).

From glutamate the reverse pathway is equally possible: with α-keto acids from other sources the corresponding amino acids and α-ketoglutarate can be formed[1] (the above diagram in reverse). The organism is able in this way to synthesize several amino acids: alanine out of pyruvate, serine out of hydroxypyruvate, and, of course, the two dicarboxylic acids already mentioned, glutamate and aspartate.

For a number of amino acids, however, neither the proper α-keto acids nor any other suitable precursors are available. These amino acids must be supplied to the organism and hence are called *essential* or *indispensable amino acids* (cf., also Chapter XXII, "Nutrition"). The list of essential amino acids varies somewhat for different animals. For man they include: valine, leucine, isoleucine, lysine, methionine, threonine, phenylalanine, and tryptophan.

The rat requires, in addition, histidine. Arginine, proline, serine, tyrosine, and cystine stimulate growth. Although the organism can synthesize them, it apparently cannot supply them fast enough to have optimal amounts available for the very intensive synthesis of protein which occurs during growth.

Plants and many microorganisms can produce all the amino acids. Ultraviolet irradiation followed by selection permits the isolation of mutants of microorganisms (bacteria, yeasts, molds) that have lost the capacity to synthesize certain amino acids and thus have become auxotrophic with respect to these particular amino acids. (A strain is called auxotrophic for a certain substance if that substance must be provided before the strain can grow.) These biochemical mutants have become invaluable both for genetics and biochemistry because many biosynthetic pathways have been elucidated through the study of such mutants (cf. also Chapters VII,7 and XVI,4).

7. Deamination

Oxidative Deamination. In addition to transamination, oxidative deamination is another pathway to produce α-keto acids. Here two hydrogen atoms are removed to produce an imino acid, which is then hydrolyzed to ammonia and a keto acid:

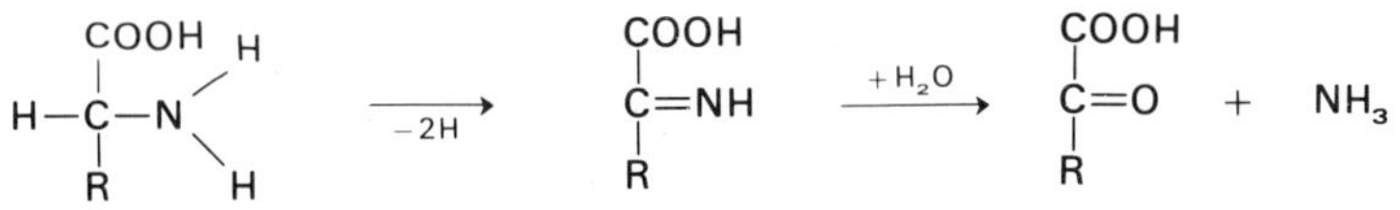

Glutamate Dehydrogenase. This enzyme occurring in the mitochondria of all tissues transfers hydrogen to nicotinamide-adenine dinucleotide. The reaction may be formulated:

$$\text{Glutamate} + \text{NAD}^{\oplus} + H_2O \rightarrow NH_3 + \alpha\text{-Ketoglutarate} + \text{NADH} + H^{\oplus}$$

[1] The transition of amino acids into keto acids and back had already been discovered by Knoop as early as 1910 from feeding experiments and later was corroborated by model reactions. Transamination was discovered by Braunstein; the role of pyridoxal phosphate, by Gunsalus and Snell.

The enzyme, specific for L-glutamate, is of exceptional importance for metabolism. Transamination is the route by which the amino group of most amino acids is transferred onto α-ketoglutarate. The glutamate produced in this transfer is converted back to α-ketoglutarate by the action of the dehydrogenase. Reduced nicotinamide coenzyme can release its hydrogen to the respiratory chain or to other systems. The ammonia is bound by the formation of urea (see below) and excreted as such.

Amino Acid Oxidases. These are flavoproteins that catalyze the same general reaction, but that transfer hydrogen onto O_2 instead. The product is H_2O_2. Some oxidases are specific for D-amino acids, while some specifically oxidize the L-form. The physiological significance of the former is obscure since D-amino acids are practically never found in the organism. L-Amino acid oxidase prepared from rat kidneys contains *flavin mononucleotide* and is of remarkably low molecular activity; one molecule of enzyme converts only six molecules of substrate per minute. It is doubtful therefore whether this enzyme has any significance in the metabolism of amino acids.

Nonoxidative Deamination. This enzymatic process eliminates NH_3 with the formation of a double bond in the remaining carbon skeleton. The systematic title of the enzymes is amino acid-ammonia lyase. In mammalian metabolism the best example is histidine-ammonia lyase (see Section 13). Bacteria and plants contain other enzymes that convert aspartate to fumarate and phenylalanine to cinnamate.

8. Urea Cycle

With the formation of ammonia in the dehydrogenation of glutamate nothing appears to have been gained, since ammonia is an unsuitable form of nitrogen for excretion. Only animals living in water, crabs and some fishes, are able to excrete ammonia directly through their gills. In higher organisms, ammonia, a cellular poison even at relatively low concentrations, is converted to urea by a cyclic process (Krebs and Henseleit). The urea cycle takes place in mitochondria of the liver closely coupled to the dehydrogenation of glutamate. The net reaction is endergonic:

$$HCO_3^{\ominus} + NH_4^{\oplus} + NH_3 \longrightarrow O{=}C(NH_2)_2 + 2\,H_2O$$

$$\Delta G^\circ = 3.54 \text{ kcal/mole}, \qquad \Delta G_{\text{physiol}} \sim 14 \text{ kcal/mole}$$

Thus, at physiological concentrations (for the dependence of ΔG on concentration, cf. Chapter V,2 and 5), the net expenditure of energy amounts to about 14 kcal/mole. We should expect, therefore, that energy-rich intermediates participate.

The most important energy-rich intermediate is a derivative of carbamic acid $H_2N{-}COOH$ namely *carbamyl phosphate* formed in the following manner:

$$NH_4^{\oplus} + CO_2 + 2\,ATP \longrightarrow H_2N{-}\overset{\overset{\displaystyle O}{\|}}{C}{-}O\sim \textcircled{P} + 2\,ADP + \textcircled{P}$$

Two moles of ATP are consumed. One donates the phosphate group in carbamyl phosphate; the other is cleaved to ADP and Ⓟ; the reaction is virtually irreversible.

Carbamyl phosphate synthetase is the enzyme catalyzing this reaction. It occurs predominantly in mammalian liver and binds $NH_4^{\oplus}$ ions even at very low concentrations ($10^{-4}M$) with the assistance of the allosteric activator *N-acetylglutamate*. It is thought that initially ATP reacts with CO_2 to form enzyme-bound carboxyl phosphate whose phosphate group is then exchanged for an H_2N group. The resulting enzyme-bound carbamic acid can then react with ATP to form carbamyl phosphate, ADP, and enzyme.

Cytosol contains another enzyme, carbamoyl-phosphate synthase II, which catalyzes the following reaction: Glutamine + CO_2 + 2 ATP ⟶ Glutamate + carbamyl phosphate + 2 ADP + Ⓟ. Glutamine donates the H_2N group. Carbamyl phosphate required for pyrimidine systhesis (Chapter VII,2) arises in this manner.

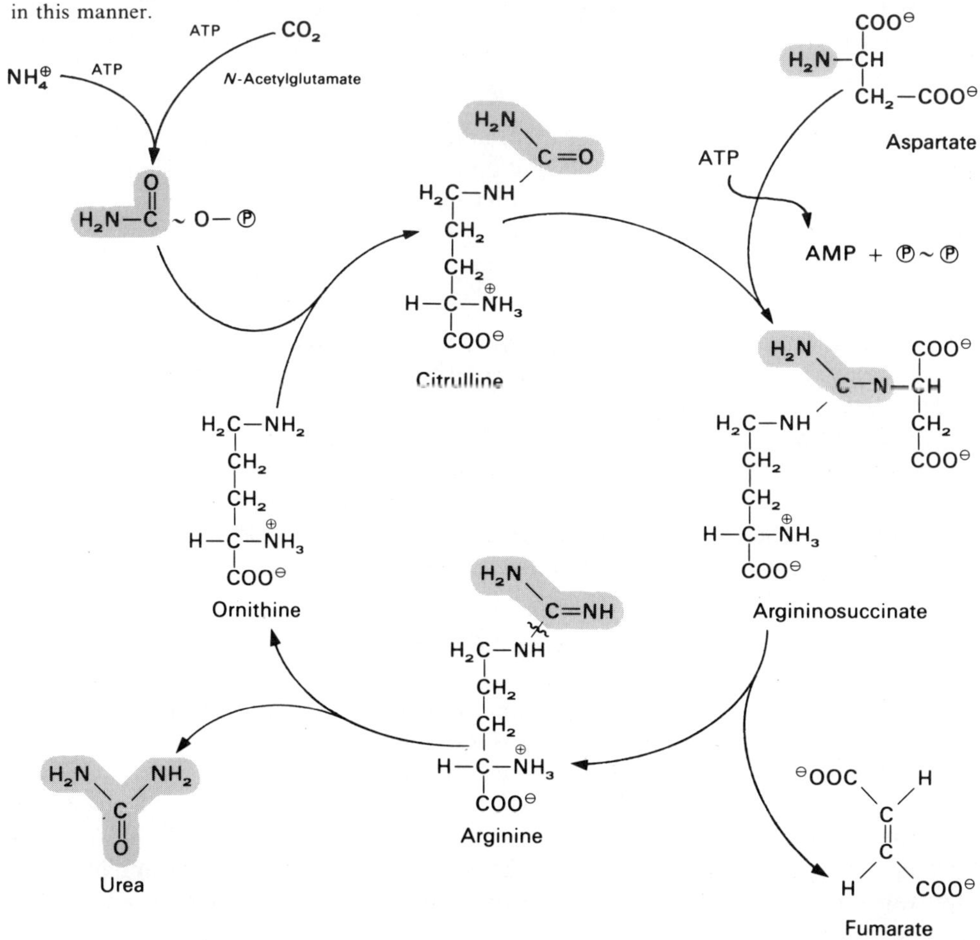

Carbamyl phosphate then reacts with the δ-amino group of *ornithine* to produce *citrulline* (α-amino-δ-ureidovalerate), and citrulline is transformed in two steps to *arginine*. The NH_2 group necessary for this is provided by aspartate which, with the consumption of 1 mole of ATP and the action of argininosuccinate synthetase, joins with citrulline to give *argininosuccinate*. ATP in this case is cleaved to AMP and pyrophosphate. The further hydrolysis of pyrophosphate prevents the reverse reaction. An enzyme helps to break the intermediate product into arginine and fumarate and, finally, arginine is cleaved by the well-known enzyme *arginase* into urea and ornithine. Urea is actually formed in its tautomeric form, isourea, but it rearranges spontaneously. This closes the cycle. Ornithine is then ready to accept another carbamyl phosphate.

In summary, 2 NH_3 (from glutamate and aspartate) join with CO_2 to give urea; in the process 3 moles of ATP are required. If the cleavage of pyrophosphate is included in the calculation, then four energy-rich bonds are consumed. The formation of urea is, therefore, from an energy viewpoint, a luxury in which the cell apparently indulges in order to escape the deleterious effect of high concentrations of free ammonia.

Glutamine Synthesis. One other metabolic reaction of ammonia is the formation of the acid amides asparagine and glutamine (formulas, Chapter II-1 and 2). The enzyme *glutamine synthetase* occurs primarily in the brain and liver and catalyzes the reaction

$$\text{Glutamate} + \text{ATP} + NH_4^{\oplus} \longrightarrow \text{Glutamine} + \text{ADP} + P_i$$

The mechanism of this reaction presumably involves the intermediary formation of γ-glutamyl phosphate and the subsequent exchange of the phosphate group for the $-NH_2$ group.

Glutamine acts as NH_2 donor in general metabolism, e.g., in the synthesis of purines (see Chapter VII-2) and of glucosamine. It is also another form for transporting ammonia in the organism. To this end, glutamine is hydrolyzed easily to glutamate and ammonia. The corresponding enzyme *glutaminase* abounds in the kidney and is largely responsible for the ammonium salts in urine.

9. Fate of the Carbon Skeleton of Amino Acids

The α-keto acids arising from transamination are shunted into general metabolism and eventually are burned to CO_2 and H_2O. Since we will not study until later the final catabolic steps of any of the foodstuffs (Chapters XI and XII), we will consider here catabolism only up to the points where common intermediates are formed.

These common intermediates are, first, saturated monocarboxylic acids (often arising in their activated form, bound to coenzyme A); second, pyruvate and hydroxypyruvate; and, third, the C_4-dicarboxylic acids: succinate, fumarate, and oxaloacetate.

First we will mention *acetoacetate*, a C_4-compound which is a pathological constituent of urine. It is one of the "ketone bodies." Those amino acids that can give rise to acetoacetate are called *ketogenic*. They are *phenylalanine*, *tyrosine*, and *leucine*. To a lesser extent isoleucine also is *ketogenic* since it gives rise to acetyl-CoA during breakdown, which in turn can condense to form acetoacetyl-CoA. Isoleucine at the same time is also glucogenic, since it produces succinate via methylmalonate (see Chapter XII,4).

The rest of the amino acids, which can be broken down to one of the C_4-dicarboxylic acids enumerated above or to pyruvate, can be converted to carbohydrate. They are therefore called "*glucogenic* amino acids." The generation of glucose from amino acids is of considerable physiological importance. Only three carbon atoms of the amino acid are used for the synthesis of glucose. In addition to this conversion (along with the ever-present terminal oxidation to CO_2), it is metabolically important to produce from amino acids C_1 fragments at the oxidation level to formate or formaldehyde. These C_1 fragments arise from serine, glycine, and histidine. Finally, some amino acids are interconvertible, without deamination, merely by changes on the carbon skeleton. Excellent examples of the multiplicity of conversions are provided by the metabolism of serine (Section 12) and of phenylalanine (Section 11).

10. Degradation to Activated Fatty Acids: Oxidative Decarboxylation

This pathway is followed by alanine, valine, isoleucine, and leucine. The prototype for all is the metabolism of *alanine*, which goes to activated acetate (acetyl-CoA) in two steps. Acetyl-CoA, of course, then can undergo a multitude of reactions. The

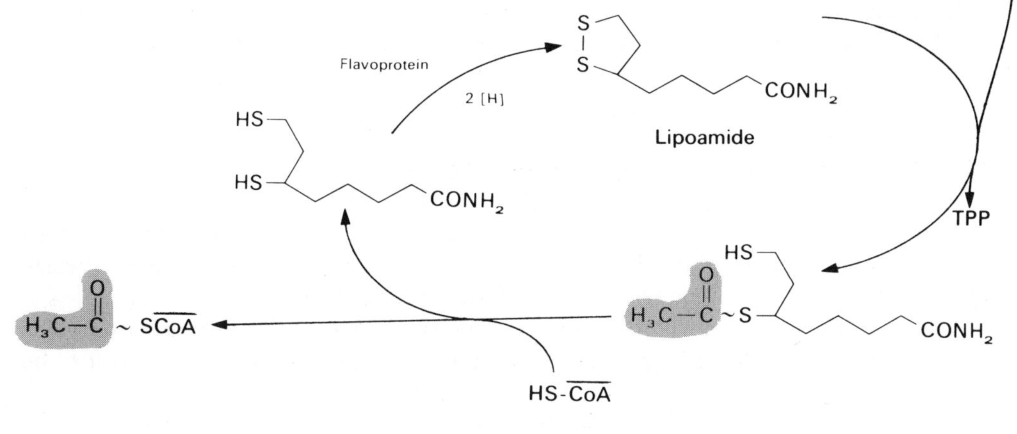

first step is transamination, yielding in the usual manner the α-keto acid, pyruvate. The latter is decarboxylated and at the same time oxidized by a complicated enzyme system (oxidative decarboxylation of the α-keto acid). The cofactors for the reaction are thiamine pyrophosphate, lipoamide, and coenzyme A. In the initial step, CO_2 is split out and "active acetaldehyde" is formed (see Chapter VI,7); it is bound to C atom 2 of the thiazole ring of TPP. In a second step, the aldehyde is transferred to lipoamide, the disulfide ring is opened, and the aldehyde separated into $H_3C—CO—$ and —H. This last step is really the oxidizing (more precisely, the dehydrogenating) step, because here lipoamide becomes a derivative of dihydrolipoic acid and the aldehyde actually becomes the derivative of an acid. The free energy of the aldehyde oxidation is retained by the product with the acetyl residue being held in an energy-rich bond (thioester bond); it is then easily transferred to coenzyme A. This is the "activated fatty acid;" in the case of alanine, it is activated acetate, which enters the general metabolic pool. A flavoprotein again dehydrogenates the dihydrolipoamide; the hydrogen then may end up on NAD (cf. Chapters VI,4 and XI,2).

Valine, *isoleucine*, and *leucine* are degraded in a quite analogous manner. Activated fatty acids shortened by one C atom are treated in metabolism in essentially the same way as ordinary fatty acids. The only problem is raised by amino acids with a methyl group as a side chain of the carbon skeleton. We will discuss their degradation in connection with β-oxidation of fatty acids (Chapter XII,4). As mentioned already, leucine is converted to acetoacetate, isoleucine partially so. Valine, however, becomes methylmalonate, and is changed further by a rearrangement of the carboxyl group to succinate thereby opening a pathway to carbohydrate synthesis.

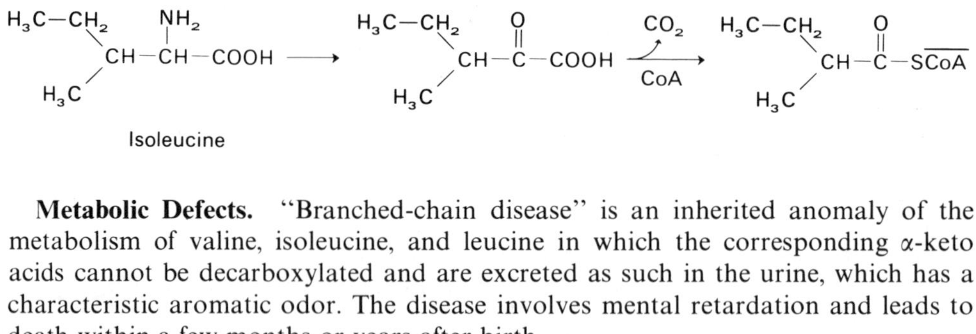

Isoleucine

Metabolic Defects. "Branched-chain disease" is an inherited anomaly of the metabolism of valine, isoleucine, and leucine in which the corresponding α-keto acids cannot be decarboxylated and are excreted as such in the urine, which has a characteristic aromatic odor. The disease involves mental retardation and leads to death within a few months or years after birth.

11. Metabolism of Aromatic Amino Acids

Aromatic compounds are generally not synthesized in the animal organism. This explains the indispensability of phenylalanine. Plants and microorganisms synthesize the aromatic ring out of carbohydrates (cf. Chapter XVI,5).

The aromatic ring is broken down by oxidation. One of the double bonds of a Kekulé structure is forced open enzymatically and satisfied with oxygen (for the mechanism see Chapter X,7). An analogous reaction would be the ozonolysis of benzene.

Metabolism of Phenylalanine and Tyrosine. These amino acids are ketogenic. The degradation of phenylalanine first of all yields tyrosine: A powerful oxygenase (phenylalanine-4-hydroxylase) introduces a hydroxyl group in the para-position of the benzene ring (the oxygen is from air; for the reaction mechanism see Chapter X,7).

Tyrosine—whether formed from phenylalanine or arising directly from ingested proteins—has various pathways available. Quantitatively, the most important is the degradation to acetoacetate + fumarate. The melanins, the dark hair and skin

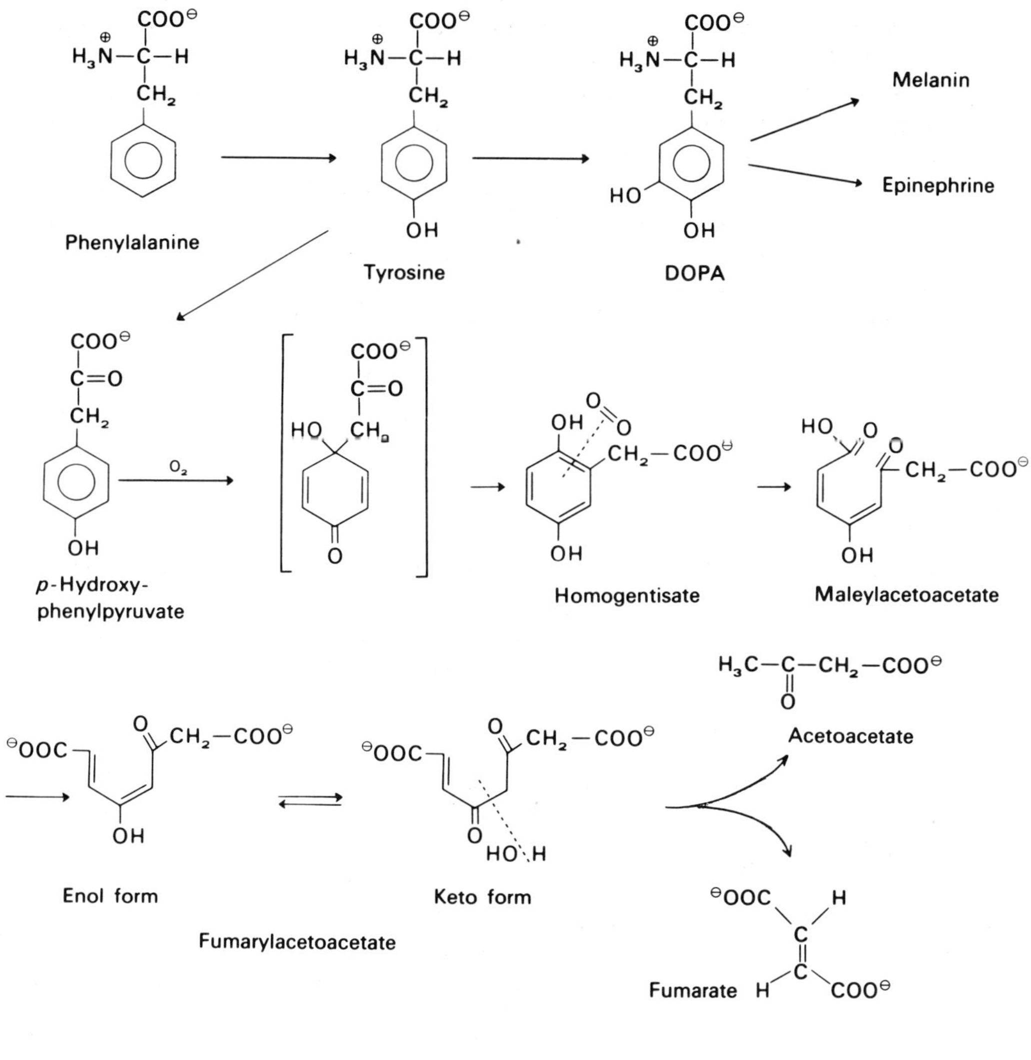

pigments, also come from tyrosine, as do the hormones epinephrine (see Section 5 and Chapter XX,4) and thyroxine (cf. Chapter XX,6).

The first of the enumerated pathways proceeds via transamination to *p*-hydroxyphenylpyruvate. *Tyrosine-α-ketoglutarate transaminase* is one of the inducible enzymes. The substrate tyrosine as well as the adrenocortical hormone cortisol are inducers. The subsequent oxidation of *p*-hydroxyphenylpyruvate is accompanied by a rearrangement of the side chain to the neighboring *o*-position. Rearrangement, oxidative decarboxylation, and hydroxylation, all three steps, are catalyzed by the single enzyme *p-hydroxyphenylpyruvate dioxygenase*. Ascorbate is a required cofactor. *p*-Quinol has been postulated as intermediate product (bracketed in the diagram) in the formation of *homogentisate*, but this mechanism has not yet been proved.

Homogentisate is further oxidized by an oxygenase (see Chapter X, 6). The double bond is split open and one oxygen is attached to each end-carbon atom. The resultant compound is *maleylacetoacetate* which has the *cis*-configuration around its double bonds. Interestingly enough, an enzyme catalyzes the rearrangement of the *cis*- to the *trans*-compound. In this reaction glutathione is a cofactor. The resulting *fumarylacetoacetate* is hydrolyzed into fumarate and acetoacetate. Tyrosine, therefore, provides four C atoms for acetoacetate and four C atoms for fumarate. The latter might possibly be used for glucose formation, but ultimate oxidation seems more likely.

The steps leading to melanin are of no quantitative significance for the degradation of tyrosine but are of interest because of the biogenesis of that pigment. The whole sequence of reactions is catalyzed by a single enzyme, *o-diphenol oxidase* (formerly catechol oxidase or tyrosinase) and follows essentially the scheme shown here:

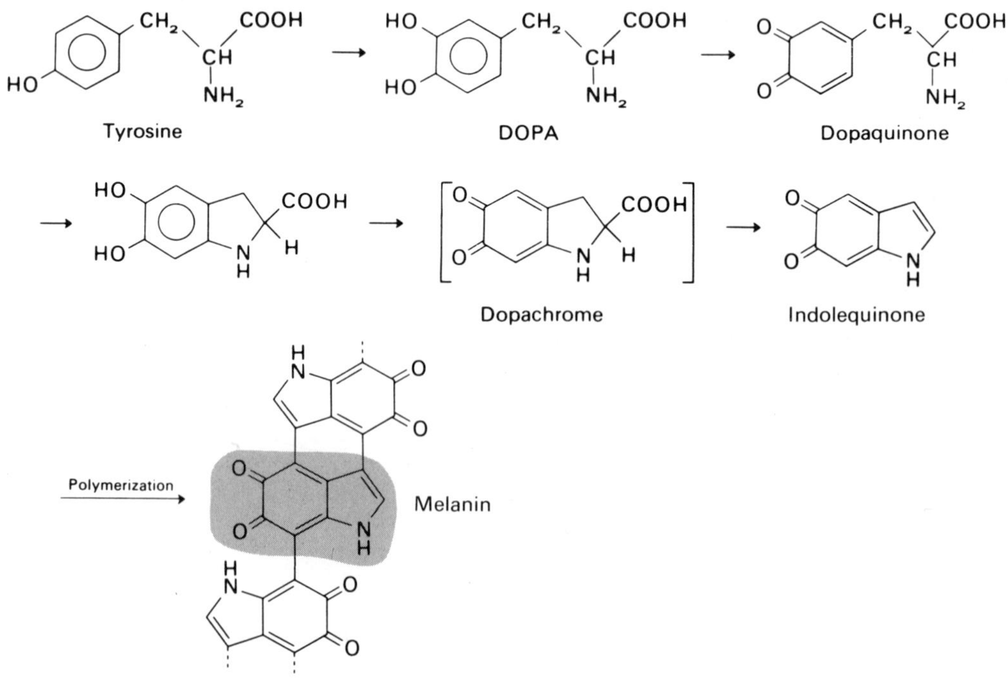

The first and second steps are associated intimately; the mechanism is discussed in Chapter X,7. Ring closure to indole, an addition to the quinonoid system, takes place spontaneously and so probably does the polymerization of the indolequinone in the positions indicated. Melanin seems to retain several radical positions.

For the biosynthesis of thyroxine compare Chapter XX,6; the steps in the formation of epinephrine are discussed in Chapter XX,4.

Metabolic Defects. Several hereditary errors in metabolism of phenylalanine and tyrosine are known. The most prominent is *phenylketonuria* caused by the absence of the hydroxylase that converts phenylalanine into tyrosine. Phenylalanine instead is shunted by transamination to *phenylpyruvate* (and in part to phenylacetate), which is excreted in the urine. The disease again is associated with mental retardation which can be prevented, however, by the feeding in infancy of a diet very low in phenylalanine.

Tyrosinosis is a rare disease caused by the absence of *p*-hydroxyphenylpyruvate oxidase; as a result *p*-hydroxyphenylpyruvate is excreted. Still another disease is *alkaptonuria*, in which homogentisate cannot be degraded further and is excreted as such. The urine quickly turns black upon standing due to air oxidation of homogentisate to the corresponding *p*-quinone, which polymerizes spontaneously to a dark pigment. This disease is free of any serious consequences.

A defect in the biosynthesis of melanin results in *albinism*, characterized by the absence of pigments in skin, hair, and eyes.

Metabolism of Tryptophan. The catabolism of typtophan is interesting since it leads to the biosynthesis of a vitamin, nicotinamide. It appears contradictory to state that a vitamin is formed by an organism, but nicotinamide deficiency can indeed be demonstrated only with a tryptophan-poor diet and the concurrent deficiency of pyridoxine. Tryptophan largely replaces the vitamin even in man.

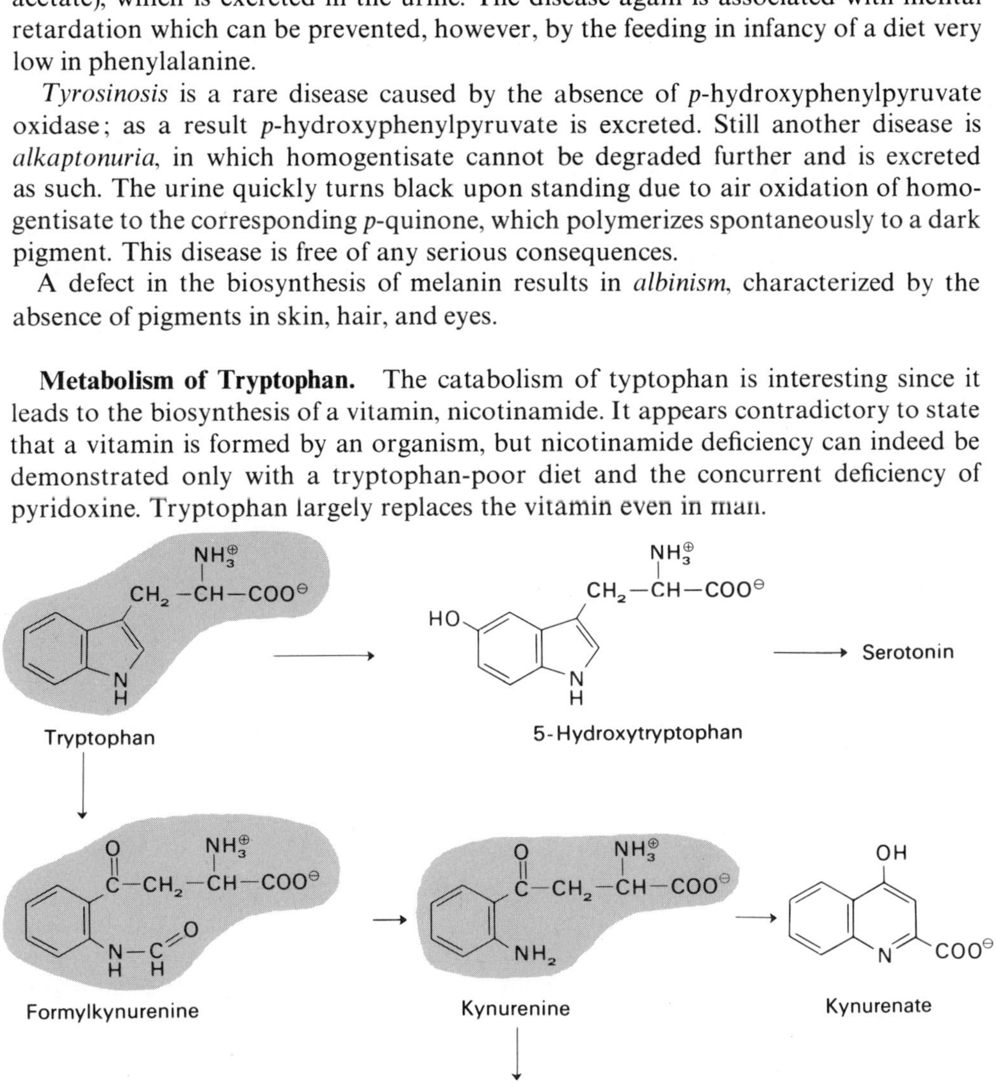

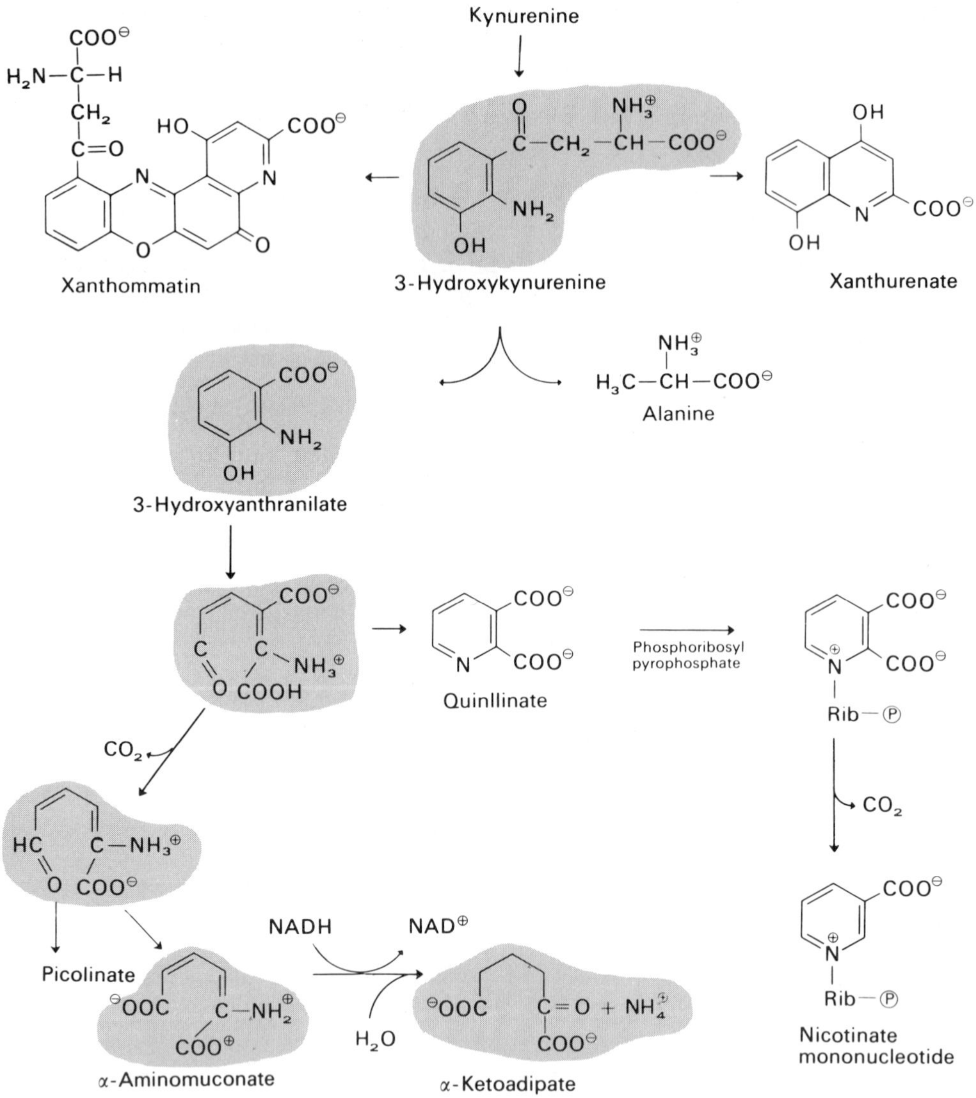

The catabolism of tryptophan seems to have many steps in common with the formation of nicotinamide. The amino nitrogen is not attacked initially, but instead the indole ring is opened up by *tryptophan pyrrolase* (an iron-porphyrin enzyme). Just as in other cases of oxidative ring opening (cf. Chapter X,7), the double bond is split and each new end receives one oxygen atom. In this way *N-formylkynurenine* is formed; it is then hydrolyzed to formate and *kynurenine*.

Kynurenine is oxidized (again by atmospheric oxygen) to yield *3-hydroxykynurenine* and, subsequently, is cleaved to 3-hydroxyanthranilate and alanine. The fate of alanine has already been presented. Sidepaths from kynurenine and 3-hydroxykynurenine via transamination and spontaneous ring closure lead to *kynurenate* and *xanthurenate.*

Initially, 3-hydroxykynurenine was not discovered as an intermediate for nicotinic acid formation but rather as an intermediate for ommochrome formation (Butenandt and co-workers). *Ommochromes* are pigments found chiefly among insects and crabs. The simplest representative, *xanthommatin* (for formula see above), is easy to prepare *in vitro* by careful oxidation of hydroxykynurenine. The synthesis of ommochromes is disrupted in several mutants of the fruit fly *Drosophila* (and other insects) (cf. Chapter VII,7).

3-Hydroxyanthranilate is subjected to another oxidative ring opening adjacent to the hydroxyl group. The unsaturated aldehyde is very unstable and easily condenses with the amino group to form the pyridine derivative *quinolinate.* This product is linked enzymatically with phosphoribosyl pyrophosphate (analogously to the formation of orotic acid, see Chapter VII,2). The same enzyme then effects decarboxylation to yield nicotinate mononucleotide, which becomes a dinucleotide with the aid of ATP. Finally, glutamine donates an NH_2 group for the formation of nicotinamide-adenine dinucleotide (NAD).

Although the production of the coenzymes NAD and NADP is obviously valuable physiologically, it is quantitatively but a minor pathway of tryptophan degradation. The main pathway proceeds from the product of ring cleavage via decarboxylation and dehydrogenation of the aldehyde group to aminomuconate. Being a labile enamine, the latter hydrolyzes readily and releases ammonia. Hydrogenation of the double bond with NADH yields α-ketoadipate, which undergoes oxidative decarboxylation resulting finally in glutaryl-CoA (cf. breakdown of lysine, Section 13).

Microorganisms (including intestinal bacteria) break down the side chain of tryptophan. The following compounds may be formed: indolepyruvate, indoleacetate, skatole (methylindole), indole, and indoxyl. Some of the latter is reabsorbed by the gut and esterified in the liver with sulfuric acid to form *indoxyl sulfate* which is excreted in the urine and also known as indican. Indoleacetate is a plant hormone (*auxin*; Chapter XX,17).

Tryptophan also can be hydroxylated to *5-hydroxytryptophan*, the parent compound of serotonin (Chapter XX,13). The enzyme tryptophan 5-monooxygenase requires tetrahydropteridine as cofactor for this conversion.

12. Amino Acids Supplying C_1 Fragments

We must distinguish here between methyl groups provided by methionine, on the one hand, and formaldehyde and formate arising from serine and glycine, on the other.

Methionine and Cysteine. Methyl groups play an important role in the biosynthesis of many substances, and methionine has turned out to be the methyl donor *par excellence.* "Active methyl" arises directly out of methionine and ATP; it is a sulfonium compound (formula in Chapter VI,6). We shall return to the active methyl group on other occasions (cf. also Chapters VI,6, XIII,2, and XX,5), but now will discuss the metabolism of demethylated methionine *homocysteine*, so called because its carbon chain is longer than that of cysteine by one CH_2 group. It can provide the sulfur for cysteine by condensing with serine; the resultant thioether (*cystathionine*) breaks into homoserine and cysteine (see diagram). The cleavage involves the pyridoxal phosphate enzyme cystathionase (also called homoserine dehydratase). γ-Elimination frees cysteine from the complex, and the remaining carbonium ion either can accept $OH^{\ominus}$ to form homoserine or it can release $H^{\oplus}$ to form an unsaturated compound which isomerizes to the imino acid and hydrolyzes to yield α-ketobutyrate. Oxidative decarboxylation finally produces propionate, whose further breakdown is described in Chapter XII,4.

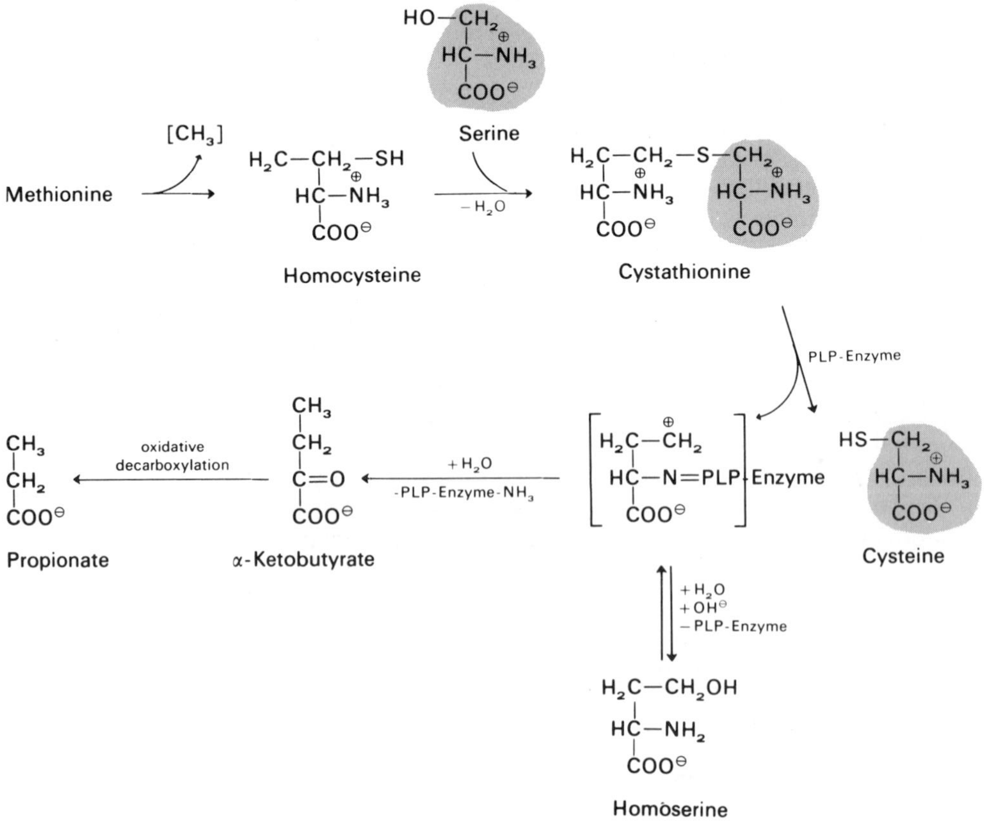

Cysteine, whose synthesis we have just seen, is degraded chiefly through these two reaction sequences: Either to pyruvate + H_2S + NH_3 by action of a desulfhydrase, or through oxidation to cysteic acid and subsequent decarboxylation to taurine:

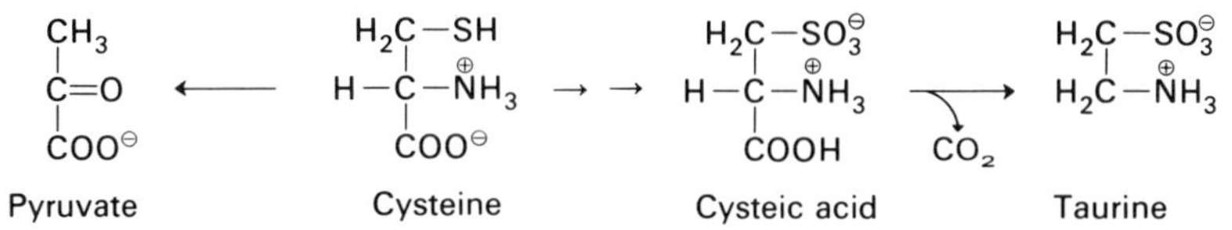

The oxidation of taurine involves several steps, which we will not discuss here. The disulfide cystine also can be oxidized in similar manner.

Serine and Glycine. Serine can be converted to glycine with the loss of active formaldehyde. This reaction is one of the most important suppliers of the C_1 fragment. Two coenzymes are necessary, tetrahydrofolate (Chapter VI,6) and pyridoxal phosphate. The elimination of the β-C atom is a pyridoxal-catalyzed reaction involving the resonance structure mentioned before in Section 4, but simultaneously serine is also bound to tetrahydrofolate. The reaction is reversible: serine can be formed from glycine and active formaldehyde.

Both amino acids can be metabolized through many other pathways as shown below by the diagram of formulas.

Serine can be converted to pyruvate (elimination of water forms the unsaturated amino acid which goes to the imino acid and then the α-keto acid; cf. breakdown of threonine, below) as well as hydroxypyruvate or phosphoserine. The phosphorylated form, which could come as well from 3-phosphoglycerate, occurs in a few proteins and phosphatides (Chapter XIII). Other bases present in the phosphatides, i.e., ethanolamine and choline, also originate from serine.

Glycine can provide an active C_1 fragment (active formate) by oxidation or transamination (both pathways seem possible) to glyoxylate and further by "oxidative decarboxylation" to activated formate. With high concentrations of glyoxylate, the body can produce oxalate, but this is probably not normal.

Glyoxylate can also come from sarcosine (*N*-methylglycine) by the action of an amino acid oxidase; the side product methylamine is oxidized to active formate during the reaction. Finally, the degradation of betaine is important. It is the oxidation product of choline and proceeds via dimethylglycine and monomethylglycine (sarcosine) to glycine. The methyl group becomes active formaldehyde, which then can join with glycine to form serine. The end effect is that sarcosine is converted to serine. Finally, to mention a few more uses for glycine, the biosyntheses of the blood pigment, of purines, and of creatine all require glycine and arginine (by amidino group transfer).

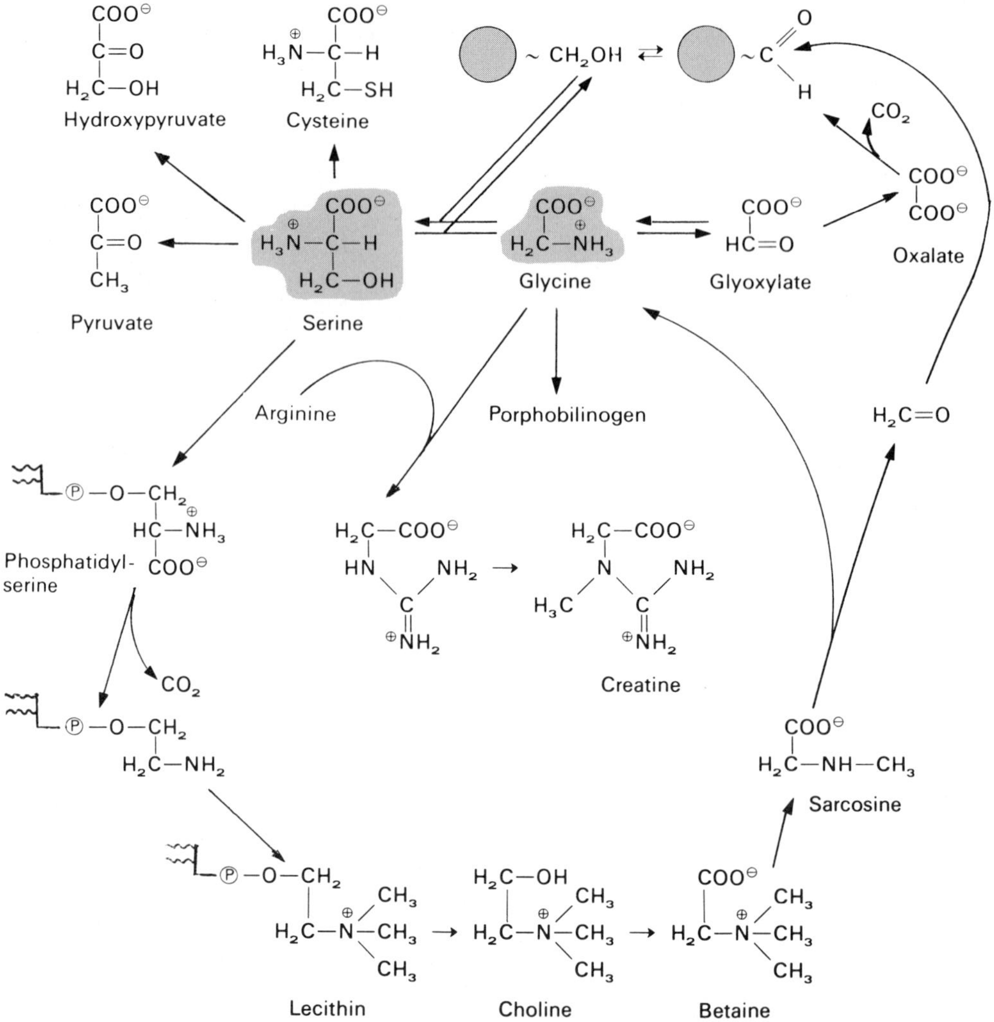

Threonine. As an essential amino acid, threonine is not involved in transamination. Its breakdown is related to the metabolism of glycine. Threonine is cleaved by the action of threonine aldolase to acetaldehyde and glycine. The chief degradative pathway probably is dehydration (removal of a molecule of water) catalyzed by "serine dehydratase." A comparison of K_m values of this enzyme indicates that in the cell the real substrate should be threonine and not serine. The reaction product is an enamine which undergoes rearrangement and hydrolysis to α-ketobutyrate:

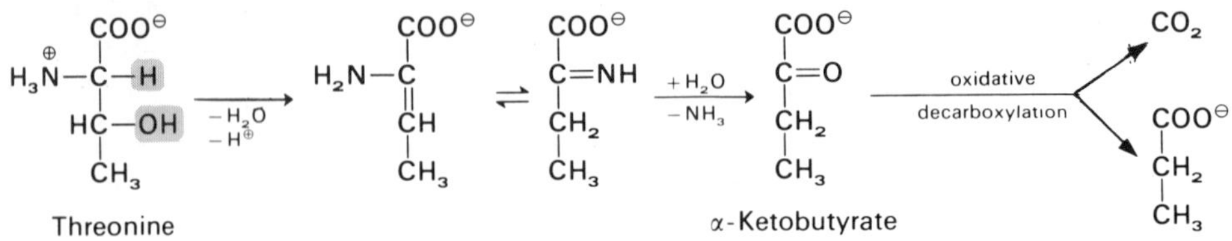

Another sequence of reaction results first in aminoacetone and then methylglyoxal and lactate:

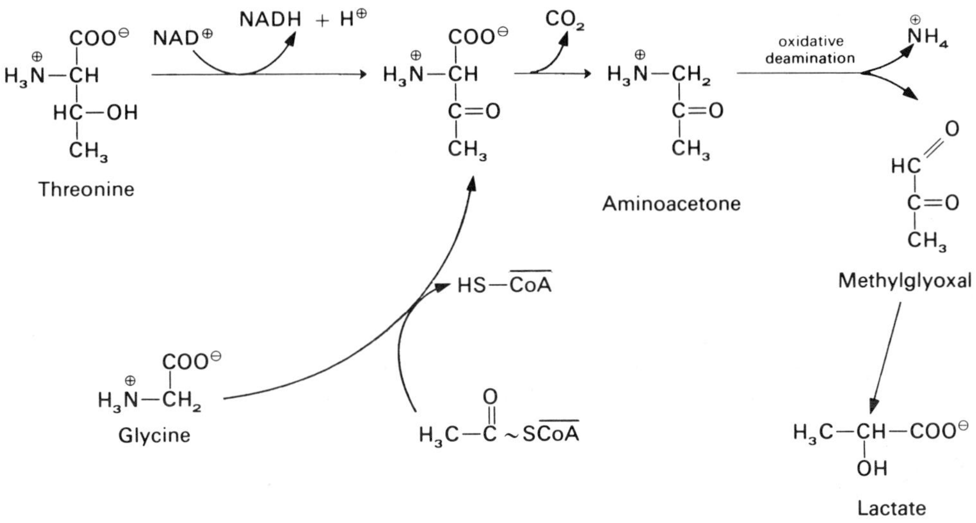

The diagram also shows the formation of aminoacetone from the condensation of glycine and acetyl-CoA, but the quantitative significance of this pathway is not known.

13. Ketoglutaric Acid or C_4-Dicarboxylic Acid-Supplying Amino Acids

The following belong under this heading: phenylalanine, tyrosine, (tryptophan), histidine, glutamate, aspartate, lysine, proline, arginine.

The degradation of phenylalanine and tyrosine has already been discussed. Besides fumarate, acetoacetate also occurs as a cleavage product, making these amino acids ketogenic. Tryptophan belongs to this group only to the extent that during its breakdown it passes through hydroxyanthranilate, the product of ring cleavage, to glutarate: hence, we place it in parentheses.

Aspartate. As we have seen, aspartate not only can transfer its amino group to keto acids, but also can supply one nitrogen directly to the urea cycle involving the formation of argininosuccinate and fumarate, which in the citrate cycle becomes malate by the addition of one water molecule. Malate, in turn, is dehydrogenated to oxaloacetate. The latter is also the transamination product of aspartate.

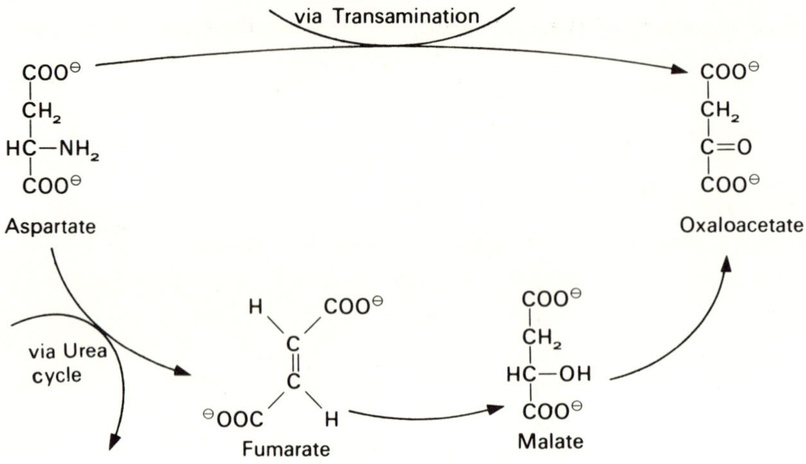

Oxaloacetate is the key intermediate both for gluconeogenesis, i.e., glucose formation (cf. Chapter XV), and for terminal oxidation. It can be thought to actually spark the citrate cycle. For this reason it will be discussed again in Chapter XI.

Lysine. An essential amino acid, lysine is metabolically relatively inert. With unusually large concentrations it can be broken down via glutarate to *β*-hydroxyglutarate and finally to *β*-ketoglutarate.

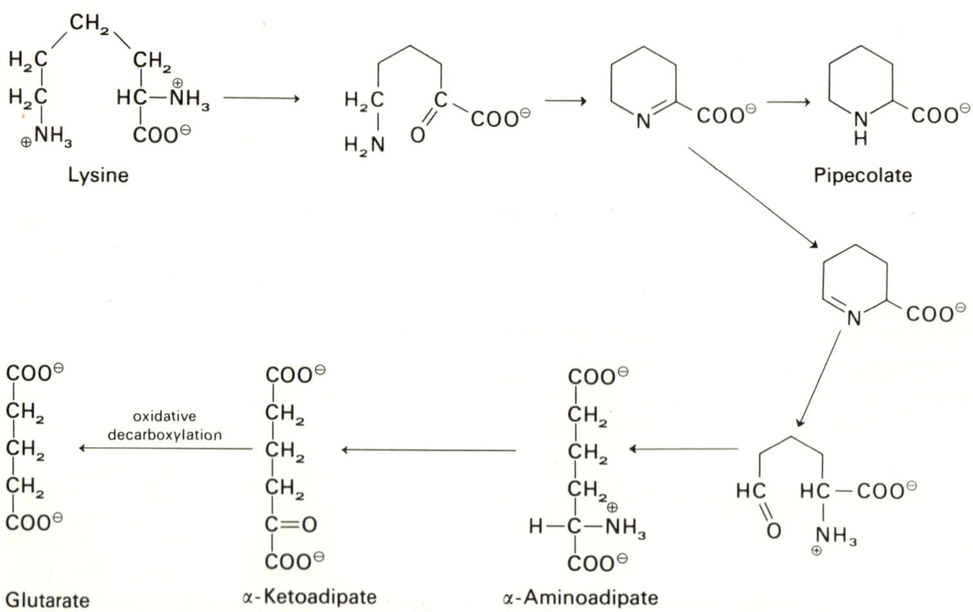

The reaction pathway (see diagram) goes through Δ^1-piperidine-2-carboxylic acid, which arises from the α-keto acid by ring closure (the two compounds are in equilibrium with each other). Δ^1-Piperidine-2-

carboxylic acid can be reduced to pipecolic acid (an extretion product in the rat); but it can also shift its double bond and open the ring by hydrolysis to give an α-amino δ-aldehyde acid which is easily converted to α-aminoadipate (aminoadipate is a rarely occurring amino acid found in proteins from corn, Chapter II,2).

The degradation of aminoadipate follows well-known steps; transamination to give the α-keto acid and oxidative decarboxylation to form glutarate, which is further degraded as the CoA derivative, just like a fatty acid. From the unsaturated compound β-hydroxy- and then β-ketoglutarate are formed; the latter in turn is split to acetyl-CoA and malonyl-CoA.

Histidine. First histidine is deaminated by the enzyme histidine-ammonia lyase with the insertion of a double bond between the α- and β-carbon atoms. The unsaturated product is called *urocanate* since it was initially found in dog urine. The 1,4-addition of water and shift of the double bond affords *imidazolone propionate.* This product then can be metabolized in two ways: by the action of a flavoprotein, it can be oxidized to *hydantoin propionate* and be excreted as such in the urine. The chief pathway, however, involves enzymatic hydrolysis of the imidazolone ring to form *formiminoglutamate.* The formimino group can be transferred as an "active C_1 fragment" to tetrahydrofolate. Glutamate thus emerges as the product of the breakdown of histidine.

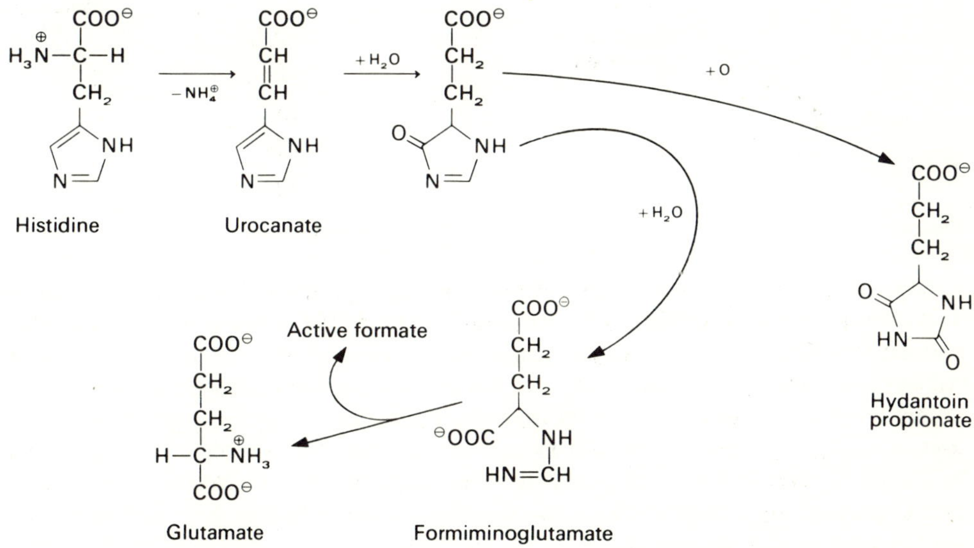

Arginine and Proline. These amino acids are also converted to glutamate. *Arginine* is cleaved by arginase into ornithine and urea (one of the reactions of the urea cycle, Section 8), and ornithine is attacked during transamination with pyridoxal phosphate primarily at the δ-amino group; it can be seen that the result is *glutamate semialdehyde* which is either oxidized to glutamate or cyclized with the elimination of water to form pyrroline carboxylate. *Proline* is produced by hydrogenation. These reactions are reversible, as shown by isotope experiments, so that the degradation of proline also goes through glutamate semialdehyde (see diagram below).

Proline is also parent substance of the collagen component *hydroxyproline* (cf. Chapter IV,4). The oxidation of proline takes place while it is part of a peptide chain through the action of a hydroxylase that requires *ascorbate* (vitamin C) as hydrogen donor.

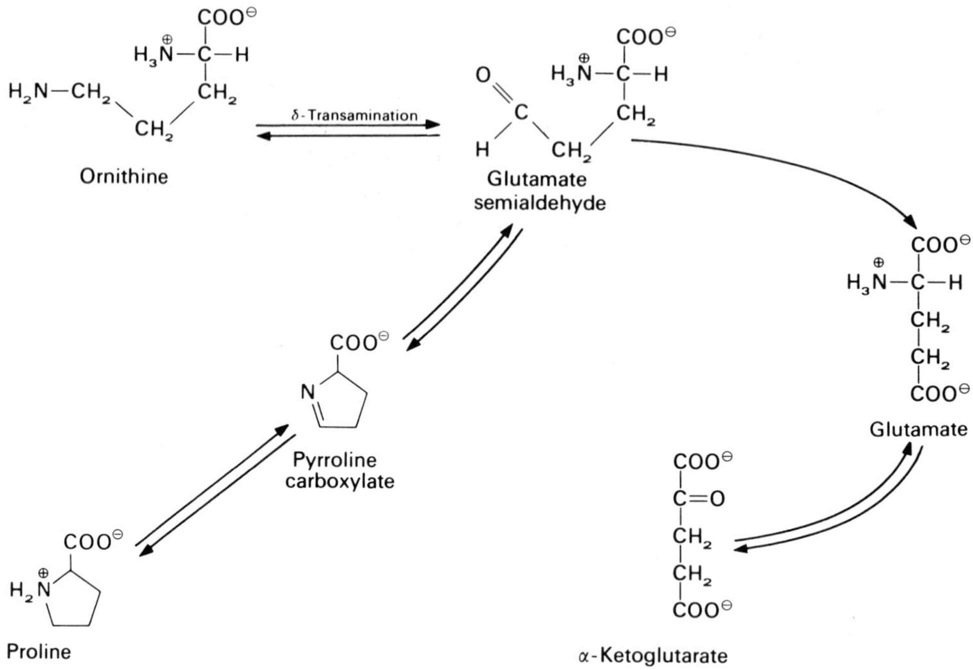

Glutamate. Its appearance as a central junction point in the network of amino acid metabolism, glutamate is dehydrogenated with NAD to the imino acid and then converted to ketoglutarate. The ammonia liberated is used directly for urea formation (via carbamyl phosphate; Section 8). On the one hand, α-ketoglutarate is the universal acceptor of amino groups in transamination reactions; on the other hand, as an α-keto acid it can also be decarboxylated oxidatively. In complete analogy to the oxidative decarboxylation of pyruvate, this reaction leads to activated succinate; this is part of the citrate cycle, to be discussed later (Chapter XI). The important feature here is that we now have a link to the terminal oxidation of all foodstuffs. Activated succinate can also condense with glycine to form a β-keto acid which decarboxylates spontaneously to give *δ-aminolevulinate.* This compound is the precursor for the biosynthesis of the blood pigment and cell hemins, as we shall discuss in the next chapter.

BIBLIOGRAPHY

D. M. Greenberg (editor), "Metabolic Pathways," 3rd ed., Vol. III, "Amino Acids and Tetrapyrroles." Academic Press, New York, 1969.

J. P. Greenstein and M. Winitz, "Chemistry of the Amino Acids," 3 vols. Wiley, New York, 1961.

A. Meister, "Biochemistry of Amino Acids," 2nd ed., 2 vols. Academic Press, New York, 1965.

CHAPTER IX

Porphyrins and Hemins

The porphyrin system is a complicated structure consisting of four pyrrole rings connected by methine groups (=CH—). It is the basis of the red blood pigment, the green leaf pigment, and of the cytochromes (redox enzymes). Efforts to determine its chemical structure occupied many decades (Küster, H. Fischer), and finally were crowned by the synthesis of the porphyrin molecule (H. Fischer). We will not follow the tortuous route of the research chemist here, but instead will construct the ring system from simple components, just as the cell does. Thus the biosynthesis will be discussed at the same time.

1. Biosynthesis of the Porphyrin System

Porphobilinogen. *δ-Aminolevulinate* has already been mentioned briefly (Chapter VIII,13) as an important precursor. It is formed from *succinyl-CoA* (activated succinate), which arises either from the metabolism of amino acids (glutamate →α-ketoglutarate →succinyl-CoA) or from the citrate cycle (cf. Chapter XI), and which condenses with glycine (with elimination of CoA). The enzyme instrumental in this condensation, called δ-aminolevulinate synthetase, contains pyridoxal phosphate as prosthetic group. Glycine is bound probably as a Schiff's base and then undergoes reaction with succinyl-CoA, yielding the very labile α-amino-β-ketoadipate, which immediately is decarboxylated to produce α-aminolevulinate. Condensation of two molecules of δ-aminolevulinate produces the pyrrole derivative porphobilinogen. The ring closure is catalyzed by an enzyme which has been purified from bovine liver.

Isotopic Labeling. The type of linkage was first studied through experiments with isotopes (^{15}N, ^{14}C). The technique of labeling or tagging plays a great role in modern biochemistry. The principle will be discussed with the example at hand.

It will be remembered that *isotopes* are atoms of different atomic weight, but with the same nuclear charge (and the same number of electrons). Chemically they are the same; physically they can be distinguished by their mass (e.g., ^{15}N versus ^{14}N) or by radioactive properties, such as the emission of β rays (e.g., ^{14}C).

An isotopic atom can be inserted into a molecule by chemical synthesis. Thus one can prepare, e.g., glycine, which has in its α-position the β-emitting carbon of atomic weight 14. The α-position of glycine is then said to be "labeled with ^{14}C." When one uses radioactive isotopes it suffices if only one out of ten thousand or a hundred thousand molecules contains the isotope, although with tritium (^3H) much higher *specific activities* are often used.

If one introduces glycine labeled in this way into biochemical experiments, the pathway of the glycine can be followed readily. All substances subsequently produced from this glycine in enzymatic reactions or in the intact organism must still contain the isotope and can be recognized by their radioactivity.

If such a glycine is used in our experiment, we find radioactivity in porphobilinogen; by stepwise chemical degradation it can be seen that the radioactivity is localized in the atoms marked with a solid square in the formulas below. If we use glycine that is labeled in the carboxyl group, the porphobilinogen will not be radioactive. This is proof that the carboxyl group is lost during biosynthesis. If, on the other hand, the carboxyl groups of succinic acid are labeled (indicated by a solid triangle), then radioactive ^{14}C is found partly in the side chain and partly in the pyrrole ring in the positions marked by the triangle. Suitable chemical degradation procedures permit this conclusion. The path of a given carbon atom can be traced in this way to the blood pigment. For example, it has been established that the methine groups stem from glycine, whereas all the carboxyl groups derive from succinate.

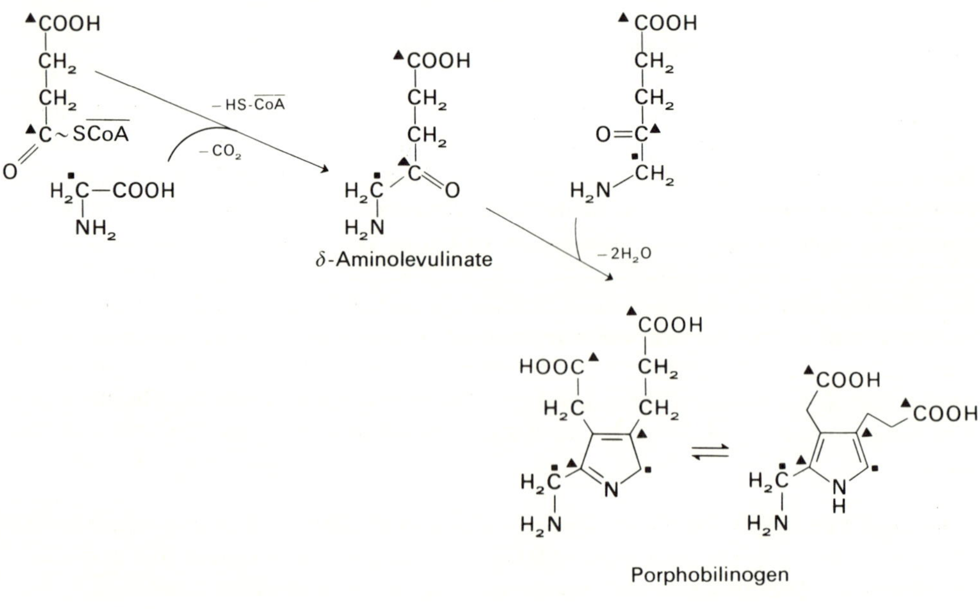

Formation of the Porphyrin System. By arranging four porphobilinogen molecules in a suitable way (see formulas below) and linking them, with elimination of ammonia, one obtains one of the possible porphyrin systems, namely *uroporphyrin I*. Uroporphyrin I, however, tends to be a pathological product and is the result of a "derailed" porphyrin synthesis. The actual precursor of the physiological substances hemoglobin and cytochrome is *uroporphyrin III*. In type III, the substituents in positions 7 and 8 are interchanged as compared to type I. Why nature prefers the isomers of type III, we do not know. In general many a structure or reaction must be looked upon simply as a fact rather than as being causally determined. One must imagine that this structure or that path of synthesis arose many millions of years ago and has survived through the generations.

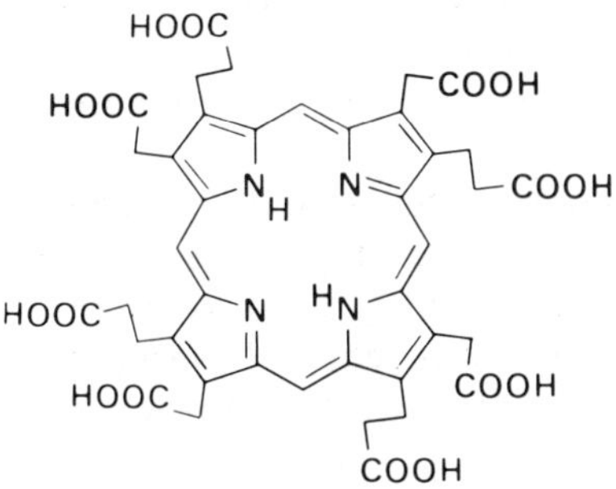

Uroporphyrin I
pathological

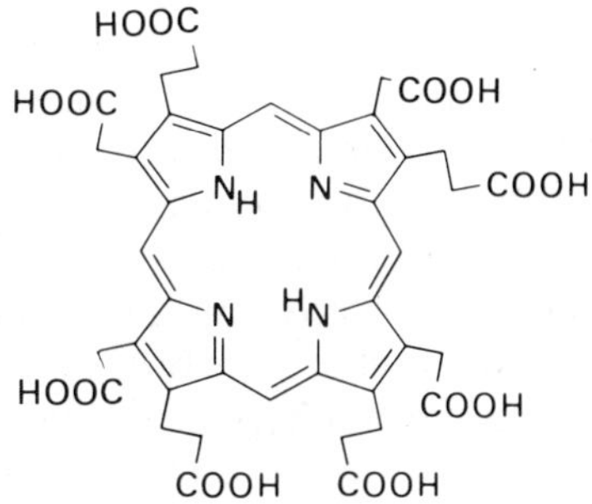

Uroporphyrin III
(precursor to heme)

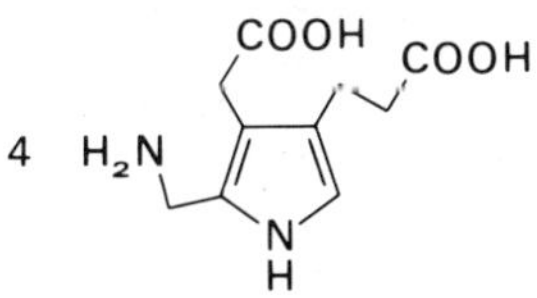

4 Porphobilinogens

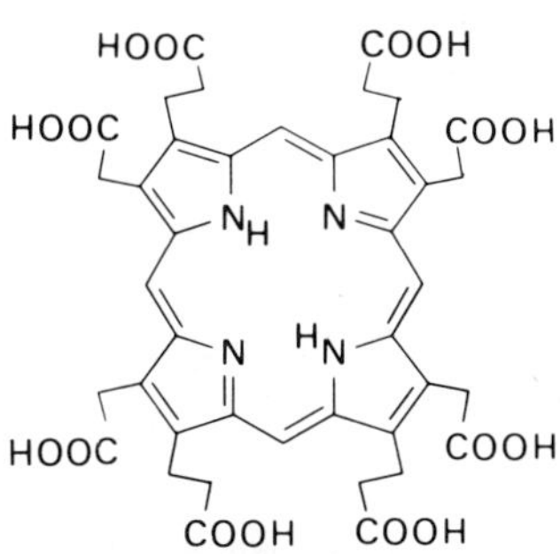

Uroporphyrin II
(available only synthetically)

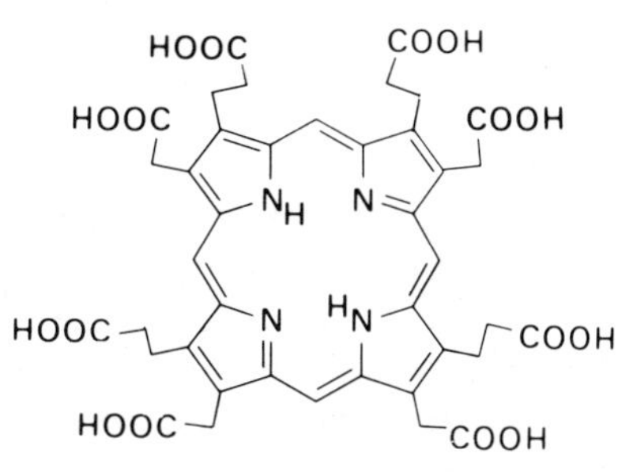

Uroporphyrin IV
(available only synthetically)

For further clarity, mention must be made about the possibilities of isomerism in the porphyrin system. The positions on the pyrrole rings bearing substituents are numbered consecutively from 1 to 8. If each pyrrole ring is to have one short and one long side chain (one acetic acid $-CH_2COOH$, and one propionic acid $-CH_2-CH_2-COOH$), then there are four possibilities, designated as types I–IV (cf. diagram of formulas).

In type I there is a regular sequence of short and long side chains all around the porphyrin. In the isomer important to us, type III, the fourth ring appears to be turned around or to be condensed in an inverted manner. Types II and IV have been synthesized but lack biological significance.

Complete decarboxylation of all side chains converts each uroporphyrin to the corresponding *etioporphyrin*, also characterized by short ($-CH_3$) and long ($-CH_2-CH_3$) side chains. Many textbooks derive types I–IV from etioporphyrins. We have discussed the isomeric forms using uroporphyrins as examples because these are the naturally occurring substances.

Two enzyme systems are involved in the biosynthesis of uroporphyrin III from porphobilinogen: a deaminase and an isomerase. The initially produced colorless *porphyrinogen* is dehydrogenated to give the porphyrin system. The deaminase acting alone catalyzes the formation of uroporphyrin I; but the isomerase suppresses the formation of type I in favor of type III.

Much discussion has centered around the problem of how the particular arrangement of the side chains realized in type III can possibly arise during synthesis. The problem is really how a ring can be turned around. Experimentally it was found that no side products are formed and that no further C_1 fragment is required as methine group. Several hypotheses have been developed to explain the sequence of reactions. Commonly an intermediate is postulated in which one pyrrole ring (I or IV) is attached by only one corner.

The following scheme of reactions (Fig. IX-1) elucidates one of the hypotheses in detail. Accordingly a deaminase catalyzes the condensation of four porphobilinogen molecules to form a tetrapyrrole derivative that can fold with its particular arrangement of double bonds (shown in Fig. IX-1) such that the group $-CH_2NH_3^{\oplus}$ of the last ring comes to rest on top of the first ring (later to be ring IV). This spatial arrangement favors ring closure with the α'position of the first ring. In the resulting ring system one pyrrole participates only with one corner.[1] This intermediate undergoes ring cleavage at the position marked; the ring rotates around the CH_2-C(pyrrole) axis and condenses with the free α-position to form *uroporphyrinogen III*. This second reaction is ascribed to the isomerase. The rotation of ring IV results in the "correct" substitution, naturally occurring in hemoglobin.

Another hypothesis has the deaminase synthesize a linearly linked tripyrrole derivative (rings I–III) and the isomerase condense ring IV in the reverse order. Here, too, initially one corner of ring IV is attached by two CH_2 groups, and the porphyrin structure comes about from the migration of one of these methine groups.

The first product of the enzymatic synthesis is *uroporphyrinogen*. Dehydrogenation (removal of 6 H) readily affords *uroporphyrin III*. In the hereditary disease porphyria synthesis takes place on a much larger scale and excess uroporphyrin of types I and III appears in the urine. The appearance of uroporphyrin I should be explained by a relative deficiency of the isomerase.

[1] A similar mechanism can be proposed to explain the formation of cobalamin (vitamin B_{12}; formula in Chapter VI,9): Even more extensive folding of the linear tetrapyrrole permits ring closure between the pyrrole rings themselves and not through the methine bridges. An analogous isomerization yields a stable precursor of cobalamin.

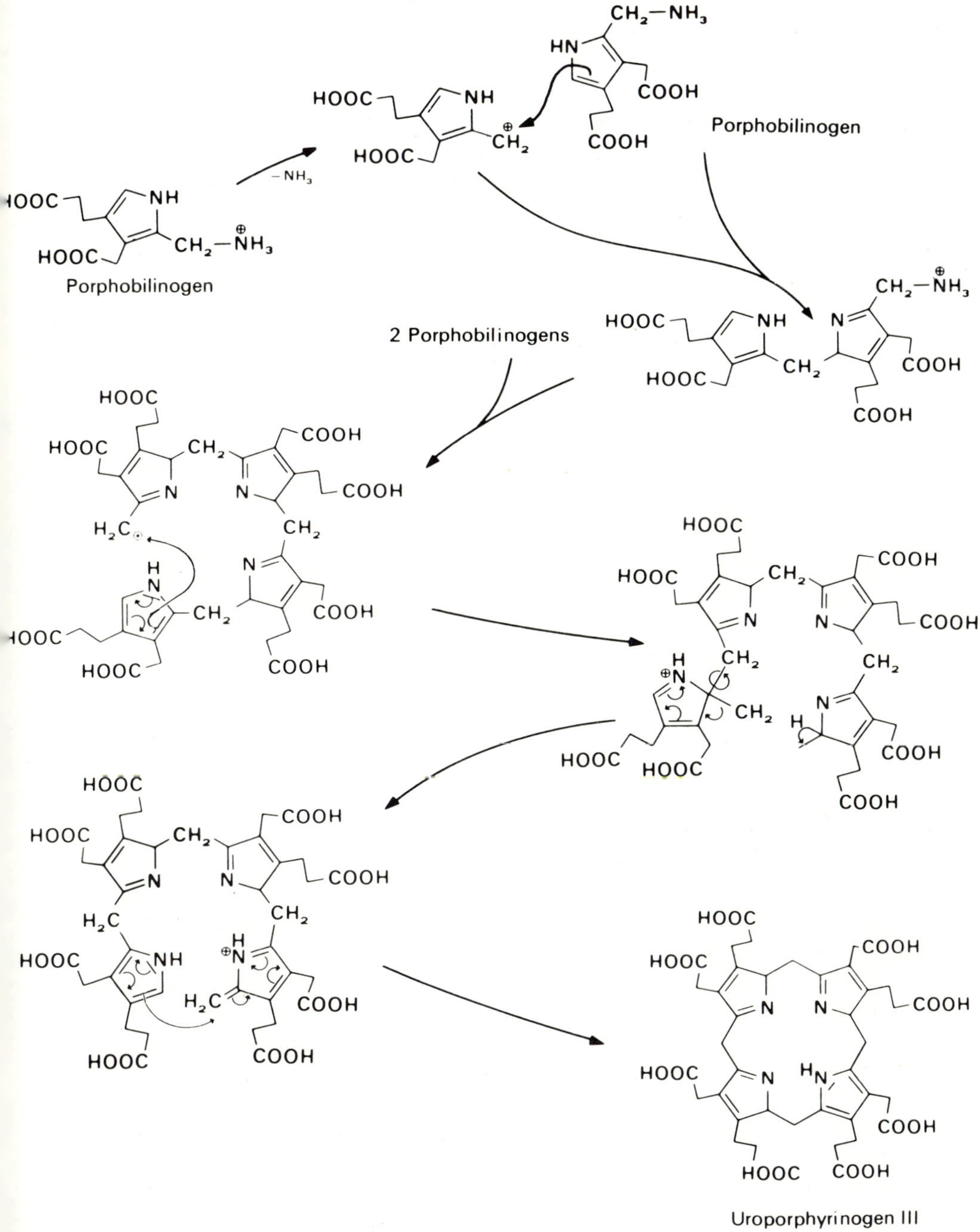

Fig. IX-1. Biosynthesis of porphyrin of type III. The small arrows denote the shift of electron pairs. See text for further explanation.

The Pathway from Uroporphyrinogen III to Heme. The pigment of blood, is no longer difficult to describe, at least on paper. First, the acetic acid side chains are shortened to methyl groups by the stepwise loss of 4 moles of CO_2. The resulting compound is called *coproporphyrin III* and is usually present in small amounts in feces. In some diseases and in congenital anomalies it may appear in greatly increased amounts. (In that case, there are also found appreciable amounts of coproporphyrin of type I.) Next, the propionic acid side chains on rings I and II must be converted to vinyl groups $-CH{=}CH_2$ by dehydrogenation and decarboxylation, yielding protoporphyrinogen. Evidently the dehydrogenation to the porphyrin system takes place only after the ferrous ion has been inserted by enzymatic catalysis (*ferrochelatase*). The nonplanar construction of the porphyrinogen facilitates the insertion of the central atom. From avian erythrocytes extracts have been obtained that catalyze the entire sequence of ractions from δ-aminolevulinate to the heme.

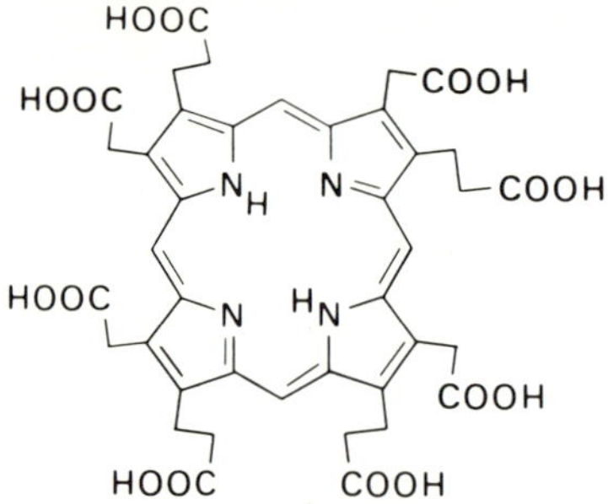

Uroporphyrin III

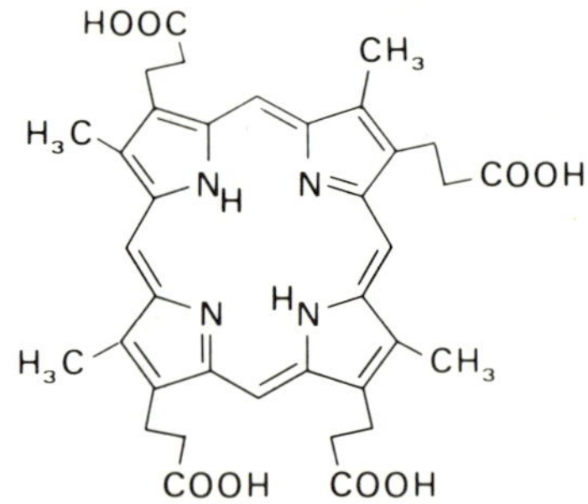

Coproporphyrin III

2. Chemical Constitution of Heme

Heme, the colored component of hemoglobin, is ferroprotoporphyrin with the formula shown below. The ring system as such (called porphine) contains nine double bonds in uninterrupted conjugation (plus two off to the side) and in the language of organic chemistry has aromatic character. The formula can be written equally well with the primary valence bonds of iron going to rings II and IV instead of rings I and III. As a resonating structure, porphine is planar. In heme, only four of the coordination valence bonds of iron (ferrous) are occupied by the porphine skeleton; the two remaining ones, lying above and below the plane of the porphine ring, are both attached to the biochemically important protein (globin) by components of the protein, usually the imidazole ring of histidine (see diagram in Section 4). Bivalent iron of heme can be oxidized to the trivalent stage; the $Fe^{3\oplus}$-porphyrin is called *hemin*.

The type and sequence of substituents on the porphine ring is determined during its biosynthesis: The propionic acid side chains of rings III and IV have been retained, but those of rings I and II have been converted to vinyl groups. In a few cytochromes the side chains are converted further by a few more steps.

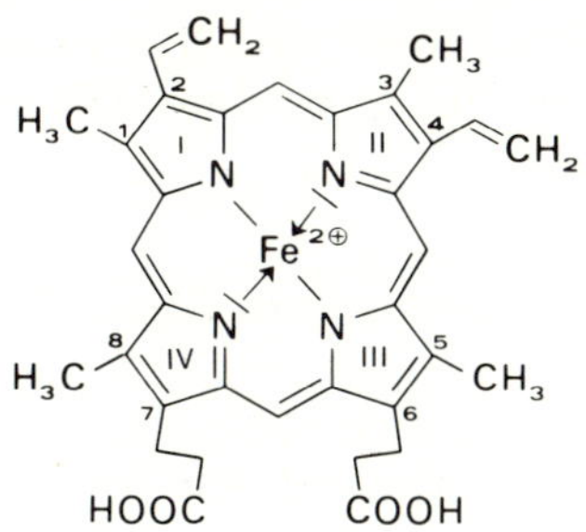

Heme

3. Multiplicity of Porphyrin Catalysis

The ferrous porphyrin molecule, called heme, and its close relatives are capable of catalyzing very many different reactions. As is generally known, hemoglobin transports oxygen in blood (this role is described in detail below). Iron does *not* change its valence in the process. A number of heme enzymes—catalases and peroxidases—interact with hydrogen peroxide H_2O_2, whose oxygen is transferred either to a substrate to be oxidized (peroxidase) or to H_2O_2 which becomes dehydrogenated to O_2 (catalase action, decomposition of H_2O_2). Other heme proteins, the cytochromes, participate in biological oxidation; they transport electrons (reduction equivalents) from suitable substrates all the way to oxygen. The type of reaction depends on the type of protein, which has a profound influence on the reactivity of the heme molecule.

TABLE IX-1
Function of Several Porphyrin Proteins

Protein	Function	Prosthetic group	Valence change[a]
Hemoglobin	O_2 transport	Heme ($Fe^{2\oplus}$-protoporphyrin)	−
Plant peroxidase	Oxidation by means of H_2O_2	Hemin ($Fe^{3\oplus}$-protoporphyrin)	+
Animal peroxidase	Oxidation by means of H_2O_2	Green hematin (structure unknown)	?
Catalase	Decomposition of $2\,H_2O_2 \rightarrow 2\,H_2O + O_2$	Hemin ($Fe^{3\oplus}$-protoporphyrin)	−
Cytochrome oxidase (cytochrome *a*)	Terminal oxidation (respiratory chain)	Cytohemin	+
Cytochrome *c*	Electron transport	Hemin; primary valence bond to the protein	+
Cytochrome *b*	Electron transport	Hemin	+
Oxygenase	Inserting O_2 into substrate	$Fe^{2\oplus}$-porphyrin	−
Chlorophyll	Conversion of light energy to chemical energy	Mg-containing porphyrin	−

[a] +, change; −, no change; ?, unknown.

Again we recognize that reaction specificity of an enzyme is not determined by the prosthetic group, but rather by the apoenzyme. The various possibilities are listed in Table IX-1.

4. Significance and Reactions of the Blood Pigment

The blood pigment *hemoglobin* (often abbreviated as Hb) is a combination of the protein globin with heme, the prosthetic group.

Treatment with glacial acetic acid easily removes the protein component, and the prosthetic group can be isolated, e.g., in the form of the easily crystallized *chlorohemin* (in hemin iron is trivalent, in heme it is bivalent). These Teichmann's crystals are excellent for identifying blood.

Hemochromogens are heme compounds with two nitrogen bases, such as pyridine and ammonia, occupying coordination places 5 and 6. The protein moiety is denatured under the conditions of hemochromogen formation.

Hemoglobin has a molecular weight of 67,000 and is an aggregate of four peptide chains. Each chain contains a heme group, and there are two identical pairs (cf. Fig. IV-7). The subunits are held together by secondary valences.

Man carries the genetic information for four different chains designated α, β, γ, and δ. The main hemoglobin of the adult is Hb A_1 comprising 97.5% of the total. Its composition is according to the formula $\alpha_2\beta_2$. The remaining 2.5% is Hb A_2 ($\alpha_2\delta_2$). The fetus has Hb F ($\alpha_2\gamma_2$); the infant a mixture of Hb A_1 and Hb F. Besides these normal hemoglobins numerous genetic variants have been identified, some of which are the cause for serious anemias (e.g., sickle-cell anemia, Chapters IV,2 and VII,7).

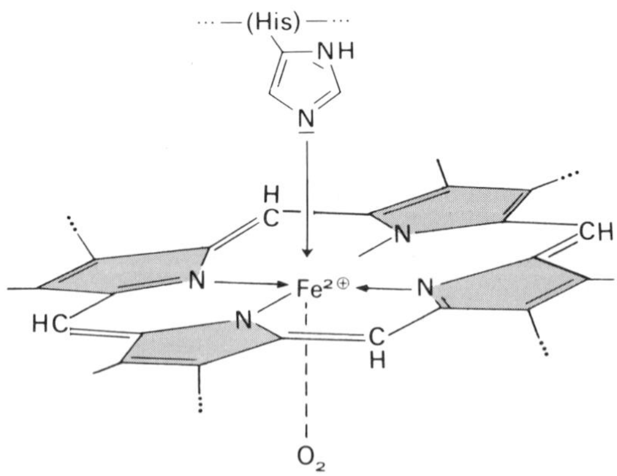

In hemoglobin (as in many other hemoproteins) one histidine residue is near the iron atom. Histidine occupies the fifth coordination position of the iron (cf. also spatial structure of hemoglobin, Chapter IV,5). Oxygen is held at the sixth place:

it is released easily. The hypothesis that a molecule of water replaces the O_2 has, however, not been borne out. During the transport of O_2, iron always remains bivalent. The amount of bound O_2, i.e., the ratio of $Hb \cdot O_2$ to Hb, depends on the concentration of O_2, in other words, on the partial pressure of oxygen. In oxygen-deficient tissues oxygen is released. This reversible binding of oxygen explains the transport phenomenon.

The curve of oxygen saturation (Fig. IX-2) is S-shaped and not—as expected—hyperbolic. The explanation lies with the cooperative action of the four chains of hemoglobin. When only one chain is charged with oxygen the conformation of the protein molecule is changed by allosteric action (cf. Chapter IV, 5) such that the three subsequent O_2 molecules are bound more tightly. This condition ensures that the hemoglobin molecule is loaded with four O_2 in the lungs, since the affinity for oxygen of partially saturated hemoglobin is greater than that of oxygen-free hemoglobin. Myoglobin in contrast, which consists only of one peptide chain and one heme group, has the expected hyperbolic form of the saturation curve (Fig. IX-2).

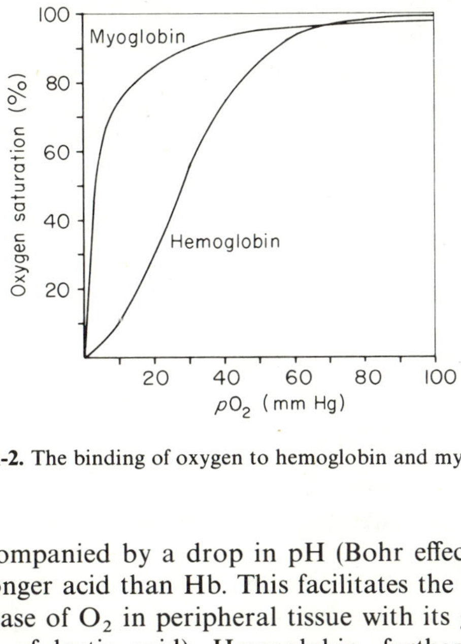

Fig. IX-2. The binding of oxygen to hemoglobin and myoglobin.

Oxygenation is accompanied by a drop in pH (Bohr effect; cf. Chapter XXI,4), since $Hb \cdot O_2$ is a stronger acid than Hb. This facilitates the elimination of CO_2 in the lungs and the release of O_2 in peripheral tissue with its greater acidity (due to anaerobic production of lactic acid). Hemoglobin, furthermore, binds CO_2 as carbamic acid (at the N-terminal amino group) and thus participates in the transport of CO_2.

Substances other than the O_2 molecule also can be bound to the coordination places, e.g., carbon monoxide, CO. The latter is bound much more firmly than oxygen; even minute amounts in the atmosphere saturate hemoglobin with CO and thus prevent oxygen transport. Since we are dealing with an equilibrium reaction, we can slowly displace the CO with copious supplies of oxygen.

Should iron become oxidized to the trivalent (ferric) level, then the new positive charge is balanced by an anion in ionic linkage. $Hb(Fe^{3\oplus})OH^{\ominus}$ is called *methemoglobin*; it cannot transport oxygen. An enzyme found in blood cells can reduce methemoglobin back to hemoglobin.

5. Degradation of Blood Pigment

The life span of erythrocytes is limited to around 4 months. The hemoglobin from dying erythrocytes is broken down in the reticuloendothelial system. At first hydroxylation and the loss of a formaldehyde group yields the green pigment *choleglobin* or *verdoglobin*, which still retains globin and $Fe^{3\oplus}$, but whose porphine ring is opened between pyrrole rings I and II. One methine carbon is lost; it is replaced with hydroxyl groups in rings I and II. Both iron and the protein component are removed easily from verdoglobin, and understandably so, because the porphine ring system no longer exists. The result is a straight, quadrinuclear pyrrole pigment, *biliverdin*. It is reduced easily to a red substance, *bilirubin*. The latter is always found in small quantities in blood, bound to albumin.

The liver traps all the bilirubin. The greater portion is "conjugated" with UDP-glucuronate to form the glucuronide and ends up in the gall bladder. Bilirubin glucuronide is the main bile pigment. With diazotized sulfanilic acid the glucuronide immediately gives a red color (so-called direct diazo reaction); the free bilirubin, only after addition of alcohol (indirect reaction).

In the gut, the bile pigments are further transformed. Vinyl groups are reduced to ethyl groups. The result is *mesobilirubin*, which through further reduction becomes *mesobilirubinogen* (= *urobilinogen*) and *stercobilinogen*. In these pigments whose name ends in *-ogen* all the pyrrole rings are connected by methylene groups ($-CH_2-$). Dehydrogenation regenerates the pigments *urobilin* and *stercobilin* in both of which the middle rings III and IV are restored to conjugation by the new double bond. For the sake of clarity the added hydrogen atoms are marked by an asterisk in the diagram of formulas.

Part of the bile pigments are reabsorbed in the intestine and returned to the liver in the portal vein blood (enterohepatic circulation). The greater part, however, is excreted by the intestine.

Stercobilin, urobilin, and further degradation products with but two pyrrole nuclei (mesobilifuscin) are the pigments of feces. They are produced largely by the intestinal flora. Whenever the food pulp passes through the gut rapidly, the feces have the yellow color of unchanged bilirubin.

In various liver diseases (jaundice, icterus) excess bilirubin is produced and at the same time the permeability of liver cells to the pigment is increased, so that bilirubin glucuronide ("direct bilirubin") and bilirubin itself ("indirect bilirubin") pass into blood and then diffuse into skin tissue. Obstruction of the bile duct can cause the same symptoms. The determination of bilirubin, therefore, is important clinically.

Bile pigments have been found in various places in the animal kingdom: in egg shells of birds; among the pigments of wings and skin of insects; and finally in sea algae, where they are bound to the C-terminal group of a peptide chain.

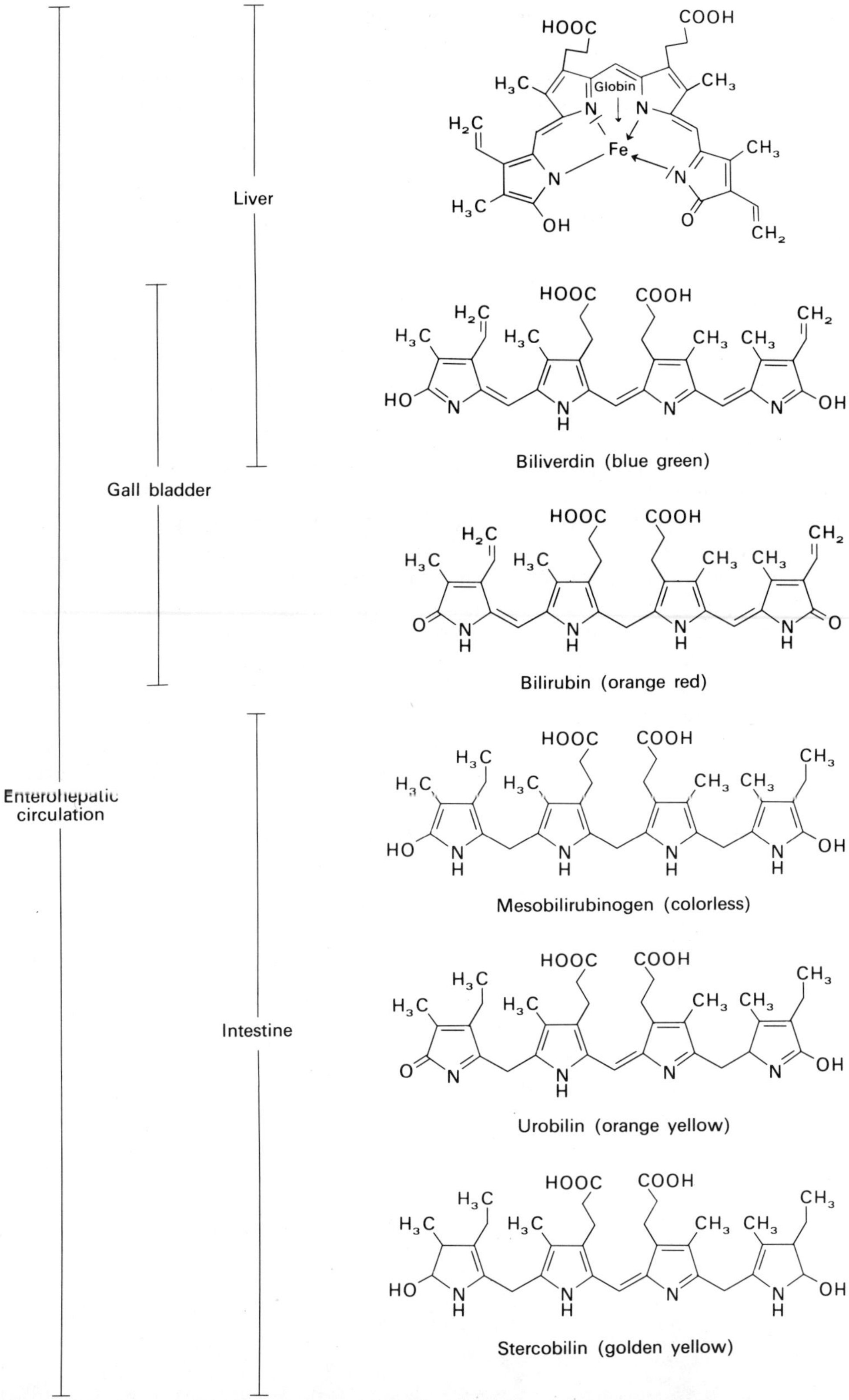
Liver
Gall bladder
Intestine
Enterohepatic circulation
Globin
Fe
Biliverdin (blue green)
Bilirubin (orange red)
Mesobilirubinogen (colorless)
Urobilin (orange yellow)
Stercobilin (golden yellow)

Phytochrome. The prosthetic group of phytochrome is closely related to biliverdin. This chromoprotein effectuates numerous light-dependent developmental processes in plants, e.g., photoperiodism, germination of seeds, and the prevention of etiolation. Phytochrome, which absorbs at 660 nm, is isomerized by bright red light to pigment P_{730}; dark red light reconverts P_{730} back to phytochrome. Consequently, it is possible to cancel the physiological effect of bright red light with dark red light.

6. Cytochromes, Catalases, and Peroxidases. Chlorophyll

Cytochromes. Discovered because of their light absorption, cytochromes occur in nearly all cells, usually bound to mitochondria or similar structures. They are the catalysts of cellular respiration (function: change in valence of iron, cf. Chapter X,4). *Cytochromes a, b,* and *c* were distinguished first on the basis of absorption spectra; later, these groups had to be subdivided further by adding subscripts. Today the cytochrome *a* group comprises all cytochromes whose prosthetic group is *hemin a* (= cytochemin). Cytochromes *b* contain ferroprotoporphyrin (just as in hemoglobin), and in cytochromes of the *c* group the porphyrin residue with its hydrogenated side chain is bound to the protein by primary valences.

The most thoroughly studied is *cytochrome c*. It is a hemoprotein with a molecular weight of 12,000 and has one heme group per molecule. The amino acid sequence and the three-dimensional structure are now known. There are primary valence bonds between the protein and the porphine ring: The SH groups from two cysteine side chains have added across the vinyl groups resulting in very stable thiolether bonds. The neighboring histidyl residue (His-18) occupies the fifth coordination position of iron; sulfur of methionine in position 80 occupies the sixth position (see diagram).

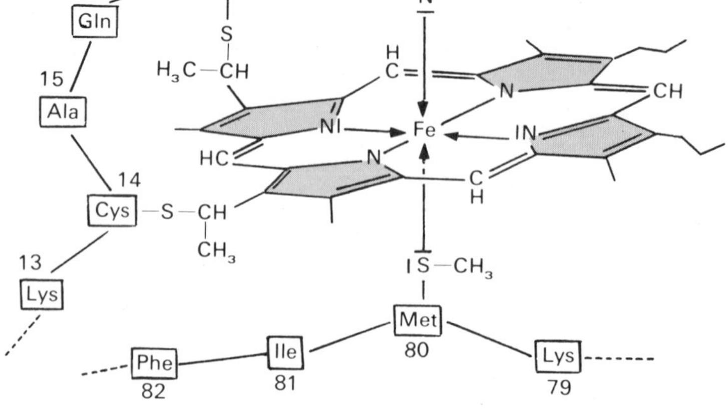

Since the coordination places are all occupied, the heme group is incapable of reacting with O_2, $CN^{\ominus}$, or CO; it thus cannot be "poisoned" with the usual inhibitors.

Cytochrome c_1 is closely related to cytochrome *c*, but has a molecular weight of 37,000. It is a component of the respiratory chain (Chapter X,5).

Cytochrome b has a molecular weight of about 30,000 and heme as prosthetic group. It is found in mitochondria, where it is part of the respiratory chain, presumably inserted between ubiquinone and cytochrome *c*, as well as in the succinate oxidation chain.

Cytochrome b_5 is closely related to the better known cytochrome *b* and has the same prosthetic group. Cytochrome b_5 is a characteristic component of microsomes, where it probably participates in electron transport or in hydroxylation reactions.

Cytochrome a and all other cytochromes of type *a* contain the prosthetic group heme *a* or cytoheme, which features a formyl group in ring IV with an accompanying shift of the absorption maximum to a longer wave length. Ring I additionally bears a lipophilic side chain biosynthetically derived from a condensation with farnesyl pyrophosphate (Lynen). Under some conditions the OH group can add across the double bond nearby with the formation of a furan or pyran ring.

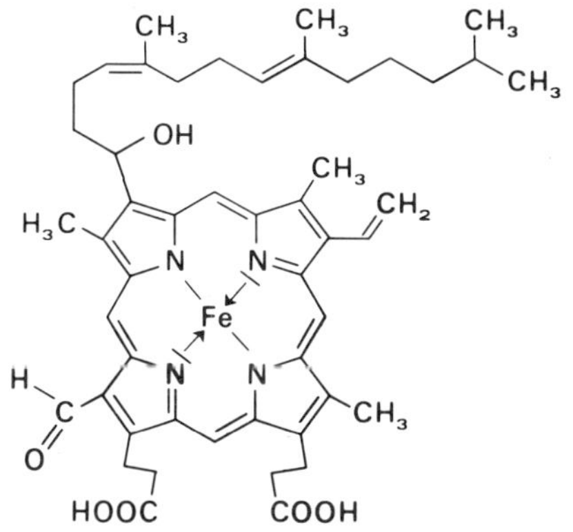

Cytoheme

The most prominent example of cytochromes of this type is Warburg's "Atmungsferment," now called *cytochrome oxidase*. It occurs in mitochondria of all animal tissue. Spectroscopic observations and the reaction of O_2 and CO gave rise to the distinction between cytochrome *a* and a_3. The actual separation of the two components, however, has never been achieved. The smallest isolable unit has been a hemo-lipo-cupro-protein with molecular weight of 150,000 to 200,000. It contains two copper atoms and two heme *a* groups, of which one behaves as cytochrome *a* and the other as a_3. It is as yet unclear whether the two heme groups are bound differently to one protein molecule or whether there is a stable dimer with two different subunits. The complex, which easily aggregates to oligomeres with four or six heme groups, has also been called cytochrome aa_3. In the respiratory chain it functions as cytochrome *c* oxidase (cf. Chapter X,4).

Catalase. This is an enzyme that breaks H_2O_2 into H_2O and O_2. Its active group is protohemin. It has an unusually high activity (in terms of turnover number). The molecular weight of the crystalline protein comes to 240,000. There are four heme groups in each molecule.

Peroxidases. These enzymes oxidize substrates by employing H_2O_2 as an oxidizing agent. Depending on their origin, they may contain either a red heme (protoheme or closely related ones) or a green heme; if they contain the latter they may be called *verdoperoxidases.* One red enzyme from milk has a molecular weight of 82,000. Horseradish is particularly rich in peroxidase, which has been crystallized (1 protoheme per molecule). A verdoperoxidase is found in pus; molecular weight 150,000.

Chlorophyll. This pigment of green leaves is a magnesium-containing porphyrin. The main features of the structure are the two "supernumerary" hydrogen atoms in ring IV, the isocyclic ring, and the phytyl side chain imparting lipid solubility to the whole molecule. Almost all plants have the two closely related pigments chlorophylls a and b, differing only in the side chain of ring II as shown in the partial formula:

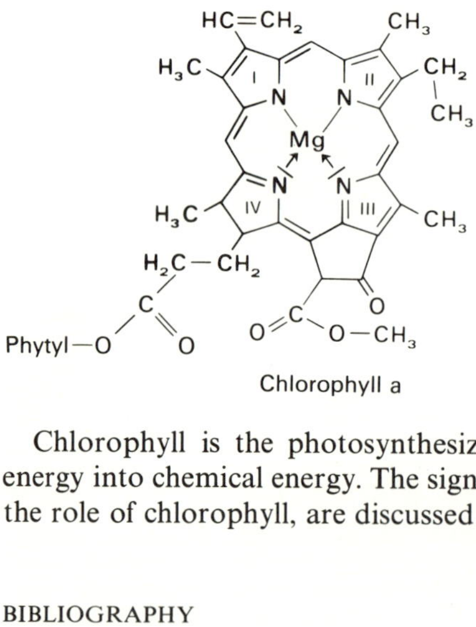

Chlorophyll a

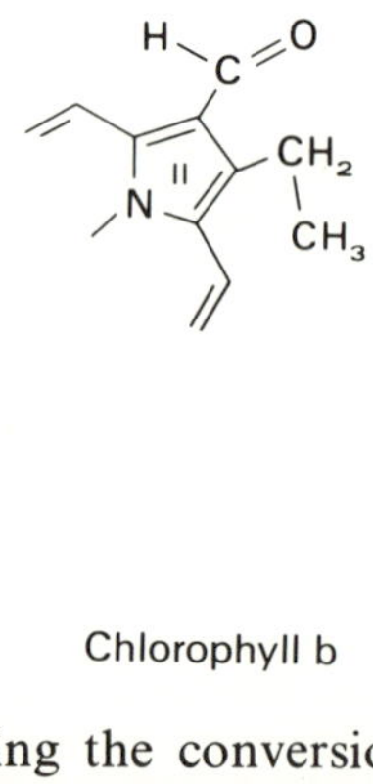

Chlorophyll b

Chlorophyll is the photosynthesizing pigment effecting the conversion of light energy into chemical energy. The significance and mechanism of photosynthesis, and the role of chlorophyll, are discussed in detail in Chapter XVI.

BIBLIOGRAPHY

W. S. Caughey, Porphyrin proteins and enzymes. *Annu. Rev. Biochem.* **36,** 611–644 (1967); contains 383 references.

B. Chance, T. Yonetani, and A. S. Mildvan (editors), "Probes of Structures and Function of Macromolecules and Membranes," Vol. 2: Probes of Enzymes and Hemoproteins, Part 2: Structural Interactions in Ligand Binding in Hemoproteins. Academic Press, New York, 1971.

T. W. Goodwin (editor), "Porphyrins and Related Compounds." Academic Press, New York, 1969.

CHAPTER **X**

Biological Oxidation—Metabolism of Oxygen

1. Combustion and Biological Oxidation

About 1780, Lavoisier concluded that combustion processes must also take place in the animal organism. Since then, biological oxidation has often been compared to combustion. There is indeed no difference in the net equations; the end products in both cases are CO_2 and H_2O, and the energy released during combustion and biological oxidation is equal, at least quantitatively. For example:

Glucose	$C_6H_{12}O_6 + 6\ O_2 = 6\ CO_2 + 6\ H_2O$	$\Delta H = -674$ kcal/mole
Fatty acid	$C_{16}H_{32}O_2 + 23\ O_2 = 16\ CO_2 + 16\ H_2O$	$\Delta H = -2379$ kcal/mole

(ΔH = heat of combustion at constant pressure)

The analogy, however, cannot be extended much beyond this. While in technology the oxidation of carbon to CO_2 is still the most important source of energy, the same process assumes a rather subordinate role in biochemistry. Furthermore, one characteristic feature of combustion processes is a drastic rise in temperature and unchecked evolution of heat. In the mammalian body, in contrast, all processes proceed at a constant temperature (around 37°C) and only part of the energy of oxidation appears as heat; the remainder is conserved as chemical energy.

The principles of biological "combustion" of foodstuffs (carbohydrates, fats, and to some extent proteins) may be summarized in four sentences:

1. Complex organic molecules are first broken down to two-carbon fragments[1] (activated acetate).

2. The further breakdown of the C_2 fragments occurs in a series of separate steps, in each of which one CO_2 or two H atoms are split off; or, the molecule is altered so as to prepare for such a step.

---

[1] In carbohydrates this step is reached via decarboxylation. A C_6 compound yields two C_2 fragments and two CO_2 (Chapter XV, 9).

3. The end product CO_2 arises by the decarboxylation of organic acids *without any considerable change in energy.*

4. The end product H_2O arises from reduced coenzymes of the respiratory chain and atmospheric oxygen with concomitant storage of some of the energy produced in the form of an energy-rich compound, adenosine triphosphate.

The most important CO_2-yielding processes will be treated in the subsequent chapter. Here, however, we will discuss the formation of water (point 4). This not only is the chief energy-producing process, but it also represents the main pathway for utilizing respiratory oxygen and requires the participation of cytochromes as mentioned in Chapter IX. In addition, other reactions involve the direct insertion of oxygen into organic molecules. Enzymes catalyzing such reactions are called oxygenases.

2. Oxidation as Loss of Electrons

Oxidation is defined today very generally as a loss of electrons ($e^{\ominus}$). Oxidation of molecular hydrogen can therefore be formulated as follows:

$$H_2 - 2\,e^{\ominus} \rightleftarrows 2\,H^{\oplus} \tag{1}$$

The electrons must be accepted by an oxidizing agent. If we use, for example, a ferric salt, the equation becomes:

$$\begin{array}{rrcl} & H_2 - 2\,e^{\ominus} & = & 2\,H^{\oplus} \\ & 2\,Fe^{3\oplus} + 2\,e^{\ominus} & = & 2\,Fe^{2\oplus} \\ \hline \text{Sum:} & H_2 + 2\,Fe^{3\oplus} & = & 2\,H^{\oplus} + 2\,Fe^{2\oplus} \end{array} \tag{2}$$

Molecular oxygen can act as an oxidizing agent similarly by picking up either two or four electrons:

$$O_2 + 2\,e^{\ominus} = O_2^{2\ominus} \underset{}{\overset{+2H^{\oplus}}{\rightleftarrows}} H_2O_2 \tag{3}$$

$$O_2 + 4\,e^{\ominus} = 2\,O^{2\ominus} \overset{+4H^{\oplus}}{\rightleftarrows} 2\,H_2O \tag{4}$$

The coupling of reactions (1) and (4) constitutes the formation of water from its elements:

$$\begin{array}{rcl} 2\,H_2 - 4\,e^{\ominus} & = & 4\,H^{\oplus} \\ O_2 + 4\,e^{\ominus} & = & 2\,O^{2\ominus} \\ \hline 2\,H_2 + O_2 & = & 2\,H_2O \end{array} \qquad \Delta G^{o} = -57\ \text{kcal/mole}\ H_2O$$

It will be remembered that the reaction is strongly exergonic.

The oxidation of organic molecules can also be interpreted as a removal of electrons since the chemical bond is a pair of electrons. The molecule undergoing oxidation in most cases donates H ions in addition to the electrons, as illustrated in the following example:

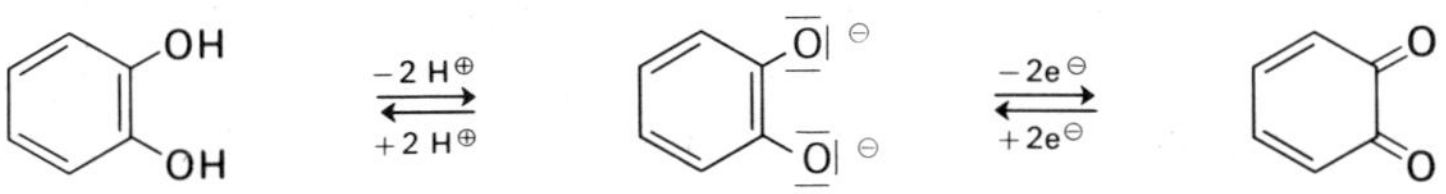

Catechol Catecholate *o*-Quinone

Oxidation of catechol consists of the removal of electrons from the catecholate ion, which thus becomes quinone.[2] Previously we have similarly formulated the oxidation of glutathione-SH to the disulfide (Chapter III,3).

Dehydrogenation of a primary alcohol (oxidation by removal of hydrogen) also can be seen to be a removal of electrons (formulas below). It should be noted that the initial step, the dissociation of a proton with formation of the alcoholate ion, does not involve a change in oxidation level. In the second step, the molecule becomes even more polarized, as symbolized by the wedge-shaped bond, and finally the binding pair of electrons comes off together with the proton.

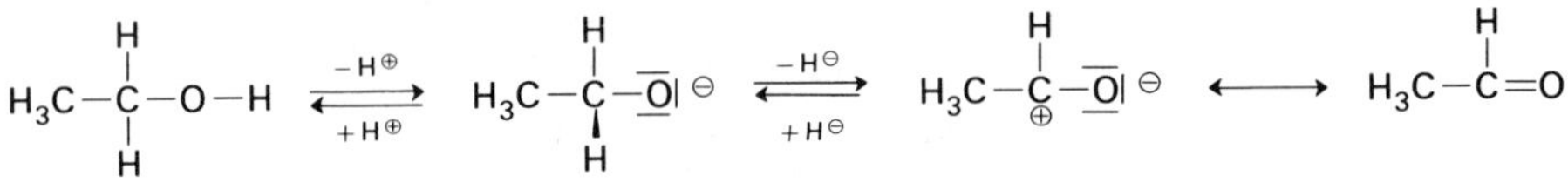

Alcohol Alcoholate Aldehyde

This last step, the removal of electrons, is the oxidation proper (or dehydrogenation). In the process, a hydride ion[3] $H^{\ominus}$ is transferred to the dehydrogenating agent, and the remaining molecule stabilizes itself by the formation of the C=O double bond. If a nicotinamide nucleotide should serve as the dehydrogenating agent, as is the case in enzymatic dehydrogenations, then we have arrived at the mechanism of nicotinamide nucleotide catalysis, as outlined briefly in Chapter VI,4. The loss of both $H^{\oplus}$ and $H^{\ominus}$ really amounts to nothing more than the dissociation of two protons and the loss of two electrons.

The examples mentioned are all two-electron transfers. One-electron transfers are also observed in organic chemistry. Here the products are free radicals characterized by a single (unpaired) electron. They are stable only in exceptional cases (e.g., some semiquinones, including those of flavoproteins). Free radicals can be demonstrated by electron spin resonance (ESR) spectroscopy. This technique has also become useful in biochemistry for the elucidation of reaction mechanisms.

In the formation of quinones, oxygen can accept electrons directly not only in chemical experiments but also biochemically, with the aid of the enzyme catechol oxidase. As far as the reaction of the substrate is concerned, it is immaterial whether the electrons are transferred directly to oxygen or to some prosthetic group. It suffices that they are removed and the substrate is oxidized.

[2] We have chosen the system catechol/*o*-quinone; hydroquinone/*p*-quinone would exemplify the reaction just as well.

[3] If a proton picks up two electrons, then a negatively charged hydrogen ion, the hydride ion, is formed. Compounds containing the hydride ion are familiar from inorganic chemistry: the hydrides of alkali and alkaline earth metals. They are decomposed instantly by water, since the $H^{\ominus}$ ion reacts with the $H^{\oplus}$ ion to release H_2.

3. The Redox Potential

Exchange of Electrons through Wires. Once more we return to the oxidation of molecular hydrogen by $Fe^{3\oplus}$ ions (Section 2). Instead of the electrons being exchanged in a direct interaction between partners, they can also be transported through a wire. This constitutes an electric current. The generation of electricity in galvanic elements (e.g., dry cells) depends on such processes.

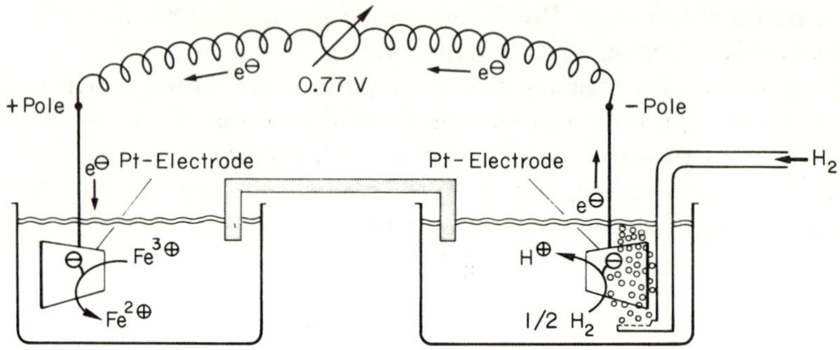

Fig. X-1. Electrochemical cell.

We may construct two half-cells (Fig. X-1). In the left solution are both $Fe^{2\oplus}$ and $Fe^{3\oplus}$ salts; in the right half-cell are $H^{\oplus}$ ions and molecular hydrogen. The electrodes would ordinarily be platinum. The electrons leave the molecular hydrogen and temporarily build up a negative charge, but after closing of the circuit they can travel to the iron salt solution, which is positively charged as compared to the hydrogen half-cell. The electric potential between the two half-cells can be measured directly.

To permit a comparison of the various half-cells or redox systems with one another —e.g., the pair $Fe^{2\oplus}/Fe^{3\oplus}$ with the pair $Ti^{3\oplus}/Ti^{4\oplus}$ or $Sn^{2\oplus}/Sn^{4\oplus}$—the concept of a redox potential was invented. Arbitrarily, the hydrogen half-cell has been chosen as the reference cell for all other systems. The redox potential then is designated simply as that electric potential (in volts or millivolts) measured against the hydrogen reference electrode.[4]

Redox Potential and Free Energy. The concept of redox potential, derived from the above experimental setup, has been an invaluable aid in chemistry. The concept is intimately associated with that of free energy of an oxidation–reduction reaction, because the reaction in a galvanic cell is reversible and electric energy is made available for useful work. Thus the redox potential becomes a direct measure of the free energy (cf. Chapter V,2), except that it is expressed in different units. It must always be remembered, however, that the redox potential invariably refers to the reaction with gaseous hydrogen. That is the zero point of the redox scale.

[4] The half-cell potentials of metal ions, e.g., $Cu/Cu^{2\oplus}$, are of great significance in inorganic chemistry. The arrangement of metals according to increasing potentials results in the familiar electromotive series.

The reaction with gaseous hydrogen is usually less interesting than that with some other oxidant or reductant. The difference in redox potentials determines the change of free energy. To convert one to the other, we use the expression:

$$\Delta G^0 = -n\mathscr{F}\Delta E_0$$

(n = number of electrons transferred; $\mathscr{F}$ = amount of charge per mole = 96,500 coulombs). With the turnover of 1 mole of electrons (= change in valency by 1, and 1 mole of substance reacted) and a potential difference of 1 V, 23.07 kcal are developed. In two-electron transfers the available work would be double that, or 46.1 kcal. As another example, the free energy of ATP hydrolysis (−7.0 kcal/mole under standard conditions) corresponds to 0.152 V potential difference with a two-electron change.

The standard values of free energy and also the redox potentials refer to standard concentrations of oxidized and reduced reactants, including the concentration (more precisely, the activity) of $H^{\oplus} = 1$, pH = 0. The dependency of half-cell potentials on concentration is given by the following expression:

$$E_n = E_0 + \frac{RT}{n\mathscr{F}} \ln \frac{C_{\text{oxidized}}}{C_{\text{reduced}}}$$

This equation is entirely analogous to Eq. (5) of Chapter V, 2.

Zero Point on the Biochemical Redox Scale. Since hydrogen ions participate in most biochemical redox reactions, their concentration should also be set at 1*M* in order to measure the standard potentials. Of course, this is not feasible because enzymes are not active at such a low pH. Furthermore, it seems more reasonable to choose "physiological" conditions. In biochemistry, therefore, it is customary to use standard potentials E_0' referring to pH of 7. At that pH, the hydrogen electrode has a potential difference of −0.42 V with respect to the hydrogen electrode of pH 0.

For reactions in living cells the relative amounts of oxidized and reduced substrates or coenzymes obviously are important. A drastic shift in concentration may actually reverse the sign, so that the more powerful oxidizing agent, according to the normal potentials, is in effect oxidized by the weaker one.

It is generally very difficult to measure directly the redox potential in biological systems. Instead, the concentration of several key metabolites (e.g., lactate and pyruvate) can be measured, and from the knowledge of their standard potentials the effective redox potential can be calculated.

Biological Redox Systems. The cytochromes discussed in Chapter IX are truly biological redox systems *par excellence*. The iron atoms that they contain change their valence with the release or uptake of electrons. Electron transport is their physiological function (see below). The redox potentials have been measured (cf. Table X-1). It should be noted that the values depend very much on the protein components, the protein-free system heme/hemin having a much more negative potential (−0.115 V) than the cytochromes.

In Chapter VI we have discussed the chemical nature of several hydrogen-transferring coenzymes. They are redox systems, because hydrogen transfer is equivalent to electron transfer (Section 2). Other redox systems consist of substrates which react with enzymes and their cosubstrates (= coenzymes); for example, the mixture of alcohol and acetaldehyde (equimolar) can be assigned a standard potential, as can the system succinate-fumarate, and so forth. Redox potentials of this kind are listed in the right-hand half of Table X-1.

TABLE X-1

Redox Potentials (E_0') for Some Biochemical Redox Systems at pH 7 and 25°

	(a) Coenzymes		(b) Substrates	
ΔG^0 (kcal)	E_0' (V)	Substance	E_0' (V)	Substance
	−0.43	Ferredoxin	−0.57	Acetaldehyde/acetate
			−0.42	$H_2/2\,H^{\oplus}$
	−0.31	NADH + $H^{\oplus}$/$NAD^{\oplus}$		
			−0.20	Ethanol/acetaldehyde
−14	−0.21	Riboflavin-P · H_2/riboflavin-P		
			−0.185	Lactate/pyruvate
			−0.166	Malate/oxaloacetate
	0.00	Flavoprotein (ubiquinone reductase)		
			−0.03	Succinate/fumarate
	+0.10	Ubihydroquinone/ubiquinone	+0.01	Methylene blue/leuko pigment
−14	+0.12	Cytochrome *b*		
			+0.20	Ascorbate/dehydroascorbate (pH 3.3)
	+0.26	Cytochrome *c*		
−24	+0.29	Cytochrome *a*		
			+0.81	$\frac{1}{2}O_2/O^{2\ominus}$

4. The Respiratory Chain

Energy of Water Formation. In Section 1 we emphasized that the formation of water is the decisive energy-yielding reaction of metabolism. The potential difference between the hydrogen electrode and the oxygen electrode amounts to 0.81 − (−0.42) = 1.23 V and is equivalent to a change of free energy of $\Delta G^0 = -57$ kcal/mole. The cell, however, ordinarily oxidizes NADH, and not molecular hydrogen, since this coenzyme accepts hydrogen from substrates. The potential difference compared to O_2 then is only 1.12 V, and $\Delta G^0 = -52$ kcal.

This amount of energy (52 kcal) is too large for a single biochemical reaction and is subdivided into small individual energy parcels. Thus NADH does not react with oxygen directly but rather in a series of intermediate steps. A comparison has been made with a cascade where running water reaches a lower level over several steps. The total potential energy of the water fall is released in parts. The biological significance of such a chain of redox systems stems from the possibility of *coupling* each of the steps *with other reactions*. Thus, much of the free energy of water formation is conserved as chemical energy and not dissipated as heat. The carrier of chemical energy is again adenosine triphosphate (ATP): The respiratory chain is coupled with *oxidative phosphorylation*, i.e., with the formation of ATP from ADP and inorganic phosphate.

Mitochondria as Site of Cellular Respiration. The respiratory chain and the coupled system of phosphorylation is localized in certain cell organelles, the *mitocondria*. The coupled systems are components of the inner membrane and thus are bound to the structure of mitochondria. The structure will be described below.

Respiration can be studied on mitochondria isolated from cell homogenates. Though much valuable information was gained from such studies, it has not been possible to learn the chemical structure of the enzymes involved. When the redox systems are isolated from mitochondria and studied separately the ordered structure is destroyed with substantial loss of the main functions. Furthermore, new nonphysiological enzymatic properties appear such as reactivity toward certain pigments. Consequently, any conclusions about the physiological state have to be limited appropriately. All this makes research on the respiratory chain very difficult.

Sequence of Redox Systems in the Respiratory Chain. A logical arrangement of the component enzymes of the respiratory chain can be devised on the basis of the redox potentials of their prosthetic groups. The most negative potential is that of nicotinamide-adenine dinucleotide, the $(NADH + H^{\oplus})/NAD^{\oplus}$ system. Reduced NAD can be oxidized again by a flavoprotein (FP_{NAD}) which transfers hydrogen to ubiquinone (Q) which itself becomes reduced. Ubiquinone is also the point of entry of hydrogen from the dehydrogenation of succinate and fatty acids. The reoxidation of reduced ubiquinone is the job of the cytochrome system. Since the latter functions by change of valence of iron, this is the latest point where hydrogen can be ionized to $H^{\oplus}$; from here on only electrons are transferred. Again we can order the cytochromes by their redox potentials in the sequence shown, where cytochrome aa_3 transfers the electrons to oxygen ($O_2 + 4\,e^{\ominus} \rightarrow 2\,O^{2\ominus}$). The resulting $O^{2\ominus}$ ion, is not stable (its equilibrium concentration in water is $10^{-52}\,M$); it immediately picks up two $H^{\oplus}$ ions to produce H_2O. The formation of water has thereby been accomplished.

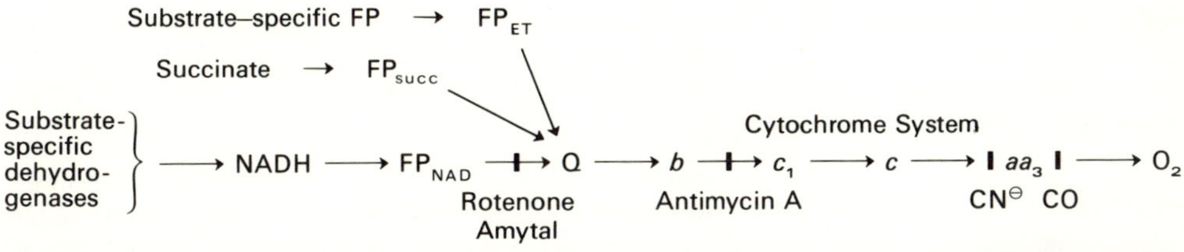

This diagram as well as Fig. X-2 does not take account of the stoichiometry of the redox system. It takes 4 molecules of cytochrome c-$Fe^{2\oplus}$ to deliver the 4 electrons which cytochrome aa_3 transfers to 1 molecule of O_2. At other steps the situation is comparable. The above diagram also shows the site of action of certain inhibitors which were of inestimable value in the study of the respiratory chain. Rotenone and amytal (isoamylethylbarbiturate), the antibiotic antimycin, and finally cyanide and carbon monoxide block the redox chain as shown.

Experimental studies of the respiratory chain have corroborated this arrangement, in principle at least, although numerous questions remain unanswered.

Auxiliary Substrates of the Respiratory Chain. In the chapter on coenzymes (Chapter VI) we attempted to differentiate the concepts of "prosthetic groups" and "coenzymes," or more correctly "cosubstrates." NADH in this context certainly is a cosubstrate, indeed, the most important substrate of the respiratory chain. As transport metabolite for hydrogen it mediates between substrate-specific dehydrogenases and the redox systems.

NAD[5] occurs both in mitochondria and in the cytoplasm, but only the coenzyme reduced within the mitochondrion is oxidized by the respiratory chain; extramitochondrial NADH cannot pass across the mitochondrial membrane. The exchange of hydrogen between cytoplasm and mitochondria appears to be carried out by certain substrates (glycerophosphate, β-hydroxybutyrate, malate) functioning as transport metabolites (see end of Chapter XVIII,3).

Figure X-2 indicates that ubiquinone and cytochrome *c* also are auxiliary substrates. *Cytochrome c* can be extracted relatively easily from mitochondria with salt solutions. It is the substrate for cytochrome *c* oxidase (= cytochrome aa_3) and is reduced again by another enzyme system described extensively below.

Ubiquinone constitutes a further pool for hydrogen, supplied in part by NADH, in part by succinate or fatty acids (cf. Chapter XII,3). Its structure is described in Chapter XIV,9. Because of the lipophilic character imparted by the long isoprene side chains, ubiquinone is inserted into the lipid phase of mitochondrial membranes (see below).

The reduction of ubiquinone may be looked upon as two one-electron transfers:

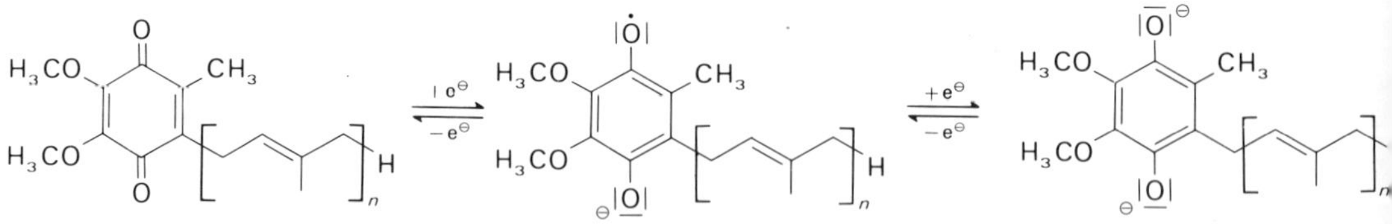

The intermediate step hereby is semiquinone. Ubihydroquinone anion can accept 2 $H^{\oplus}$ ions and assume the undissocated form.

[5] We remind the reader again that the Committee on Enzymes of the International Union of Biochemistry has suggested the designations NAD (nicotinamide-adenine dinucleotide) and NADP (nicotinamide-adenine dinucleotide phosphate) for what have previously been known as DPN and TPN (di- and triphosphopyridine dinucleotide). This nomenclature has now been adopted widely. Strictly speaking, the more correct abbreviation for the oxidized form is $NAD^{\oplus}$ (or $NADP^{\oplus}$), and for the reduced form NADH + $H^{\oplus}$ (or NADP + $H^{\oplus}$); for convenience, these are shortened in the text to NAD (or NADP) and NADH (or NADPH), respectively. In balanced equations the more complete form is used.

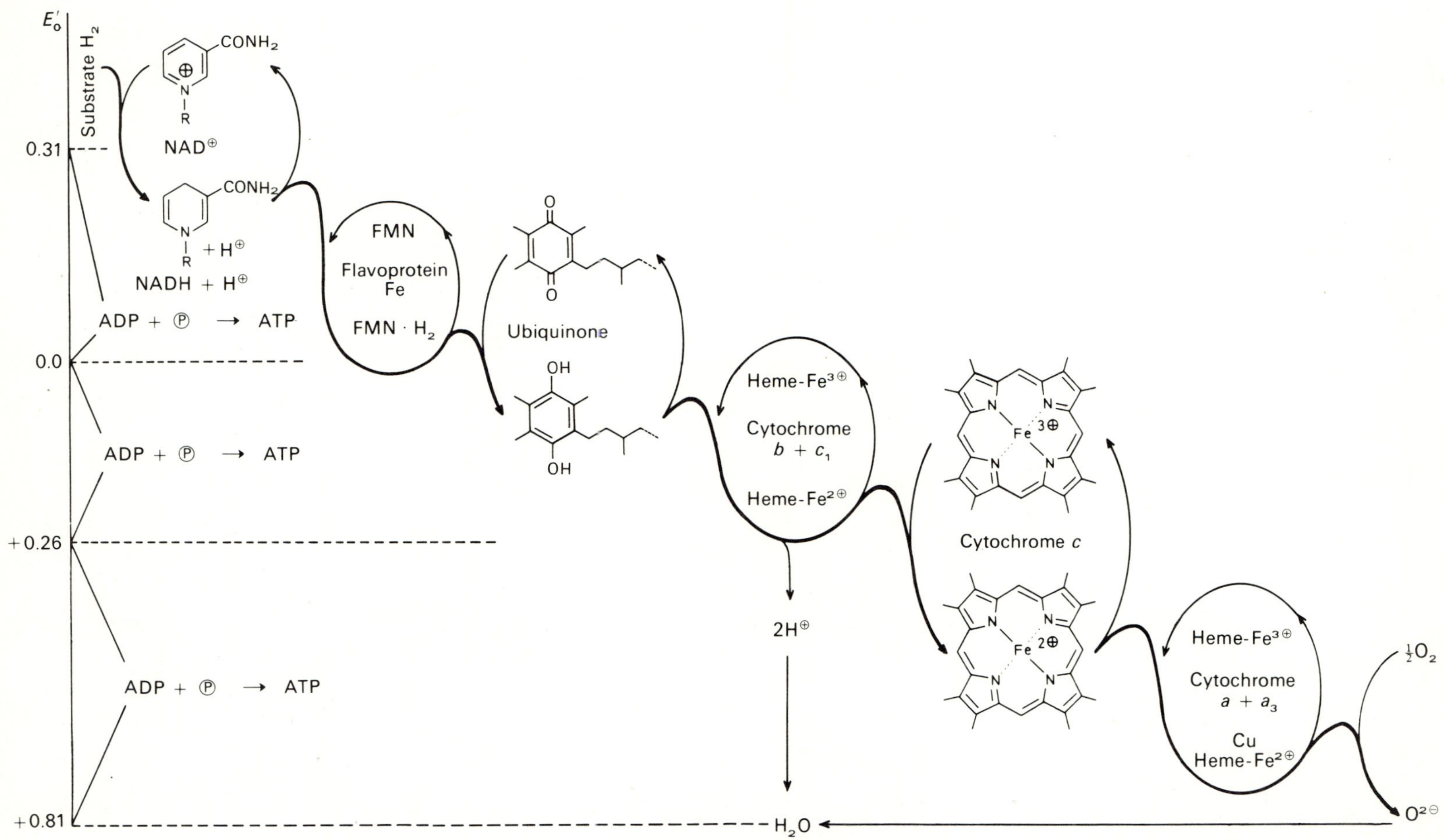

Fig. X-2. Diagram of the respiratory chain for the oxidation of NADH. The diagram is intended to illustrate the flux of electrons through the various redox systems without regard for the stoichiometry involved.

The rest of the redox systems are enzymes, or rather enzyme complexes, which will be discussed below.

Flavoproteins of the Respiratory Chain. From among some sixty different flavin-containing enzymes known today, only three flavoproteins are important for the respiratory chain. They are NADH dehydrogenase, succinate dehydrogenase, and the electron-transferring flavoprotein.

NADH dehydrogenase in the respiratory chain accepts the hydrogen from reduced nicotinamide-adenine dinucleotide and passes it along to the next redox system—probably ubiquinone. The enzyme is complex and consists of one flavoprotein with FMN (flavin mononucleotide) as prosthetic group, iron ions, and a second protein that contains iron and loosely bound sulfur.

A number of similar iron-sulfur proteins have recently been isolated and characterized; they are also termed "nonheme-iron proteins." Ionic iron is bound partially by sulfhydryl groups of the protein and partially by hydrogen sulfide. One of these redox systems is ferredoxin (cf. Chapter XVI,2). Iron participates in redox catalysis by a change in valency. It is thought that iron facilitates the transition from the two-electron transport (or H_2 transport) to the one-electron transport of the cytochromes. Furthermore, the separation of 2(H) into $2\,H^{\oplus}$ and $2\,e^{\ominus}$ may occur here.

The flavoprotein *succinate dehydrogenase*[6] is very closely associated with the respiratory chain. It acts on its substrate without the mediation of the nicotinamide coenzymes[7] and can transfer the hydrogen to unphysiological acceptors such as methylene blue. The physiological acceptor is ubiquinone or cytochrome *b*. The flavin group of succinate dehydrogenase is very similar to FAD, but the isoalloxazine nucleus is bound to the protein by primary valence bonds. The enzyme also contains 4 atoms of Fe per flavin group. The link of the dehydrogenation of succinate with the respiratory chain is of great physiological significance, since succinate is an intermediate of the citrate cycle (Chapter XI,2).

The *electron-transferring flavoprotein* contains FAD and has a molecular weight of 70,000. It is the connecting link between the many substrate-specific flavin enzymes e.g., the acyl-CoA dehydrogenases of the fatty acid breakdown (see β-oxidation, Chapter XII,3) or α-glycerophosphate dehydrogenase, and the respiratory chain. The hydrogen acceptor probably is ubiquinone.

The Cytochrome System. Cytochromes are hemoproteins that transfer electrons in redox chains. By their chemical structure and their spectra they belong to one of

---

[6] Since at the pH of the cell (pH ~ 7) succinic acid exists in its salt form, succinate dehydrogenase is the more correct name (rather than succinic dehydrogenase).

[7] Collaboration of the NAD system in the dehydrogenation of succinate is thermodynamically impossible because the nicotinamide system with a redox potential of $E_0' = -0.31$ V is too weak an oxidizing agent. The flavoprotein ($E_0' = 0.00$), in contrast, can attain equilibrium with the system succinate/fumarate ($E_0' = -0.03$). It is a fairly general rule that the introduction of a double bond into a saturated carbon chain requires flavoproteins.

the three groups *a*, *b*, and *c*. Individual cytochromes within a group are distinguished further by subscripts.

Cytochromes b and c_1 according to Fig. X-2 mediate redox catalysis between ubiquinone and cytochrome *c*, the "auxiliary substrates" of the respiratory chain. The structure and properties of these cytochromes have already been discussed in Chapter IX,6.

A complex with a molecular weight of about 280,000 can be isolated from the respiratory chain. The complex consists of 2 cytochrome *b*, 1 cytochrome c_1, 1 iron-sulfur protein ("nonheme-iron protein"), and lipid (including bound ubiquinone). This complex has the catalytic property of ubiquinone-cytochrome-*c* oxidoreductase, i.e., of the enzyme that mediates the redox reaction described above.

Some authors look upon this complex (also called complex III) as a preformed multienzyme system; others see it more or less as a random fragment of the inner mitochondrial membrane in which the respiratory chain is localized (see below). This question is unresolved. In addition, several authors regard cytochrome *b* as H_2 acceptor of succinate dehydrogenase.

Cytochrome c, which follows the b–c_1 complex in Fig. X-2 and is classified as an auxiliary substrate, has been described in Chapter IX,6.

Cytochrome aa_3, often called cytochrome *c* oxidase or simply cytochrome oxidase, is identical with Warburg's "Atmungsferment." It is a complex hemoprotein containing cytoheme, copper, and lipid; copper participates in the catalysis by a change in valency (chemical structure, Chapter IX,6). The respiratory chain seems to contain predominantly a hexameric form.

Components "*a*" and "a_3" are distinguishable by their spectra and reaction with inhibitors; the chemical basis for this difference is not clear. It is established, however, that only a_3 reacts with oxygen and is poisoned by CO and $CN^{\ominus}$.

This complex is the so-called *terminal oxidase*, that enzyme of the electron transport chain which reacts directly with respiratory oxygen. The enzyme binds 1 molecule of cytochrome *c* per heme group and transfers 4 electrons to 1 molecule of O_2. Since the change of valence of cytochrome iron releases only one electron ($Fe^{2\oplus} \rightarrow Fe^{3\oplus}$), the cooperation of several heme groups appears to be necessary; Cu is assumed to cooperate, but the exact mechanism is still not known. The oxidized form of cytochrome aa_3 is reduced again by cytochrome *c*.

The Respiratory Chain as a Dynamic Equilibrium. The respiratory chain is an excellent example of the fact that in the organism true chemical equilibria do not exist. The respiratory chain fulfills its job only if it continuously accepts hydrogen from the substrates and with it reduces the oxygen supplied by hemoglobin. A steady state is reached, which is a *dynamic equilibrium* and which is influenced much more by the supply of oxygen, the concentration of substrates, and above all by the phosphorylating system (Section 6) coupled to it, than by redox potentials. The nicotinamide nucleotides may be present largely in the reduced form, and the successive coenzymes along the chain are increasingly oxidized. This has been supported by actual measurements. Some of the values found are given in Table X-2.

TABLE X-2

Dynamic Equilibria in the Respiratory Chain

	Percentage of active groups in "reduced form"					
				Cytochromes		
Experimental condition	NAD	Flavoprotein	Ubiquinone	*b*	*c*	*a*
"Controlled respiration": Much substrate; ADP limiting	78%	42%	36%	50%	3%	0%
"Active state": Maximal O_2 consumption	5%	15%	6%	20%	3%	2%

The Respiratory Chain as Component of the Mitochondrial Membrane. The study of the ultrastructure of mitochondria gives some clues as to where the respiratory chain is located. Figure X-3 is a schematic presentation of the mitochondrial structure as it would appear in high-resolutions electron microscopy (see also Chapter XVIII,3).

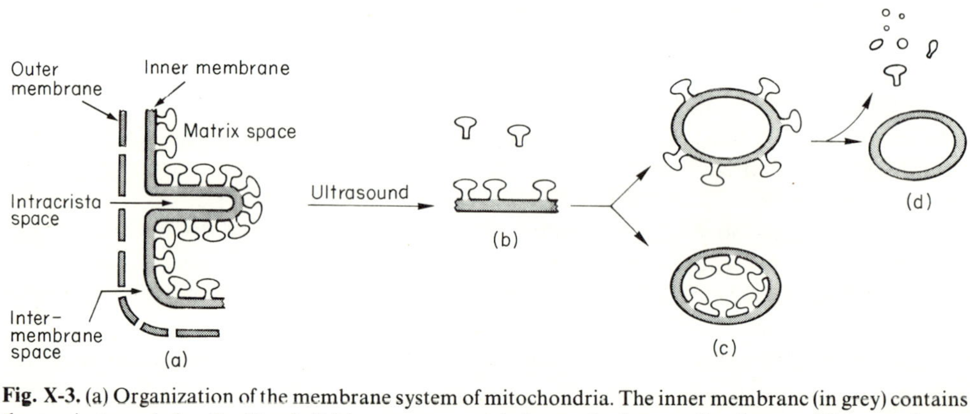

Fig. X-3. (a) Organization of the membrane system of mitochondria. The inner membrane (in grey) contains the respiratory chain; the "knobs" (elementary particles) contain the coupling factors. Ultrasound produces membrane pieces (b) that ball up to phosphorylating electron-transport particles (c) in some of which the matrix side faces inward, in some outward. In (d) the knobs are set free and no longer phosphorylate.

Of the two membranes, the outer membrane is very permeable; the real limit of permeability is set by the inner membrane. It envelopes the matrix containing numerous enzymes, among others, those of the citrate cycle and fatty acid oxidation (cf. Chapters XI and XII,3). The inner membrane usually has many inward folds called cristae that protrude into the matrix thus greatly enlarging the surface area of the membrane. On the inner or matrix side of the inner membrane one can see structures consisting of a stem and a spherical head. These knobs contain coupling factors for oxidative phosphorylation.

The component enzymes and auxiliary substrates of the respiratory chain are localized in the *inner membrane*. Although the composition of the outer membrane resembles that of endoplasmic reticulum (about 45% lipid and 55% protein), the inner membranes consist of 80% protein and 20% lipid. A large portion of the total membrane protein has been estimated to consist of enzyme protein. The existence of an additional structural protein (Green) is debatable. As is true for all membranes, the mitochondrial membrane also contains much phospholipid, especially lecithins and cardiolipins with unusually high proportions of unsaturated fatty acids. The redox systems discussed above are thought to be embedded in this lipid phase which appears to be important functionally because the extraction of the lipids inactivates the redox systems. Addition of lipid can restore some of the activity.

Ultrasonic treatment disintegrates mitochondrial membranes; the fragments roll up to spherical vesicles (see Fig. X-3c), which can be purified by centrifugation. These are called *electron-transport particles* and they are still able to oxidize NADH or succinate with the consumption of oxygen and the formation of ATP (see below). If the structure is damaged further, the capacity for oxidative phosphorylation disappears first.

Certain methods can break these particles up further and enable the isolation of complexes. These have been designated with roman numerals and are listed below.

Complex I: NADH:ubiquinone oxidoreductase; contains flavoprotein with FMN, nonheme-bound iron, and ubiquinone. It catalyzes the transport of hydrogen from NADH to ubiquinone; it contains mainly FP_{NAD}.

Complex II: Succinate:ubiquinone oxidoreductase; contains the flavoprotein succinate dehydrogenase, nonheme-bound iron, and cytochrome *b*.

Complex III: Ubihydroquinone:cytochrome *c* oxidoreductase; contains cytochrome *b* and c_1, as well as nonheme-bound iron. It mediates the transport of hydrogen and electrons from the ubiquinone system to cytochrome *c*.

Complex IV: Cytochrome c:O_2 oxidoreductase (= cytochrome oxidase); contains cytochrome aa_3 and copper.

The respiratory chain (or parts thereof) can be pieced together again from these complexes and the interposed auxiliary substrates ubiquinone and cytochrome *c*. The reconstituted particle even exhibits the susceptibility to certain inhibitors.

It is still not clear whether the complexes just listed are really independent multienzyme systems or whether they are merely fragments that are always obtainable in reproducible form. Many authors claim that the individual components are built into the membrane in a fairly orderly fashion rather than that the whole complexes are inserted. Ubiquinone and cytochrome *c* have the key function as auxiliary substrates of mediating the transfer of hydrogen and electrons between the individual enzyme components in the membrane.

Ubiquinone as lipophilic component significantly is interlaced in the membrane lipids, possibly in a well-ordered structure. The transition to the semiquinone can propagate a certain redox state from one molecule to the next without the need for any individual ubiquinone molecule to leave its place. In this manner it is possible for hydrogen atoms and for electrons to be transferred from the various flavoproteins to the cytochrome *b* system.

The organization of this membrane structure has additional significance for oxidative phosphorylation. This is the subject of the following section.

5. Oxidative Phosphorylation (Respiratory Chain Phosphorylation)

Energy Balance and P/O Ratio. The importance of the respiratory chain resides in the fact that the free energy of oxidation of each of the individual steps is trapped and stored in the form of ATP. This process has been termed oxidative phosphorylation or respiratory-chain phosphorylation. The energy difference calculations have been outlined in the scheme in Section 4 and are listed in Table X-1. Whenever 1 mole of NADH is oxidized by $\frac{1}{2}O_2$, 52 kcal are set free. The synthesis of 3 moles of ATP from ADP and orthophosphate (= "inorganic phosphate") requires about 21 kcal; thus, from an energy standpoint, the synthesis is entirely feasible and has an "efficiency quotient" of 40% under standard conditions (with physiological concentrations the efficiency would tend to be greater). The table of energies reveals that 1 ATP can arise for each pair of electrons (or hydrogens) transferred at each of the three sites, between NADH and flavoprotein, ubiquinone and cytochrome *c*, and cytochrome aa_3 and oxygen.

Measurements in various laboratories have repeatedly confirmed this value of 3 ATP per $\frac{1}{2}O_2$ (or per 1 mole H_2O). The quotient *ATP formed/O consumed* is called the *P/O ratio.* A P/O ratio of 3, however, applies only to substrates that are dehydrogenated by NAD. Succinate, oxidized directly by flavoprotein, yields only 2 moles of ATP per mole of H_2O, and cytochrome *c* only 1 mole.

Control of Respiration. Normally the respiratory chain and respiratory-chain phosphorylation are coupled tightly. Even with substrate and O_2 available, the redox reaction nevertheless cannot proceed unless ADP is phosphorylated at the same time. ADP thus becomes the limiting reagent of the overall reaction including O_2 uptake. The phenomenon can be called *control of respiration by ADP* (see Table X-2, line 1). The greater the supply of ADP, the greater the rate of transport of electrons to oxygen. Oxygen consumption and ATP production increase until with excess ADP in the "active state" maximal velocity of electron flow is attained (Table X-2, line 2).

This mechanism regulates itself. When large amounts of chemical energy (ATP) are consumed, for example, by a working muscle, the accumulating cleavage product ADP stimulates respiration and is rephosphorylated to ATP. This example of biochemical regulation will be discussed again in Chapter XIX.

Uncouplers and Inhibitors of Phosphorylation. 2,4-Dinitrophenol (DNP), substituted phenylhydrazones such as carbonylcyanide-*m*-chlorophenylhydrazone (CCCP), and other compounds are called uncouplers of oxidative phosphorylation because they stimulate respiration of mitochondria and at the same time prevent the formation of ATP. The respiratory chain is free-wheeling. Inhibitors, as *oligomycin*, on the other hand, in intact mitochondria (tight coupling) prevent both respiration and phosphorylation. Respiration cannot be affected in the absence of coupling.

In certain specialized tissues, such as the brown fat of some mammals, and under certain physiological states, such as upon awakening from hibernation or during cold stress, oxidative phosphorylation can be uncoupled *in vivo*. In that case the free energy of the reaction is released as heat. This is called chemical thermogenesis.

Coupling Factors. The study of membrane fragments obtained by ultrasonic disintegration (see end of Section 4) led to the discovery of coupling factors. When the proteins appearing as knobs in the electron microscope are stripped off the phosphorylating electron-transport particles then these particles cease to phosphorylate although they still respire. The addition of the proteins restores phosphorylation.

Racker has isolated two coupling factors from the protein fraction of these membrane fragments, factor F_0 involved in oligomycin inhibition and factor F_1, a protein complex of high molecular weight that catalyzes the reaction ATP + $H_2O \rightleftharpoons$ ADP + phosphate. Since equilibrium favors the right side of the equation, this isolated factor functions as an ATPase (abbreviation for adenosine triphosphatase), i.e., it hydrolyzes ATP. In intact mitochondria, however, it presumably acts as ATP synthase and at the expense of the energy released by the respiratory chain catalyzes the synthesis of ATP from ADP and phosphate. The detailed study of the "mitochondrial ATPase" is expected to reveal the mechanism of oxidative phosphorylation and how the released energy is harnessed for this synthesis.

Carefully isolated intact mitochondria possess no ATPase activity. It appears only in aged or damaged preparations or after addition of the uncoupler dinitrophenol.

The coupling factors undoubtedly are involved in the generation of an energy-rich intermediate conventionally symbolized by $X \sim X$. It is capable of providing the energy for other endergonic processes of the mitochondria, such as the accumulation of $Ca^{2\oplus}$ ions by active transport; the transfer of hydrogen from succinate to NAD opposite a redox gradient—probably a nonphysiological model reaction (Klingenberg); the contraction of mitochondria from the swollen to the condensed state; and last—by transfer at a high energy level—the formation of ATP, surely both quantitatively and physiologically the most important process.

The essential question, by what reaction mechanism are the redox reactions of the respiratory chain linked to the formation of energy-rich bonds, still cannot be answered. Basically two hypotheses have been advanced:

1. *Chemical coupling* via energy-rich intermediate products, in analogy to substrate chain phosphorylation (Slater, 1953). A variant of this hypothesis postulates the primary formation of an *energy-rich conformation* of the inner membrane with whose assistance another energy-rich bond is established (Boyer, 1965).

2. *Chemiosmotic theory* postulating an electrochemical concentration gradient (Mitchell, 1961).

The Theory of Chemical Coupling. This process is best illustrated by a diagram, which also shows the postulated action of inhibitors and uncouplers (see below).

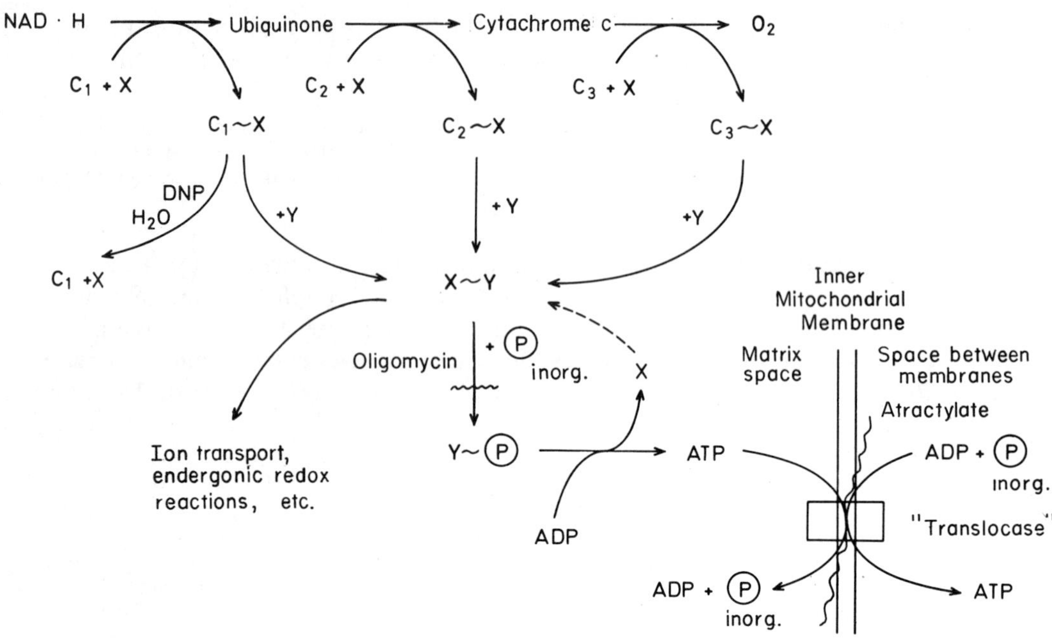

Initially the redox reaction is coupled with the formation of an energy-rich product C ~ X. For the three phosphorylation sites we must postulate three different "coupling factors" C_1, C_2, and C_3 (possibly they are parts of redox enzymes). These initial energy-rich intermediate C ~ X can be destroyed (i.e., hydrolyzed) by dinitrophenol, for example, thereby explaining its uncoupling effect.

The intermediate C ~ X now transfers ~ X, bound with high energy, to the ATP-forming enzyme Y (the ATPase) with the formation of X ~ Y. The latter is an energy-rich product which can be utilized within the mitochondrion for a variety of energy-consuming processes (see above). It is also able to react with inorganic phosphate to form X ~ P; it is this step that is blocked by oligomycin. Last, ~ P is transferred from X to ADP with the formation of ATP.

Since the respiratory chain of mitochondria is the chief producer of ATP, and since most ATP is consumed in the cytoplasm, ATP has to be channeled out of the mitochondria. A *translocase system* in the internal mitochondrial membrane is responsible for this transportation. The plant *N*-glycoside *attractylate* inhibits the translocase.

One major objection to this otherwise plausible hypothesis is that it has not been possible to isolate and characterize the postulated high-energy intermediates despite considerable effort.

The Chemiosmotic Theory. This theory proposes that in the course of the redox reaction each of the membrane-bound enzyme complexes exports 2 $H^{\oplus}$ ions per coupling site, while $OH^{\ominus}$ ions remain in the matrix space, thereby establishing a pH gradient (see Fig. X-4).

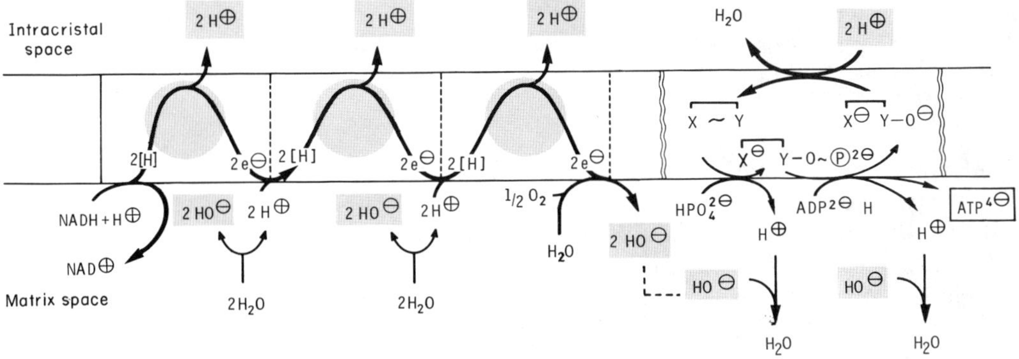

Fig. X-4. Schematic representation of the chemiosmotic coupling theory. In each "loop" the redox systems pump two protons outside into the intracristal space; $HO^{\ominus}$ remain inside and elevate the pH. At the coupling site (only one is shown on the right) the pH gradient is harnessed for the formation of an energy-rich bond X$\sim$Y. Simultaneously two negative charges diffuse outside. In the diagram these negative charges arise from the removal of one H_2O from ADP and phosphate. The $H^{\oplus}$ ions help to maintain neutrality in the interior; oxygen functions in the transfer of charges at a protein. Formally the coupling reaction resembles the neutralization of an acid by a base.

The enzyme accepts the hydrogen from the substrate (NADH or succinate) on the matrix side ("inside") and releases protons to the outside while the electrons react on the inside of the membrane with the auxiliary substrate (e.g., Q) which then picks up $2\,H^{\oplus}$ from the interior to yield the reduced auxiliary substrate (QH_2). The latter is oxidized by the next enzyme and the cycle repeats itself. By oxidation of the auxiliary substrate $2\,H^{\oplus}$ are exported again, while electrons inside of the membrane trap $H^{\oplus}$ ions. In the third step, electrons are transferred to O_2 and H_2O whereby $HO^{\ominus}$ ions are generated. A consequence of this charge separation is a membrane potential borne by $H^{\oplus}$ and $OH^{\ominus}$.

Prerequisite for this mechanism is a vectorial property of the enzyme complexes involved as well as an appropriate localization in the mitochondrial membrane. It must furthermore be postulated that the membrane is impermeable to $H^{\oplus}$ and $OH^{\ominus}$ ions, for otherwise the gradients would collapse.

According to this scheme, the energy of the redox reaction initially is spent on setting up a concentration gradient of $H^{\oplus}$ ions across the membrane and the gradient is able to perform work. At certain coupling sites (Fig. X-4, right side) this gradient is equalized provided it is coupled with the formation of an energy-rich bond. Each combination of redox enzyme, electron carrier, and coupling site Mitchell has called a *loop*. The hypothetical energy-rich primary product X$\sim$Y is then employed in an exchange reaction for the synthesis of ATP. For every $2\,H^{\oplus}$ exported one ATP is formed; thus the oxidation of NADH nets a P/O ratio of 3. The reading of the coupling reaction in the reverse direction results in the hydrolysis of ATP in the manner of the catalysis of the coupling factor F_1 discussed above.

Many experimental results agree very well with this chemiosmotic theory: The mitochondrial membrane is indeed impermeable to protons; respiratory chain phosphorylation is observed only with particles that have an intact membrane structure; and the predicted shift of pH has actually been measured. The uncoupling effect of dinitrophenol can be explained if it is assumed that this lipophilic weak acid diffuses into the membrane and acting as proton carrier equalizes the pH gradient. Furthermore there is extensive similarity with photophosphorylation in the thylakoid membrane of chloroplasts for which a chemiosmotic mechanism has also been postulated (see Fig. XVI-2).

The difficulty with this chemiosmotic theory is the inability to explain the actual coupling step. There is no known reaction mechanism that could be cited as model. Formally it is the reversal of ATP-dependent active transport, whose detailed mechanism is not known either.

Completely different primary mechanisms of energy conservation are postulated by the two theories discussed above. It can be seen, nevertheless, that both mechanisms depend on the existence of a nonphosphorylated, energy-rich intermediate product X~Y, which is involved directly in the endergonic processes of the (artificial) "uphill" transport of electrons, the active transport of ions (particularly $Ca^{2\oplus}$), and the formation of ATP. With all the differences, this important concept is at least common to both theories.

6. Other Oxygen-Activating Enzymes

The respiratory chain contains only one enzyme that reacts directly with oxygen, namely cytochrome oxidase. Another name for it is terminal oxidase because it is found at the end of the electron transport chain. Cytochrome oxidase may be the most important enzyme to react with oxygen, but it is not the only one.

Classification and Nomenclature. In recent years it has been recognized more and more that oxygen is not only an acceptor of electrons (or hydrogen), but that it can also enter organic molecules directly. Three groups of enzymes react with O_2 and catalyze the following reactions:

1. *Oxidases* (electron-transferring oxidases). The general equations are:

$$O_2 + 4\,e^{\ominus} \rightarrow 2\,O^{2\ominus} \overset{+4H^{\oplus}}{\rightleftarrows} 2\,H_2O$$

or

$$O_2 + 2\,e^{\ominus} \rightarrow O_2^{\,2\ominus} \overset{+2H^{\oplus}}{\rightleftarrows} H_2O_2$$

2. *Dioxygenases* (= oxygen transferases). The general equation is:

$$A + O_2 \rightarrow AO_2$$

3. *Monooxygenases* (= hydroxylases). General equation:

$$AH + DH_2 + O_2 \rightarrow AOH + D + H_2O$$

(AH = substrate; DH_2 = hydrogen donor)

Further Oxidases. Several enzymes which contain copper just as cytochrome oxidase, but which do not contain any porphyrin group, are able to transfer electrons from *o*- or *p*-hydroquinones or from ene-diols to oxygen. Water (not H_2O_2) is the other product of the reaction. The enzymes therefore transfer *4 electrons* to 1 O_2. To this group belong *p-diphenol oxidase* (or laccase), *o-diphenol oxidase* (catechol oxidase, catecholase), and *ascorbate oxidase*. The latter is found in plants, and it has often been suggested that it can serve as terminal oxidase in cellular respiration. *o*-Diphenol oxidase will be discussed at the end of this section.

Two-electron transferring oxidases (= H_2O_2-forming enzymes) practically without exception contain flavin as prosthetic group. The flavin enzymes are also capable of transferring hydrogen (or its electrons) to acceptors other than oxygen. Methylene blue and other quinoid dyes in *in vitro* experiments serve as unphysiological acceptors, whereas *in vivo* either electron-transferring flavoprotein, ubiquinone, or the cytochromes of the microsomes generally would seem to assume the role of the redox dye; thus the formation of H_2O_2 is circumvented.

Oxygen Transferases. The typical reaction of these enzymes is the cleavage of an aromatic ring; the product has two C=O groups at the points left after C=C scission. The mechanism calls for the addition of an oxygen molecule forming, as an intermediate, a cyclic peroxide:

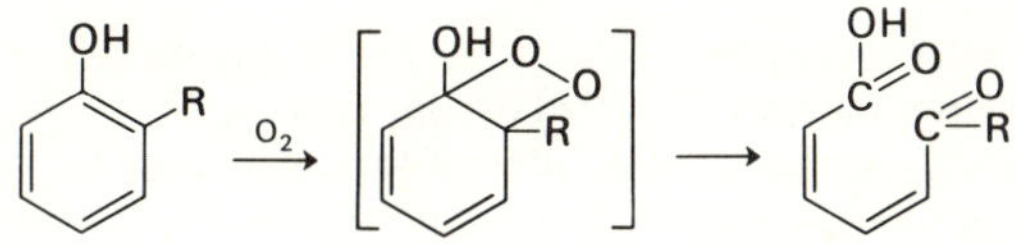

The metabolism of the aromatic amino acids provides examples of this reaction; e.g., the oxidation of homogentisate (Chapter VIII,11), the oxidation of 3-hydroxyanthranilate (Chapter VIII,11), and the oxidation of tryptophan to formylkynurenine (Chapter VIII,11). The enzyme for this last oxidation, contains a ferrophorphyrin system and the oxidation of $Fe^{2\oplus}$ to $Fe^{3\oplus}$ inactivates the enzyme. The other iron-containing enzymes have no porphyrin.

Enzymatic Hydroxylations. A distinction must be made between oxidative ring cleavage and hydroxylations. In the latter only one oxygen atom is introduced into the substrate molecule; the other oxygen is reduced to water. A hydrogen donor is required for this, which may be one of several different kinds. The most common donors are α-ketoglutarate, NADPH, flavoprotein, ferredoxin, and tetrahydropteridine.

Microsomal Hydroxylases. The most important hydroxylation system is that of the "microsomes," which must be taken as fragments of the endoplasmic reticulum (cf. Chapter XVIII-2). True hydroxylases in this case are enzymes that contain *cytochrome* P_{450}. Furthermore, "microsomes" contain an electron-transport chain evidently for the purpose of supplying electrons to cytochrome P_{450}. The electrons originate from NADPH and are transported via a flavoprotein. The following diagram—still hypothetical in several respects—represents the mechanism of hydroxylation as postulated by Ullrich.

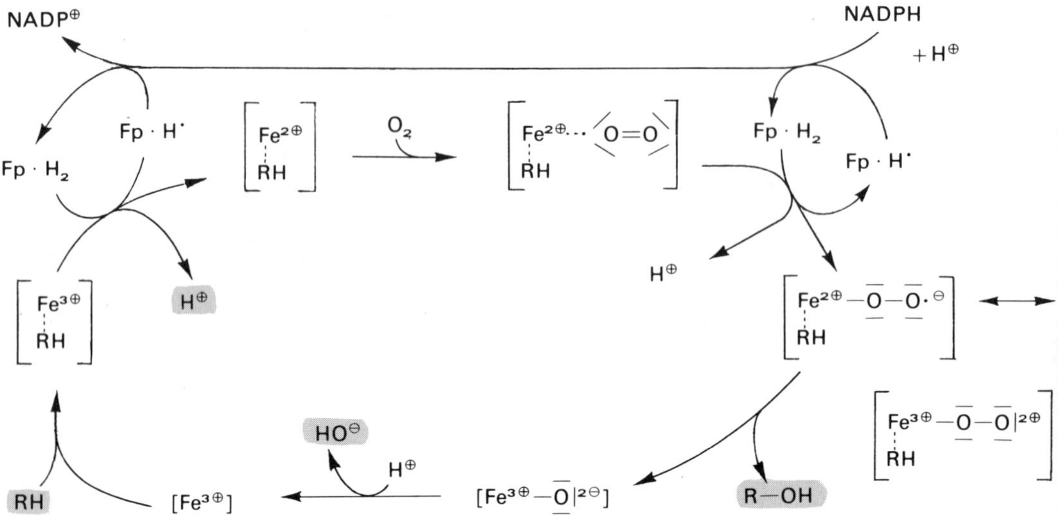

During the introduction of oxygen into the substrate (R–H) it is the task of cytochrome P_{450} to bind the O_2 molecule and to activate oxygen. Activated oxygen can transfer one oxygen atom to the substrate; the remaining oxygen atom is released as $HO^{\ominus}$. Both electrons required for the activation of oxygen are supplied by a flavoprotein, which is oxidized from the dihydro form to a semiquinone and subsequently reduced again by NADPH. The semiquinone step facilitates the transition to the one-electron transport.

This same mechanism is also employed by the liver in the hydroxylation of exogenous lipophilic substances, such as drugs and steroids, preparatory to excretion by the kidneys. C-21 hydroxylation of steroids proceeds analogously in the endoplasmic reticulum of the adrenal cortex by means of a steroid-specific cytochrome P_{450}.

Mitochondria of the adrenal cortex also contain steroid hydroxylases that differ from the ones just described by having *ferredoxin* (also called *adrenodoxin*) inserted between flavoprotein and cytochrome P_{450}. Hydroxylation proceeds entirely analogously. There are very specific hydroxylases for the 11β, for the 18 position etc., as well as for the cholesterol side chain, which is then shortened (cf. Chapter XIV,6).

Further Hydroxylases. In addition to the more complicated hydroxylation systems of the endoplasmic reticulum, there are hydroxylases with simpler structure that can be studied in solution. We will mention a few that differ in their requirements for hydrogen donors.

In the hydroxylation of phenylalanine to form tyrosine, the hydrogen donor is the pteridine derivative *tetrahydrobiopterin*, although *in vitro* tetrahydrofolate can serve as H_2 donor. The corresponding dihydropteridine derivative is formed in the reaction and is reduced again by action of the *dihydropteridine reductase* to the tetrahydro level by NADPH. This enzyme system is localized in the cytoplasm.

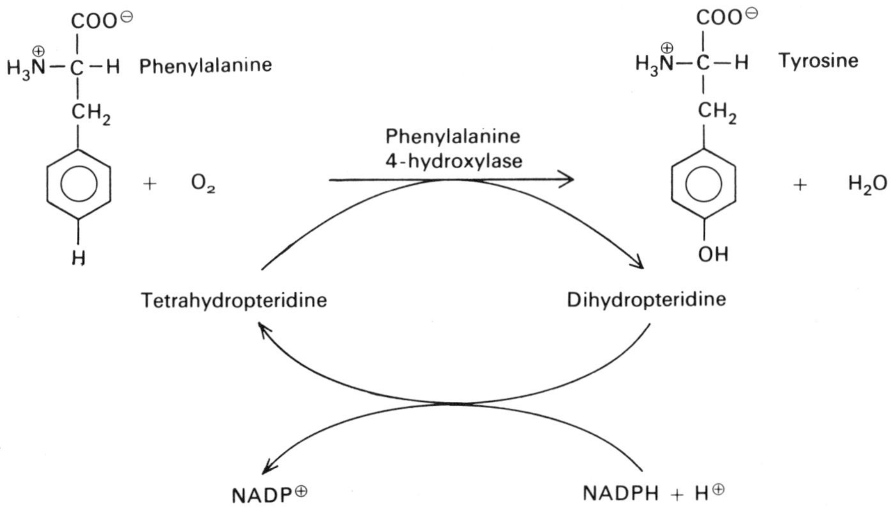

α-Ketoglutarate is hydrogen donor for the hydroxylation of prolyl residues in the protocollagen molecule, which proceeds according to the following scheme:

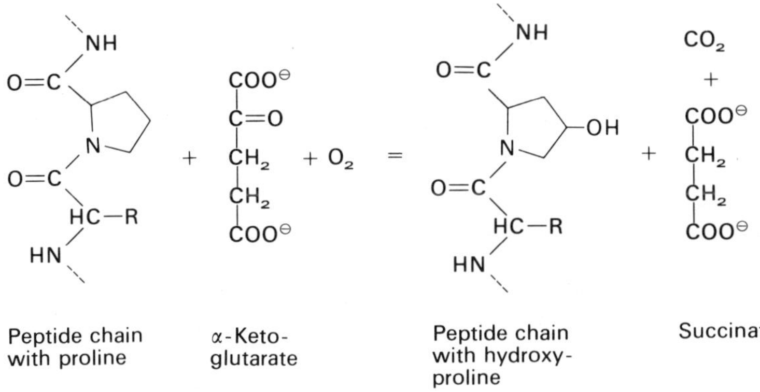

Protocollagen hydroxylase, which catalyzes this reaction, requires the cofactors $Fe^{2\oplus}$ ions and ascorbate. The hydroxylation of lysyl side chains proceeds in the same fashion. In the oxidation of *p*-hydroxyphenylpyruvate to form homogentisate—a similar reaction in principle—hydrogen donor is the α-ketoacid group of the side chain, which in the process is oxidatively decarboxylated (cf. Chapter VIII,11).

The o-Diphenol Oxidase Complex. *o*-Diphenol oxidase (catechol oxidase, phenolase, tyrosinase) converts tyrosine to DOPA (=dihydroxyphenylalanine) and oxidizes the dihydroxy derivative further to the quinone stage. Through a series of subsequent reactions, some of which occur spontaneously and without enzymatic catalysis, black or brownish black melanin is finally formed (for a schematic representation of the reactions see Chapter VIII,11). *o*-Diphenol oxidase is an oxidase with mixed functions; the product of hydroxylation, the hydroquinone derivative, simultaneously acts as hydrogen donor. The mechanism of reaction can be written as follows:

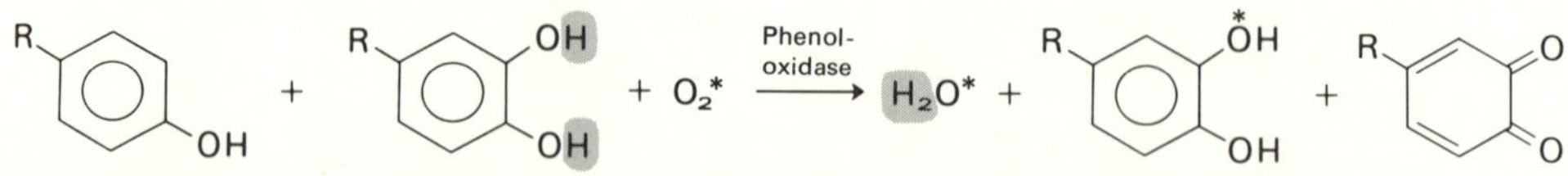

Experiments with isotopically labeled oxygen (O*) have supported this mechanism. The majority of diphenol oxidases can carry out the oxidation of hydroquinone derivatives by themselves without coupling. This is also termed the catechol oxidase effect (catechol = 1,2-dihydroxybenzene). Here H_2O (and not H_2O_2) is formed.

o-Diphenol oxidases (tyrosinases) are widely distributed in the plant kingdom (champignon mushrooms, potatoes, bananas). They are responsible for the darkening of freshly cut surfaces of plants or fruits. For insects, *o*-diphenol oxidases are important both for melanin formation and for browning and hardening (sclerotization) of the cuticle.

BIBLIOGRAPHY

H. Baltscheffsky and M. Baltscheffsky, Electron Transport Phosphorylation, *Annu. Rev. Biochem.* **43**, 871–897 (1974).

P. D. Boyer, *in* "Biological Oxidation" (T. P. Singer, ed.), Wiley, New York, 1967.

M. Florkin and E. H. Stotz (eds.), "Biological Oxidations" (Comprehensive Biochemistry, Vol. 14). Am. Elsevier, New York, 1966.

P. Mitchell, Translocations through natural membranes. *Advan. Enzymol.* **29**, 33–87 (1967).

E. Racker, The membrane of the mitochondrion. *Sci. Amer.* **218,** 32–39 (1968).

E. Racker, Function and structure of the inner membrane of mitochondria and chloroplasts. *In* "Membranes of Mitochondria and Chloroplasts," E. Racker (editor). Van Nostrand Reinhold, New York, 1970.

K. van Dam and A. J. Meyer, Oxidation and energy conservation by mitochondria. *Annu. Rev. Biochem.* **40**, 115–160 (1971).

CHAPTER **XI**

Carbon Dioxide Formation in the Citrate Cycle

1. Significance of the Citrate Cycle

The end products of aerobic metabolism are chiefly carbon dioxide and water (cf. Chapter X,1). Carbon dioxide arises metabolically from carboxylic acids mainly by two pathways of decarboxylation:

1. Decarboxylation of β-keto acids, a reaction which proceeds spontaneously even *in vitro.*

2. "Oxidative decarboxylation" of α-keto acids, whose mechanism has already been introduced in Chapters VI,4 and VIII,10.

The ultimate purpose of intermediary metabolism (i.e., the energy-yielding or catabolic aspect of metabolism) may be thought to consist in converting the various foodstuffs into compounds that can undergo one of these two general reactions. It is a welcome simplification that in almost all organisms breakdown invariably follows the same path. The starting material for this final degradation is "activated acetate," in which the acetyl group is bound to *coenzyme A*. The origin of activated acetate from the amino acid alanine has already been discussed. How fats and carbohydrates are broken down to activated acetate will be described in Chapters XII and XV.

The breakdown of activated acetate itself takes place in the citrate cycle (Krebs' cycle; tricarboxylic acid cycle). In this cycle, the pathways of protein fat, and carbohydrate catabolism are united. Furthermore, the cycle provides many of the necessary components for the synthesis of endogenous substances. The citrate cycle therefore encompasses a large pool of common intermediates, which can be used either for synthesis of new material or for degradation to gain energy. The full significance of these interrelationships will be demonstrated in Chapter XIX.

The net reaction for the breakdown of acetyl-CoA can be written as follows:

$$CH_3-\overset{\overset{\displaystyle O}{\|}}{C}\sim S\overline{CoA} + 3\,H_2O \rightarrow 2\,CO_2 + 8\,[H] + HS\overline{CoA}$$

No oxygen is taken up in this oxidation; instead, water is added and dehydrogenated several times. The hydrogen is picked up by the appropriate coenzymes and brought to reaction with oxygen in the respiratory chain.

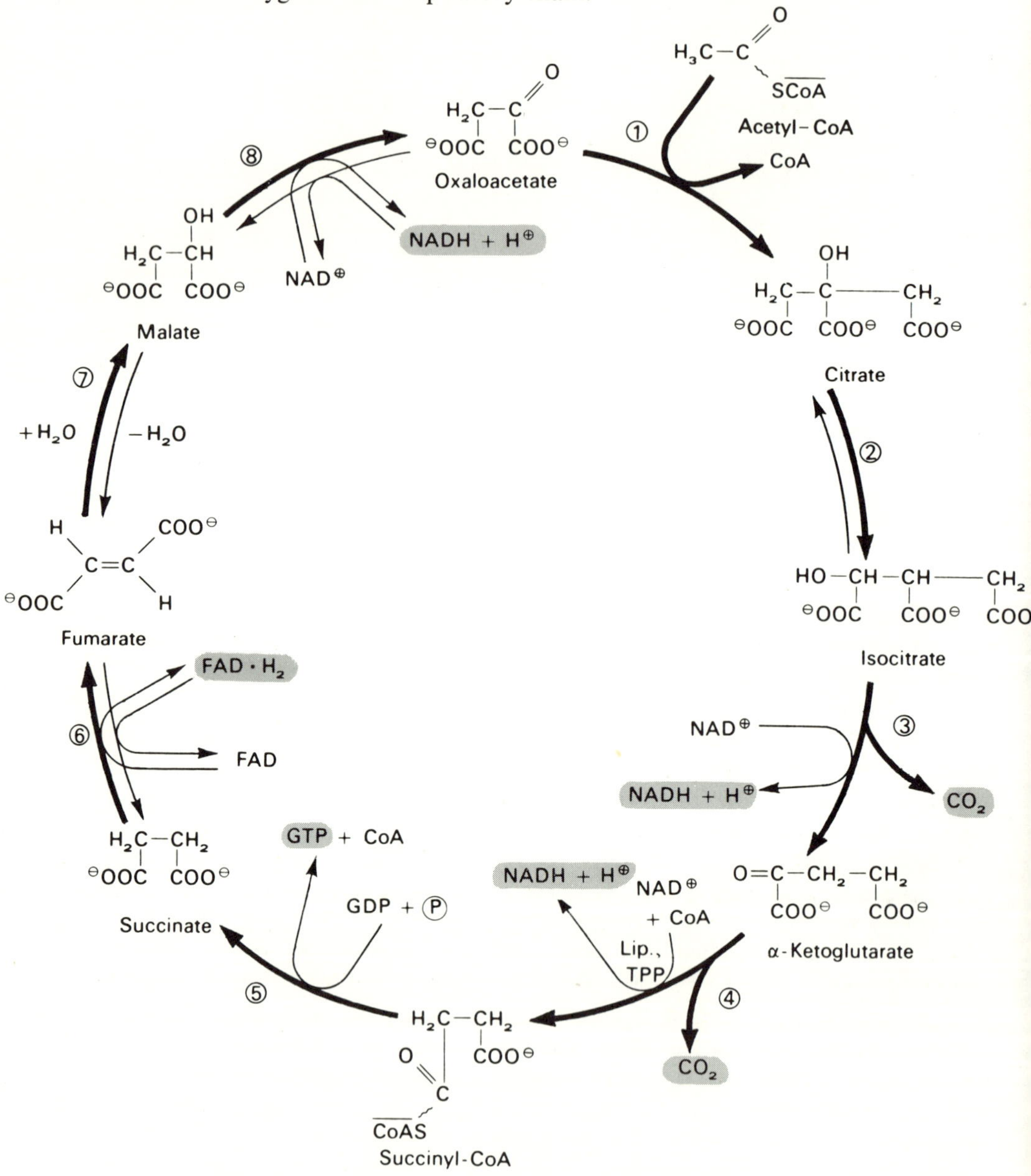

The process, however, does not involve C_2 and C_1 fragments. On the contrary, the condensation of acetyl-CoA with a C_4 compound yields the C_6 compound citrate (whence the name of the cycle is derived), and by a series of steps the C_4 compound is eventually regenerated, whereby the cycle is completed. The whole sequence of reactions is illustrated above. Since at the pH of the cell organic acids are ionized, we have drawn the salt forms. The steps have been numbered for more convenient reference.

The citrate cycle was discovered independently at about the same time in 1937 by Krebs and by Martius and Knoop. The discovery permitted the correlation of many isolated facts.

2. The Individual Steps

The formation of citrate (reaction 1) is catalyzed by *citrate synthase*. Substrate in addition to activated acetate is *oxaloacetate*, which can arise from aspartate by transamination (Chapter VIII,13). The keto group of oxaloacetate can react with acetyl-CoA by a modified aldol condensation. In this reaction (1) the activating CoA group is split off; for the reaction mechanism see Chapter VI,7. The product of the condensation reaction is a monohydroxytricarboxylic acid with the trivial name *citric acid*.

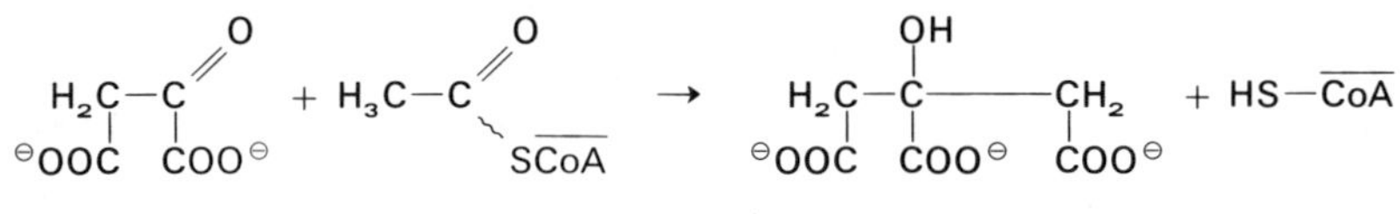

The series of transformations of citrate begins with the isomerization to *isocitrate* (step 2). The enzyme *aconitase* (aconitate hydratase) catalyzes the attainment of the equilibrium among citrate (89 %), *cis*-aconitate (3 %), and isocitrate (8 %). It is now believed that the reaction proceeds via a carbonium cation as the common intermediate for all three tricarboxylic acids:

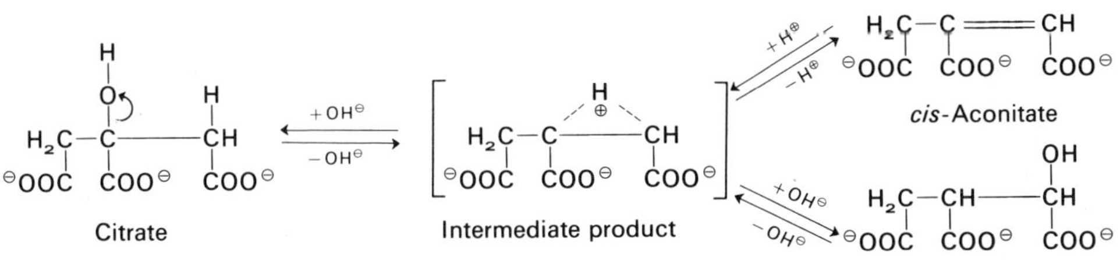

Citrate is a symmetrical compound, and yet the enzyme converts it to isocitrate in strict stereospecific fashion. The hydroxyl group appears only on the C atom originally contributed by oxaloacetate. Ogston has explained the phenomenon of asymmetric treatment of symmetrical substrates by postulating a *three-point attachment* of the substrate to the enzyme surface. The symmetry of the citrate ion is lost by its adherence to an asymmetric surface (of the protein) and the approach of the $OH^{\ominus}$ ion from one particular side may be favored enormously.

The isomerization to isocitrate gives rise to a secondary hydroxyl group, which is easily oxidized to a keto group. *Isocitrate dehydrogenase* catalyzes that reaction as well as the reversible decarboxylation of the β-keto acid oxalosuccinate (reaction 3). The product is *α-ketoglutarate*, with which we have already become acquainted in connection with the degradation of glutamate.

There are two different isocitrate dehydrogenases: The NADP-dependent enzyme has been known for some time. It occurs both inside and outside of mitochondria and makes hydrogen available for syntheses. More recently an NAD-dependent isocitrate dehydrogenases was found. It is activated by ADP, inhibited by ATP, and occurs exclusively in mitochondria, where it transfers hydrogen to the respiratory chain. This enzyme plays the major role in terminal oxidation. Both enzymes require $Mg^{2\oplus}$ or $Mn^{2\oplus}$ for full activity.

α-Ketoglutarate is an important branching point. In the cycle it is decarboxylated oxidatively; this means that CO_2 is split off with a concomitant loss of hydrogen, resulting in a dicarboxylic acid (succinate) shorter by one C atom (reactions 4 and 5). This last step is irreversible and is catalyzed by a multienzyme complex consisting of three enzymes: (1) *α-ketoglutarate decarboxylase*, which contains thiamine pyrophosphate as prosthetic group and carries out the decarboxylation yielding thiamine-bound succinic semialdehyde; (2) *lipoyl reductase-transsuccinylase*, which with lipoate as prosthetic group catalyzes the transfer of the succinyl residue to lipoate involving the opening of the disulfide ring and which also can catalyze the transfer of the succinyl residue to coenzyme A forming the product *succinyl-CoA*; and finally (3) *dihydrolipoyl dehydrogenase* instrumental in returning the lipoyl residue to the disulfide form. The reactions proceed as shown below and are analogous to oxidative decarboxylation of pyruvate (cf. Chapters VI,4 and VIII,10).

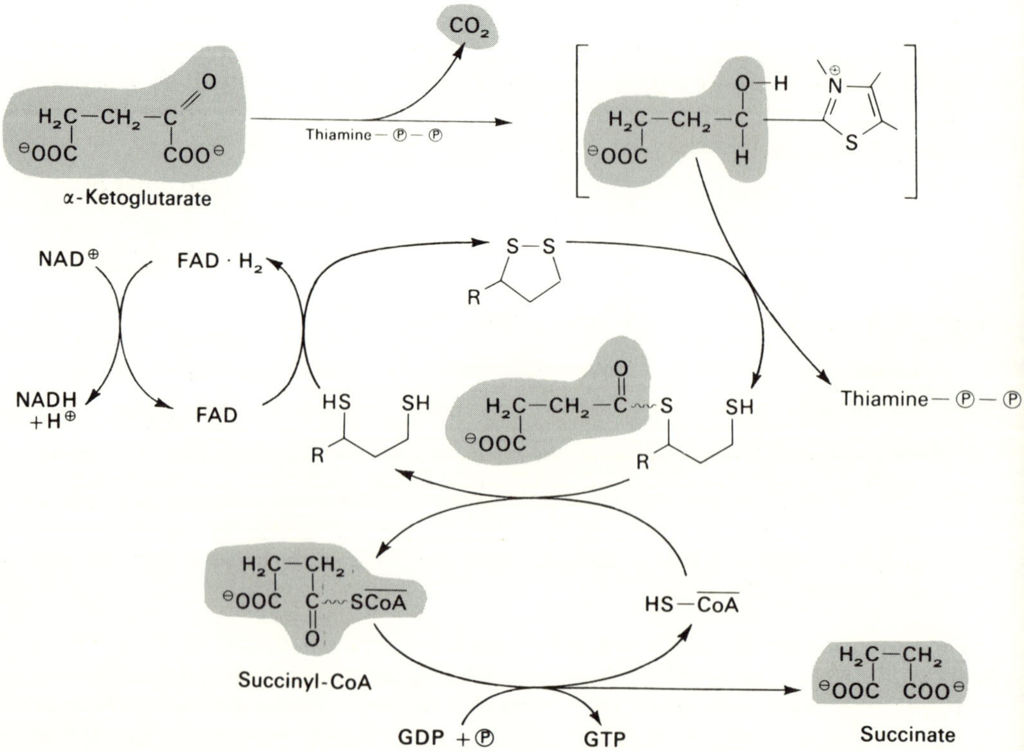

S-Acyl dihydrolipoate is a compound with a high potential for group transfer (a type of thioester); the free energy of the aldehyde oxidation has been trapped as chemical energy. The succinyl residue is transferred to CoA, forming succinyl-CoA and dihydrolipoate, which must be dehydrogenated again. A flavoprotein, *lipoyl dehydrogenase* (formerly called diaphorase), takes care of the hydrogen transfer to NAD. The NADH produced is reoxidized in the respiratory chain. We can now summarize the train of redox catalysts in the following way:

$$\text{Lipoamide} \longrightarrow \underset{E'_0 = -0.34}{\text{flavoprotein}} \longrightarrow \text{NAD} \longrightarrow \underset{E'_0 = 0.00}{\text{flavoprotein}} \longrightarrow \text{cytochromes} \longrightarrow O_2$$

It is remarkable that two flavoproteins with opposite functions—the reduction of NAD and the dehydrogenation of NADH—should be involved here.

Succinyl-CoA either is used in synthetic reactions (cf. Chapter IX,1) or is converted to succinate; the energy content is preserved as chemical energy by the coupled transformation of *guanosine diphosphate* (GDP) and orthophosphate to the energy-rich *guanosine triphosphate* (GTP). Evidently, the adenylic acid system does not always have to participate directly; in this case, it is in a secondary equilibrium relation with the GDP-GTP system.

Dehydrogenation of succinate introduces a C—C double bond to give rise to the *trans*-compound *fumarate* (step 6). The enzyme *succinate dehydrogenase* is a flavoprotein and is a member of the respiratory chain (cf. Chapter X,4). This reaction is inhibited by malonate—the classical example of competitive inhibition: The "wrong substrate" is attached to the enzyme, but cannot undergo the reaction for simple chemical reasons.

Water is added across the double bond of fumarate in a reaction (step 7) which is catalyzed by the enzyme *fumarate hydratase* (fumarase). The reaction is reversible although slightly exergonic. At equilibrium there is 82% L-*malate*. In the eighth and last step of the cycle, the secondary hydroxyl group of malate is dehydrogenated. The enzyme *malate dehydrogenase* transfers the hydrogen to NAD. The product of this reaction is *oxaloacetate*, the primer for the whole chain of reactions. The cycle is closed and we have finished one trip around it.

In summary we can see that one C_2 fragment, activated acetate, forms a C_6 compound which is then broken down via a complicated series of seven steps to the same C_4 compound that initially entered the cycle as a carrier. Two carbon atoms are removed in steps 3 and 4. In addition, four times 2 H atoms are taken up by co-dehydrogenases and oxidized to H_2O in the respiratory chain.

Actually it is not quite correct to say that the same oxaloacetate molecule is introduced merely as a carrier and appears again. It has been found, in fact, that two C atoms of oxaloacetate are split off and the acetate added originally is retained in the new oxaloacetate molecule.

3. Energy Yield of the Citrate Cycle

In the citrate cycle, water is taken up and one molecule of activated acetate is broken down into $2\,CO_2$ and hydrogen ($4 \cdot [2\,H]$) bound to the *coenzyme*. The combined steps yield a relatively small decrease of free energy ($\Delta G^0 = -25$ kcal), accounted for essentially by the cleavage of the energy-rich CoA bond (-8 kcal) and by the two decarboxylation reactions. Thermodynamically, this is an advantage because at equilibrium all the reactants can be present in comparable concentrations.

For the cycle to run in this fashion—i.e., anaerobically—there would always have to be a fresh supply of NAD to enter the cycle as a true substrate. But actually the coenzyme is present only in catalytic amounts and is regenerated continuously. The citrate cycle runs only in conjunction with the respiratory chain, and the calculation of energy yields must be revised. Indeed, of a total of 216 kcal of chemical energy released 191 are due to the respiratory chain. This should make it quite clear that the energy is really derived from the formation of water and not from CO_2 production!

The organism utilizes this energy in part as heat—to maintain its body temperature—and partly as chemical energy in the form of ATP gained from oxidative phosphorylation. The net result is:

from steps 3, 4, and 8, three ATP for each NADH oxidized, or a total of	9 ATP
from step 6, through flavin oxidation	2 ATP
from step 5, through transfer of the energy-rich bond	1 ATP
Total:	12 ATP

Under standard conditions this corresponds to a total amount of energy of $12 \times 7 = 84$ kcal/mole of activated acetate or an efficiency of about 40% of the theoretically available free energy.

In general, the storage of energy in the form of ATP is probably more important for the body than any incidentally produced heat. For most purposes it suffices to know the yield of ATP in moles per mole of oxidized acetate (through the cooperation of the citrate cycle and the respiratory chain).

Regulation of the Citrate Cycle. Under physiological conditions the citrate cycle is in dynamic equilibrium, just as the respiratory chain: The concentrations of the intermediate products remain approximately constant (at the order of magnitude of 10^{-4} *M*), but the turnover of substrates or, less precisely, the flow rate has to adapt to the physiological requirements. The adaptation is achieved by the regulation of enzyme activities with ADP, ATP, and NADH as effectors.

Prominent in this regulation are the allosteric properties of *isocitrate dehydrogenase*, regarded by many authors as the pacemaker enzyme of the citrate cycle. It requires ADP as allosteric activator; the enzyme is inhibited by ATP and NADH. *Citrate synthase*, too, is subject to allosteric control. ATP sharply elevates the Michaelis constant of this enzyme for acetyl-CoA and consequently throttles the rate of reaction. Since the citrate cycle operates in close conjunction with oxidative phosphorylation of the respiratory chain, the combination of effects achieves a stationary level of ATP.

4. Relations to Synthetic Tasks. Glyoxylate Cycle

The citrate cycle is not only a degradative scheme but furnishes also a general pool of intermediate products. It accepts the C_2 fragments from carbohydrate, fat, and protein catabolism; it is involved in the synthesis of glucose (which will be discussed in Chapter XV, the key substance being oxaloacetate); and it provides the raw material for the synthesis of several amino acids—aspartate and glutamate, for example. Finally, the blood pigment also arises from succinyl-CoA, another intermediate of the cycle.

The involvement of such nitrogen-containing intermediates can alter the citrate cycle. In the brain, almost all α-ketoglutarate is converted to *glutamate*, which is decarboxylated to *γ-aminobutyrate*. By transamination γ-aminobutyrate affords succinic semialdehyde, which is oxidized further to succinate. The last three reactions replace oxidative decarboxylation. It is not yet clear what physiological significance this metabolic pathway has for the brain.

As another variation, Shemin has postulated the following pathway:

$$\begin{matrix}\text{Succinyl-CoA} \\ +\ \text{Glycine}\end{matrix} \longrightarrow \delta\text{-Aminolevulinate} \longrightarrow \gamma\text{-Ketoglutaraldehyde} \longrightarrow \begin{matrix}\text{Succinate} \\ +\ C_1\ \text{fragment}\end{matrix}$$

The formation of δ-aminolevulinate has already been presented in context of the biosynthesis of porphobilinogens (see formulas in Chapter IX,1). By action of a transaminase δ-aminolevulinate is converted to γ ketogutaraldehyde, which is then split to succinate and a C_1 fragment. The C_1 fragment may be used for the synthesis of serine or of purines. We do not know yet whether these reactions proceed to a sufficient extent to warrant their designation as separate metabolic pathway.

The cycle demands for its proper functioning sufficient reserves of oxaloacetate. If the oxaloacetate is converted to aspartate, or if some other intermediate is diverted, the cycle falters; acetyl-CoA cannot enter the cycle without the acceptor molecule oxaloacetate. It is highly significant, therefore, that oxaloacetate may also be formed from pyruvate, the key substance of carbohydrate breakdown (via malate, cf. Chapter XV,10).

In some microorganisms an interesting variation of the citrate cycle is realized. There, the emphasis is not on the degradation of activated acetate, but rather on the synthesis of succinate, malate, and oxaloacetate (and eventually of carbohydrates) from acetyl-CoA. Two new reactions have been added:

1. Cleavage of isocitrate into succinate and *glyoxylate*, the reverse of an aldol condensation.
2. Condensation of glyoxylate with acetyl-CoA, similar to the formation of citrate (which means addition of the methyl group of acetate to the carbonyl group), resulting directly in the formation of malate.

With the inclusion of some reactions of the citrate cycle, we may draw the *glyoxylate cycle* shown here:

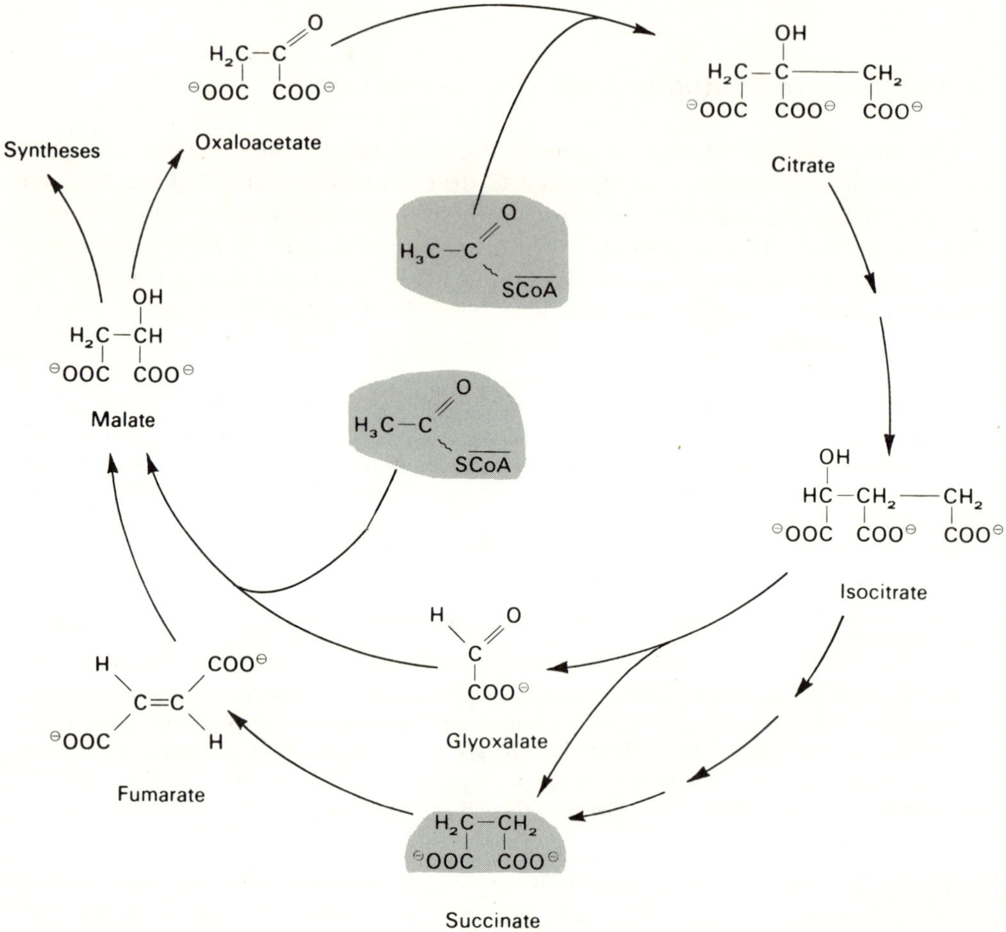

Oxaloacetate can be regenerated from both cleavage products of isocitrate: From glyoxylate by condensation with acetyl-CoA to form malate which is subsequently dehydrogenated, and from succinate through the usual citrate cycle. As a net result, 2 moles of activated acetate have been converted to succinate, which can undergo further reactions by familiar pathways. Other synthetic pathways branch off these secondary products.

It is of historical interest that the linkage of 2 moles of acetate with elimination of 2 H to form succinate was postulated already by Wieland and Thunberg around 1920, and had long been considered a central reaction of metabolism.

The glyoxylate cycle does not exist in the mammalian organism. The search for the enzymes in question has been fruitless. The cycle plays a dominant role, however, in microorganisms that grow on fatty acids or acetate as sole carbon source, as well as in plant seedlings, which by this device can use their fat reserves for the synthesis of carbohydrate. The cycle in such organisms is localized in special cell organelles, the *glyoxysomes.*

BIBLIOGRAPHY

T. W. Goodwin (editor), "Metabolic Roles of Citrate," Academic Press, New York, 1968.

H. L. Kornberg, Anaplerotic sequences in microbial metabolism. *Angew. Chem., Intern. Ed. Engl.* **4,** 558–565 (1965); contains a detailed description of the glyoxylate cycle.

P. A. Srere, The molecular physiology of citrate, *Nature* (*London*) **205,** 766–770 (1965).

CHAPTER **XII**

Fats and Fat Metabolism

The fats proper and fatlike materials are classified together as "lipids" mainly because of their similar solubilities. Without exception, they are almost insoluble in water but very soluble in common organic solvents such as benzene, ether, chloroform, or chloroform–methanol mixtures. Fatlike materials will be discussed in subsequent chapters without emphasizing their similarities to fats.

1. Chemical Composition of Fats

Neutral fats are composed of several simpler compounds. Chemically they are esters (cf. Chapter I,2) of unbranched monocarboxylic acids, the *fatty acids*. The alcohol component is invariably *glycerol*, which has three hydroxyl groups (a trihydric alcohol):

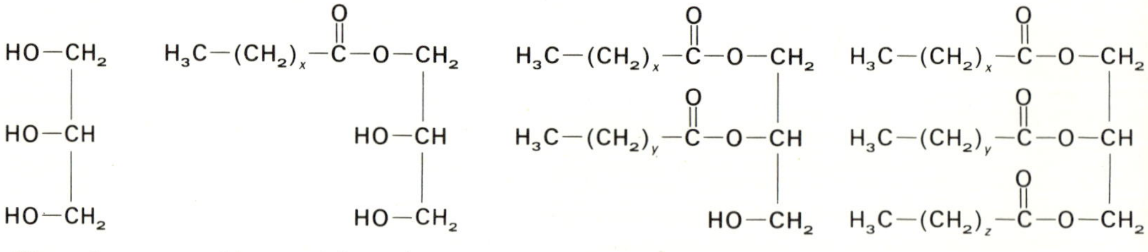

Glycerol — Monoacylglycerol — Diacylglycerol — Triacylglycerol

Being a trihydric alcohol, glycerol can form mono-, di, or triesters. Such esters have long been known as mono-, di-, or triglycerides. The new nomenclature for these compounds is monoacyl-, diacyl-, and triacylglycerols. Triacylglycerol generally contains two or three different fatty acids. Naturally occurring fats are always mixtures of numerous triacylglycerols, which fact makes the isolation of a single, homogenous triacylglycerol component inordinately laborious.

Hydrolysis of the ester linkages releases the components. This is accomplished quite readily in alkaline solutions, in which case the alkali salts, the so-called *soaps*, rather than the free acids are released. The technique of preparing soaps is a very old one, and the term saponification has been applied in organic chemistry not only to the hydrolysis of fats but also to that of other esters, amides, and so forth.

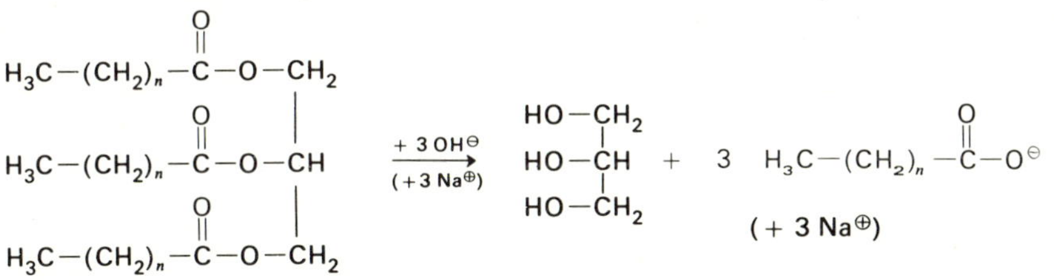

Fatty Acids. All the acids of naturally occurring fats have an even number of carbon atoms. This phenomenon is quite plausible since they are all synthesized from C_2 units (acetyl radicals) (cf. Section 6). It is not so clear, however, why acids with 16 or 18 C atoms, e.g., palmitic acid ($C_{16}H_{32}O_2$) and stearic acid ($C_{18}H_{36}O_2$), are most commonly found in nature.

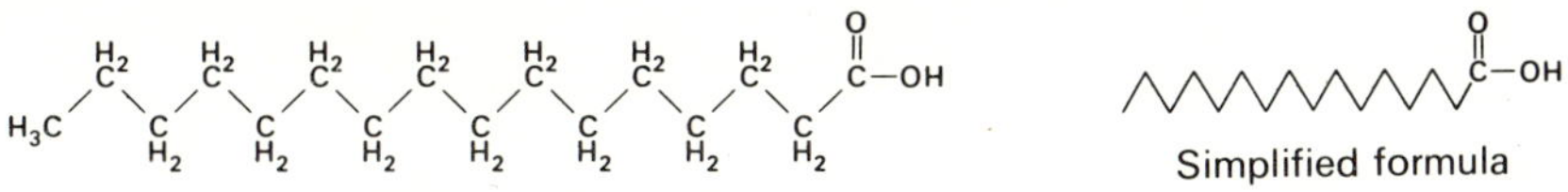

Palmitic acid — Simplified formula

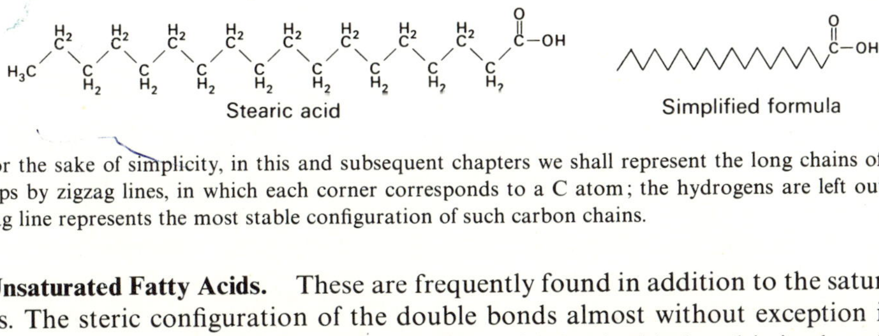

Stearic acid — Simplified formula

For the sake of simplicity, in this and subsequent chapters we shall represent the long chains of CH_2 groups by zigzag lines, in which each corner corresponds to a C atom; the hydrogens are left out. The zigzag line represents the most stable configuration of such carbon chains.

Unsaturated Fatty Acids. These are frequently found in addition to the saturated ones. The steric configuration of the double bonds almost without exception is *cis* (cf. *cis-trans*-Isomerism, Chapter I,4). In oleic acid the double bond is in the middle of the carbon chain, between C-9 and C-10; the molecule is almost symmetrical. In multiply ("poly-") unsaturated fatty acids the double bonds are generally isolated, i.e., separated by a CH_2 group. Their π-electrons cannot interact as they can in conjugated double bonds (cf. carotenoids, Chapter XIV,7). Some examples are linoleic, linolenic, and arachidonic acid, whose formulas are shown here.

For the description of the more common fatty acids a kind of shorthand has been adopted consisting of the number of carbon atoms and double bonds. Oleic acid thus becomes 18:1 (see formula). For the analysis of mixtures of fatty acids, as derived from the saponification of natural fat or lipid fractions, the technique of gas-liquid chromatography is generally used.

Linoleic and *linolenic acids* abound in certain vegetable oils, e.g., linseed oil. More highly unsaturated fatty acids occur in fish liver oils.

Linoleic and linolenic acids are essential food components ("*vitamin F*," see Chapter XXII,2). They cannot be synthesized by the mammalian organism; they are metabolized, however, and converted to even more highly unsaturated acids (often with chain elongation). Any double bonds introduced enter on the side toward the carboxyl end. Unsaturated C_{20} fatty acids furthermore are biosynthetic precursors for *prostaglandins* (see Chapter XXII,4).

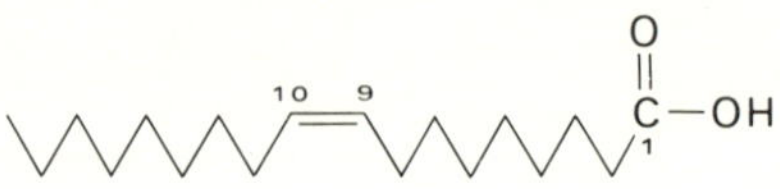

Oleic acid $C_{18}H_{34}O_2$ 18:1

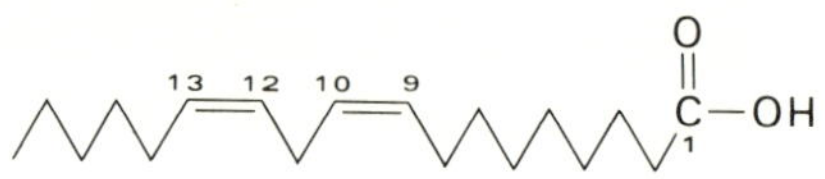

Linoleic acid $C_{18}H_{32}O_2$ 18:2

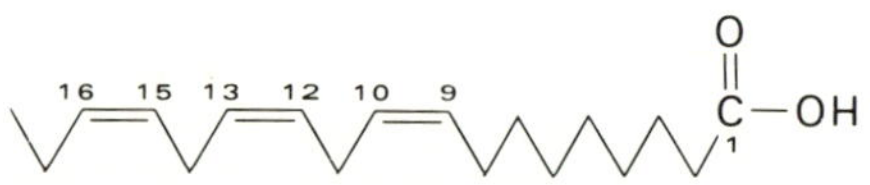

Linolenic acid $C_{18}H_{30}O_2$ 18:3

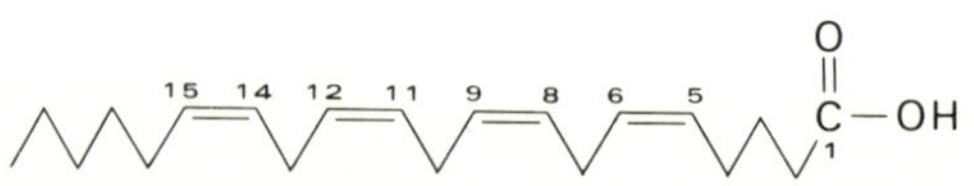

Arachidonic acid $C_{20}H_{32}O_2$ 20:4

Hardening of Fats. Fats that contain a large proportion of saturated fatty acids (there is always a variety of fatty acids in a fat) have a higher melting point and are solid at room temperature, whereas fats with a high content of unsaturated fatty acids are liquid or oily (the term *oil* denotes consistency and not chemical structure). The unsaturated fats can be hydrogenated by adding catalytically activated hydrogen to the double bonds. The melting point of the fats is raised by this procedure so that previously oily fats become solid at room temperature; hence, the term "hardening" of fats. The process plays a significant role in the production of oleomargarine and shortening and consequently in the nutrition of man.

Waxes. Natural waxes (beeswax, spermaceti, plant wax) are mixtures of different materials. The main component is esters of long-chain monohydric alcohols and higher fatty acids. From beeswax, myricin, an ester of palmitic acid and myricyl alcohol, $C_{30}H_{61}OH$, has been isolated; from spermaceti (or "head oil" of the sperm whale), cetyl palmitate, $CH_3(CH_2)_{14}COOC_{16}H_{33}$. Besides these esters, waxes also contain higher molecular weight, e.g., unbranched hydrocarbons (arisen from the decarboxylation of fatty acids), esters of sterols, free fatty acids, and hydroxy fatty acids.

2. Fats as Depot Material

The fats play a major role in human nutrition (and in the nutrition of many animals). Fats are nutrients of high caloric value (cf. also Chapter XXII,1), but their real biological significance lies in the fact that they constitute reserve stores for the organism. Any foodstuffs ingested beyond the normal requirements are largely converted to fat and deposited in appropriate tissues. Then in times of need the fat is again made available. The fat of the liver has a much more rapid turnover than that of the adipose tissue. In the rat the biological half-life for liver fat is 1–2 days; for depot fat it is 15–20 days.

In certain tissues, fats are structural components. They are also found in lipoproteins of blood, which in addition to phosphatides and free fatty acids also contain a certain percentage of triacylglycerols.

Whenever the organism utilizes dietary fats or calls on its fat reserves, it must first degrade them. The initial step is the hydrolytic cleavage catalyzed by lipases to release glycerol and fatty acids.

Lipases. These enzymes are hydrolases, subgroup esterases. They split ester bonds; with the uptake of one molecule of water a free fatty acid and a diacylglycerol are produced; with continued action the second and third acid residues are also removed. The true lipases act only on emulsified fats, not on soluble fats.

Lipases are often cited as examples of synthetic hydrolases. With the proper choice of concentrations *in vitro* (very high proportion of glycerol), appreciable amounts of esters are found at equilibrium. Within the cell or in the intestine, however, the reactant water is present in great excess and, therefore, biosynthesis by lipases is impossible. The equilibrium under *in vivo* conditions is entirely on the side of cleavage (for fat synthesis see Section 6).

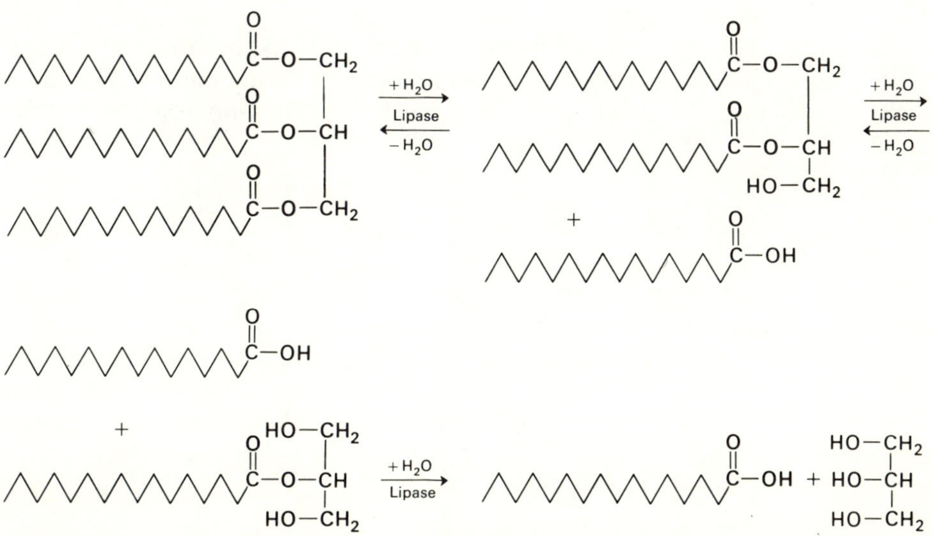

Lipases are widely distributed. Unusually high concentrations are found in the pancreas, in the intestinal wall, and the liver. Pancreatic lipase hydrolyzes only fatty acids in α- or α′-position, whereas the enzyme of the intestinal wall attacks the β-position as well. At least 75% of dietary fat is absorbed as β-monoacylglycerol. The muscosa rebuilds neutral fat, which is transported away through the lymphatic system.

The mobilization of reserves from adipose tissue entails intracellular fat hydrolysis. The liberated "nonesterified" fatty acids are transported to the liver in the blood bound to serum proteins.

The cleavage products follow separate pathways of metabolism. Glycerol, closely related to carbohydrates, is either utilized for the biosynthesis of fructose and glucose or, after phosphorylation, is broken down in the same way as the carbohydrates (Chapter XIX,1). Fatty acids are decomposed according to the rules of β-oxidation to C_2 units (activated acetate) which may then either be used for biosynthesis or be oxidized to CO_2 and H_2O through the citrate cycle (Chapter XI) and the respiratory chain.

3. β-Oxidation of Fatty Acids

Activation of Fatty Acids. Fatty acids are relatively inert chemically. Their reactivity is enhanced when they become converted to thioesters. Thioesters have a great potential for group transfer; they are "energy-rich." The thiol group is provided by coenzyme A, which contains β-mercaptoethylamine (formula Chapter VI,7). The formation of the energy-rich bond requires an expenditure of 1 ATP:

$$CH_3-(CH_2)_n-COOH + ATP + HS-\overline{CoA} \longrightarrow CH_3-(CH_2)_n-CO-S\overline{CoA} + AMP + PP$$

This net equation can be rewritten in two parts. Initially the fatty acid reacts with ATP and an enzyme and by loss of pyrophosphate forms the intermediate acyl adenylate (Chapter VI,5), which is then transformed by $\overline{CoA}$—SH to acyl-S$\overline{CoA}$ and AMP.

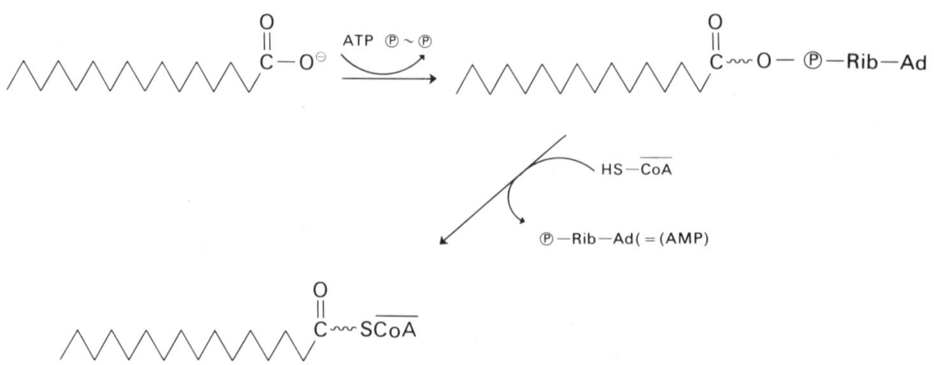

The CoA derivative of fatty acids is in equilibrium with another energy-rich derivative, the *ester of carnitine*:

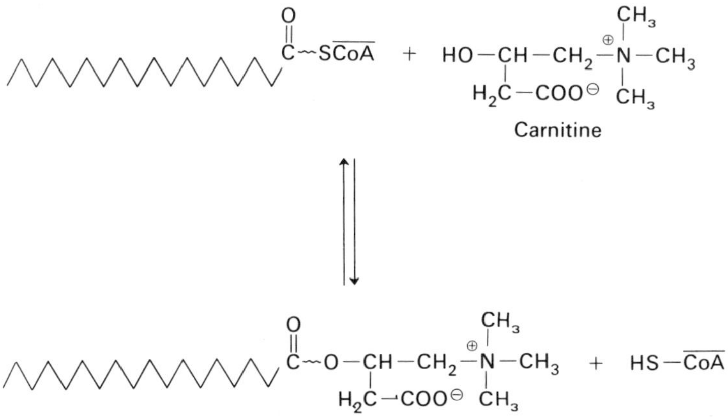

Acyl carnitine

Acyl carnitine is an important intracellular transport metabolite of fatty acids since it can pass through mitochondrial membranes. Inside the mitochondria, the fatty acyl residue is transferred back to HS-$\overline{\text{CoA}}$, and all further reactions take place on CoA derivatives.

Degradation of Fatty Acids. The first step of breakdown (see spiral diagram of formulas below) is a dehydrogenation reaction; the product is the corresponding α,β-unsaturated acyl-CoA compound. The enzyme *acyl-CoA dehydrogenase* contains flavin-adenine dinucleotide (FAD, cf. Chapter VI,4) as the active group. There are three such enzymes, not just one, with differing substrate specificities; the first prefers to act on long-chain fatty acids (C_{18}, C_{16}, C_{14}) the other two on medium- or short-chain fatty acids. The flavoproteins channel the hydrogen via electron-transfering flavoprotein (see Chapter X,4) to the respiratory chain so that they can dehydrogenate more fatty acids.

Water is then added across the double bond of the unsaturated fatty acyl-CoA compound to yield a β-hydroxy acid. Reaction (2) is catalyzed by *enoyl-CoA hydratase.* In the next step, the β-hydroxy group is dehydrogenated to the keto group (reaction 3). The enzyme for this step, called *β-hydroxyacyl dehydrogenase*, transfers hydrogen to NAD, where it is transported along the respiratory chain to become oxidized to water. The dehydrogenated compound, the β-keto acid, is not very stable; with another molecule of $\overline{\text{CoA}}$—SH it is cleaved in the manner indicated below. The process is termed *thioclastic cleavage*, and the enzyme involved is *acetyl-CoA acyltransferase* (β-ketothiolase). The prominent feature of the whole degradation is this reaction (4), the cleavage of a C—C bond. The two scission products are acetyl-CoA (called activated acetate because of its high potential for group transfer) and the remaining acyl-CoA molecule. Since the latter is bound to CoA it need not be activated by ATP by the above scheme in order to react further. The free energy of the cleavage reaction (4) has been preserved as chemical energy in the enzymatic reaction with CoA (instead of water).

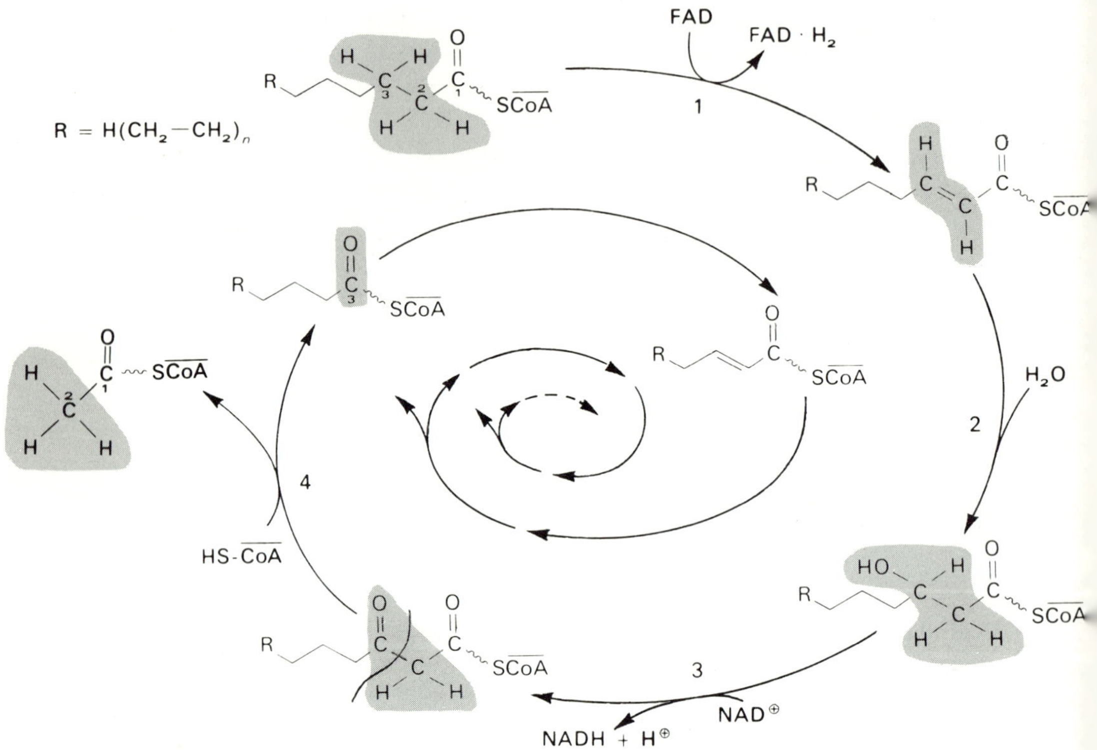

Note that the newly formed acyl-CoA compound can undergo reaction (1) and can run through the whole series of subsequent reactions without the need for another molecule of ATP. It suffices to have a fatty acid, be it 12, or 16, or 22 C atoms long, activated once with CoA (with consumption of 1 ATP) in order for it to be broken down step by step into C_2 units (acetyl-CoA units). Since without exception the chain is shortened by two C atoms at a time, and the dehydrogenation, hydration, and further dehydrogenation afford β-hydroxy and β-keto acids, the entire course of reactions is simply called β-oxidation.[1]

β-Oxidation and the Respiratory Chain. We should stop and ask ourselves what has been achieved up to this point. Energy has been gained first of all by funneling a number of hydrogen atoms through the respiratory chain: From each C_2 fragment

[1] The principle of β-oxidation (degradation by two C atoms) was discovered more than 60 years ago by F. Knoop. The individual steps have been elucidated more recently, chiefly by F. Lynen, F. Lipmann, D. E. Green, and others.

two H atoms release 2 ATP via flavoprotein, and the other 2 H release 3 ATP via NAD, presuming maximum yield from phosphorylation. However, at this point we have not yet seen the production of CO_2—only that of acetyl-CoA. β-Oxidation consequently must be linked to the final breakdown of acetate in the citrate cycle, in order to proceed unimpeded and to derive an optimal yield of energy from the catabolism of fat. Furthermore, the reduced coenzymes must be reoxidized by the respiratory chain. The cooperation of all three processes in the mitochondria therefore is of signal importance.

In certain pathological conditions the collaboration of β-oxidation and operation of the citrate cycle is disturbed. Consequent overproduction of acetyl-CoA causes the abnormal appearance of acetoacetate and acetone in the urine. The formation of these "ketone bodies" will be discussed in Section 5.

4. Metabolism of Unsaturated and Branched-Chain Fatty Acids

Unsaturation. Besides the simple saturated fatty acids, most fats and lipids also have unsaturated fatty acids with one or more double bonds, of the type oleic, linoleic, and linolenic acid (formulas, Section 1). The unsaturated acids are activated and subjected to β-oxidation just as the saturated ones are. The double bonds, however, call for two additional enzymes.

Linoleic acid, for example, is degraded by losing successively three C_2 units. The product is a β,γ-unsaturated acyl-CoA compound. A special enzyme isomerizes it to an α,β-unsaturated compound with *trans*-configuration. Now the enzyme system of β-oxidation can remove two more C_2 units, when another double bond—originally the Δ^{12} double bond—begins to interfere. Since this is a *cis* double bond, the enzymatic hydration yields the D-(−)-hydroxy compound. Another special enzyme inverts it to the antipode L-(+)-hydroxyacyl-CoA. Degradation then can proceed as usual. All unsaturated fatty acids are broken down by this scheme.

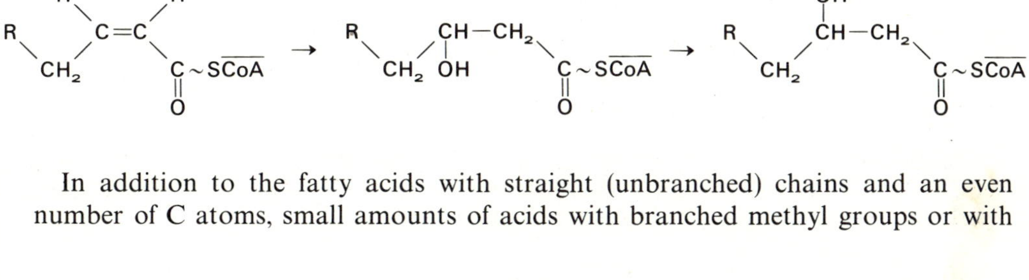

In addition to the fatty acids with straight (unbranched) chains and an even number of C atoms, small amounts of acids with branched methyl groups or with an odd number of C atoms are also found in nature. Their metabolic fate in β-oxidation is of considerable interest and can best be studied on the examples of the methyl branched C_4- and the C_5-carboxylic acids. These two acids arise from the catabolism of the amino acids leucine, isoleucine, and valine by transamination and oxidative decarboxylation, as described in Chapter VIII,10. Basically only two situations need to be discussed: one with a methyl group in the α-position, or potentially in α-position (i.e., by repeated shortening of the chain by two C atoms the methyl group eventually ends up in a α-position); the other with a methyl group in the β-position, or at least potentially in β-position.

α-Branching. *Isoleucine* is initially broken down to α-methylbutyryl-CoA (Chapter VIII,10). The methyl group does not impede the action of the acyl dehydrogenase; indeed, the breakdown continues to proceed normally through the stage of the unsaturated acid, the β-hydroxy acid, and the β-keto acid. The specificity of the enzymes evidently is not strict enough to permit a methyl group to inhibit the reactions. Thioclastic cleavage yields two products: acetyl-CoA and propionyl-CoA. The methyl group was responsible for the formation of the next higher homolog.

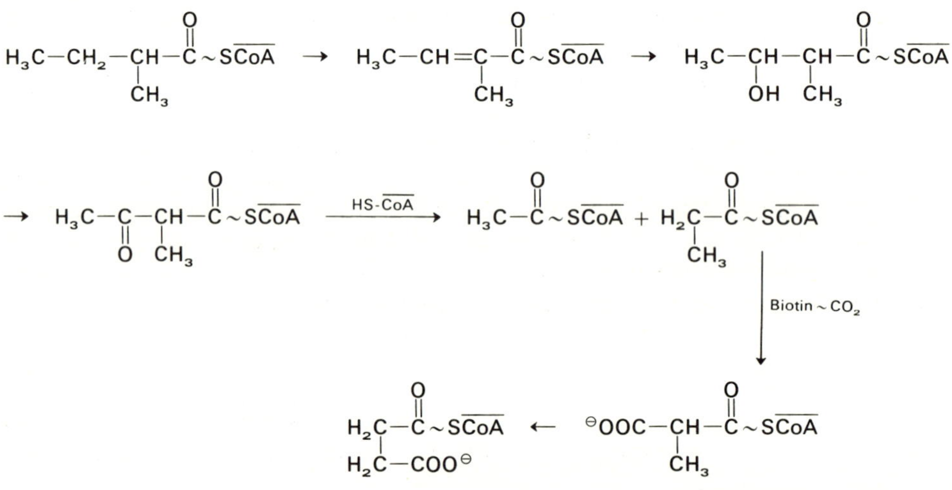

Propionyl-CoA is carboxylated (with the cooperation of a biotin enzyme and ATP; cf. Chapter VI,6) and converted to methylmalonyl-CoA. This derivative of a C_4-dicarboxylic acid can be isomerized (with the aid of a cobalamine coenzyme) to form succinate. The problem of this type of methyl branching is solved for the organism with this last step, because succinate can be metabolized along pathways discussed earlier (cf. citrate cycle, Chapter XI,1).

Unbranched fatty acids with an odd number of carbon atoms similarly end up as propionyl-CoA (instead of acetyl-CoA) after β-oxidation. They are oxidized completely in the same way as the branched-chain acids.

Oxidative decarboxylation of *valine* produces isobutyryl-CoA which is then subjected to a modified fatty acid degradation: Dehydrogenation is followed by addition of water to the double bond. The product is not a secondary, but a primary, alcohol group which apparently is not attacked by the enzyme β-hydroxyacyl dehydrogenase. After the hydrolytic removal of the CoA group, dehydrogenation to methylmalonate semialdehyde can take place. This β-aldehyde acid is easily decarboxylated to yield propionic aldehyde, which subsequently is oxidized to the acid. Propionic acid reacts in the manner indicated above. This scheme points out that, although the degradative pathway is the same in principle, certain detours are followed occasionally. In this case, methylmalonate semialdehyde first is decarboxylated and propionate is carboxylated to give back methylmalonate.

β-Branching. Branching in the form of a methyl group in β-position, for example, as in isovaleryl-CoA, the degradation product of *leucine*, poses a more formidable problem. The sequence of formulas shown reveals that by dehydrogenation the unsaturated compound can be formed but the addition of water would yield a tertiary alcohol which could not be dehydrogenated to form the keto compound. The regular pathway of breakdown is blocked there.

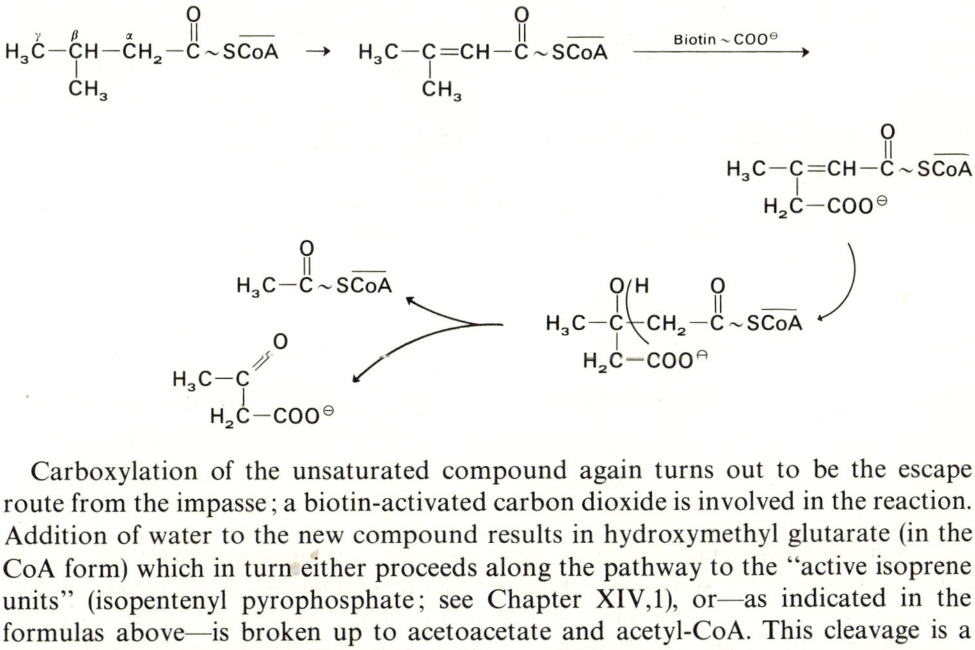

Carboxylation of the unsaturated compound again turns out to be the escape route from the impasse; a biotin-activated carbon dioxide is involved in the reaction. Addition of water to the new compound results in hydroxymethyl glutarate (in the CoA form) which in turn either proceeds along the pathway to the "active isoprene units" (isopentenyl pyrophosphate; see Chapter XIV,1), or—as indicated in the formulas above—is broken up to acetoacetate and acetyl-CoA. This cleavage is a retro aldol condensation and is catalyzed by a special enzyme. We can now appreciate why leucine is particularly ketogenic; it is degraded directly to acetoacetate (and not to acetoacetyl-CoA).

5. Formation of Acetoacetate—"Ketogenesis"

Acetoacetate is a normal metabolic product. As seen in Section 4, it arises from the breakdown of leucine, as well as from phenylalanine and tyrosine (Chapter VIII,11), but the bulk is derived from the catabolism of fat. It has been known for a long time that during starvation and in diabetes mellitus acetoacetate is at a much higher concentration in the blood and spills over in the urine.

Common to these pathological conditions is the accelerated breakdown of fat and consequent elevated production of activated acetate, even exceeding the capacity of the citrate cycle to utilize. Two molecules of acetyl-CoA form *acetoacetyl-CoA*. Most probably this does not take place by a reversal of the thioclastic cleavage, but rather proceeds via malonyl-CoA (discussed in detail in Section 6 below). Acetoacetyl-CoA then can give rise to free acetoacetate by two mechanisms:

1. In the kidneys and in the muscle, by transacylation with succinate:

$$\text{Acetoacetyl—CoA} + \text{Succinate} \rightleftharpoons \text{Succinyl—CoA} + \text{Acetoacetate}$$

2. In the liver, through the "attempt at synthesis," i.e., via *hydroxymethylglutaryl-Co-A*, in the manner shown in the diagram:

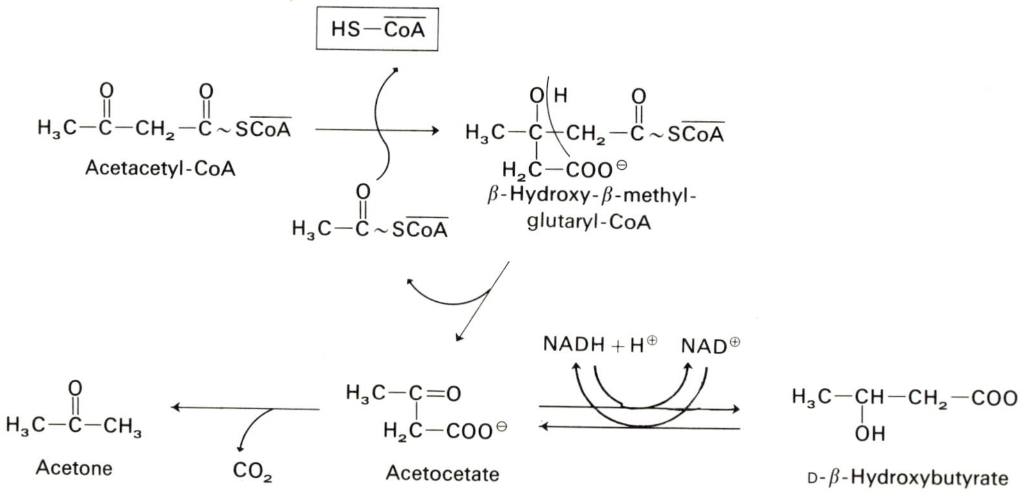

Acetoacetyl-CoA condenses at its carbonyl group with one more molecule of acetyl-CoA. The C=O double bond is opened and a new C—C bond is formed. By the hydrolytic loss of one molecule of CoA, β-hydroxy-β-methylglutaryl-CoA is produced. This branched-chain acid is a precursor of cholesterol (cf. Chapter XIV,1). In our context, β-hydroxy-β-methylglutaryl-CoA is split again to release acetyl-CoA and free acetoacetate. The cleavage is very similar to the reverse of the synthesis. In this cycle a total of two molecules of acetyl-CoA are converted to free acetoacetate and two molecules of CoA. The cycle is a mechanism for regenerating the coenzyme. Free acetoacetate can either undergo spontaneous decarboxylation to form acetone, or it can be reduced by NAD-dependent β-hydroxybutyrate dehydrogenase to *β-hydroxybutyrate* (setting up an equilibrium). Acetoacetate, β-hydroxybutyrate, and acetone (termed collectively "ketone bodies") enter the blood and finally the urine. Their appearance in the urine is significant for pathognomy.

6. Biosynthesis of Fatty Acids

Fat can be synthesized in the mammalian organism from various other foodstuffs, as is amply demonstrated by the hog; when fed with potatoes or bran (i.e., a diet rich in carbohydrates) it puts on much fat.

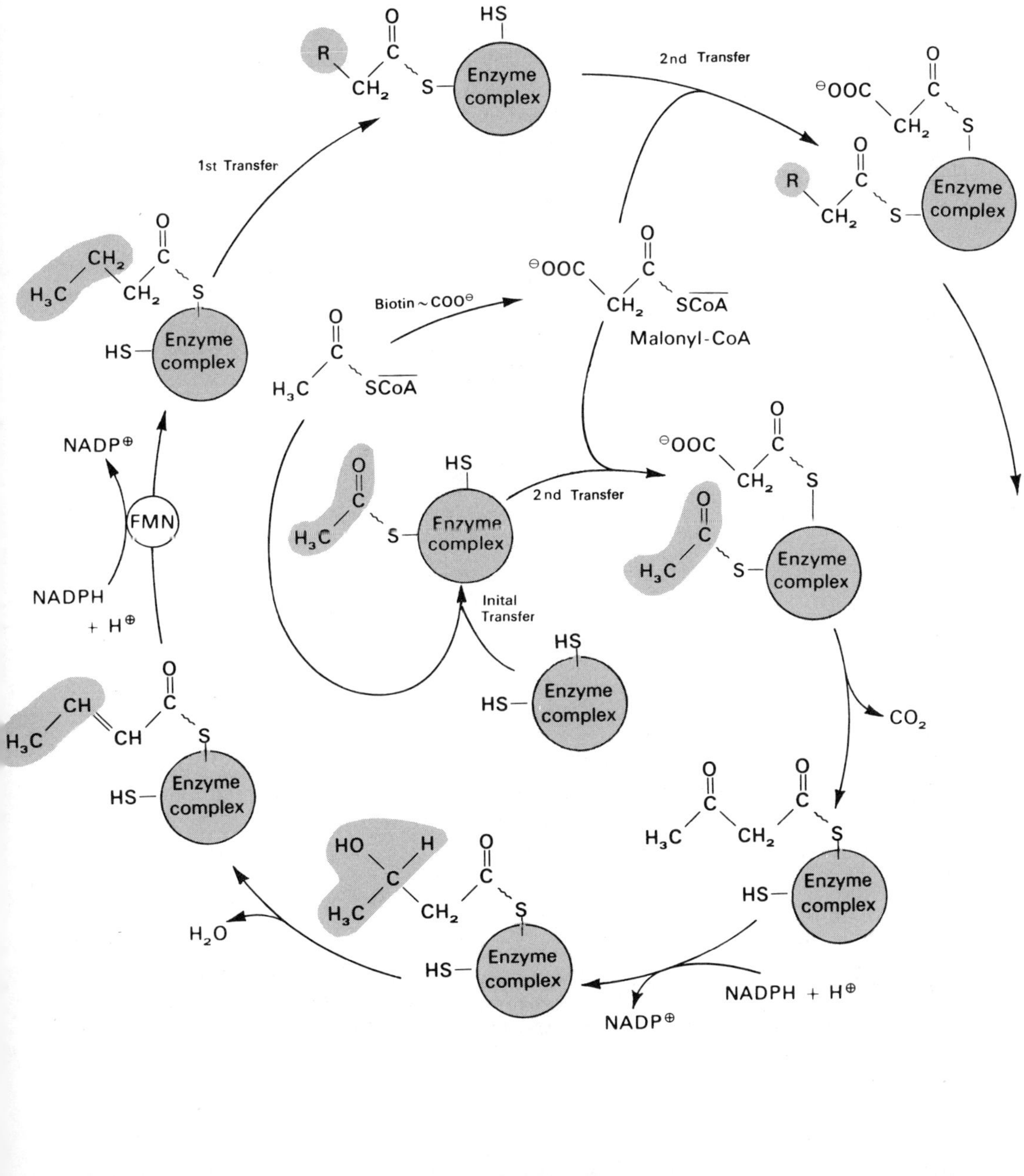

Starting material for the synthesis of fatty acids is *acetyl-CoA*, derived from the aerobic breakdown of carbohydrates by oxidative decarboxylation of pyruvate (see Chapter VIII,10). At first acetyl-CoA is carboxylated and is converted to *malonyl-CoA*, a particularly reactive compound. The carboxylation reaction requires ATP and involves "active carboxyl," *carboxybiotin* (Chapter VI,6). According to Lynen, all subsequent synthetic reactions take place on a multienzyme complex (see Figs. XII-1 and 2). All intermediate products remain attached to the enzymes in the complex.

The reaction sequence begins when an acetyl residue is transferred from acetyl CoA to an SH group of the enzyme complex (transfer 1). A malonyl residue then is transferred from malonyl-CoA to another reactive SH group of the complex (transfer 2). At this point the acetyl group condenses with the highly reactive methylene group (CH_2) of the malonyl residue to form a β-keto acyl derivative. This condensation elongates the C chain by a C_2 unit. CO_2 is eliminated in the process. The subsequent steps of reduction to the hydroxy group, removal of a molecule water, hydrogenation of the double bond, all proceed formally as a reversal of β-oxidation, with the exception of employing NADPH as reducing agent for the keto group. The enzyme instrumental in the second reduction reaction contains flavin mononucleotide (FMN) as prosthetic group.

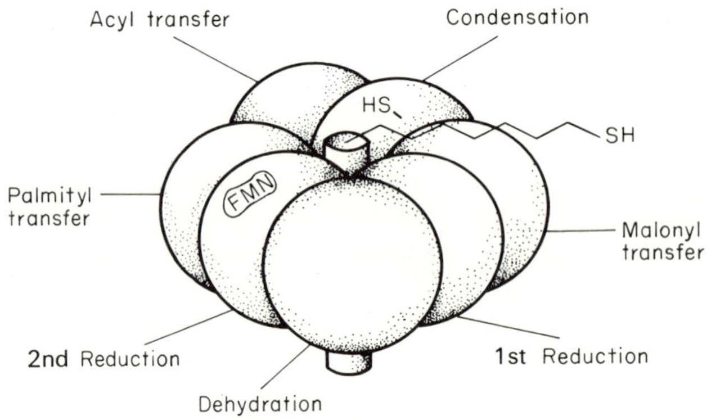

Fig. XII-1. Diagram of the multienzyme complex of fatty acid synthesis.

The elongated acyl residue now is transferred back to the first SH group of the enzyme complex, so that the central SH group is free to accept another malonyl residue. With this, the reaction sequence can start again. Finally the acyl residue is transferred back to coenzyme A and released as activated fatty acid, but only after having reached the proper chain length of C_{16} or C_{18}. Thus neutral fat contains predominantly palmitic and stearic acid.

In bacteria (*E. coli*), a low-molecular weight protein, itself without enzymatic activity, combines with the acyl group, and the various soluble enzymes of fatty acid synthesis act on this complex (Vagelos and coworkers). This *acyl-carrier protein* (ACP) contains 4′-phosphopantetheine as prosthetic group.

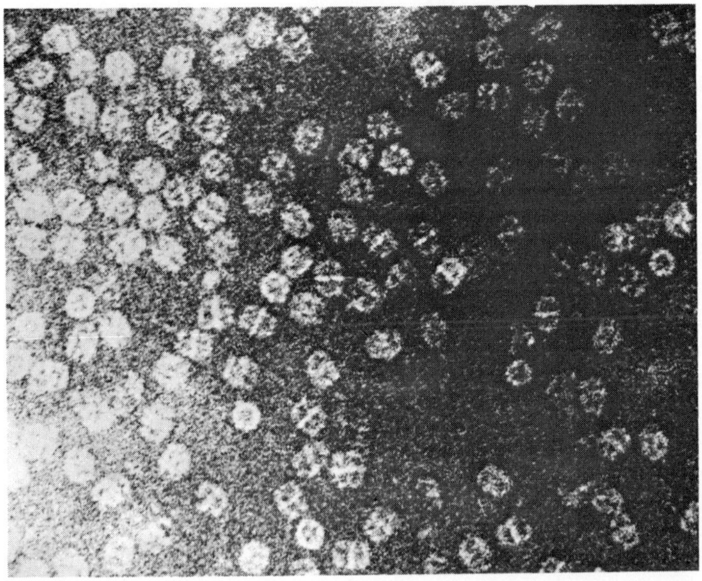

Fig. XII-2. Complex of fatty acid synthetase from yeast. Note the subunits of the structures. (Electron micrograph by Dr. P. Hofschneider, enlarged 200,000-fold.)

It should be pointed out that fatty acid synthesis proceeds more or less as the reverse of β-oxidation, but deviates in at least one significant step (see footnote 4 in Chapter XIX,2, for other differences). A very similar situation will be found with the carbohydrates. The detour is more favorable thermodynamically (synthesis is favored by the equilibrium) but does require one extra ATP per C_2 unit. For *separate control of the two pathways* it is also crucial to realize that the entire synthetic sequence proceeds with all intermediates firmly bound to an enzyme complex (or in bacteria bound to the acyl-carrier protein). None of the intermediates thus can shuttle between synthetic and degradative pathways. Last, β-oxidation takes place in the mitochondria while fatty acid synthesis proceeds in the cytosol.

Synthesis of Fats from Fatty Acids. The above biosynthesis releases fatty acids in the form of the acyl-CoA derivatives. They are then stored as glycerol esters, i.e., as neutral fats. Esterification does not take place on the glycerol molecule itself, but rather on glycerol phosphate, which could have arisen, for example, from reduction of dihydroxyacetone phosphate (cf. Chapter XIII,2). There is an enzyme which is responsible for the formation of diacylglycerol phosphate (also called phosphatidic acid) from glycerol phosphate and 2 moles of activated fatty acid. CoA—SH is released in the process.

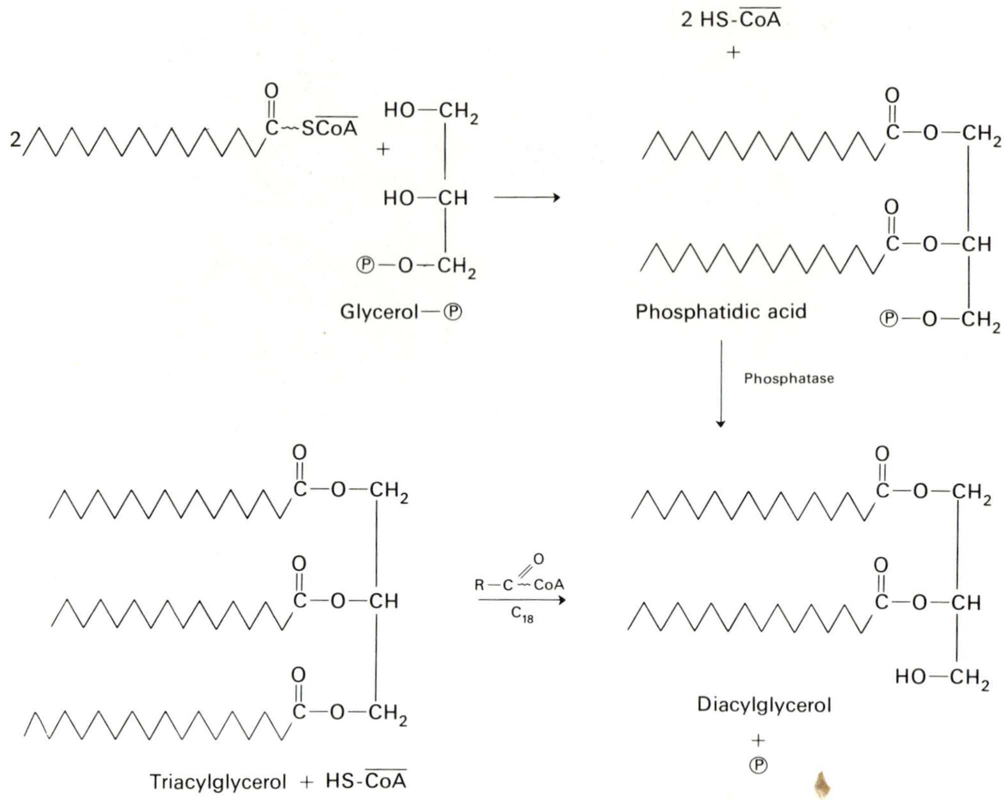

The enzyme is not very specific for definite chain lengths of the activated acids; it reacts more swiftly, however, with C_{16}-, C_{17}-, and C_{18}-fatty acids. In the last step, the diacylglycerol phosphate is dephosphorylated by the action of a phosphatase and is brought to reaction with another mole of acyl-CoA. So-called neutral fat (triacylglycerols) have been formed and can now be transported and stored as depot fat (or organ fat).

The resynthesis of triacylglycerols in the intestinal wall differs somewhat. Here the monoacylglycerol absorbed by the intestine are used directly and converted to triacylglycerol by the reaction with two molecules of acyl-CoA. These triaclyglycerols then are transported to the liver in form of chylomicrons in colloidal dispersion.

BIBLIOGRAPHY

F. C. Gran (editor), "Cellular Compartmentalization and Control of Fatty Acid Metabolism," Academic Press, New York, 1968.

F. Lynen, The role of biotin-dependent carboxylations in biosynthetic reactions. *Biochem. J.* **102,** 381–400. (1967).

J. J. Volpe and P. R. Vagelos, Saturated fatty acid synthesis and its regulation. *Annu. Rev. Biochem.* **42,** 21–60 (1973).

CHAPTER **XIII**

Phospholipids, Glycolipids, and Membranes

Structure and function of biological membranes is an area of biochemistry that in recent years has become increasingly prominent. Many cell properties are determined by the surface of cell membranes. Moreover, within cells numerous membranes form various compartments of the cell (see Chapter XVIII) and many of these membranes also contain structure-bound enzymes, as already seen in the case of the respiratory chain and the microsomal hydroxylation system. The intensive interest in the study of biological membranes thus is well justified.

Membranes are formed of proteins and lipids, particularly phospho- and glycolipids. The latter possess both hydrophobic and hydrophilic groups in one molecule and, therefore, in aqueous media spontaneously form ordered structures called micelles. These properties are essential also for the structure of biological membranes.

Originally the term "lipid" was applied to all substances with solubility properties similar to those of fats. In addition to phospholipids and glycolipids, lipids also include sterols and several isoprenoid lipids, which will be discussed in Chapter XIV. First we will present the biochemistry of phospho- and glycolipids and derive the various characteristic structures on the basis of their biosynthetic pathways.

1. Structural Features

Phospholipids, often called *phosphatides*, chemically are phosphodiesters. Phosphoric acid is esterified on the one side with a derivative of either sphingosine or glycerol (usually diacylglycerol) and on the other side with either choline, ethanolamine, serine, inositol, or another glycerol. The three components choline, ethanolamine, and serine contain basic nitrogen which bears a positive charge at physiological pH. And since the phosphate group has a negative charge, phosphatides are amphoteric (zwitterions).

Glycolipids in contrast do not contain phosphate,[1] but rather a mono- or oligosaccharide residue which very often is linked to sphingosine. As a result, *glycerophosphatides* have been distinguished from *sphingolipids*; although a useful distinction, it is not without problems. We prefer the classification of phospholipids on the one hand and glycolipids on the other. Within these classes the further distinction between glycerol and sphingosine derivatives can be made.

It is additionally important to realize that almost all phospho- and glycolipids, also known by such trivial names as lecithin, cerebroside, etc., are complex mixtures of closely related compounds: Similar to neutral fats, these also contain fatty acids with varying chain lengths and a number of double bonds. The extraction of these lipids from biological material generally yields difficult to resolve mixtures. The isolation of a pure component remains difficult even with modern methods. Consequently, new types of compounds continue to be discovered.

2. Biosynthesis and Structure of Glycerophosphatides

The basic component common to all glycerophosphatides is *sn-glycerol 3-phosphate*.[2] As seen in the diagram below, it can be esterified with the two moles of acyl-CoA to yield a phosphatidic acid, which is also an intermediate product in the synthesis of neutral fats.

Another pathway is followed in liver microsomes. It begins with the esterification of the CH_2OH group of dihydroxyacetone phosphate with a saturated fatty acid. The keto group is then reduced and the resultant secondary HO group is acylated through another enzyme that prefers an unsaturated fatty acyl-CoA as acyl donor. In this manner phosphatidic acids are generated that bear a saturated fatty acid at C-1 and an unsaturated one at C-2; this substitution pattern is typical for many phosphatides.

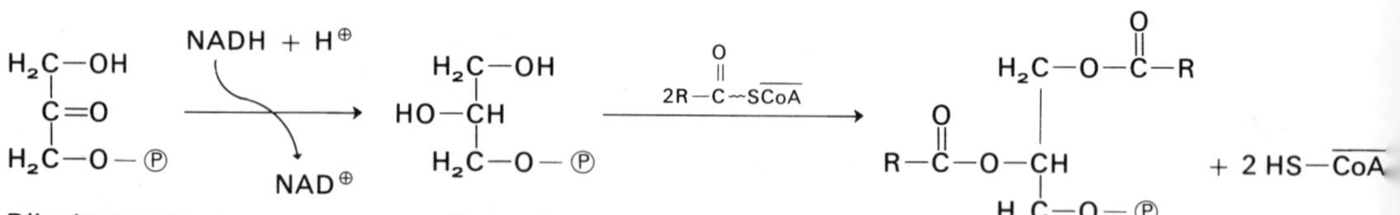

Dihydroxyacetone phosphate — L-α-Glycerol 3-phosphate — Phosphatidic acid

Linkage of the Phosphodiester bond. Since phosphatides are diesters of phosphoric acid the phosphatidic acid must be linked to a second alcohol component

[1] Certain bacteria, however, contain lipids that consist of phosphate as well as a sugar moiety; these could be termed phospho-glycolipids.

[2] The prefix "*sn*" stands for stereospecific numbering according to which convention glycerol has the L-configuration (HO is to the left in Fischer projection formulas) because of the close relationship to L-α-glycerolphosphate. See "The Nomenclature of Lipids," *J. Biol. Chem.* **242**, 4845–4849 (1967).

to complete the molecule. By reaction with cytidine triphosphate (CTP) a compound with a high potential for group-transfer is formed; it is *cytidine diphosphate diacylglycerol.* A specific transferase then transfers the phosphatidyl residue onto the HO group of serine and by removal of CMP produces *phosphatidylserine.*

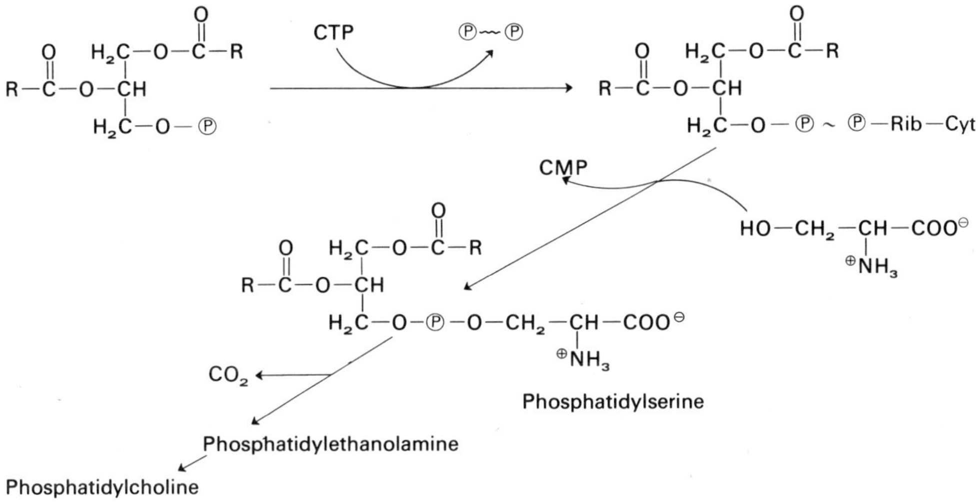

Phosphatidylserine thus is the first phosphatide in this sequence. In a pyridoxal phosphate-dependent reaction it can then be decarboxylated to yield *phosphatidyl ethanolamine,* from which finally *phosphatidylcholine* (= *lecithin*) is produced by the transfer of three methyl groups from *S*-adenosylmethionine to the $H_3N^{\oplus}$ group.

In an alternate pathway of forming lecithin, choline is first phosphorylated with ATP and subsequently activated with CTP. The resulting *CDP-choline* (see formulas in Chapter VI,8) undergoes reaction with a diacylglycerol to form DMP and phosphatidylcholine. This pathway is significant for the reutilization of free choline.

Phosphatidylserine and phosphatidylethanolamine together have also been called *cephalins.* Lysolecithins and lysocephalins are the cleavage products of the enzyme phospholipase A (see below); the 1-position of glycerol still has a fatty acid attached, but the hydroxyl group of C-2 is free.

Figure XIII-1 summarizes the structural relationship of the three main glycerophosphatides. The fatty acid composition resembles that of the neutral fats. Generally, the fatty acid linked to C-1 is saturated (predominantly palmitate), while at C-2 it is usually unsaturated (oleate, linolenate, or higher homologs thereof). Mixtures of these combinations are invariably obtained from extracts of biological materials.

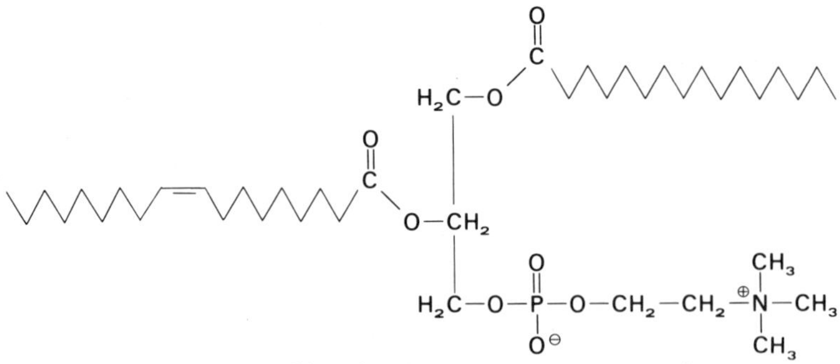

Phosphatidylcholine (Lecithin)

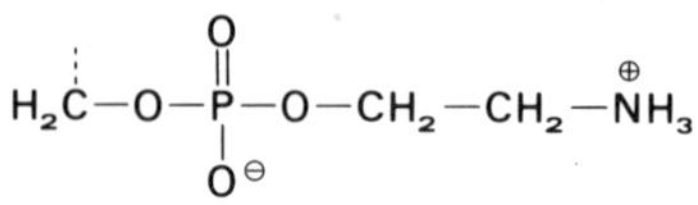

Phosphatidylethanolamine

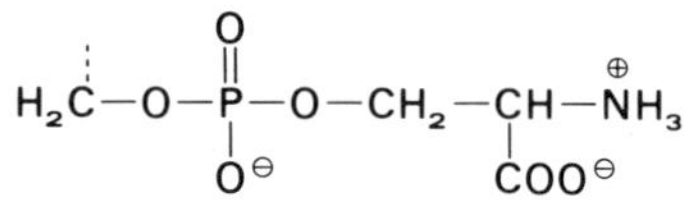

Phosphatidylserine

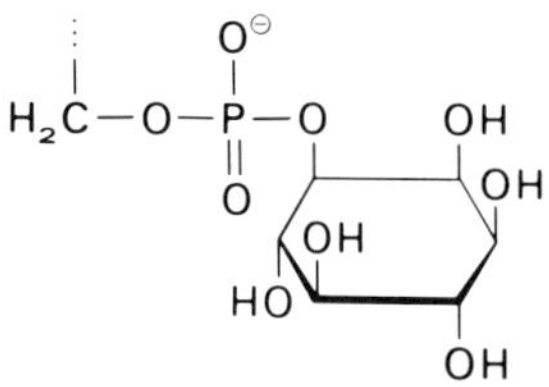

Phosphatidylinositol

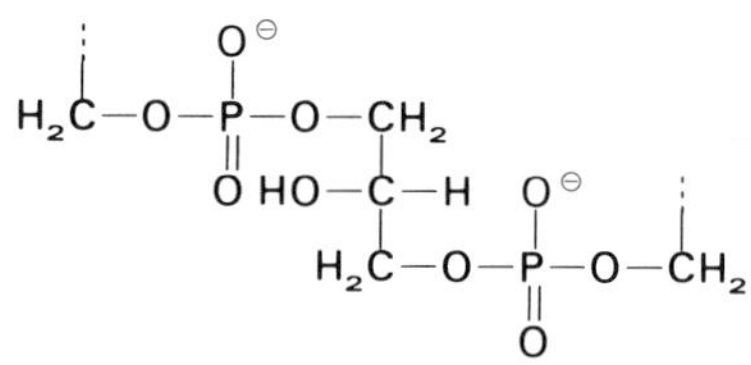

Bisphosphatidylglycerol
(Cardiolipin)

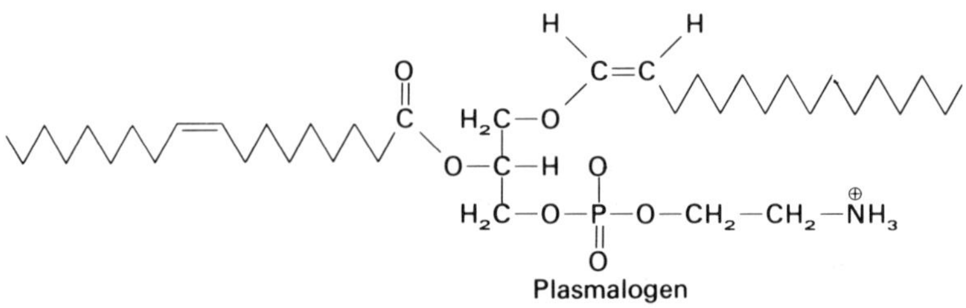

Plasmalogen

[(1-Alk-1′-enyl-2-acylglycerol-3-phosphoryl)ethanolamine]

Fig. XIII-1. Comparison of structures of common glycerophosphatides.

The prime importance of glycerolphosphatides is their incorporation in biological membranes (see Section 5). They play an additional significant role in metabolism: Fatty acids synthesized *de novo* in the liver are used there to form phosphatides (chiefly phosphatidylcholine plus some phosphatidylethanolamine) which in turn are incorporated into lipoproteins and in this form fatty acids are transported into the blood. Adipose tissue can trap the phosphatides and store the fatty acids as triacylglycerols. Disturbances of hepatic phosphatide biosynthesis interferes with the exportation of fatty acids and results in possibly pathological fat accumulation in the liver.

Phosphatidylinositol. Besides serine, another acceptor of the activated phosphatidyl residue during biosynthesis often is *myo*-inositol, which is a cyclic, hexahydric alcohol (i.e., a carbocyclic ring with six HO groups). The product is phosphatidylinositol.

The six-membered ring defines a plane in reference to which the substituent HO groups are oriented either "above" or "below." There are nine possible stereoisomers (7 meso and 2 optically active forms). *myo*-Inositol occurring free in muscle tissue is one of the symmetric, optically inactive meso forms, as can be seen clearly in the formula on the left. The right formula stands for the same *myo*-inositol but in chair form; five of the six HO groups evidently are oriented equatorially.[3]

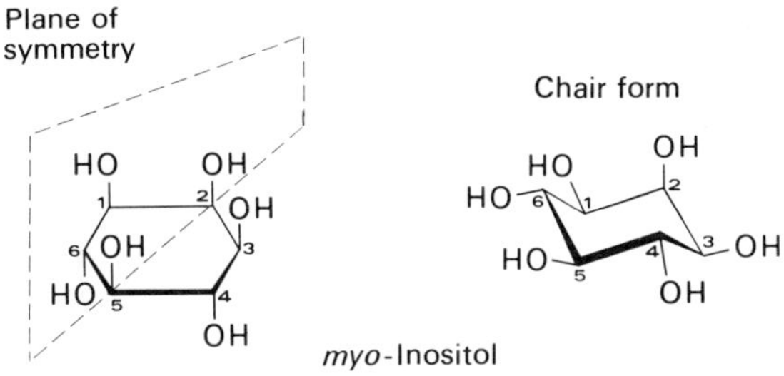

myo-Inositol

In phosphatidylinositol hydroxyl 1 of inositol is esterified with phosphatidic acid (see formula). This type of compound is formed in many tissues. In brain the phosphorylated derivatives 1-phosphatidylinositol 4,5-bisphosphate have also been identified. Finally, plants also contain an inositol hexaphosphate called *phytic acid.*

Cardiolipin. Analogous to the reaction mechanism discussed above, the linkage of two phosphatidyl groups with the two HOH_2C groups of glycerol results in cardiolipin originally isolated from heart muscle. It is component of the mitochondrial membrane. The formula is shown here:

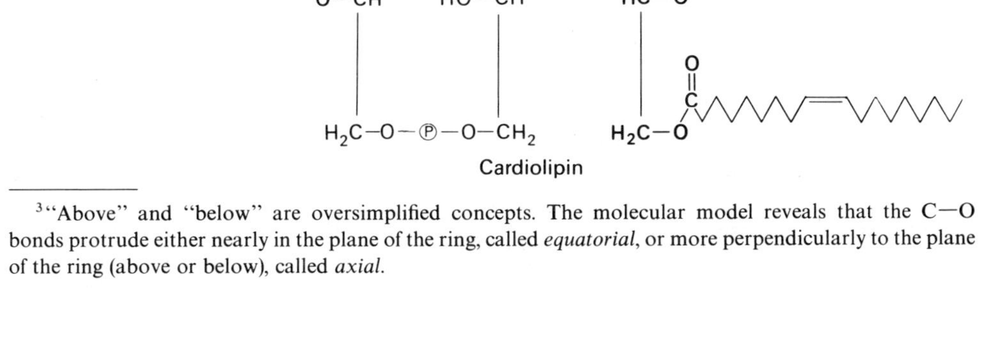

Cardiolipin

[3]"Above" and "below" are oversimplified concepts. The molecular model reveals that the C—O bonds protrude either nearly in the plane of the ring, called *equatorial*, or more perpendicularly to the plane of the ring (above or below), called *axial*.

Plasmalogens. These phosphatides are distinguished by an enol ether group in 1-position of glycerol. Their biosynthesis proceeds via a glycerol ether; the following sequence of reaction steps has been postulated:

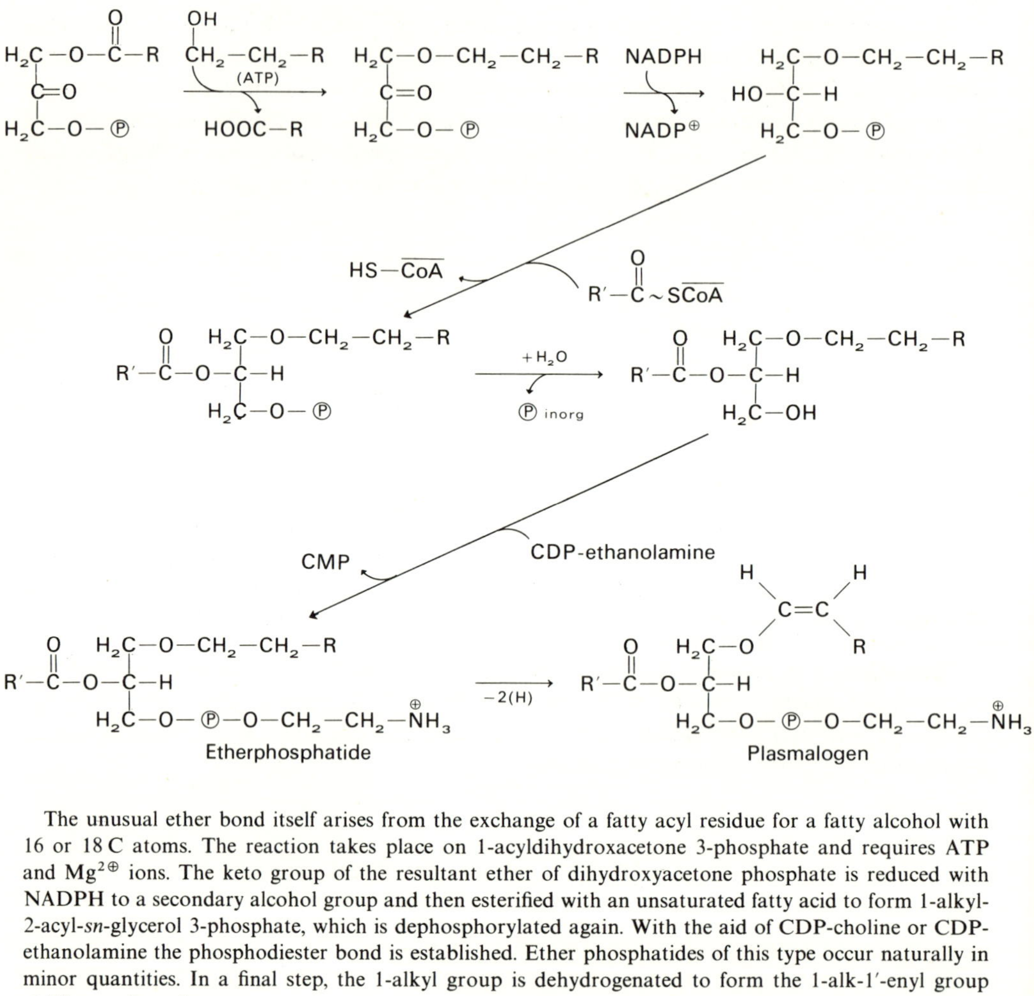

The unusual ether bond itself arises from the exchange of a fatty acyl residue for a fatty alcohol with 16 or 18 C atoms. The reaction takes place on 1-acyldihydroxacetone 3-phosphate and requires ATP and $Mg^{2\oplus}$ ions. The keto group of the resultant ether of dihydroxyacetone phosphate is reduced with NADPH to a secondary alcohol group and then esterified with an unsaturated fatty acid to form 1-alkyl-2-acyl-*sn*-glycerol 3-phosphate, which is dephosphorylated again. With the aid of CDP-choline or CDP-ethanolamine the phosphodiester bond is established. Ether phosphatides of this type occur naturally in minor quantities. In a final step, the 1-alkyl group is dehydrogenated to form the 1-alk-1′-enyl group yielding a plasmalogen.

Plasmalogens contain only unsaturated fatty acids; the basic group consists of ethanolamine or choline. The treatment with mineral acid cleaves the enol ether bond and releases aldehydes that can be detected histochemically. This aldehyde reaction in cytoplasm (= plasmal reaction) gave rise to the term *plasmalogen.*

Phosphatide-Cleaving Enzymes (Phospholipases). Both the carboxylate and phosphate ester bonds of phosphatides can be cleaved hydrolytically (the hydrolyses are exergonic). The applicable enzymes are, in part, highly specific. The following diagram illustrates the points of attack of several phospholipases using a lecithin as an example.

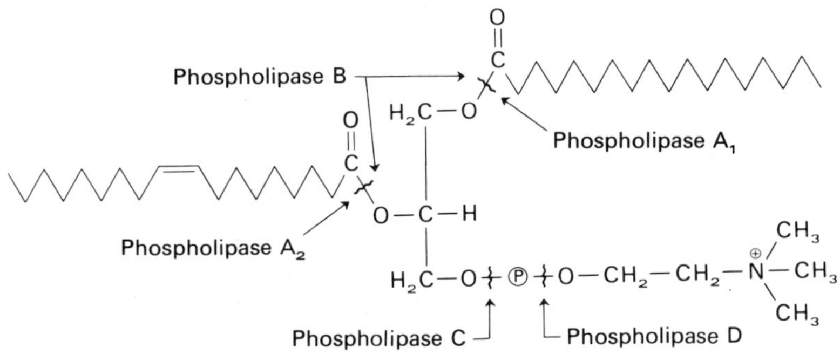

The phospholipases A are carboxyesterases by their specificity; they are mainly found in the venom of snakes and bees. The products of the enzymatic action are the *lysolecithins*, so called because they hemolyze erythrocytes. It was once believed that the poisonous quality of the venoms could be traced to that effect, but instead snake venom contains potent neurotoxins without phospholipase activity. Phospholipase B splits off both fatty acids. Both phosphatidase C (from gas gangrene bacilli) and D (from plants) are classified as *phosphodiesterases*.

Phospholipases of type A and B were discovered in the pancreas and other animal organs; nerve tissue contains a specific glycerol 3-phosphorylcholine diesterase, which further breaks down the cleavage products of the action of phospholipase B.

3. Sphingosines and Sphingomyelins

Sphingolipids contain the aminodiol *sphingosine* instead of glycerol. The structure of sphingosine is derived best from its biosynthesis. The enzymatic, pyridoxal phosphate-dependent condensation of serine and palmityl-CoA with simultaneous decarboxylation yields a 3-keto compound that can be reduced by NADPH to *sphinganine* (dihydrosphingosine).

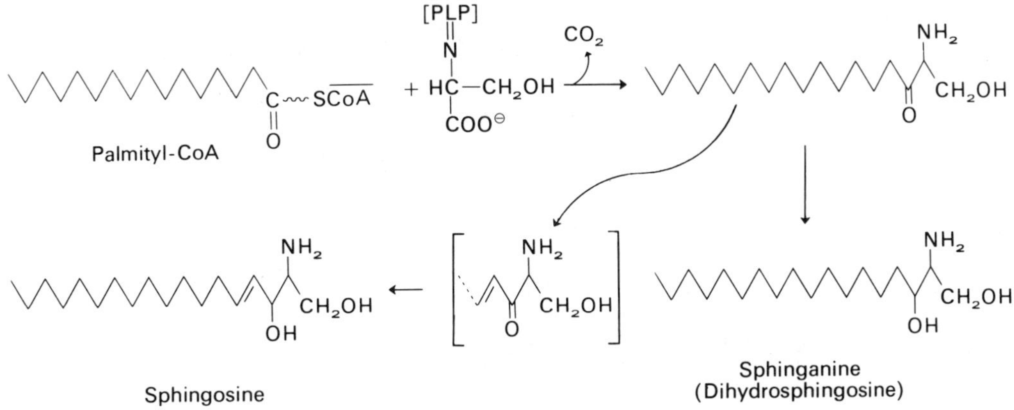

If a double bond is inserted next to the keto group before the reduction step, then the final product is sphingosine. *Sphingosine* thus features a long-carbon chain (C_{18}), a double bond with *trans*-orientation, one amino group, and two hydroxyl groups. Its steric configuration is that of a D-*erythro* compound.

Dihydrosphingosine, just mentioned as a side product, also appears in small amounts in sphingolipids. Furthermore, in plants there is *phytosphingosine* (4-hydroxydihydrosphingosine). In animals one can find in addition to (C_{18}) sphingosine the homologous C_{20} compound (eicosasphingosine), especially as a component of gangliosides. It may well arise analogously from serine and stearyl-CoA.

The *breakdown of sphinogosine* begins with a phosphorylation of the hydroxyl group at C-1 with ATP. A specific aldolase then splits the product into phospho-ethanolamine and the unsaturated fatty aldehyde hexadec-2-en-1-al. The aldehyde can either be reduced further to the corresponding fatty alcohol and utilized for the synthesis of ether phosphatides (see above) or be oxidized to palmitate and thus enter the general pool of fatty acids.

Ceramides and Sphingomyelins. Naturally occurring sphingolipids invariably bear a fatty acyl residue on their amino group. Such amides of sphingosine and a fatty acid are called *ceramides* and occur as such in small amounts.

Sphingosine phosphatides are called *sphingomyelins*, because of their occurrence in the myelin sheaths of nerves. Sphingomyelin has a fatty acid residue in amide linkage with its nitrogen and phosphorylcholine on its terminal hydroxyl group. The formula is shown below. It has an obvious similarity to the glycerophosphatides. The biosynthesis, too, proceeds analogously from CDP-choline and ceramide.

Sphingomyelins are also isolated as mixtures with different fatty acid components. Predominant among these are the acids with 24 C atoms lignoceric acid (saturated) and nervonic acid (one double bond).

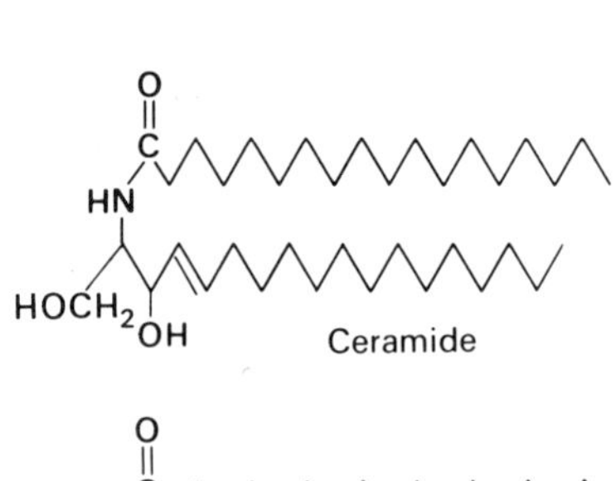

Ceramide

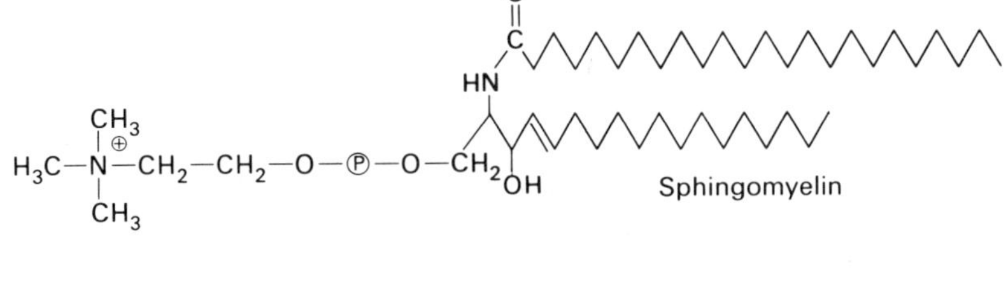

Sphingomyelin

4. Glycolipids

Glycolipids have become increasingly prominent in the biochemical literature in recent years. They embody both a lipid and a carbohydrate moiety; the latter constitutes the hydrophilic pole of the molecule.

Glyceroglycolipids have a relatively simple structure. They contain 1,2-diacylglycerol and a mono- or oligosaccharide bound glycosidically in position 3 of the glycerol. They are particularly abundant in bacteria, although they have been found in mammalian systems.

Glycosphingolipids are much more important. The main component is *ceramide*, which we have already encountered above. They can be categorized in one of three classes depending on the nature of the carbohydrate moiety: Neutral glycosphingolipids, sulfatides, and gangliosides.

Cerebrosides are simple neutral glycosphingolipids. Their carbohydrate is a monosaccharide. In cerebrosides derived from brain the monosaccharide is predominantly galactose (see formula); in those derived from parenchymatous organs (liver, spleen, and others) it is mostly glucose. The fatty acids are saturated C_{24} acids that occasionally feature a hydroxyl group in α position.

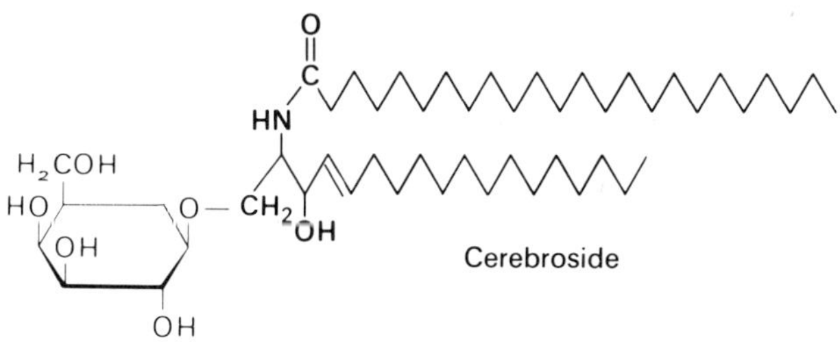

Cerebroside

Besides cerebrosides there are also compounds consisting of a ceramide and a di-, tri-, or tetrasaccharide. Such higher neutral glycosphingolipids are widely distributed but exist only in very low concentrations. By their chemical structure they can be considered to be the neuraminate-free basic form of gangliosides; there are also biogenetic relationships.

Sulfatides are esters of sulfuric acid and neutral glycosphingolipids. One very common type is derived from a galactocerebroside with a sulfate group at C-3 of galactose. In a certain storage disease (see below) it occurs in greatly increased amounts (Yatzkewitz).

Gangliosides are characterized by the presence of one or more sialic acid residues in glycosidic linkage. *Sialic acid* is the generic term for *N*-acetyl- and *N*-glycolyl-neuraminate. (Neuraminate is the condensation product of mannosamine and

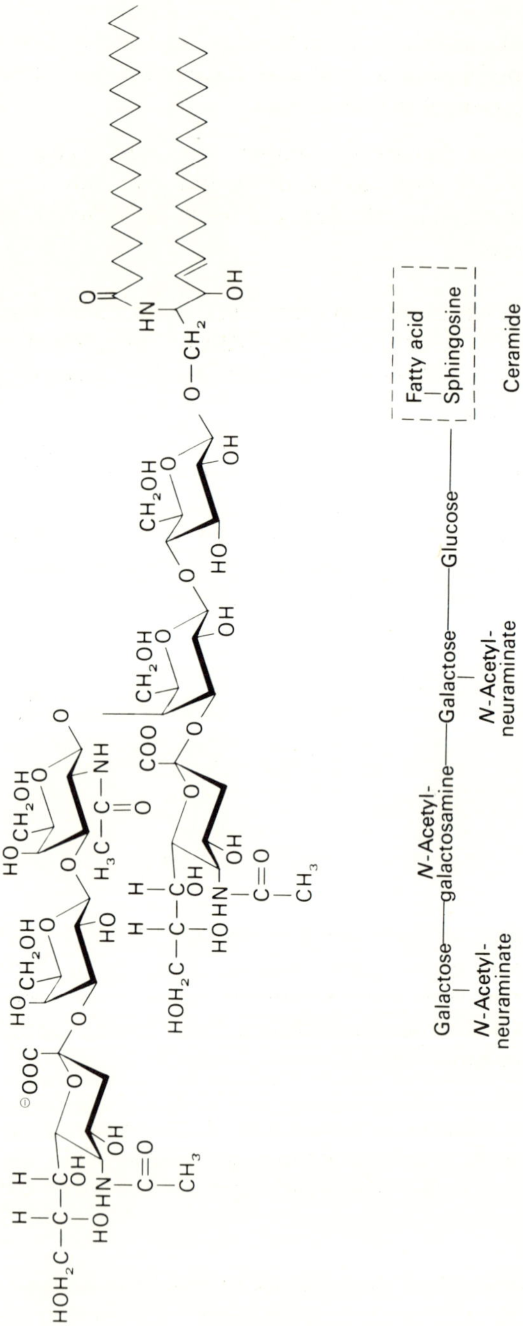
Galactose
N-Acetyl-neuraminate
N-Acetyl-galactosamine
Galactose
N-Acetyl-neuraminate
Glucose
Fatty acid
Sphingosine
Ceramide

pyruvate, see Chapter XV,4.) Gangliosides consist then of a ceramide portion, a neutral oligosaccharide with generally two to four sugar residues, primarily glucose, galactose, and *N*-acetylgalactosamine, and one or more sialic acid residues. The structural formula shown here is that of a commonly found disialoganglioside. The enzymatic cleavage of the terminal *N*-acetylneuraminate residue frees a common monosialoganglioside.

Particularly high concentrations of gangliosides occur in the gray matter of the brain, but they are also found in other organs, especially in cellular surface membranes. Gangliosides can neutralize tetanus toxin.

Neuraminate-containing components of cellular membranes play a role in the attachment and penetration of virus particles into the cell. The virus receptors of the cellular membrane can be destroyed by an enzyme from *Vibrio cholerae*. The *receptor-destroying enzyme* is a *neuraminidase* which hydrolytically splits off neuraminate. Neuraminidases are found in several viruses, e.g., in the influenza virus.

Sphingolipidoses are inherited metabolic defects in which certain sphingolipids are stored in the organism in greater than normal amounts. These diseases appear in infancy or in early childhood and usually are rapidly fatal. The most common symptom is mental retardation as a result of brain damage. The diseases are caused by the absence of certain enzymes that normally accomplish the hydrolytic degradation of the corresponding sphingolipids. Anabated *de novo* synthesis in the presence of deficient breakdown results in the accummulation of sphingolipids. The following diagram (Fig. XIII-2) shows which of the breakdown steps can be blocked. In each instance the sphingolipid occurring just before the blocked step accummulates abnormally.

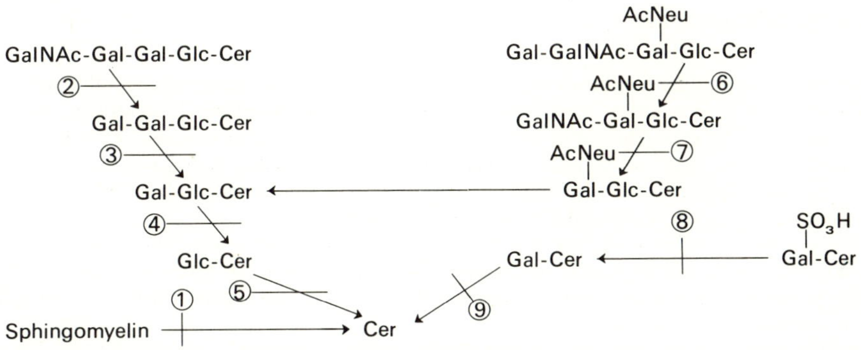

Fig. XIII-2. Breakdown of sphingolipids. When enzyme defects obstruct degradative steps designated by numbers the following storage diseases appear: 1. Niemann–Pick's disease; 2. Sandhoff's disease; 3. Fabry's disease; 4. Lactosyl ceramidosis; 5. Gaucher's disease; 6. Generalized gangliosidosis; 7. Amaurotic idiocy of type Tay-Sachs; 8. Metachromatic leucodystrophy; 9. Leucodystrophy of type Krabbe. *Abbreviations used:* Cer, ceramide; AcNeu, acetylneuraminate; Gal, galactose; GalNAc, *N*-acetylgalactosamine; Glc, glucose.

5. Biological Membranes

Research into ultrastructure has shown that cells of higher organisms are not only enclosed in a cellular membrane but are also filled with a multitude of plasma membranes. It has become clear in recent decades that plasma membranes with their structure-bound enzymes play an important role in metabolism. The biochemical role of individual cell organelles that are delineated by plasma membranes will be detailed in Chapter XVIII. Here we will discuss the molecular structure of these membranes.

Components of the Membranes. All membranes consist of protein and lipid. The protein component, depending on the type of membrane, makes up on the average between 40 and 60%. Extremely protein rich (80%) is the inner membrane of mitochondria (see Chapter XVIII,3); in contrast, the myelin sheaths of nerves have only a 25% protein and a 75% lipid content. These proteins generally are sparingly soluble in water and tend to aggregate. Some have the property of lipoproteins, or rather apolipoproteins.

The lipids are predominantly *phosphatides* (lecithin, phosphotidylethanolamine); *cholesterol* is present especially in relatively rigid membranes. Quantitative analysis has shown that the composition is characteristic for the type of membrane and for the particular organ. Even the content of unsaturated fatty acids of the phosphatides is characteristic, although the composition can be influenced by nutrition. There is no constant composition in the sense of a stoichiometric relationship.

In order to understand the molecular architecture of membranes, it is important to consider the behavior of lipids in aqueous media. We can distinguish three modes of behavior: 1. phase separation without molecular order; 2. formation of micelles; and 3. formation of lamellae, which corresponds to the type of order predominating in biological membranes.

The first case, typical for hydrocarbons (mineral oil, paraffin oil) and largely also for oily triacylglycerols is not relevant to this discussion. The *formation of micelles* occurs whenever lipids arrange themselves as illustrated in Fig. XIII-3a: They form a structure such that hydrophobic residues lie on the inside while hydrophilic residues face outside. Such micelles can form long tubules or they can order themselves to spheres or disc-shaped elipsoids, in whose interior neutral fats can be encapsulated (for example, during the cleaning action of soaps). Glycolipids form such micelles.

Lamellae, on the other hand, are double layers which theoretically are infinite in two dimensions (Fig. XIII-3b). These can be stacked in sandwich fashion, whereby the hydrophilic poles of the lipids begin to interact. Phosphatides generally form lamellae in water; "the myelin structures" obtainable from lecithin and water show such a structure.

Lamellae consisting of but one double layer can be prepared artificially from phosphatides. They appear black in frontal view in the electron microscope and consequently have been designated "black membranes." For many investigations into the behavior of biological membranes these are suitable model systems.

Structure of Membranes. In the electron microscope many biological membranes reveal a surprisingly uniform picture. One usually sees two dark stained lines separated by a lighter space. The total thickness is 70–100 Å. Together with the properties of

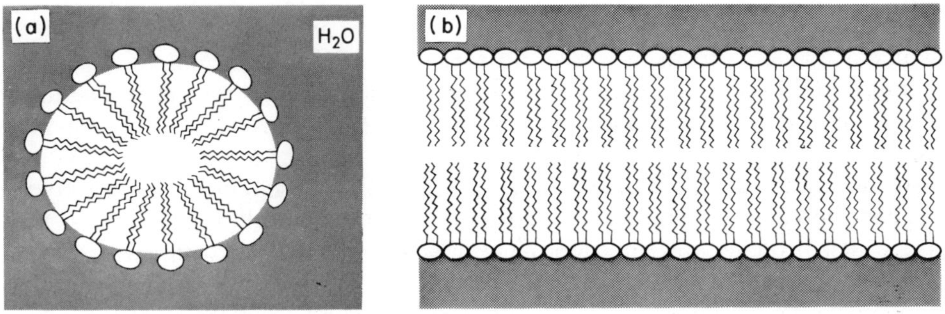

Fig. XIII-3. Arrangement of lipids in aqueous medium. (a) Formation of micelles; the size of the micelle often is determined by the spatial requirement of the molecule and thus limited. (b) Formation of lamellae; the flat surface has no natural boundaries. The surfaces can bend and finally close to form vesicles.

phospholipids just described this picture has led to the concept of a "*unit membrane*," according to which the membrane contains in its interior a lipid double layer to which protein molecules in pleated sheath conformation are attached on both sides.

Subsequently it has not been possible, however, to substantiate such a simple model. Especially physical-chemical investigations have revealed that the proteins of

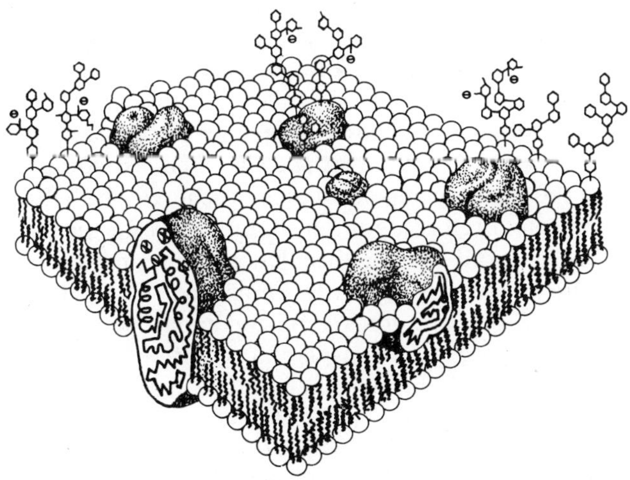

Fig. XIII-4. Diagram of the structure of biological membranes according to the fluid mosaic model. The gray shaded bodies represent proteins; the alpha-helical structure is seen in cross-section. Some proteins coagulate. Large protein molecules protrude on both sides of the membrane, while smaller proteins are embedded entirely in the hydrophobic portion of the lipid phase. Both types of proteins can move around within the phospholipid phase. Carbohydrate side chains of glycolipids are indicated near the right and left margins; they also come out of the membrane surface. (Modified diagram according to Singer and Nicolson, The Fluid Mosaic Model of the Structure of Cell Membranes, *Science* **175**, 720–731, 1972.)

membranes are present essentially as globular proteins which occupy more space than was available to them in the old model. On the basis of numerous physical-chemical and biological observations, a new model for the structure of membranes is emerging. It is called the model of a "*fluid mosaic*" and is much more plausible (see Fig. XIII-4).

According to this model the protein molecules, which might be up to 100 Å tall, are distributed in a bimolecular phospholipid film. Phospholipids form a matrix into which proteins are embedded. The lipid film is not rigid but rather has the property of a very viscous fluid. The proteins can move about within this film and indeed have been found to do that under certain conditions. Some of the proteins penetrate the membrane and are exposed on both sides of the membrane to the aqueous medium. These proteins have been designated "tunnel proteins" and most probably are very important for the transport of substrates through the membrane. Other proteins probably are merely embedded in the membrane facing outside or inside with their hydrophilic side. It is a necessary condition that in the tertiary structure of proteins both hydrophilic *and* hydrophobic regions are present on the surface of the molecule. By means of its hydrophobic regions the protein is able to interact with the lipids of the membranes. As a result, the lipid matrix in the immediate vicinity of the protein becomes ordered and semi-crystalline in nature; the individual lipid molecules are no longer able to shift around as easily as in other locations.

Many membranes, especially those in surface membranes of cells, are characterized by having carbohydrate groups located on their surfaces. These carbohydrates are part of *glycoproteins* or of *glycolipids*. Gangliosides have a special role in this connection, because their sialic acid residues bear a negative charge. These charges have actually been demonstrated.

The carbohydrate residues extend into the aqueous medium and impart a specificity to the cell surface because of their own specific chemical composition. The most familiar example is the blood group substances anchored on the surface of erythrocytes. Many phenomena of cell recognition so important for the formation of tissues and organs can now be traced to the molecular structures on cell surfaces. However, their *chemical* nature is still insufficiently known (cf. Chapter XVII,8). Such cell specificities are equally decisive for the histocompatibility which plays such a significant role in surgical organ transplantation. If there is incompatibility the transplanted tissue is rejected through an immune reaction. (For the function of membranes, see Chapter XVIII, 5.)

Function of Membranes. The most important and most general function of biological membranes is to restrict the exchange of substances among various areas of reaction or compartments. Selective permeability is the key feature for this purpose: Certain substances are prevented from passing through; others must be let through easily; and others have to be transported even against a concentration gradient (see Chapter XXI,3) to the inside or outside. The membrane thus possesses a certain asymmetry or one-sidedness since it distinguishes between "inside" and "outside."

Mitochondrial membranes appear to be components as well of multienzyme complexes, especially in the respiratory chain. "Microsomes" (fragments of the endoplasmic reticulum), too, seem to have such an arrangement of the electron transport chain. Last, there is evidence that several enzymes of the cytoplasm in fact are bound to membranes *in vivo*.

Finally, there are highly specialized membranes capable of maintaining large electrical potential differences. They are electrically excitable, i.e., they respond to an electrical stimulus by a change in permeability and consequently by a change in potential. This phenomenon is the basis for the conduction of a stimulus in nerves (see Chapter XXIII,8).

BIBLIOGRAPHY

S. Gatt and Y. Barenholz, Enzymes of complex lipid metabolism. *Annu. Rev. Biochem.* **42,** 61–90 (1973).
W. C. McMurray and W. L. Magee, Phospholipid metabolism. *Annu. Rev. Biochem.* **41,** 129–160 (1972).
W. Stoffel, Sphingolipids, *Annu. Rev. Biochem.* **40,** 57–82 (1971).
H. van den Bosch, Phosphoglyceride metabolism. *Annu. Rev. Biochem.* **43**, 243–277 (1974).

CHAPTER XIV

Isoprenoid Lipids: Steroids and Carotenoids

Isoprene is a branched-chain, unsaturated hydrocarbon of five C atoms. The carbon skeleton

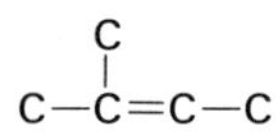

called a *prenyl radical* after Popják's proposal of a nomenclature, is the basis of a great many plant and animal products (terpenes, camphor; carotenoids; steroids, etc.). In 1922, Ruzicka suggested that all these natural products are derived from one common precursor. His "*isoprene rule*" has proved very useful; the nature of the biological isoprene, however, has been elucidated only in comparatively recent years through the efforts of Lynen, Bloch, Popják, and Folkers. The biosynthesis of cholesterol, the most important steroid for the animal organism, will be described here.

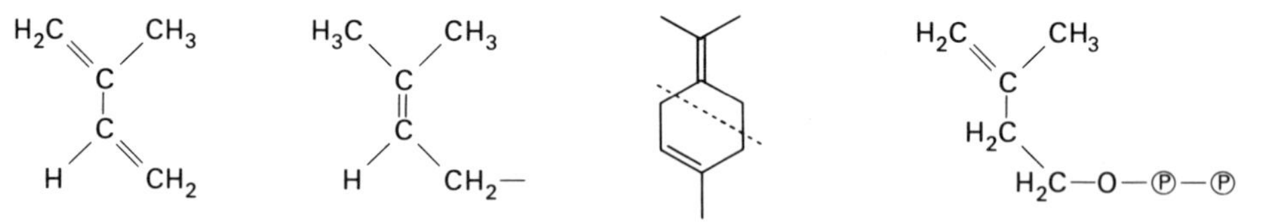

Isoprene | Prenyl residue | Terpinolene (2 prenyl residues) | Biological isoprene (isopentenyl pyrophosphate)

1. Biosynthesis of Cholesterol

Formation of Mevalonic Acid. Starting material for the formation of the isoprenoid precursor of cholesterol is activated acetate, *acetyl-CoA*, which we have already met several times. We have also seen that two molecules of activated acetate condense to form *acetoacetyl-CoA* and a third acetyl-CoA may condense onto it as a side chain. This reaction, formulated in Chapter XII,5, is entirely analogous to the formation of citrate from oxaloacetate and acetyl-CoA (cf. citrate cycle, Chapter XI,1). *β-Hydroxy-β-methylglutaryl-CoA* is the product of this condensation.

Next, the CoA-bearing carboxyl group is reduced to the alcohol in two steps by the action of hydroxymethylglutaryl-CoA reductase. HS-CoA is freed and two NADPH are consumed in the process. In this way, we have arrived at *mevalonate*, the key intermediate for the biosynthesis of isoprenoids.

Isopentenyl Pyrophosphate, the "Active Isoprene Unit." Mevalonate is subjected to several changes before it can condense to form larger compounds. First it is phosphorylated to form the pyrophosphate, with the consumption of 2 ATP. In a complex reaction, whose detailed mechanism is not understood, H_2O and CO_2 are split off, ATP again being required. The product is *isopentenyl pyrophosphate* (Lynen; Bloch),

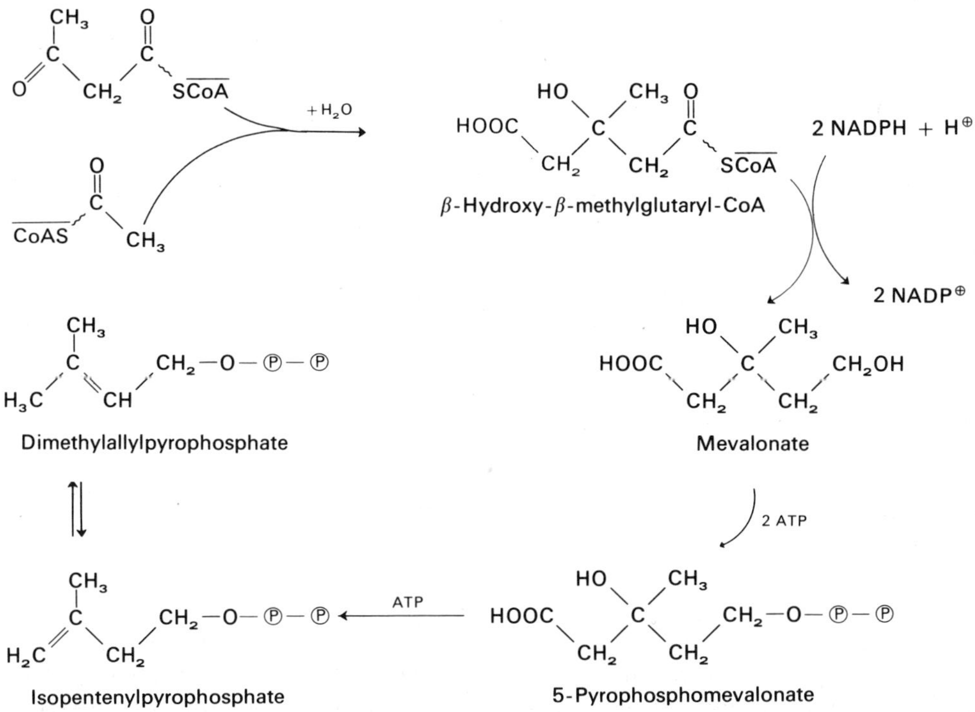

which is very similar to isoprene (the loss of pyrophosphoric acid would yield isoprene). The active isoprene unit (isopentenyl pyrophosphate) now undergoes a double bond migration (from Δ^3 to Δ^2) catalyzed by a specific isomerase giving *dimethylallyl pyrophosphate* ("prenyl pyrophosphate"). A pyrophosphate anion is eliminated easily from the allyl derivative, and the remaining carbonium cation attaches itself to the double bond of an isopentenyl pyrophosphate. The loss of a proton stabilizes the molecule yielding *geranyl pyrophosphate.* A repeat of this condensation reaction (i.e., between geranyl pyrophosphate and another isopentenyl pyrophosphate) finally produces the C_{15} compound *farnesyl pyrophosphate.*

From this compound, the series of reactions either proceeds with more condensations of the same type (Section 7) or is terminated by a head-to-head condensation of two C_{15} units. The mechanism again involves an allylic rearrangement, farnesyl

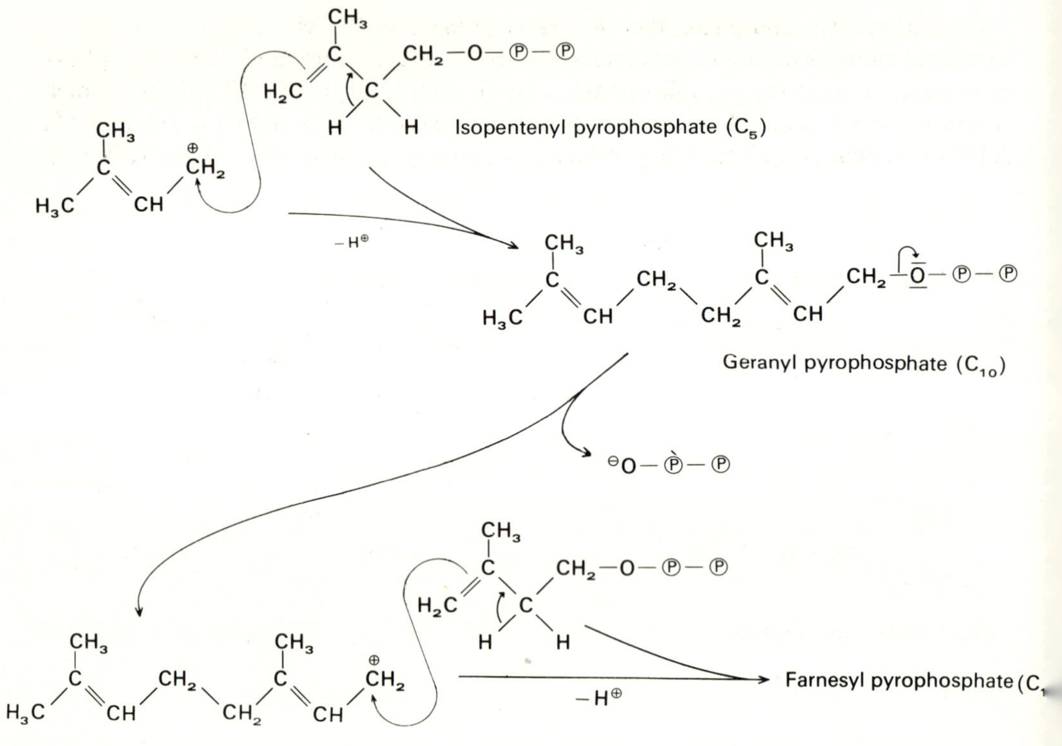

pyrophosphate to a nerolidol derivative, and an attack by a carbonium ion. The newly discovered intermediate presqualene (not shown) contains a cyclopropane ring. This compound by rearrangement, dephosphorylation, and a reduction step with NADPH finally forms *squalene.* The last condensation step ($C_{15} + C_{15}$) is catalyzed by a microsomal enzyme, whereas the enzymes of all preceding steps are soluble.

Between pyrophosphomevalonate and squalene there are no fewer than fourteen stereospecific transformations. The detailed stereochemical course of all but one of these steps is now known (Popják and Cornforth), as well as the absolute configurations of all intermediates.

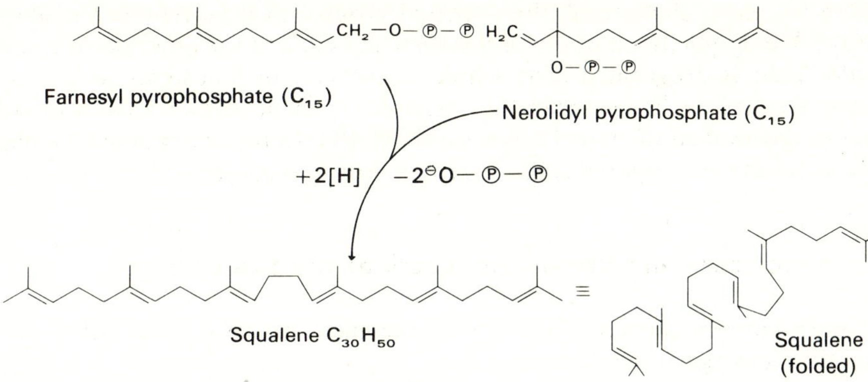

Squalene. This is a hydrocarbon $C_{30}H_{50}$, which was first isolated from shark liver but is also found in small amounts in mammals. As is evident from the biosynthesis just discussed, squalene is an isoprenoid composed of two symmetrical halves. Another way of writing the formula of squalene which better shows the possibility of cyclization is shown next to the linear formula.

This amazing cyclization of squalene to lanosterol proceeds via the epoxide intermediate, *squalene 2,3-oxide* involving more than one enzyme (Corey; van Tamelen and Clayton). The oxygen atom later becomes part of the C-3 hydroxyl group. As a mechanism, it is assumed that a proton opens up the epoxide ring and that the resulting carbonium cation continues to cyclize through the molecule in the fashion indicated. Two methyl groups migrate during cyclization; they are the two methyl groups that end up between rings C and D in the first isolable product, *lanosterol*.

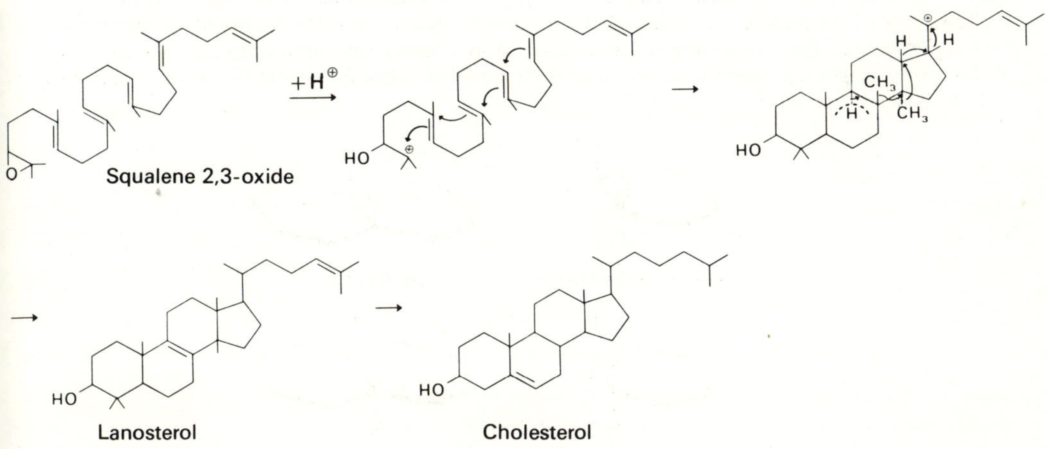

Before we reach cholesterol, three methyl groups must be removed oxidatively (as CO_2), one double bond in the ring system shifted, and the other double bond in the side chain hydrogenated. The whole conversion of lanosterol to cholesterol proceeds over at least fifteen separate steps, most of which are now known and which require the equivalent of 10 to 15 moles of NADPH. In the diagram of formulas we merely show the end product *cholesterol*, the principal zoosterol.

2. Nomenclature and Stereochemistry of the Steroids

We must concern ourselves briefly with the chemistry of steroids before we can discuss their biological significance.

The Ring System and its Possibilities for Isomerism. According to systematic chemical nomenclature, cholesterol is a derivative of cyclopentanoperhydrophenanthrene. This alicyclic, saturated hydrocarbon (without double bonds or aromatic character) bears the trivial name *sterane*. The C atoms are numbered in the manner indicated.

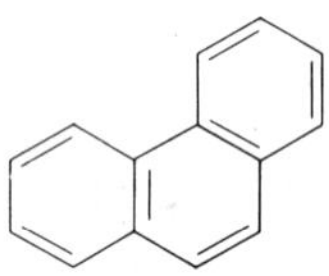

Phenanthrene (aromatic)

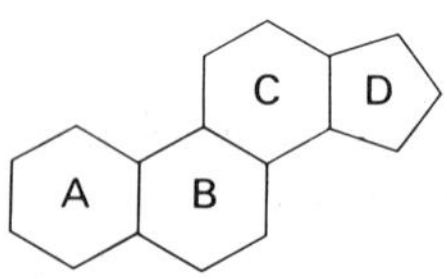

Sterane (alicyclic)

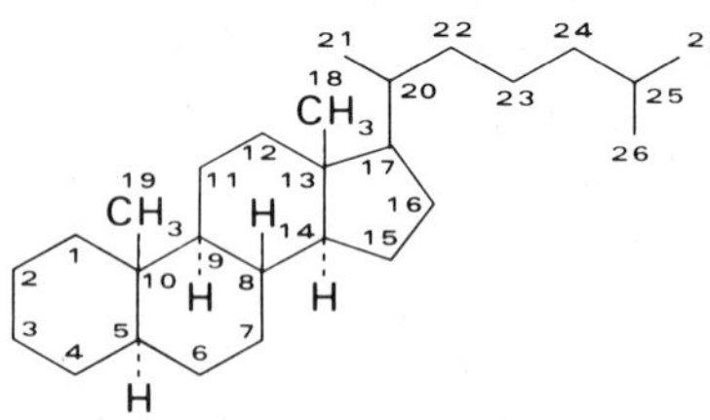

Cholestane

Two substituents of a cyclohexane ring (in *ortho*-position in the formulas below) can be either both on the same side of the plane of the ring (*cis*) or on opposite sides (*trans*). The possibility of isomerism is retained even when the two substituents are connected by C atoms (see formula) to form a second ring. If this second ring is a six-membered ring then the compound is Decalin, whose two isomers have been isolated separately (Hückel).

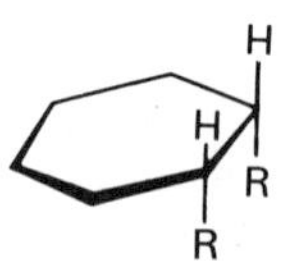

cis-Configuration

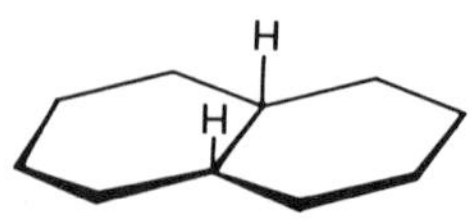

cis-Decalin

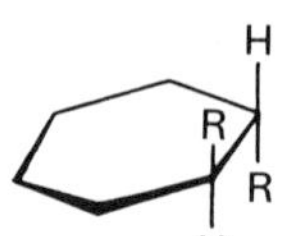

trans-Configuration

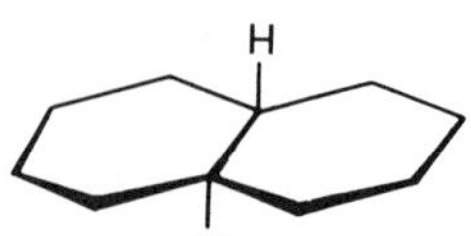

trans-Decalin

Two saturated rings, therefore, can be fused in two ways: *cis* or *trans*. The sterane system has four fused rings which have been designated A, B, C, and D. It is easy to remember that in almost all naturally occurring steroids, rings B and C as well as rings C and D are attached *trans*; only in the attachment of rings A and B are both isomers found. The fundamental hydrocarbon with *trans* connection is called *5α-androstane* (formerly androstane); with *cis* connection, *5β-androstane* (formerly etiocholane). As can be seen from the formulas, in 5α-androstane (and in *cholestane*) the hydrogen atoms and the methyl groups are alternatively above or below the plane of the rings. In cholesterol itself the isomerism of the ring attachment A/B is abolished, since there is a double bond at C-5.

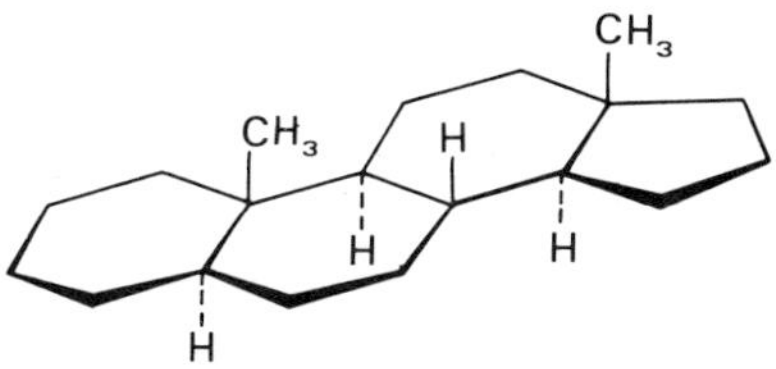

5α-Androstane (androstane)

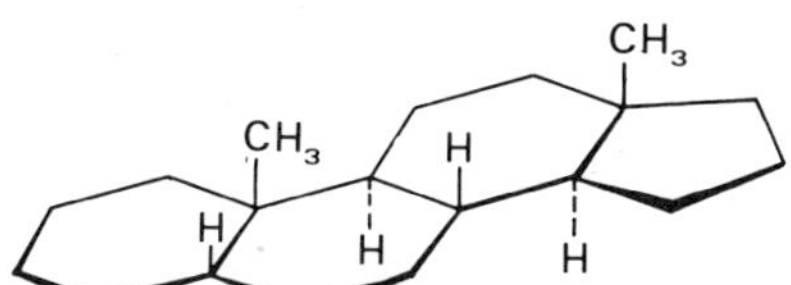

5β-Androstane (etiocholane)

The Position of Other Groups. By convention, the spatial orientation of hydrogen at the points of ring juncture, of hydroxyl groups, or side chains, i.e., of every substituent, is *referred to the methyl group at C atom 10*. In the formulas above, this particular methyl group sticks out above the flat ring system.[1] Thus, one talks about the β-position of a OH group if it is *cis* to the methyl group on C atom 10; the β—OH group sticks out above the ring system. In formulas this orientation of the bond is represented by a solid or heavy line.

If a substituent is oriented *trans* to the 10-methyl group and points down in the flat model, then such a substituent is said to be in α-position, symbolized in formulas by a dotted line as valence bond. 5α-Androstane, with *trans*-fused rings, has its hydrogen on C atom 5 in *trans*-position and therefore is called the 5α-compound. The same convention holds for hydroxyl groups. Cholesterol is a 3β-hydroxy compound. These designations will be very important in the field of steroid hormones, where isomers are frequently found together.

Conformation. For a better explanation of chemical properties, especially of reactivity, the concept of conformation has proved more valuable than the assignment to the α- or β-series. One must realize that cyclohexane rings actually are not planar, but exist rather in the so-called *chair form*, in which C atoms *para* to each other are pushed out of the plane above and below. (The *boat form* of cyclohexane does not exist normally.) All substituents are either *equatorial*, i.e., away from the center of the ring and approximately in the plane of the ring, or they are *axial*, i.e., approximately perpendicular to the plane of the ring (see drawing). In this way, *trans*-oriented substituents (in the drawing the hydroxyls on C-1 and C-2) can come together just as closely as *cis*-oriented groups (e.g., those on C-4 and C-5).

---

[1] The convention had been agreed upon before the absolute configuration was determined. This arbitrary choice later turned out to be the correct one.

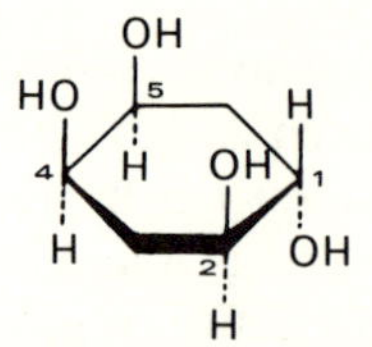

Simplified representation

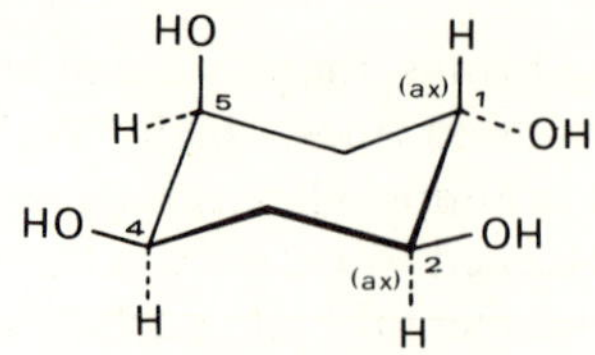

Chair form

Such considerations are pertinent for the "molecular anatomy" of membranes. The formulas of cholestanol and coprostanol (below) illustrate how space is filled very differently when at one time rings A and B are fused *trans* and at another *cis* (respectively). A clear concept of these relationships can be acquired only by using molecular models.[2]

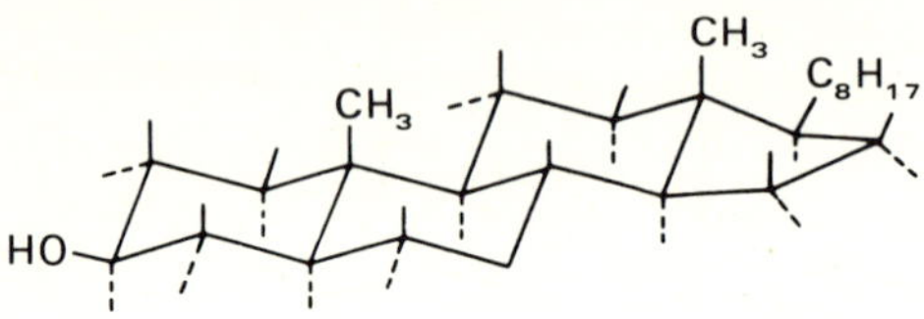

Cholestanol

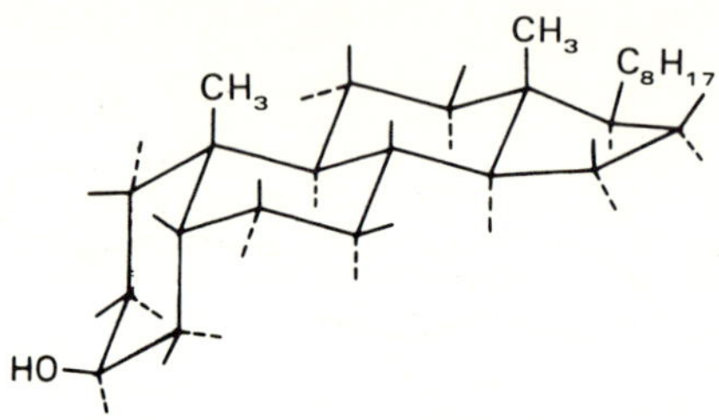

Coprostanol

The Great Variety of Steroids. In nature, steroids are converted to substances that execute the most varied tasks. An outline of some of these functions is provided by Table XIV-1.

TABLE XIV-1

Some Steroids and Their Biological Role

Group	Representative	Formula	Occurrence and function
Sterols			
C_{27}—C_{30}	Cholesterol	$C_{27}H_{46}O$	Ubiquitous; structural component and precursor
	Δ^7-Dehydrocholesterol	$C_{27}H_{44}O$	Skin; provitamin D
	Ergosterol	$C_{28}H_{44}O$	Yeast; provitamin D
Bile Acids			
C_{24}	Cholic acid	$C_{24}H_{40}O_5$	Gall bladder and intestines; absorption of fats
Hormones			
C_{21}	Progesterone	$C_{21}H_{30}O_2$	Corpus luteum hormone
	Deoxycorticosterone	$C_{21}H_{30}O_3$	Adrenocortical hormone
	Cortisol	$C_{21}H_{30}O_5$	Adrenocortical hormone
C_{19}	Testosterone	$C_{19}H_{28}O_2$	Testicular hormone
C_{18}	Estradiol	$C_{18}H_{24}O_2$	Follicular hormone

[2] For a more extensive treatment of conformational analysis see L. F. Fieser and M. Fieser, "Steroids." Reinhold, New York, 1959.

3. Sterols and Plant Steroids

Sterols characteristically have a hydroxyl group on C atom 3. The sterol of vertebrates is cholesterol, which is found in all cells. It participates, in conjunction with phospholipids, in the construction of membranes; the shape of the molecule, a small flat disk, seems to be particularly suited for this purpose. In nerve tissue, cholesterol is a component of the myelin sheath (cf. Chapter XXIII,7). The excessive appearance in vascular walls (atherosclerosis) and in gallstones (where it was discovered as early as 1769) is pathological.

Besides being a structural component, cholesterol also serves the organism as a starting material for the synthesis of numerous other steroids such as bile acids, the steroid hormones of the adrenal cortex and the gonads, and vitamin D. Intestinal bacteria reduce cholesterol to *coprostanol*, in which rings A and B are *cis*.

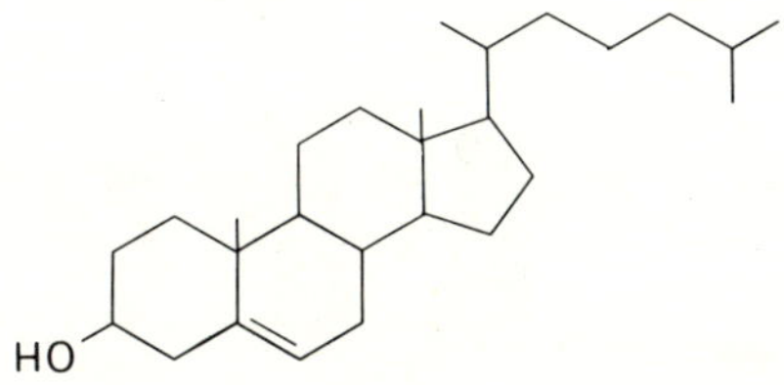

Cholesterol

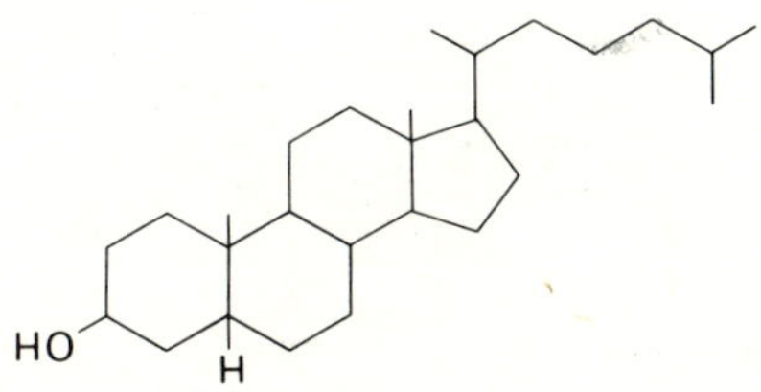

Coprostanol

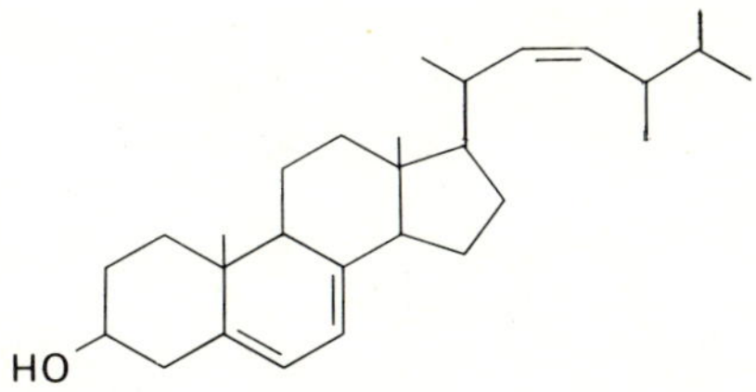

Ergosterol

In addition to cholesterol, a whole series of other sterols has been found in plants and lower animals—but the higher in the animal kingdom, the more uniform the pattern of sterols.

Lanosterol has already been mentioned as a biosynthetic intermediate; it is abundant in the wool fat of sheep. The related *zymosterol*, which lacks the three extra methyl groups, was first isolated from yeast. The most important yeast sterol, however, is *ergosterol*, a C_{28}-steroid with three double bonds; it is a provitamin D (see below). Plants contain several sterols with additional C substituents in the side chain, such as ergosterol has; examples are stigmasterol and sitosterol. Insects are incapable of synthesizing the sterol skeleton; they require sterols as essential food components.

Saponins, Digitaloids, and Steroid Alkaloids. These substances are plant products. They are characterized by numerous hydroxyl groups, ether and lactone linkages, and, in the case of alkaloids, by the content of nitrogen. Many of them exist as glycosides. They are secondary plant products with greater pharmacological than physiological interest. The formulas included here demonstrate the great variety of steroids rather than point out or catalog individual substances.

Saponins in solution foam strongly. One typical saponin is *digitonin* (from digitalis seeds). With 3β-hydroxy steroids (e.g., cholesterol) digitonin forms insoluble addition compounds; the reagent is useful for the determination of cholesterol. The reaction with cholesterol is probably the cause of its hemolytic (erythrocyte-disrupting) action. Digitonin is a glycoside consisting of a pentasaccharide and the steroid digitogenin. The formula is shown below.

As an example of a cardiac glycoside we mention *strophanthin*, a glycoside with therapeutic applications. Noteworthy features of the formula are that the methyl group on C-10 is oxidized to the aldehyde (the methyl group is intact in the digitalis compounds) and that the side chain is in the form of a five-membered, unsaturated lactone ring. The lactone ring is common to all glycosides with cardiac activity. In glycosides from *squill* and from *toad poisons* the ring is six-membered.

From the vast number of steroid alkaloids we have chosen *solanine* as a representative. As the formula shows, it also is a glycoside. It occurs in potato shoots. The steroid alkaloids cannot all be derived from one common structure, as is possible with the cardiac drugs. For other representatives, see special reference works.

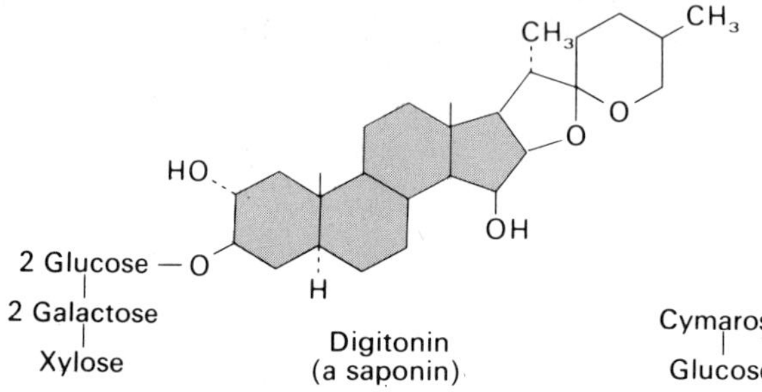

Digitonin
(a saponin)

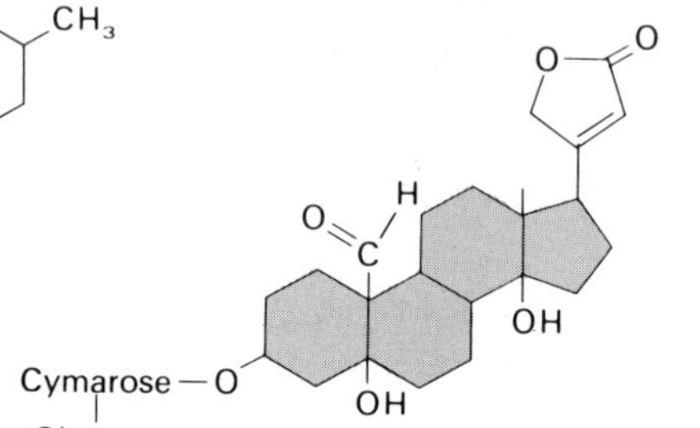

Strophanthin
(a cardiac stimulant)

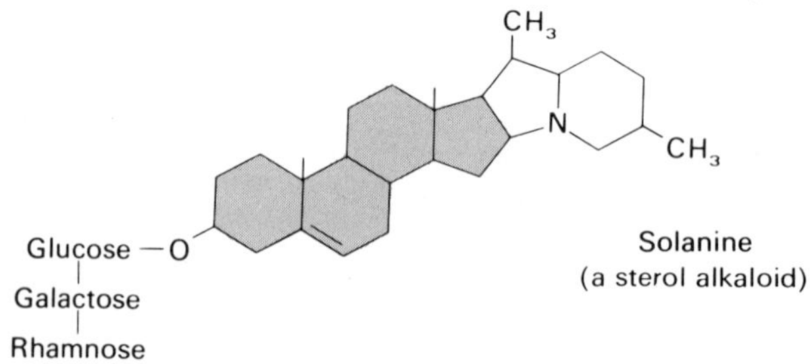

Solanine
(a sterol alkaloid)

4. Vitamin D

The letter D was used to designate the antirachitic vitamin of liver oil (for its physiological action see Chapter XXII,4). Several active substances are known (vitamin D_2, D_3, etc.[3]), and one correctly speaks of the vitamins of the D group, they are also called *calciferols*. Windaus first prepared vitamin D_2 by irradiating ergosterol and determined its constitution.

Vitamins D are not really steroids themselves since they do not possess the four-ring system. Ring B is split open when the provitamins D, namely ergosterol and 7-dehydrocholesterol, with two conjugated double bonds in ring B are irradiated with ultraviolet light (see accompanying formulas). Not long ago it was found that the active forms of the vitamin are the hydroxylation products of the calciferols, particularly 25-hydroxycholecalciferol and *1,25-dihydroxycholecalciferol*.

[3] D_1 was the designation for the first crystalline material, which later turned out to be a mixture.

Windaus had assumed a series of intermediate steps, but Velluz could show that the photochemical reaction produces precalciferol (*previtamin D_2*), which is in equilibrium with vitamin D_2 through a thermal rearrangement reaction; vitamin D_2 is the major component of the equilibrium. Both *lumisterol* and *tachysterol* are side products, not intermediates. As can be seen from the formulas, the transition to precalciferol entails the opening of ring B with the consequent formation of a new double bond. The rearrangement to vitamin D_2 consists in a shift of the double bonds by one place.

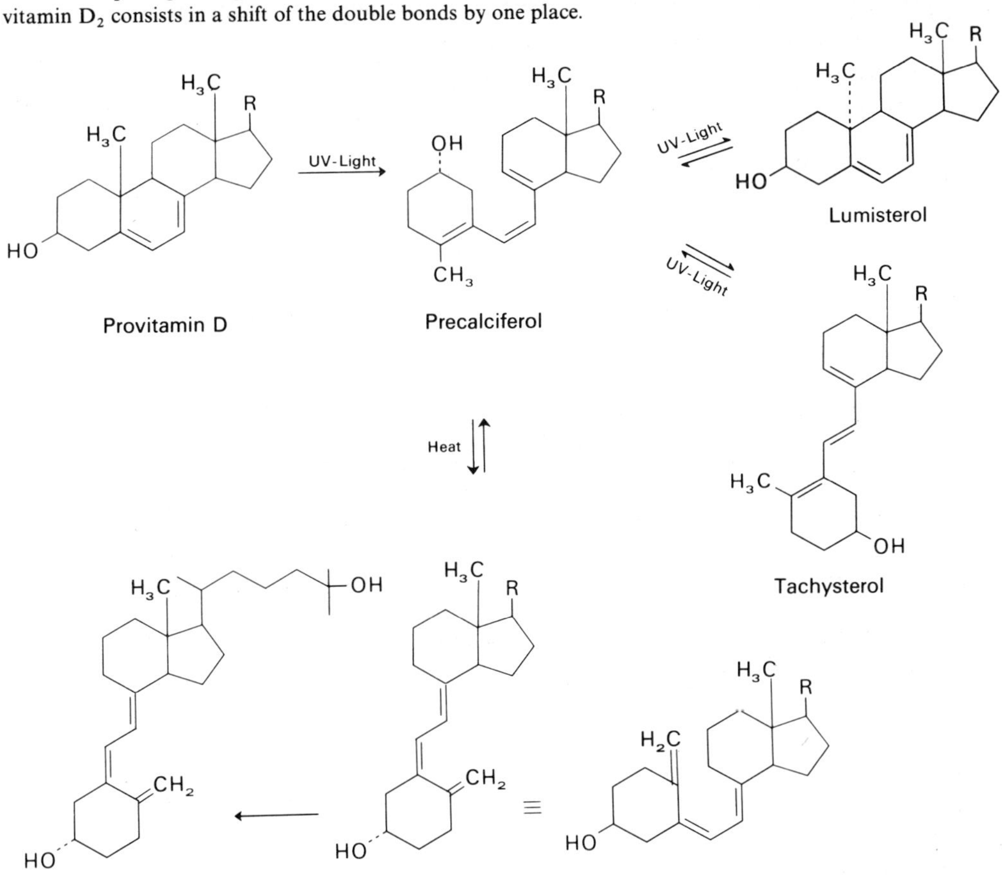

The temptation is great to represent vitamin D_2 in a way which emphasizes its relationship with the steroids, but such a representation does not correspond to the true configuration. In writing such formulas it can be noticed that the methylene group really has not enough room. The representation at the left corresponds much better with the true situation. The two formulas for calciferol stand for the same substance, not for isomers.

The photochemical reaction is not limited to ergosterol. Other steroids with $\Delta^{5,7}$-double bonds undergo the conversion. Of physiological significance are vitamin D_2, *ergocalciferol*, derived from ergosterol, and vitamin D_3, *cholecalciferol*, derived from dehydrocholesterol. Cholecalciferol is found in liver oil and during exposure to direct sunlight arises also in the skin from 7-dehydrocholesterol, which is present in relatively high concentrations; this fact explains the beneficial effect of sunlight or UV-radiation on rachitis (rickets).

5. Bile Acids

The *salts of the bile acids* are the digestion-promoting constituents of bile. They are surface active agents; this means that they lower surface tension and thus can emulsify fats. They also activate lipases. The bile acids, for these reasons, play a leading role in the digestion and absorption of fats.

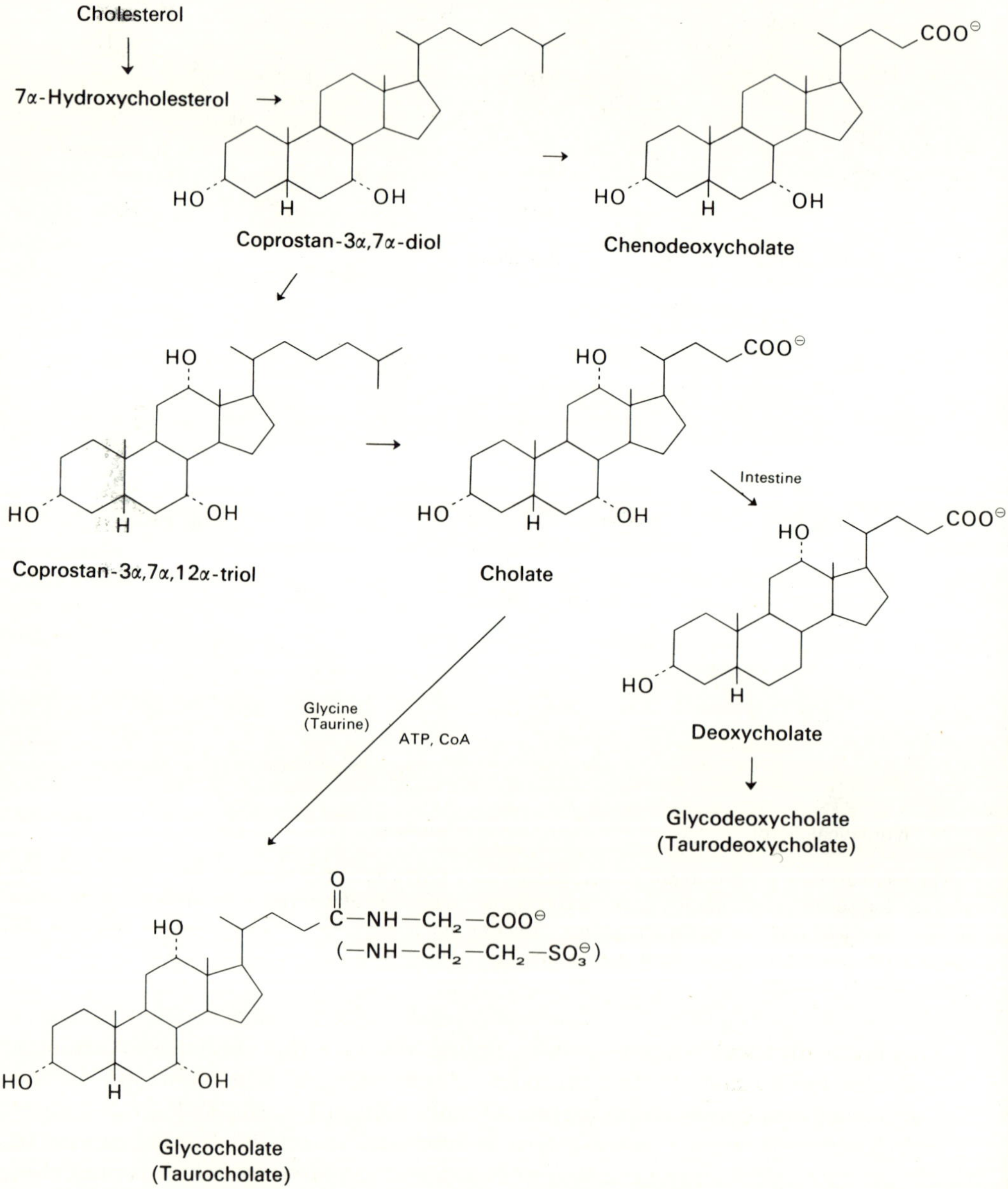

Without exception, in all bile acids, the juncture of rings A and B is *cis*, and the hydroxyl groups are α-oriented. The side chain is usually made up of five C atoms and bears the carboxyl group.

The biosynthesis of bile acids takes place in the liver. It begins with the insertion of a 7α-hydroxyl group into cholesterol; the ring is then reduced and epimerized at C-3 (the ketone must be assumed to be the intermediate). After these changes in the ring system, and occasionally after hydroxylation at C-12, the side chain is shortened oxidatively: One methyl group is hydroxylated and the CH_2—OH group is oxidized further to the carboxylic acid; the C_{27} carboxylic acid is then broken down by β-oxidation (as are the fatty acids with α-methyl branching; see formulas in Chapter XII,4). The splitting-off of propionic acid produces the "specific bile acids," which are activated with ATP and CoA and connected to the amino group of *glycine* or *taurine* in amide linkage. Deoxycholate (3α,12α-dihydroxycholanate) arises from cholate by the action of intestinal bacteria.

The chief bile salts are the taurine and glycine derivatives of *cholate* (3α,7α,12α-trihydroxycholanate), of *deoxycholate* (3α,12α-dihydroxycholanate), and of the isomeric *chenodeoxycholate* (3α,7α-dihydroxycholanate). The bile salts are one of the end products of the metabolism of cholesterol; however, over 90% of the amount secreted (20–30 gm per day) is reabsorbed in the intestine and thus stays in the enterohepatic circulation.

Deoxycholate can join with fatty acids and other lipids (cholesterol, carotene) to form molecular compounds. In the case of stearic or palmitic acid, eight molecules of bile salts are joined to each molecule of fatty acid. Sodium deoxycholate is used extensively in the laboratory as a solubilizer of lipids.

6. Steroid Hormones

Many of the hormones are steroids. The organism synthesizes them from cholesterol (or cholesterol derivatives) in special tissues—the adrenal cortex, ovaries, corpus luteum, testes, and placenta. The hormone production in the adrenal cortex has been studied best and actually is the most versatile. In this chapter, structure and biogenesis of the hormones will be discussed; physiological properties will be treated later in Chapter XX.

The C_{21}-Steroids. This group of steroids, consisting of the corpus luteum hormone progesterone and most of the adrenocortical hormones (see Table XIV-1), is formally derived from *pregnane*, a hydrocarbon with a β-oriented side chain of two C atoms. The biosynthesis starts with cholesterol, and the first important conversion product is *pregnenolone* (systematic name: 3β-hydroxypregn-5-en-20-one). The oxidative shortening of the side chain of cholesterol proceeds by hydroxylations at C-20 and C-22 and cleavage to form isocaproic aldehyde. The resulting pregnenolone then undergoes dehydrogenation and isomerization of its double bond to form *progesterone*, one of the biological key substances.

Progesterone, the hormone of corpus luteum, is at the same time a precursor of the *corticosteroids*, which contain additional hydroxyl groups. The introduction of these hydroxyl groups follows the principle of oxygenation with O_2 + NADPH (cf. Chapter X,6). The *steroid hydroxylases* are highly specific with regard to the position of the hydroxyl group introduced into the steroid skeleton. We know a *17-hydroxylase*, a *21-hydroxylase*, and an *11β-hydroxylase*; if they act in this sequence they produce *cortisol* via 17-hydroxyprogesterone (cf. formulas above).

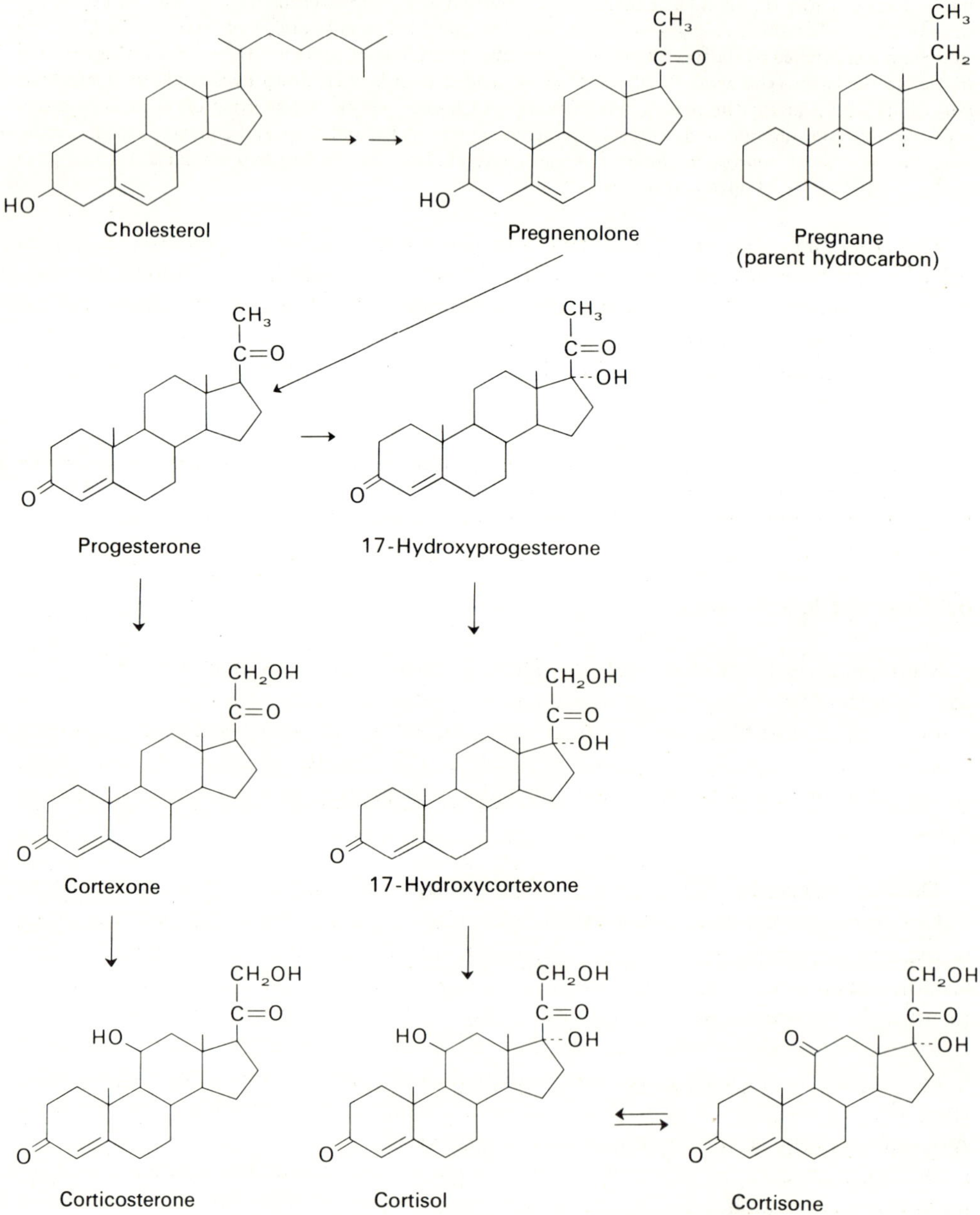

It is interesting that the 17-hydroxylase will not attack a C-21 hydroxy compound. If a hydroxyl is first introduced at C-21, then further conversion stops at *corticosterone*, and cortisol will not be produced (diagram of formulas above). The 11-hydroxy compounds can be dehydrogenated reversibly to the 11-keto compounds.

Aldosterone. Another "adrenocortical" hormone with very high activity is aldosterone, with an aldehyde function at C-18. During its biosynthesis the methyl group apparently is oxidized. The aldehyde group very easily forms a hemiacetal ring with the OH group on C-11, as is shown by the formula on the right.

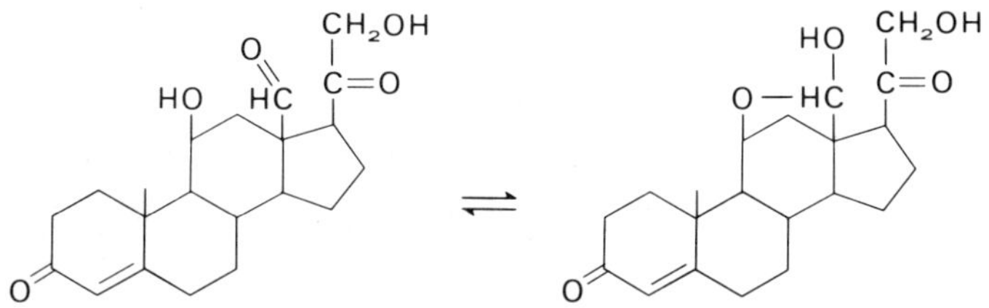

Aldosterone

Inactivation and Excretion of C_{21}-Steroid Hormones. As explained in more detail in Chapter XX,1, the inactivation and excretion of hormones is an essential link in the chain of hormonal regulation. Degradation occurs mostly in the liver; by reduction of the unsaturated ketone grouping to a C-3 hydroxyl and a new asymmetric center at C-5 (the point of ring juncture) complete or nearly complete inactivation is achieved. The majority of the excretion products are 3α-hydroxy-5β-compounds, i.e., the hydroxyl group usually has the opposite steric configuration of that of the starting materials cholesterol and pregnenolone, and rings A and B are *cis*. However, some 5α-compounds (allopregnane derivatives) are also formed.

The saturated C_{21} compounds are then either conjugated with glucuronate and appear in urine as the glucuronides, or they are further broken down. The side chain is split off particularly easily from 17-hydroxy compounds. The products are the *17-ketosteroids androsterone* (5α-androstan-3α-ol-17-one), *etiocholanolone* (5β-androstan-3β-ol-17-one), and others, also excreted as glucuronides (or as sulfates). The amount of 17-ketosteroids in urine can be estimated easily with Zimmerman's color reaction; it is useful for the diagnosis of adrenal cortical activity. One should be aware, however, that by this method only a portion of the corticosteroid excretion is measured together with testosterone inactivation products. Normally, about 10 to 20 mg of 17-ketosteroids are excreted per day.

Biosynthesis of C_{19}-Steroids. The principal C_{19}-steroid is the male sex hormone *testosterone* (17β-hydroxyandrost-4-en-3-one). Its biosynthesis in the testes proceeds essentially through the intermediates already discussed, progesterone and *17-hydroxyprogesterone*. Loss of the side chain yields *androst-4-ene-3,17-dione*, and finally testosterone by reduction at C-17.

The elimination of the side chain, which here leads to the synthesis of a hormone, is, on the other hand, a catabolic reaction for the adrenocortical hormone 17-hydroxy-11-deoxycorticosterone. As we have mentioned, the hormones are inactivated by reduction in ring A. The urinary excretion product *androsterone* is formed by the oxidative loss of the side chain at C-17 and reduction of ring A.

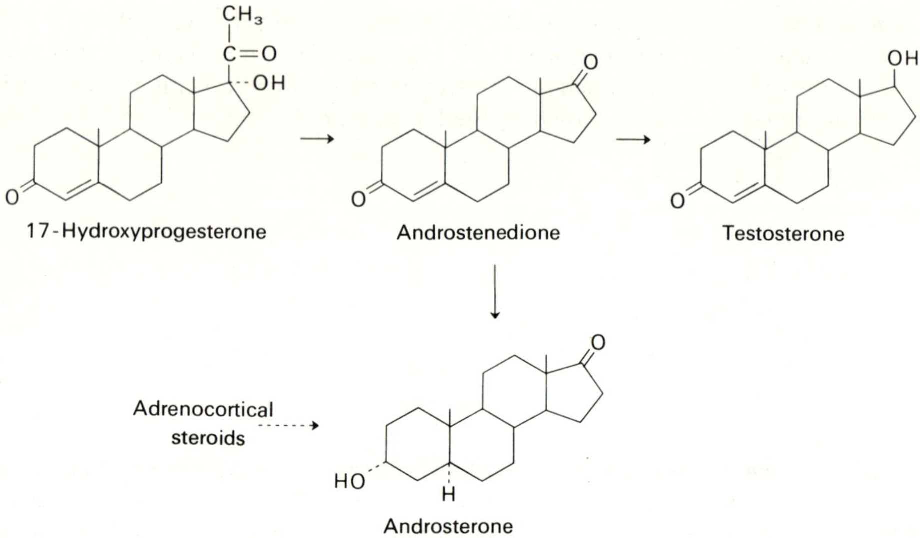

Biosynthesis of Estrogens. Estrogens (C_{18}-steroids) differ from the previously discussed steroids in one important respect: ring A is aromatic. There can be no methyl group at C-10. The aromatization of ring A, therefore, implies loss of the angular methyl group at C-10 and imparts phenolic character to the alcohol group at C-3. The biosynthesis starts with testosterone, whose C-19 angular methyl group (cf. formula of cholestane in Section 2, for the numbering) is hydroxylated and further dehydrogenated to the aldehyde. C-19 can then be removed as formaldehyde, and ring A is aromatized.

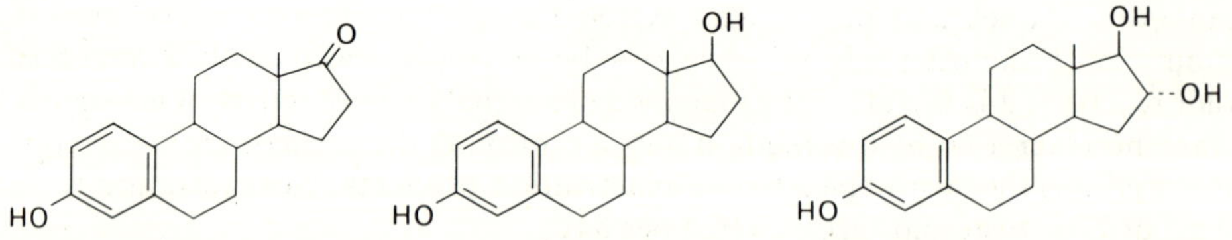

Estrone | 17β-Estradiol | Estriol-3,16α,17β

The important estrogens are *estrone*, *17β-estradiol*, and *estriol-3,16α,17β*. Estrone easily converts to estradiol, which is considered to be the true female sex hormone. For physiological properties see Chapter XX,3.

7. Carotenoids

Carotenoids enjoy wide distribution both in the animal and plant kingdoms, but are exclusively of plant origin. They are isoprene derivatives with a high degree of unsaturation. Because of the very many conjugated bonds they are colored reddish or yellow: The pigments of the tomato (lycopene) and of the carrot (α- and β-carotene) are red; many oxygen-containing carotenoids are yellow (xanthophylls). Since the double bonds permit *cis-trans*-isomerism, numerous forms are possible. Most carotenoids, however, exist in the *all-trans*-form. The long hydrocarbon chains make them lipid soluble, so that they may also be called "lipochromes."

Biosynthesis. The true carotenoids and xanthophylls have forty carbon atoms; this corresponds to eight isoprene residues. Like squalene, they are constructed symmetrically and it is assumed that they arise by head-to-head condensation of two C_{20} precursors. Indeed, appropriate plant enzyme systems convert the key precursor *mevalonate* to carotenoids. The tetraisoprenoid *geranylgeranyl pyrophosphate* can be dimerized in analogy to farnesyl pyrophosphate, to yield a C_{40}-hydrocarbon with eight or nine double bonds.

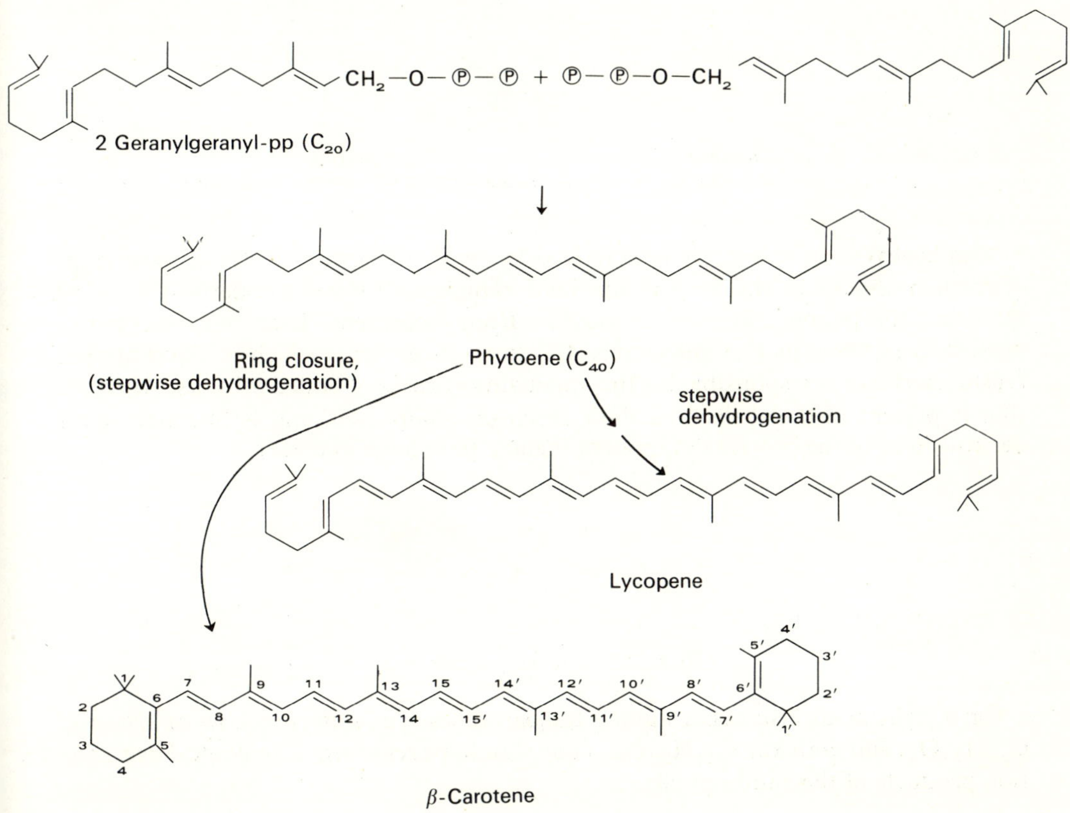

The precursors of the carotenoids, according to this scheme, are more hydrogenated than later products. The initial isolable product is still colorless *phytoene*; of its nine double bonds only three are conjugated. The conjugated system grows by gradual dehydrogenation (2 H with each step), producing successively *phytofluene*, *ζ-carotene*, *neurosporene*, and *lycopene*. In some mutants the biogenesis is interrupted at certain points; larger quantities of hydrogenated and colorless carotenoids are accumulated. From one *Neurospora* mutant, for example, the pale yellow carotenoid neurosporene was obtained.

Lycopene and the Carotenes. The formula reveals that lycopene is a long-chain hydrocarbon with thirteen double bonds, eleven of which are in conjugation. Lycopene is the main pigment of the tomato, paprika, and other fruits. We have indicated in the schematic formula that the ends of the chain can easily close up to form rings, with the disappearance of terminal double bonds. Ring closure on only one end results in γ-carotene; with ring closure on both ends, β- and α-carotene are produced. These two carotenes differ in the position of the double bonds in the rings. In *β-carotene* both annular double bonds are in conjugation with the system of double bonds to the long chain (β-ionone structure); in *α-carotene* one of the annular double bonds is removed from the system of conjugation by one position (α-ionone structure). These details have physiological significance, because vitamin A can arise only from the β-ionone structure.

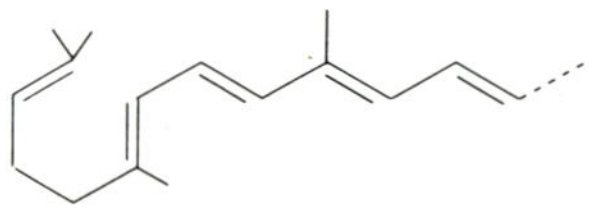

γ-Carotene
(partial formula)

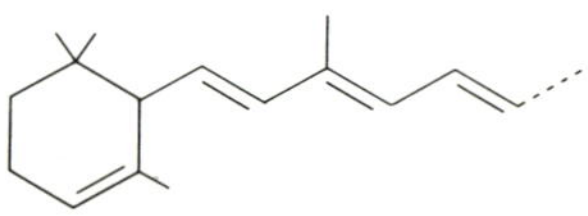

α-Carotene
(partial formula of α-ionone structure)

Xanthophylls. These are characterized by hydroxyl groups in the ionone rings of carotenes, *para* to the long chain. Leaf xanthophyll (*lutein*) is derived from α-carotene; the pigment in corn *zeaxanthin*, from β-carotene. Even more oxygen is present in *astaxanthin* (3,3′-dihydroxy-4,4′-dioxo-β-carotene) which is found among crustaceans and is responsible for the appetizing redness of boiled lobsters. Attaxanthin is present in the shells as a dark green chromoprotein and is liberated upon denaturation of the protein component (hence the change in color).

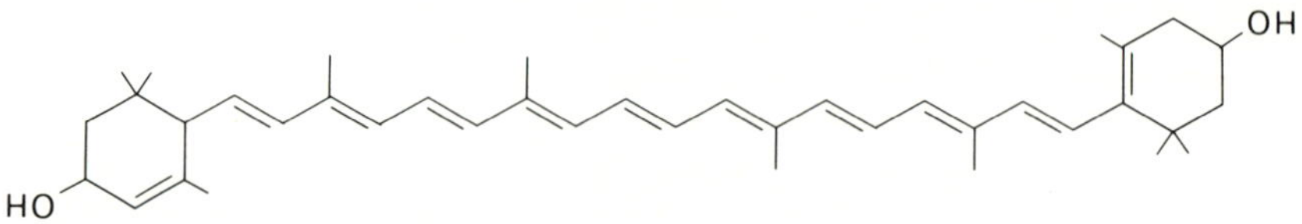

Lutein (3,3′-dihydroxy-α-carotene)

Carboxylic acids also occur among the carotenes, e.g., saffron yellow or *crocetin*, $C_{20}H_{24}O_4$, and *norbixin*, $C_{24}H_{28}O_4$. They should be regarded as biological degradation products of the xanthophylls.

Biological Significance. Being pigments, the carotenoids participate in photosynthesis. Along with chlorophyll, they are structural and functional components of the *chloroplasts* (the photosynthesizing organelles of plant cells).

For mammals, the carotenes (especially β-carotene) are *provitamins A*.

8. Vitamin A and Visual Purple

The mammalian organism oxidatively splits carotene at the plane of symmetry between C-15 and C-15′. The initial cleavage product is presumably *retinal* (vitamin A_1 aldehyde), which is then reduced to the alcohol. The alcohol *retinol*, also called vitamin A_1, can be esterified for storage. Another active substance *3-dehydroretinol* (vitamin A_2) differs from A_1 by an additional double bond in the ring in conjugation with the entire system.

Vitamin A is a growth factor. In addition to the alcohols A_1 and A_2 and the aldehydes, the corresponding acid *retinoic acid* is also active in biological assays. Perhaps the latter is the active form of the vitamin, but in the visual process the acid does not fill the role of vitamin.

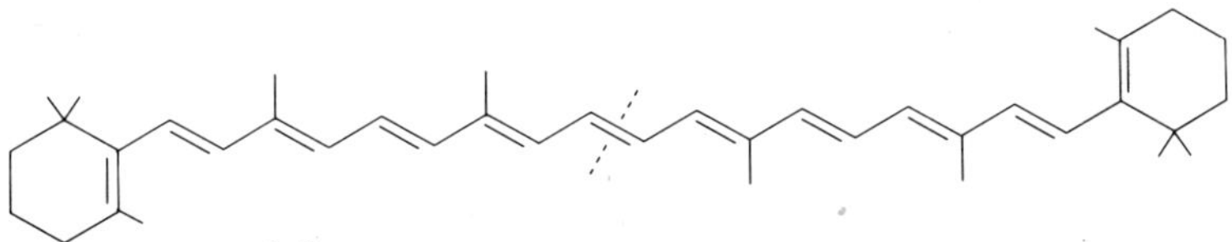

β-Carotene

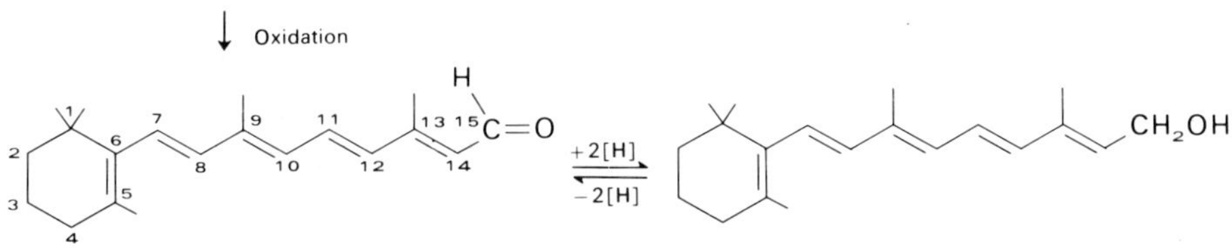

Retinal

Retinol
(Vitamin A_1)

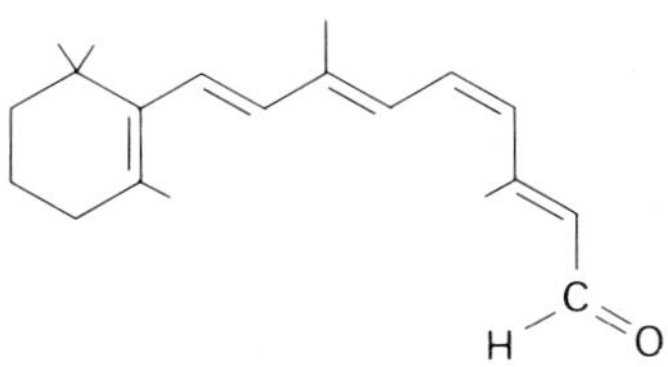

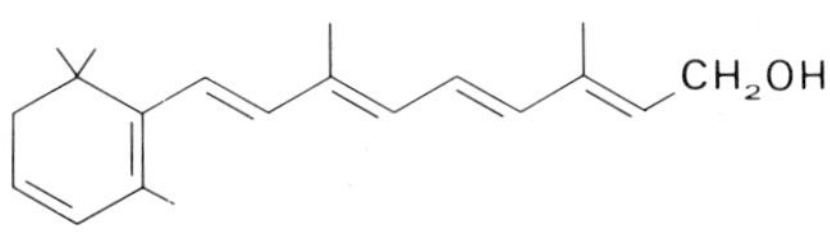

11-*cis*-Retinal
(Neoretinal b)

3-Dehydroretinol
(Vitamin A_2)

Biochemistry of the Visual Process. The photosensitive pigment of the eye is visual purple or *rhodopsin*. It consists of the protein *opsin* and *neoretinal b*, a stereo isomer of the *all-trans*-retinal (vitamin A_1 aldehyde). The double bond between C-11 and C-12 in neoretinal b is in *cis* configuration. The protein opsin attaches only to this *cis*-isomer to form the chromoprotein; the aldehyde is bound as a Schiff's base to a lysyl residue of the protein.

During the visual process—the conversion of light to nerve impulses—neoretinal b first is rearranged to the more stable *all-trans*-configuration. Only *light* triggers the process. The altered molecular shape interferes with the bond to the protein. The spatial structure of the protein probably is affected, too, and in this manner photochemical changes somehow are translated to neural excitation.

The photochemical rearrangement of the chromophore renders the visual purple rhodopsin unstable. It is then bleached over a number of intermediate steps (lumirhodopsin, metarhodopsin I and II), i.e., it is disassembled to opsin and *trans*-retinal. This bleaching process is not a prerequisite for visual excitation; it is rather a consequence. The labile intermediate product *in vivo* probably snaps back to its original form without the bleaching process, possibly in a photochemical reaction. Once bleached, however, visual purple can be regenerated only after *all-trans*-retinal is rearranged to the 11-*cis*-form. The rearrangement is possible both as a photochemical reaction and as an oxygen-dependent dark reaction. An appreciable portion of retinal is reduced by alcohol dehydrogenase with the aid of NADH to form retinol which then exchanges with retinol of the blood circulation. These processes have been represented schematically:

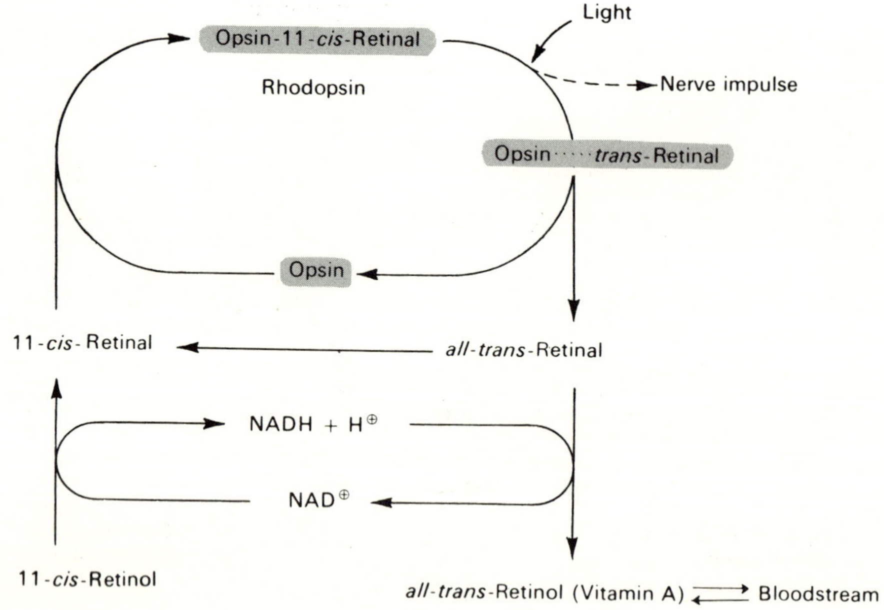

Besides opsin, there is another protein, scotopsin. Both proteins can combine either with 11-*cis*-retinal or with 11-*cis*-3-dehydroretinal. Consequently, there are four visual pigments, which cover a broad range of the spectrum. These interrelationships have been elucidated chiefly in the laboratory of G. Wald.

9. Polyprenols and Polyprenyl Quinones

In the biosynthesis of squalene (the precursor of steroids) and of phytoene (the precursor of carotenoids) the pyrophosphates, respectively, of three or four isoprene units are joined together by head-to-head condensation. The linear condensation of isopentenyl pyrophosphate, however, can readily proceed beyond the C_{20} unit. This is the origin of the widespread *polyprenols*, some of which have chain lengths exceeding 100 carbon atoms. Polyprenols with nine to thirteen prenyl residues (each prenyl group is C_5) predominate in plants, whereas *dolichols* with seventeen to twenty-two prenyl residues occur in animals. In the rat, *dolichol-18* is the predominant polyprenol (see formula here). The α-prenyl residue of dolichols invariably is saturated. Some double bonds are *trans*, most, however, are *cis*. The steric configuration most likely is set at the condensation step of the isopentenyl pyrophosphate residues. Up to 60% of the dolichols are esterified with long-chain fatty acids.

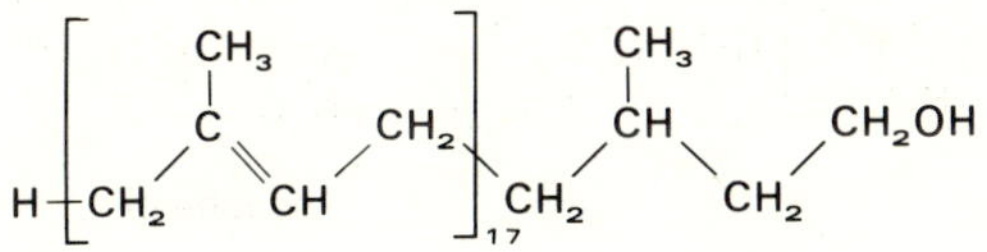

Dolichol-18, $C_{90}H_{163}OH$

The physiological function of these polyprenols in the mammalian system is still completely unknown. In bacteria they participate in the biosynthesis of the cell wall by functioning as lipid-soluble carriers for carbohydrate-peptide units. Linked to C_{55} polyprenyl pyrophosphate, these units are transported through the cell membrane and incorporated in exterior "murein" (see Chapter XVII,7).

Natural Rubber. If such a condensation process is imagined to continue until thousands of isoprene units are linked together, the natural rubber molecule is obtained. It is certain now that the biosynthesis actually proceeds in such a fashion; it has even been carried out *in vitro* with enzymes of tree sap. The milky sap of the rubber tree (Hevea) contains high-molecular weight natural rubber (molecular weight about 350,000) in the form of minute spheres. Chemically it is noteworthy that all the double bonds are exclusively in the *cis*-configuration. The shape of the molecule resulting from this configuration undoubtedly is what imparts this material its elasticity.

Polyprenyl Quinones. Widely distributed in nature, their constitution can be seen in the formulas: A highly substituted benzoquinone or naphthoquinone aromatic nucleus bears a polyprenyl side chain of variable length. The four main groups of compounds are ubiquinone, plastoquinone, tocoquinone, and tocopherols. In addition, there are also menaquinones.

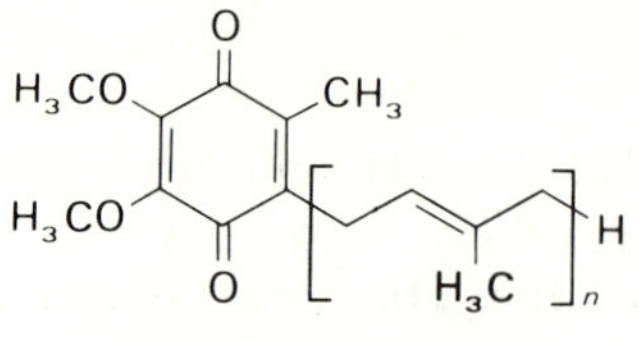

Ubiquinone ($n = 6$–10)

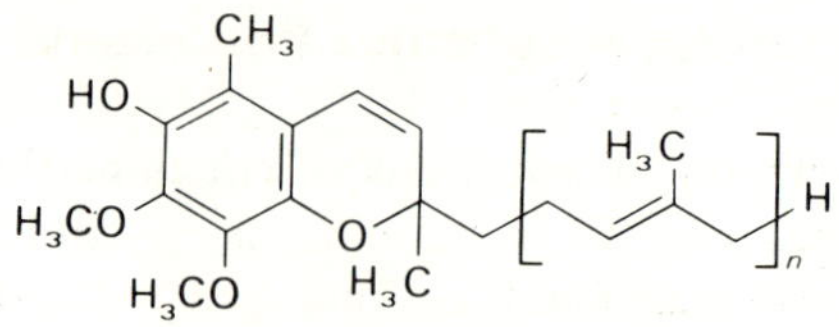

Ubichromenol

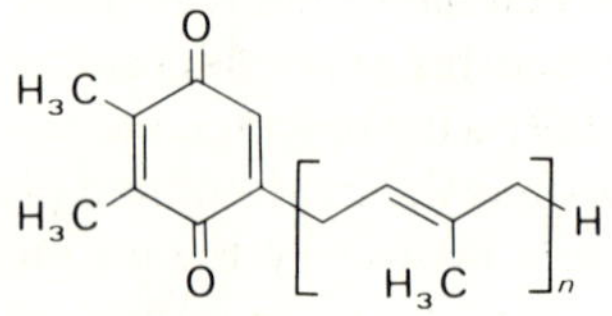

Plastoquinone ($n = 9$)

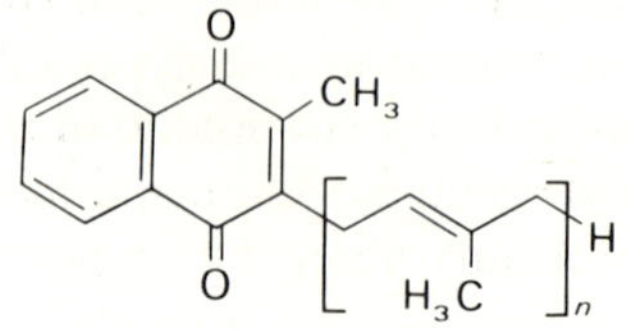

Vitamin K_2 ($n = 6$-9)

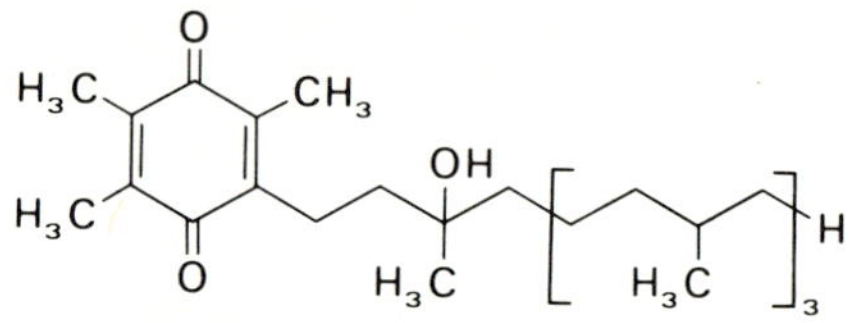

Tocoquinone

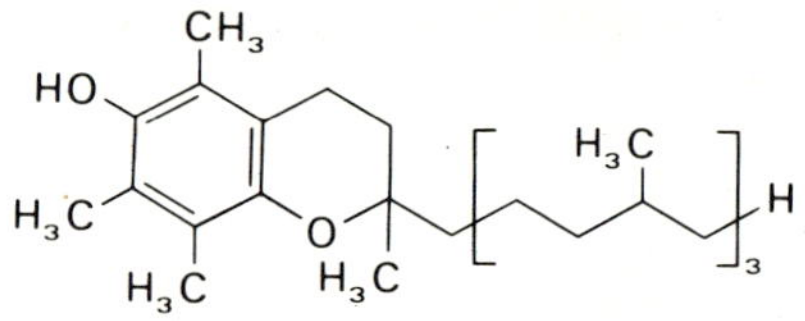

Tocopherol

Ubiquinone. Described in Chapter X, 4 as an auxiliary substrate of the respiratory chain, the biosynthesis of ubiquinone is fairly well established at present. The aromatic ring derives from tyrosine which is degraded via *p*-hydroxyphenylpyruvate to *p*-hydroxybenzoate. The polyprenyl side chain is then attached, and last, the substituents on the aromatic rings are introduced one by one. The methyl groups derive from *S*-adenosylmethionine.

Ubiquinone is, as indicated by its name, widely distributed, both in the animal and plant kingdoms. The length of the prenyl side chain varies. Ubiquinone-50 (50 C atoms in the side chain = 10 isoprene residues) has been isolated from pig heart, whereas ubiquinone-30 has been extracted from yeast.

In the equally ubiquitous, isomeric *ubichromenol* the side chain is added to the quinone system (see formula). As is the case for ubiquinones, there is a whole family of closely related compounds. Nothing is known about the function of ubichromenols.

Plastoquinone. Isolated from chloroplasts, plastoquinone differs in its structure from ubiquinone only by its methyl group substituents in the aromatic ring in place of the methoxy groups. In photosynthesis it functions as a reversible redox substrate (cf. Chapter XVI,2).

Vitamin K Group. Compounds in this group are characterized by a naphthoquinone ring system in place of the benzoquinone ring. The length of the prenyl side chain again varies. Vitamin K_2 is widely distributed with its difarnesyl side chain ($n = 6$; see formula), which is also called *menaquinone*-6. Vitamin K_1, called *phylloquinone*, contains a phytyl side chain.

In several bacteria, menaquinones appear to be components of the respiratory chain where they take the place of ubiquinone. They are vitamins for mammals and man (antihemorrhagic or coagulation vitamin) as discussed in Chapter XXII,4.

Tocoquinones. γ-Tocoquinone structurally resembles plastoquinone. It has a shorter prenyl side chain which is hydrogenated, i.e., the phytyl chain, although compounds with unsaturated side chains have also been isolated. The latter may be precursors. Isomeric β-tocoquinone differs only in the position of the methyl groups. α-Tocoquinone contains three methyl groups on the aromatic ring.

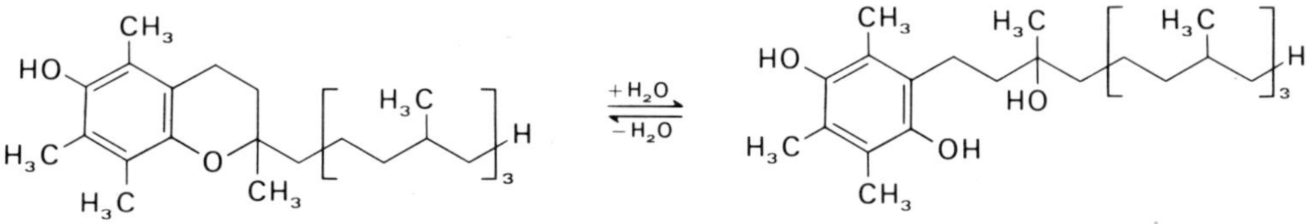

α-Tocopherol α-Tocopherol hydroquinone

Tocopherols. The interest of tocoquinones stems from their relationship to tocopherols, which became recognized as the "antisterility vitamins" of the rat (for vitamin E deficiency, see Chapter XXII,4). Structurally, tocopherols have a chroman ring that probably arose from the addition of a phenolic hydroxyl group to the double bond of the first prenyl residue. The ring can be opened by hydrolysis to yield tocopherol hydroquinone (see formulas). As with tocoquinones, the compounds designated with β, γ, and δ respectively, contain one, two, or three methyl groups fewer than the α-compound.

BIBLIOGRAPHY

M. E. Dempsey, Regulation of steroid biosynthesis. *Annu. Rev. Biochem.* **43**, 967–990 (1974).

T. W. Goodwin (editor), "Natural Substances Formed Biologically from Mevalonic Acid." Academic Press, New York, 1970.

T. W. Goodwin (editor), "Aspects of Terpenoid Chemistry and Biochemistry." Academic Press, New York, 1971.

E. Heftmann, "Steroid Biochemistry." Academic Press, New York, 1969.

J. H. Richards and J. B. Hendrickson, "The Biosynthesis of Steroids, Triterpenes, and Acetogenins." W. A. Benjamin, New York, 1964.

CHAPTER XV

Simple Sugars, Monosaccharides

The simple sugars constitute a subclass of the great class of natural substances, the carbohydrates. In sheer amount, the carbohydrates make up the bulk of organic substance on this earth. Though predominantly of plant origin, they constitute the principal component of the food of many animals and of man. Of the customary divisions of foodstuffs—protein, fats, and carbohydrates—the carbohydrates rank highest as energy suppliers.

Many simple sugars contain carbon and the elements of water in the ratio of 1:1; this fact has caused the name *carbohydrate* to be given to this class of compounds.

The designation originated at a time when real significance was attached to the empirical formula $C_x(H_2O)_n$, which represents these substances as hydrates of the element carbon. Today we are no longer bound by the empirical formula; we call deoxyribose $C_5H_{10}O_4$ and glucosamine $C_6H_{13}O_5N$ "carbohydrates," but not lactic acid $C_3H_6O_3$.

The simple sugars are either polyhydroxy aldehydes or polyhydroxy ketones. The carbohydrates ordinarily also encompass all substances that are closely related to sugars, e.g. simple derivatives (amino sugars, carboxylic acids, etc.) and polymers of them (oligo- and polysaccharides), which will be discussed in Chapter XVII.

1. Nomenclature and Definitions[1]

Aldoses and Ketoses. Carbohydrates differ from ordinary polyalcohols in that one of the alcohol groups is oxidized to a carbonyl group. We use glycerol as an example:

[1] For international rules of nomenclature see *Eur. J. Biochem.* **21,** 455–477 (1971).

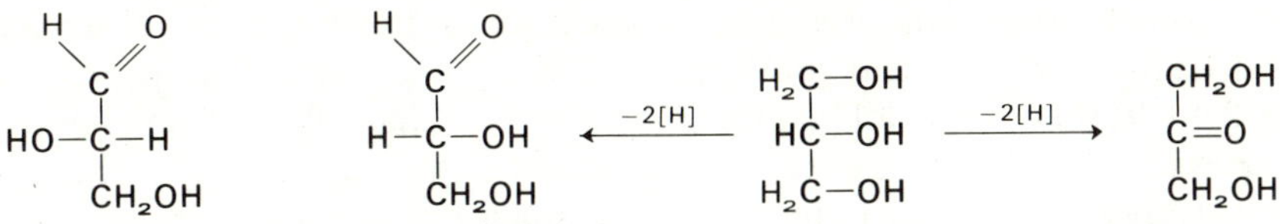

L- D-
Glyceraldehyde Glycerol Dihydroxyacetone

Obviously, two dehydrogenation products are possible, either the aldehyde or the ketone. Sugars with a keto group (e.g., dihydroxyacetone) are called *ketoses*, while those with an aldehyde group (e.g., glyceraldehyde) are called *aldoses*. The two formulas for glyceraldehyde shown above represent mirror-image isomers (cf. Chapter I,3); they are the two optically active forms, D-glyceraldehyde and L-glyceraldehyde.

By convention, D-glyceraldehyde, in whose projection formula the OH group points to the right, is taken as the basis of the entire D-series. This convention is very important for the carbohydrates. Assignment to a steric series is independent of the direction of rotation of light (cf. Chapters I,3 and II,1).

Structure of Aldoses. Mentally, the carbon chain of glyceraldehyde may be lengthened most easily by condensation with formaldehyde (the synthetic chemist has better methods available). Two compounds can arise thereby, with the following projection formulas:

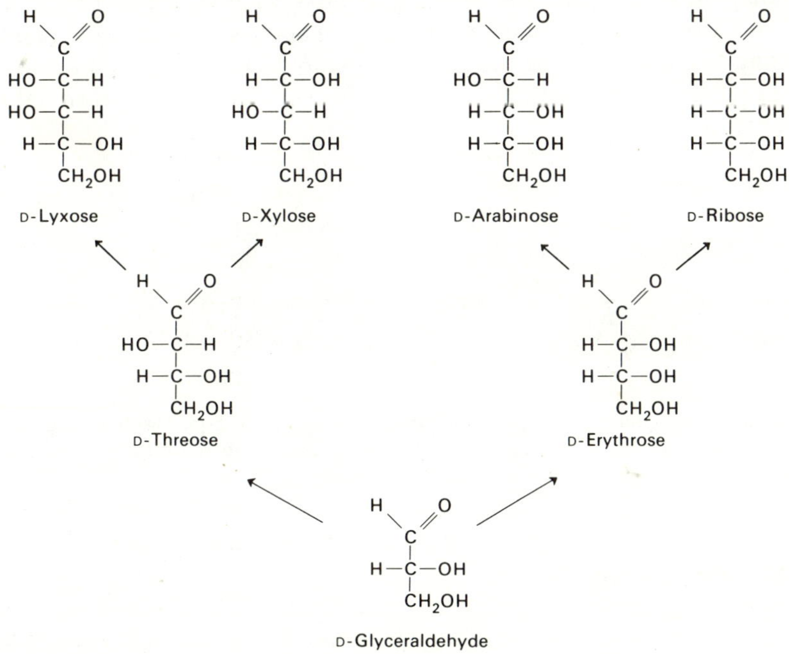

They are called *threose* and *erythrose*. A repetition of this hypothetical formaldehyde condensation would yield four pentoses. Since we started the process with D-glyceraldehyde, they would all belong to the D-series. Chain elongation of D-glyceraldehyde produces only D-sugars; the same procedure starting with L-glyceraldehyde yields exclusively L-sugars. The latter are mirror images of the former. Since by definition the aldehyde group, or more generally the most highly oxidized group, is written on top, the bottom-most asymmetric C atom determines the assignment to either the D- or L-series. It is also the asymmetric carbon with the highest number, because numbering starts with the aldehyde carbon.

The symbols D- and L- denote only membership in the corresponding configurational series; nothing is said about optical rotation, which may be either + (to the right) or − (to the left). Rotation is measured easily, whereas assignment to a steric series is usually difficult. The sugar must be broken down stepwise, converted to a compound whose configuration is already known, and then identified. The configuration of most sugars is known. For more on configurational determinations see textbooks of organic chemistry.

The number of possible aldoses grows exponentially with each additional C atom. Hence there are two glyceraldehydes (D- and L-), four sugars with four C atoms, called *tetroses*, eight *pentoses* with five C atoms (they are called D- and L-ribose, D- and L-arabinose, D- and L-xylose, and D- and L-lyxose), 16 *hexoses* and so forth. Furthermore, there is a similar series of ketoses. The large number of asymmetric C atoms and the consequent number of isomers complicates the chemistry of carbohydrates considerably.

We have employed Fischer's projection formulas for our discourse on isomerism. In the transition from projection formula to the spatial model and back, it should be noted that the projection rule (cf. Chapter I,3) applies to each individual C atom: Neighboring C atoms must be behind, while H and OH groups are in front, of the plane of projection. For a pentose (D-ribose) the correct three-dimensional model is shown in Fig. XV-1.

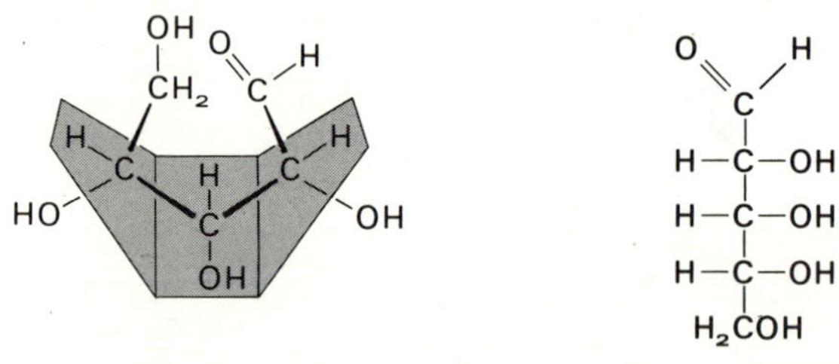

Fig. XV-1. Projection formula of a sugar. In the model the C—C—C bonds form an angle of 110°; the planes of projection (framed in the left figure), therefore, are at an angle to each other. In writing the formula (as at right), the planes are forced into the plane of the paper.

2. Hemiacetal Formulas

Figure XV-1 illustrates that the aldehyde group at C-1 and the hydroxyl group at C-5 comes rather close together. The proximity facilitates the reaction leading to hemiacetal formation. Aldehydes, in general, can add hydroxyl compounds to the C=O bond (cf. Chapter I,2). If H—O—H is added, the product is the hydrate of the aldehyde; if an alcohol is added the hemiacetal is formed. Full acetals (or simply "acetals") arise from hemiacetals plus alcohols by elimination of water (see formulas below). This reaction is the basis for glycoside formation by carbohydrates (see Chapter XVII,1).

Whenever hemiacetal formation proceeds intramolecularly, a ring is produced; it is either a five-membered ring with one O atom, or a six-membered ring with one O atom. In reference to the model compounds furan and pyran (see Table I-1) the corresponding forms of the sugars are called *furanose* or *pyranose*. The sugars of interest to us usually exist in the pyranose form; in some polysaccharides, however, the furanose form may appear (e.g., sucrose, nucleic acids).

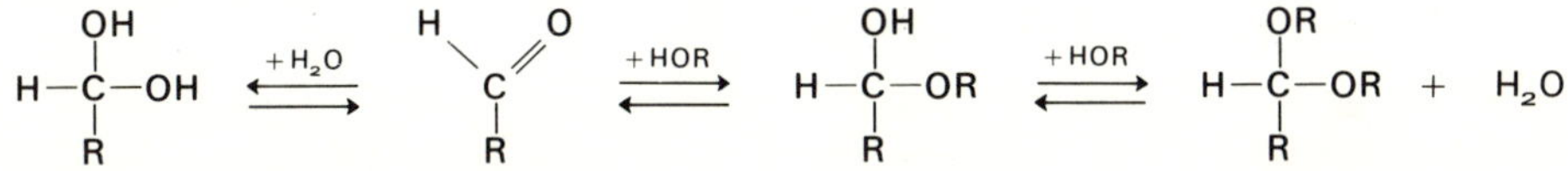

Aldehyde hydrate Aldehyde Hemiacetal (Full) Acetal

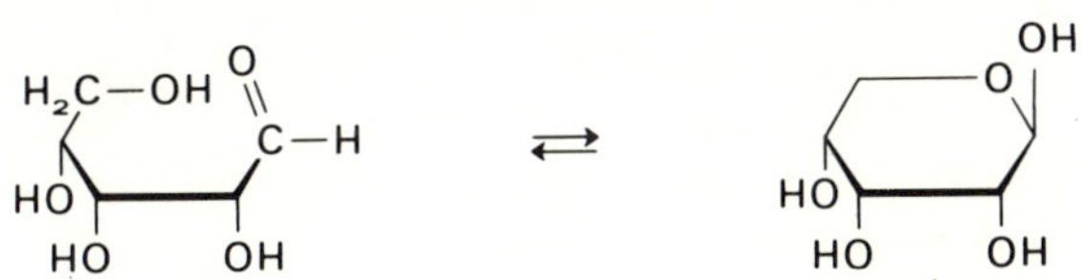

Pyranose

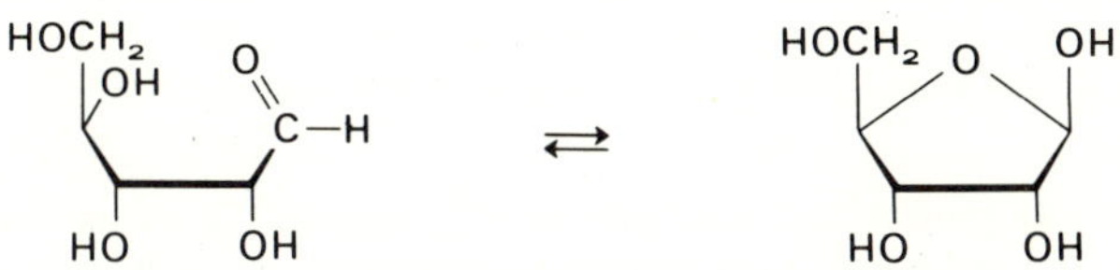

Furanose

In using the Fischer projection formulas, it is customary to indicate the ring by a long line between C-1 and C-5 [representation (b) for β-D-glucose, Fig. XV-2]. An error is thereby committed: As can be seen in the figure, ring closure becomes possible only after the OH group at C-5 is rotated to a suitable position [transition (d) → (e) → (f)]; the corresponding projection formula is shown in (c). Compared to the Fischer projection formula the H atom seems to have changed its position, a consequence of the altered projection rule.

Haworth proposed writing the ring form of sugars in perspective as shown by formula (f): Imagine the molecular model to lie on a table and draw the model looking down on it with the oxygen toward the rear and the CH_2OH group protruding on top as a side chain. (This arrangement is typical of all D-sugars in this manner of representation.) The C atoms are usually not written out. We have already used this type of representation on other occasions (inositol, Chapter XIII,2; sterols, Chapter XIV,2). The use of such *perspective* formulas is governed by other rules than the projection formulas; for one thing, the drawing may not be rotated in the plane of the paper, but must be redrawn in the new position.

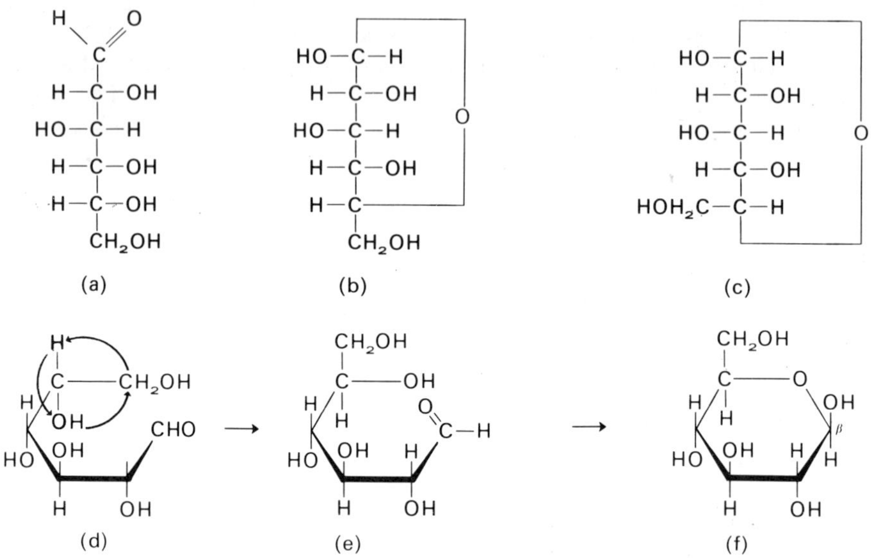

Fig. XV-2. Different ways of representing glucose. Formulas (a), (d), and (e) show the aldehyde form, which is practically nonexistent; in (b), (c), and (f), β-D-glucose is shown. The series (d) → (e) → (f) clarifies the transition to the ring form, during which the CH_2OH group reaches the top.

We have drawn the Haworth formula of the six-membered ring [drawing (f)] as a planar hexagon. This is a simplification; in reality carbohydrate molecules always exist in the energetically preferred chair form (cf. Chapter XIV,2). It can be expressed in the Haworth formula by drawing C-1 down and C-4 up.

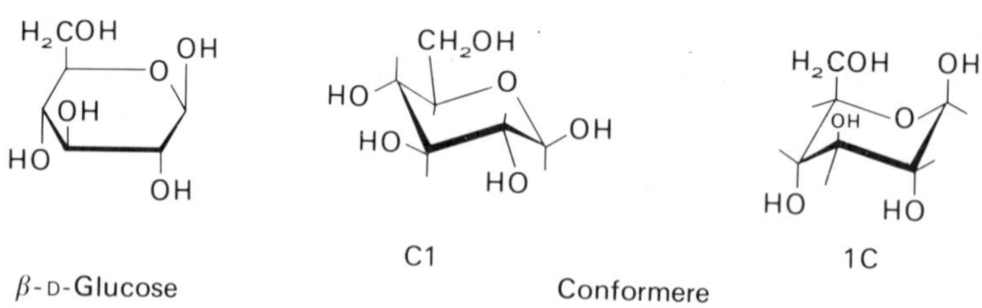

In this so-called C1 form, all hydroxyl groups and the CH_2OH group are arranged equatorially. The other possible chair form, the 1C conformere, is obtained analogously by drawing C-1 up and C-4 down. With that all functional groups become axial; this makes it an energetically unfavorable form.

Whenever possible we shall use these Haworth formulas. The following two rules should be remembered:

1. When oxygen is in back and C-1 at right—the most usual form—then in all D-sugars[2] the CH_2OH group is up.
2. In *β*-D-glucose all hydrogens are *trans*, i.e., they are alternately up and down.

α- and β-Forms. During ring closure a new asymmetric C atom is created; there are now four different substituents on the former carbonyl carbon. Therefore, two different "anomeric" forms, the α- and the *β*-forms, are possible. The *anomers* are easily interconvertible, probably through the aldehyde form, which is, however, present only in infinitesimal amounts in the equilibrium mixture. Anomers differ in their physical constants, including their optical rotation; they are, however, not mirror-image isomers, for example:

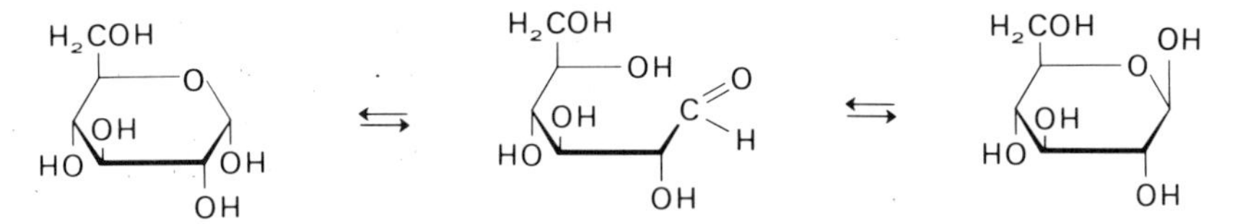

α-D-Glucose
Rotation $\alpha_D = +112°$

al-D-Glucose
(aldehyde form)

β-D-Glucose
Rotation $\alpha_D = +19°$

Immediately after α-D-glucose is dissolved in water, the degree of rotation recorded above can be observed. However, the rotation gradually diminishes and a few hours later comes to rest at $\alpha_D = +52°$, which corresponds to the equilibrium mixture. (The enzyme mutarotase, or *aldose 1-epimerase*, accelerates the reaching of equilibrium.) This same final value of rotation is observed when one starts with a solution of *β*-glucose. This phenomenon of a shift in rotation is called *mutarotation*, and the discovery of the α,*β*-anomers confirmed the correctness of the ring formulas.

The assignment of the rotational isomers to the two formulas which can be written for them has been a difficult stereochemical problem. At first, in all sugars of the D-series the more strongly dextrorotatory form was called the α-form. Later it was noticed that the α-form in the projection formula invariably has the hydroxyl group on the same side as the hydroxyl group determining the configurational series (i.e., to the right in the D-series; to the left in the L-series). The reason for this is that each C atom contributes a certain amount to the optical rotation (Hudson's rule of superimposition). In the Haworth formulas of the D-series this means that, if the OH at C-1 is down, we have the α-form; if the OH is up, the *β*-form.

3. General Reactions of Monosaccharides

The chemical properties of sugars are determined primarily by their many hydroxyl groups. Being polyhydroxy compounds they dissolve readily in water, are insoluble

[2] The presumption is that the hydroxyl which determines the D-series is a member of the ring. This is most often the case. Exceptions are aldohexoses in the rare furanose form, heptoses, and neuraminic acid.

in lipid solvents and in fats, and taste sweet (even glycerol is sweet). The aldehyde groups (or ketone groups in ketoses) are not manifest; simple color reactions for aldehydes are negative, because in solution there is practically no aldehyde present, but rather the hemiacetal.

Functional Derivatives of Hydroxyl Groups. Alcohol groups can be esterified. This reaction is used frequently in organic chemistry for the purpose of characterizing sugars or for blocking certain hydroxyl groups.

In biochemistry, the esters of phosphoric acid have special significance; the organism metabolizes sugars almost exclusively in the phosphorylated form. We symbolize the phosphate group by Ⓟ. Different phosphates of the very same sugar (e.g., glucose 1-phosphate and glucose 6-phosphate; see formulas) behave entirely differently in biochemical reactions.

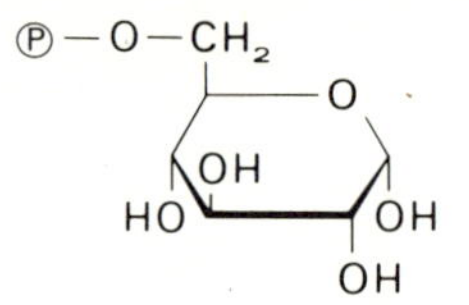

Glucose 6-phosphate

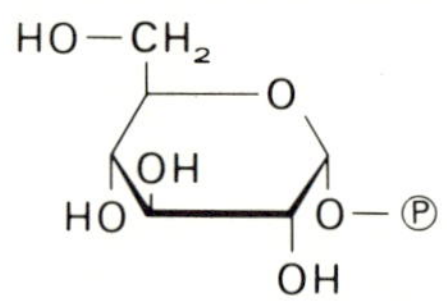

α-D-Glucose 1-phosphate

Alcoholic groups are also able to form ethers. Ether formation (e.g., methylation with dimethyl sulfate and alkali) is also important for the constitutional determination of oligosaccharides. The hemiacetal hydroxyl group on C-1 is unusually reactive; compounds derived from it are called glycosides. A separate chapter is devoted to them (Chapter XVII).

Derivatives of the Carbonyl Group. Not many reactions of the aldehyde or keto groups of sugars are observed because the carbonyl forms of sugars comprise only a fraction of 1% in the equilibrium mixture. They can be trapped, however, in a few reactions, e.g., by oxime formation. Important for the identification of sugars is the reaction with phenylhydrazine or with substituted phenylhydrazines (E. Fischer), which involves two carbon atoms and results in *osazones*:

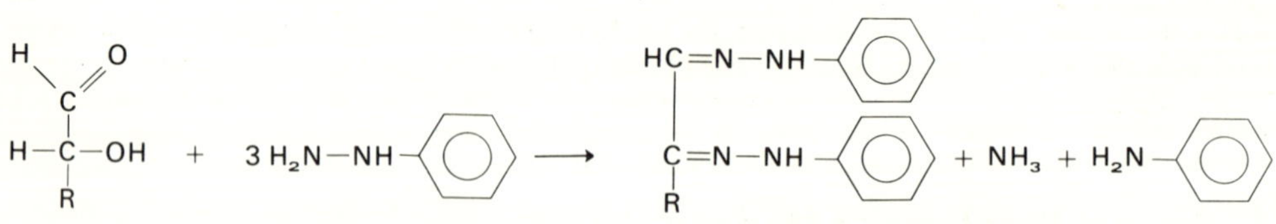

Aldose Phenylhydrazine Osazone

The course of the reaction was recognized correctly by Weygand (1941) as a series of rearrangements (cf. textbooks of organic chemistry).

Sugars possess only limited stability toward acids and alkali. In strongly acid solution, water molecules are split out; pentoses are thereby converted to furfural, while hexoses become hydroxymethylfurfural, which decomposes further to unsaturated keto aldehydes, such as $H_3C—CO—CH{=}CH—CO—CHO$ and $H_3C—CO—CH{=}CH—CHO$, that are responsible for certain color reactions of the sugars (see below). Alkali augments the formation of enols; glucose becomes fructose or mannose (or the reverse). Sugars that differ only at C-2 have been called epimeric sugars. Today the term *epimerism* is generally reserved for sterically different arrangements at one out of several asymmetric C atoms. Continued heating with more concentrated alkali destroys the molecule, and strongly reducing fragments (reductones) are generated.

Reduction of the carbonyl group produces the corresponding polyalcohol, for example, mannitol is obtained from mannose.

Cautious oxidation dehydrogenates the hemiacetal to form the lactone of an acid (in alkaline solution, the salt of that acid). The lactone may be reduced again to reform the hemiacetal. More vigorous oxidation forms the dicarboxylic acid glucarate by oxidizing the terminal CH_2OH as well.

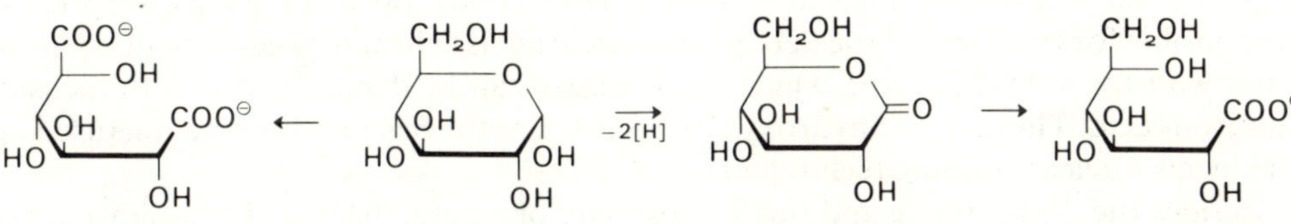

Color Reactions of Carbohydrates. There are a few color reactions of sugars with phenols (α-naphthol, anthrone, resorcinol, orcinol, etc.) and concentrated mineral acids. The acids effect loss of water, resulting in furfural derivatives, some of which break down further to form lower aldehydes. The aldehydes then condense with the phenols to produce the colored substances. The reaction mechanisms are not completely understood.

General color reactions of this type include the Molisch test with *α-naphthol* and concentrated sulfuric acid and the *anthrone reaction* with anthrone and sulfuric acid. The latter is used for quantitative colorimetric determinations. Reaction conditions can be chosen so that only ketoses react (HCl + *resorcinol*, Seliwanoff's test) or only pentoses (HCl + *orcinol*, Bial and Tollen's test). These reactions help to distinguish the various sugars.

All mono- and some oligosaccharides are reducing sugars; this property is due to the α-ketol grouping (carbonyl group adjacent to a hydroxyl group). Although tests for reducing sugars are nonspecific, they are used frequently to detect and to determine sugars. The reduction of $Cu^{2\oplus}$ ions in alkaline solution is the basis of Benedict's and Fehling's reactions. Aldoses are initially oxidized to the carboxylic acids, but may undergo further cleavage (especially in more concentrated alkali). Ketoses are split to glycolaldehyde and a tetrose that can be oxidized further.

4. The Individual Sugars

Trioses. D-*Glyceraldehyde* (formula in Section 1) is the dehydrogenation product of glycerol. More important, however, is 3-phosphoglyceraldehyde, an intermediate in the degradation of carbohydrates (Section 7), which is in equilibrium with *dihydroxyacetone phosphate*; the attainment of equilibrium is catalyzed by the enzyme triosephosphate isomerase (cf. Section 7).

Tetroses. Both threose and erythrose have already been mentioned (formulas in Section 1). Erythrose 4-phosphate is an intermediate in certain biochemical reaction sequences. The keto tetrose is called erythrulose.

Pentoses. Names and formulas of the aldopentoses may be found in Section 1. *Ribose* and ribose phosphates are components of nucleic acids and nucleotide coenzymes. The furanose form is present in these derivatives whereas free ribose exists in the pyranose form (formulas in Section 2). *Deoxyribose* (formula in Chapter VII) is responsible for the name of the deoxyribonucleic acids. The free sugars are in equilibrium with the aldehyde form, which, in this case, can be demonstrated with fuchsin sulfurous acid. The lack of a hydroxyl group at C-2 is the basis of this reaction (used in Feulgen's nuclear staining technique).

Besides the 3-phosphate and the 5-phosphate of ribose, there is also 5-phosphoribosyl pyrophosphate, an intermediate in the biosynthesis of nucleotides (cf. Chapter VII,2).

Arabinose is a component of gum arabic and otherwise widely distributed in the plant kingdom.

Ribulose is the ketose corresponding to ribose. In general, the ketoses are designated by the ending *-ulose* (unless they have trival names, such as fructose). *Xylulose* is an epimer to ribulose at C-3. Phosphates of ketopentoses play a role in biological interconversions of the sugars and in photosynthesis (Chapter XVI,4).

Hexoses. The most important and most widely distributed carbohydrate is D-*glucose* (also called "grape sugar" or "dextrose"). The β-form has already been introduced with its formula; it is easy to remember that all its hydrogen atoms are *trans* (in the Haworth formula or corresponding simplifications). Common grape sugar actually consists chiefly of α-glucose; the initial specific optical rotation of $+112°$ decreases to the final value of $+52°$ in solution (mutarotation; for explanation see end of Section 2). Free glucose occurs primarily in sweet fruits; smaller amounts are found in blood of the animal organism, between 0.06 and 0.10% in man.

The more important phosphate esters of glucose (formulas in Section 3) are *glucose 1-phosphate* (also called Cori ester), which is in equilibrium with the storage form glycogen (Chapter XVII,6), and *glucose 6-phosphate* (also called Robison ester), which is formed enzymatically from glucose and ATP (hexokinase reaction, Section 7). Glucose 1-phosphate can be converted to glucose 6-phosphate via the intermediate "coenzyme" glucose 1,6-bisphosphate.

Glucose, mannose, and the ketose fructose all form the same osazone. In alkaline solution the three are interconvertible through a common enol, but further degradation and breakdown into smaller fragments occur easily.

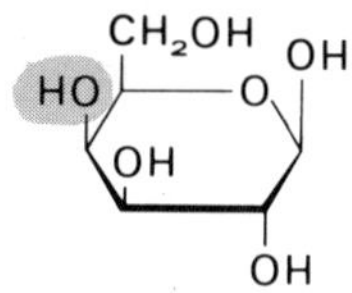

β-D-Galactose

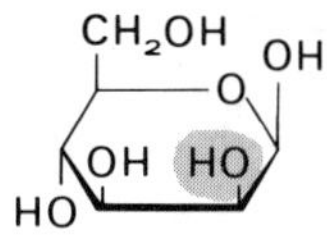

β-D-Mannose

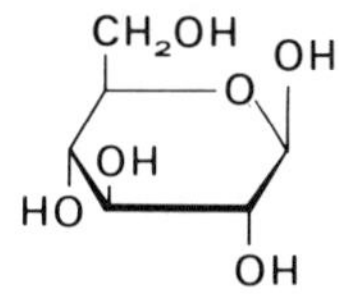

β-D-Glucose

D-(+)-*Mannose* is a C-2 epimer of glucose; i.e., the two sugars differ only at C-2. Free mannose occurs in plants, but more often is bound. In the animal organism, it is a component of glycolipids (Chapter XIII,4), glycoproteins, and blood group substances.

D-(+)-*Galactose* is a component of milk sugar (lactose) and other oligosaccharides and of a few other more complicated compounds. It differs from glucose in the steric configuration at C-4; as a consequence, three hydroxyls in the ring form are *cis*. An enzyme is able to convert galactose to glucose, i.e., it performs a Walden inversion at C-4 (glucose 4-epimerase; coenzyme: uridine diphosphate). Nitric acid oxidation of galactose produces mucic acid (tetrahydroxyadipic acid) in its *meso* form. This reaction was important in the determination of its constitution. Besides the common D-isomer, L-galactose also occurs naturally (in agar agar and elsewhere).

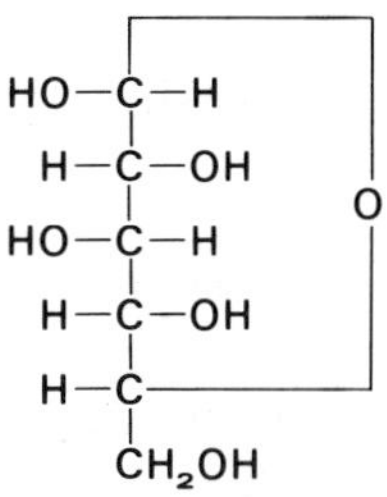

β-D-Glucose

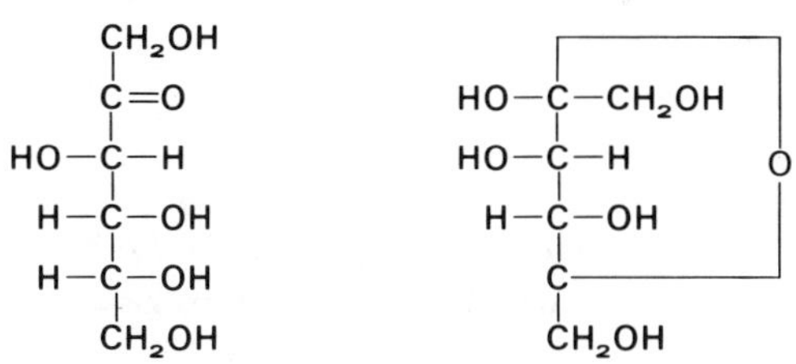

D-Fructose, projection formulas

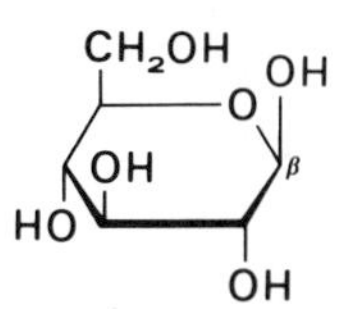

β-D-Glucose

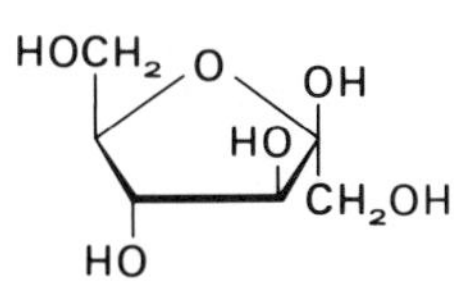

β-D-Fructofuranose

D-(−)-*Fructose*, a ketohexose, was at one time called levulose, because it rotates polarized light to the left. Nevertheless, it belongs to the D-series and is closely related to D-glucose (see formulas above). Free fructose largely exists in the pyranose form (six-membered ring); only in oligosaccharides (cane sugar), in polysaccharides (inulin), and in several phosphate esters is the furanose form realized. Fructose occurs mainly in the plant kingdom and in honey.

Two phosphate esters, fructose 6-phosphate (Neuberg ester) and fructose 1,6-bisphosphate (Harden-Young ester) are biochemically significant because they are intermediates in the breakdown of glucose according to the Embden-Meyerhof pathway (Section 7); this pathway is actually the route of breakdown of fructose 1,6-bisphosphate.

Fructose 1,6-bisphosphate formerly was known as fructose 1,6-diphosphate. New rules of nomenclature for phosphate esters reserve the expression "...diphosphate" for compounds of the type adenosine diphosphate (ADP) that contain two phosphate groups connected together in anhydride linkage. Compounds that contain two phosphate groups at two different positions in the molecule are to be called bisphosphates. We have adopted the new internationally agreed upon nomenclature rules.

Deoxyhexoses. A component of a few oligosaccharides of milk and of blood group substances (see Chapter XVII,8) is L-*fucose* (6-deoxy-L-galactose). *Rhamnose* (6-deoxy-L-mannose) occurs in glycosides, e.g., in hesperidin (Chapter XVII,1). Note that both deoxyhexoses belong to the L-series.

Heptoses. Only *sedoheptulose* is of interest to us. It is a ketose with the same structure as ribose between C-4 and C-6. Its phosphate is derived from ribose 5-phosphate during the interconversion of the monosaccharides (Section 5).

Cyclitols. These have the same formula as hexoses, $C_6H_{12}O_6$, but are not carbohydrates. Instead they are carbocyclic homologs of glycerol; for this reason they have been discussed under lipids, of which they are components (inositol, Chapter XIII,2).

Amino Sugars. Formally, amino sugars are formed by the replacement of a hydroxyl group with an amino group. In biochemistry three such compounds are important: glucosamine (chitosamine), galactosamine (chondrosamine), and neuraminic acid.

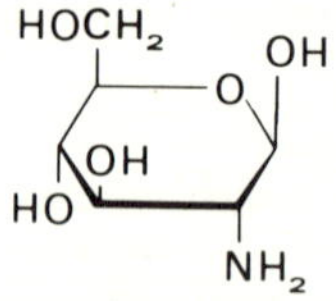

β-D-Glucosamine

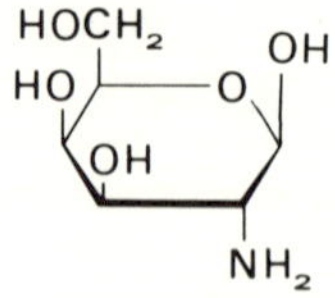

β-D-Galactosamine

Glucosamine in its structure resembles D-glucose; the amino group is at C-2 (hence the systematic name *2-amino-2-deoxy-D-glucose*). It occurs only bound, e.g., in chitin (hence the name chitosamine), in glycolipids, in blood group substances, and in other complex polysaccharides. The amino group frequently is acetylated.

Galactosamine (chondrosamine) resembles galactose; the OH group in position 2 is replaced by an amino group which usually is acetylated. Acetylgalactosamine occurs in complex glycolipids and polysaccharides.

Neuraminate has a somewhat more complicated structure, which may be derived from the product of an aldol condensation between mannosamine (an epimer to glucosamine at C-2) and pyruvate. The newly formed hydroxyl group is oriented *trans* to the amino group, and the keto group of pyruvate forms a pyranose hemiacetal ring. Its pK of 2.2 makes neuraminic acid an unusually strong acid.

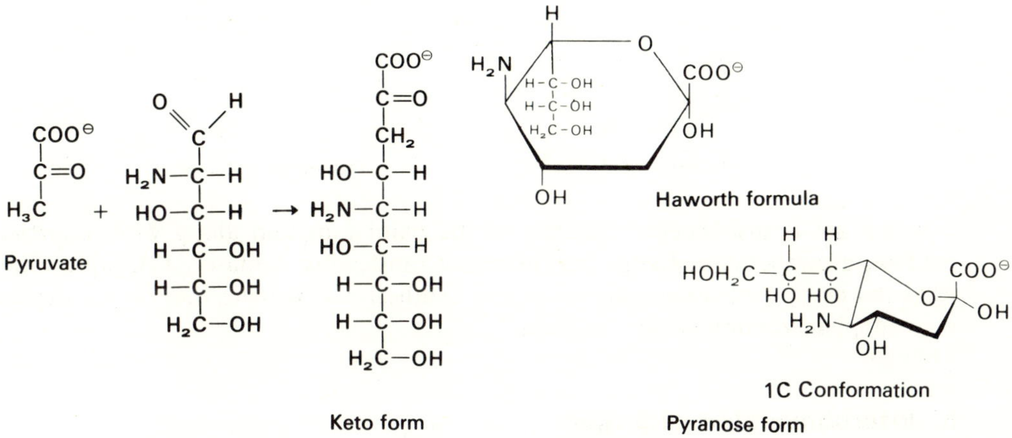

al-D-Mannosamine Neuraminate

Neuraminate does not exist free, but as *N*-acetyl- or *N*-glycolylneuraminate is a component of membrane-forming glycolipids (cf. Chapter XIII,4), as well as of blood group substances, and many glycoproteins. The latter are the origin of the generic designation *sialic acid* for *N*-acylneuraminic acids.

Uronic Acids. Oxidation of sugars at the CH_2OH group, but not at the aldehyde group, yields uronic acids. This reaction can be carried out chemically only by indirect means; the organism, however, performs the step directly (cf. Chapter XVII,4). *Glucuronate*, the prototype of the uronates, is important because it rather easily forms glycosides called *glucuronides*. Many substances are excreted in the urine after undergoing coupling with glucuronate (cf. Chapter XXIII,2). *Galacturonate* is found in pectin.

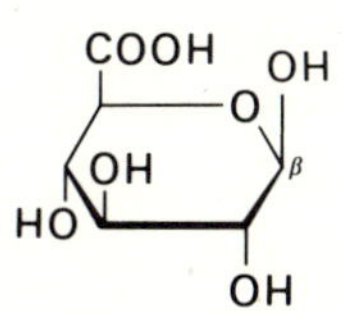

D-Glucuronate

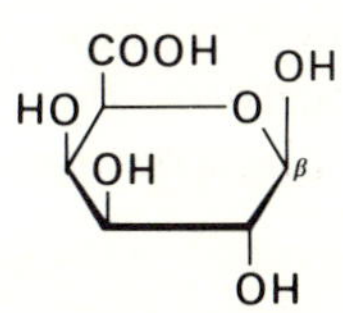

D-Galacturonate

Ascorbic Acid (*Vitamin C*). Discovered and isolated as the anti-scurvy vitamin (for physiological properties cf. Chapter XXII,5), ascorbate[3] is the lactone of 2-keto-L-gulonate. It contains an ene-diol group (a double bond between two adjacent hydroxyl groups) and is a strong reducing agent, as are all ene-diols. With the loss of hydrogen, it is converted to dehydroascorbic acid (formula at right) possibly via a semiquinonoid intermediate step. The reaction is reversible. In some biological hydroxylations ascorbate is hydrogen donor (Chapters VIII,11 and X,6).

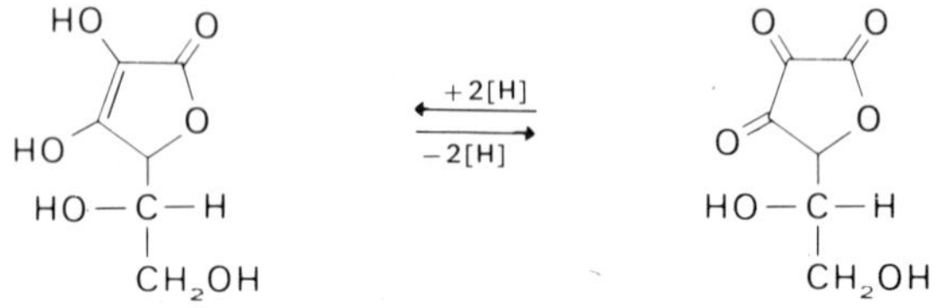

L-Ascorbic acid L-Dehydroascorbic acid

Ascorbate is distributed primarily in the plant kingdom along with ascorbate oxidase, a copper-containing, four electron-transferring oxidase (cf. Chapter X,6). Rats and many other mammals are able to synthesize ascorbate; they do not depend on a supplementation of the "vitamin."

5. Interconversion of Sugars

Most organisms can interconvert sugars; e.g., they can form ribose out of glucose whenever ribose is required, as it is for the synthesis of nucleic acids. Several types of reactions are available for such interconversions:

1. Epimerization (the inversion of steric configuration at one C atom) and isomerization (the conversion of aldose $\rightleftharpoons$ ketose). The number of C atoms does not change.

2. Oxidative removal of one C atom by dehydrogenation of the aldehyde to form the acid and subsequent decarboxylation. Hexoses become pentoses (see below); the chain is shortened by one C atom.

3. Transfer of C_3 or C_2 fragments from one sugar to another; thus hexoses can be converted to trioses, tetroses, pentoses, and heptoses. The donor of the C_3 or C_2 fragment is invariably a ketose; the acceptor is an aldose. The sum of carbohydrate C atoms remains constant.

Type 1: Biochemical *epimerization* at certain asymmetric centers has been discovered relatively recently. An example is glucose $\rightleftharpoons$ galactose; the coenzyme is uridine diphosphate, whose reactions will be discussed in another context (cf. Chapter XVII, 4). Among the pentoses, there is the conversion of ribulose 5-phosphate to xylulose 5-phosphate (epimerization at C-3).

[3] Ascorbic acid dissociates the enolic hydrogen at C-3 with a pK of 4.2 and thus exists as the enolate anion at physiological pH.

Isomerization, the shift of a carbonyl function to an adjacent carbon atom, can be achieved *in vitro* in a basic solution. The simplest example of enzymatic isomerization is the one of triose phosphate (cf. Section 7). The conversion of glucose to fructose and the reverse is also catalyzed by an isomerase.

Type 2: *Oxidation and decarboxylation* of glucose can proceed by one of two possible routes: Either C-1 or C-6 of glucose is oxidized and removed as CO_2.

Oxidation at C-1 is of greater significance for the degradation of glucose. The substrate is glucose 6-phosphate. The enzyme *glucose-6-phosphate dehydrogenase* (formerly also called "Zwischenferment") transfers hydrogen to nicotinamide-adenine dinucleotide phosphate and produces 6-phosphogluconolactone.

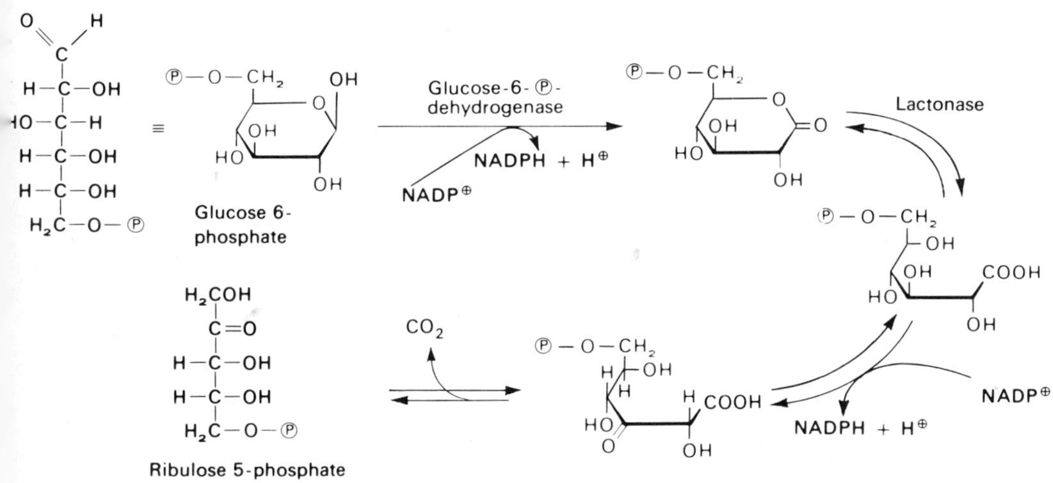

Gluconolactone is hydrolyzed easily to gluconate (the reaction is catalyzed by a lactonase but can also proceed spontaneously). *Gluconate 6-phosphate* can be dehydrogenated by another enzyme, with the simultaneous loss of CO_2. The pentose *ribulose 5-phosphate* arises from gluconate 6-phosphate presumably via the intermediate 3-ketogluconate 6-phosphate, which, being a β-keto acid, decarboxylates easily. The product ribulose 5-phosphate is in equilibrium with the aldose ribose 5-phosphate (the equilibrium reaction is catalyzed by ribose-5-phosphate isomerase). In this fashion several pentoses are formed from glucose.

The oxidation of glucose at C-6 (the other possibility mentioned above) affords *glucuronic acid*. This compound is involved in so-called detoxication reactions, and in general is coupled with foreign and endogenous substances for the purpose of excreting them in the urine. The coupling reaction is discussed in Chapter XVII,4d.

Any glucuronate still left over can be broken down further. The hemiacetal group (= aldehyde group) is reduced enzymatically with NADPH as hydrogen donor to produce *L-gulonate*. The transition to the L-configuration has been occasioned by turning the carbon chain around, i.e., the carbon atoms are numbered the other way, and C-6 of glucuronate now is C-1 of gulonate.

The *biosynthesis of ascorbate* branches off L-gulonate. The lactone formed first is dehydrogenated aerobically by L-gulonolactone oxidase (a flavoprotein) to yield ascorbate. This enzyme is missing in man and some animals (monkeys, guinea pigs) that depend consequently, on the dietary supply of vitamin C.

As discussed above with gluconate, L-gulonate can also be dehydrogenated at position 3 (with NAD as hydrogen acceptor) and be decarboxylated. The product *L-xylulose* appears in the urine in "essential pentosuria," a genetically determined metabolic anomaly. In normal metabolism, L-xylulose is reduced to form *xylitol*, a symmetrical *meso*-compound. Again by dehydrogenation (also involving a turning around of the carbon chain; C-1 becomes C-5), xylitol is converted to D-*xylulose*, which is then phosphorylated and entered into the transketolase reaction (see facing page).

Quatitatively this pathway is of subordinate significance. It is remarkable, nevertheless, that most of the intermediate products are unphosphorylated in contrast to the major pathways of carbohydrate metabolism.

Type 3: C_3 *fragments* can arise in the simplest manner from cleavage of a ketose. Fructose 1,6-bisphosphate is in equilibrium with the two trioses dihydroxyacetone phosphate and glyceraldehyde 3-phosphate (cf. Section 7). The reaction mechanism and the catalysis by *aldolase* were discussed in detail in Chapter V,9.

The aldolase reaction is the biochemical version of a reaction well known in organic chemistry, the aldol condensation. An aldehyde is attached to an active CH_2 group, whereby a new C—C bond is formed and hydrogen is accepted by the aldehyde oxygen (left part of the diagram of formulas).

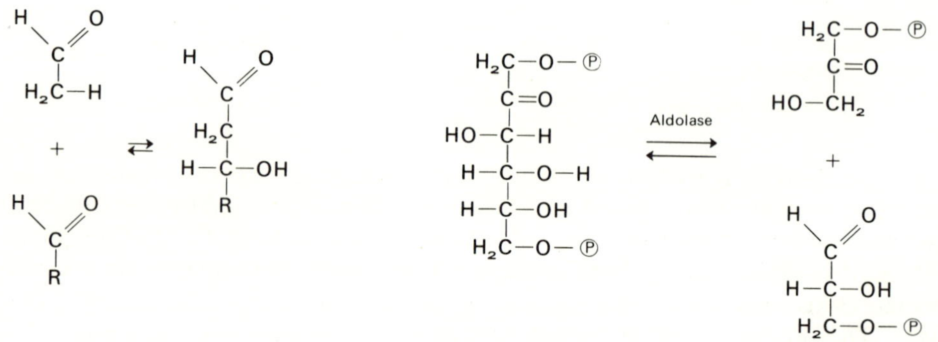

Aldol condensation Fructose-1-Ⓟ-6-Ⓟ 2 Triose-Ⓟ

Transaldolase. While aldolase produces only trioses from hexoses (and the reverse), the analogous enzyme transaldolase *transfers the dihydroxyacetone residue* onto other aldoses. The enzyme does not require a coenzyme and its reaction mechanism resembles that of aldolase. The substrate specificity is remarkable: Transaldolase splits only fructose and sedoheptulose (reverse of aldol condensation) and transfers the C_3 residue (a dihydroxyacetone, which remains bound to the ε-amino group of a lysine residue on the enzyme surface) to a corresponding aldehyde, i.e., to glyceraldehyde 3-phosphate or erythrose 4-phosphate, or possibly to ribose 5-

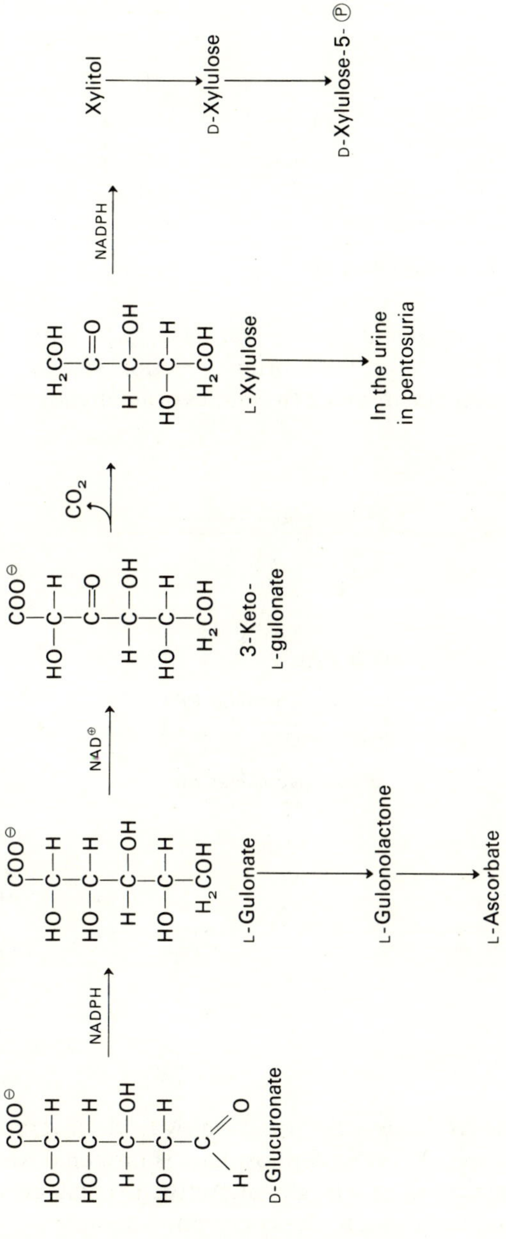
D-Glucuronate
NADPH
L-Gulonate
NAD⊕
3-Keto-L-gulonate
CO_2
L-Xylulose
NADPH
Xylitol
D-Xylulose
D-Xylulose-5-Ⓟ
In the urine in pentosuria
L-Gulonolactone
L-Ascorbate

phosphate:

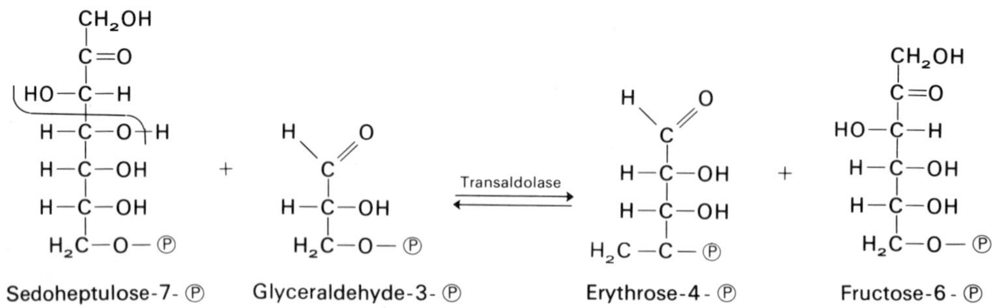

Transketolase. A C_2 fragment, *active glycolaldehyde*, is transferred during the *transketolase* reaction. This is an equilibrium reaction similar to the acyloin condensation of organic chemistry, in which two moles of aldehyde yield a hydroxyketone.

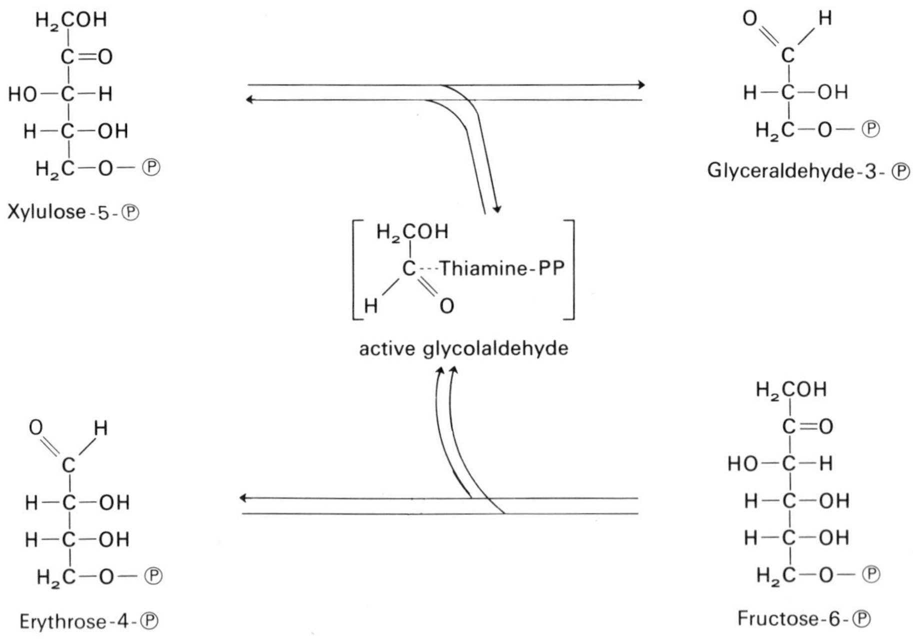

In biochemistry, the transketolase reaction was discovered as the cleavage mechanism of ketopentoses. Xylulose 5-phosphate is broken down to a C_3 fragment (glyceraldehyde phosphate) and a C_2 fragment. Active glycolaldehyde does not appear in its free form, but remains bound to the coenzyme thiamine pyrophosphate (TPP), probably in the same way as acetaldehyde does during the decarboxylation of pyruvate. In this form, glycolaldehyde is able to react with other sugars, e.g., ribose 5-phosphate or erythrose 4-phosphate. In the first case, *sedoheptulose 7-phosphate* is formed; in the second case, *fructose 6-phosphate*. The donor of the C_2 fragment evidently must be a ketose phosphate, or more precisely (because of the enzyme specificity) a ketose phosphate whose steric configuration at C-3 corresponds to that of fructose. Ribulose 5-phosphate, for this reason, can become a C_2 donor only after epimerization to xylulose 5-phosphate. There is the further possibility of obtaining active glycolaldehyde by decarboxylation (thiamine pyrophosphate assisted) of hydroxypyruvate. The acceptor molecule is always an aldose.

6. The Pentose Phosphate Cycle

The two enzymes described above, transaldolase and transketolase, function primarily in the organism to interconvert hexoses and pentoses. The conversion of hexoses (glucose and fructose) to ribose is particularly critical during the period of growth since ribose is necessary for the formation of nucleic acids.

The reaction sequence depicted below in the diagram starts with a C_2 transfer from fructose-6-Ⓟ to glyceraldehyde-3-Ⓟ. The latter sugar is an ever-present metabolite of glycolysis and arises from the cleavage of fructose 1,6-bisphosphate (see Section 7). The product of the *transketolase reaction* is xylulose-5-Ⓟ (the first of the C_5 sugars) and erythrose-4-Ⓟ, which then becomes the acceptor molecule of a C_3 fragment in a *transaldolase reaction*. The donor of the C_3 fragment is fructose-6-Ⓟ. As products from this reaction, sedoheptulose-7-Ⓟ and again glyceraldehyde-3-Ⓟ emerge. In a further *transketolase reaction* involving a C_2 transfer these two products are transformed into the two C_5 sugars ribose-5-Ⓟ and xylulose-5-Ⓟ. The latter is then epimerized to ribulose-5-Ⓟ and isomerized also to ribose-5-Ⓟ. As a result of all these interconversions three C_5 sugars have been produced out of two C_6 sugars and one C_3 sugar.

An alternate pathway of hexose (glucose-6-Ⓟ) to pentose is direct oxidation of glucose already discussed in Section 5. This pathway yields ribulose-5-Ⓟ and two molecules of NADPH, which is utilized primarily for the synthesis of fatty acids.

If the diagram of formulas is read from right to left then one sees a pathway for the reconversion of pentoses to hexoses (fructose-6-Ⓟ). Since fructose-6-Ⓟ can be isomerized to glucose-6-Ⓟ, it becomes possible to write down a full cycle in which glucose is oxidized completely. Such a cycle is called the pentose phosphate cycle or, after its discoverers, the Warburg-Dickens-Horecker degradative pathway.

The organism, in fact, does not employ this cycle to achieve complete degradation of glucose. The pathways described operate merely to produce the pentose phosphates required for the formation of nucleic acids and reduced NADP (i.e., NADPH) indispensable for the synthesis of fatty acids, cholesterol, and others. When the need is greater for pentoses than NADPH—as is the case under normal nutritional conditions—then the lower pathway in the diagram is followed. With an unusual demand for NADPH, an excess of pentoses can be reconverted to hexoses (upper pathway and return via the lower one).

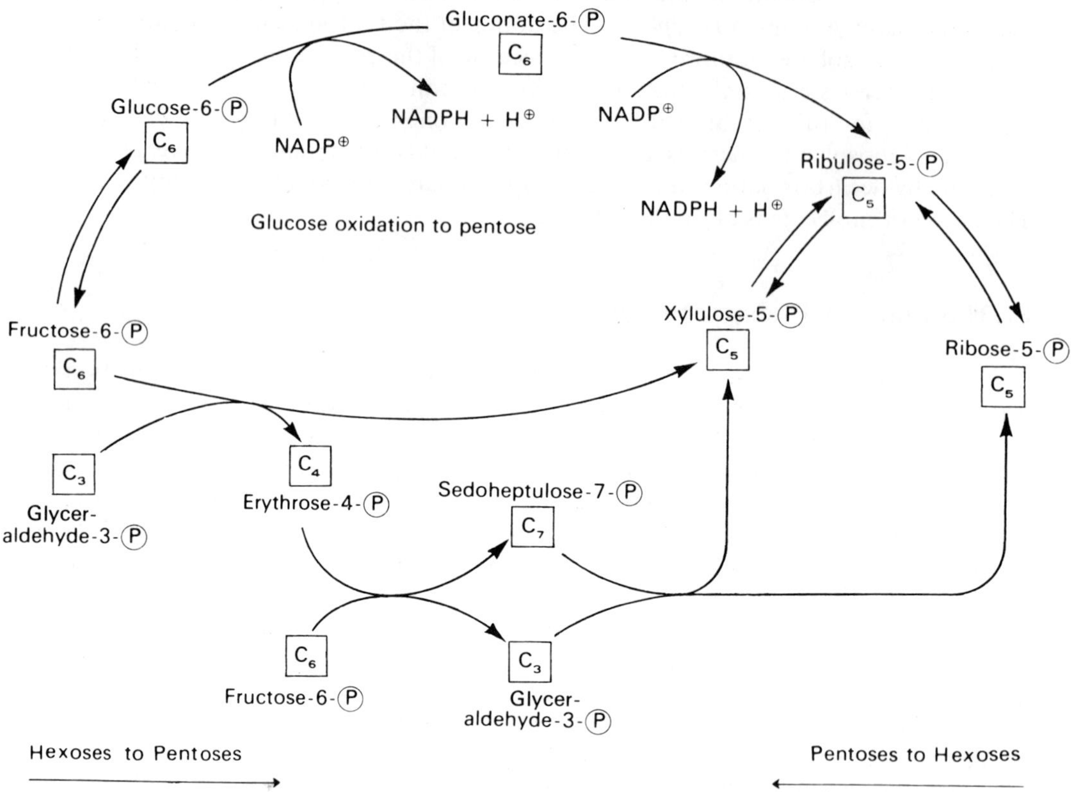

The pentose phosphate cycle plays a key part also in green plants. There the pathway prepares the CO_2 acceptor ribulose bisphosphate during photosynthesis (see Chapter XVI,3).

7. Glycolysis and Alcohol Fermentation

Glycolysis is usually defined as the anaerobic breakdown of carbohydrates (especially of glycogen) in the organism. It turns out, however, that aerobic catabolism also follows many of the steps of glycolysis. Actually, the only difference between aerobic and anaerobic breakdown is found in the utilization of the reduced coenzymes and of pyruvate. Alcoholic fermentation also branches off at the step of pyruvate utilization. It is not surprising that the different lines of research on these processes profited from each other.

Some History. Knowledge of alcoholic fermentation and particularly of its products, the intoxicating beverages, is very old. The history of the scientific investigation of alcoholic fermentation is that of biochemistry itself from its inception, at the time of Lavoisier, to the present. Gay-Lussac first presented the correct overall equation for the fermentation reaction (1815). Around 1840 it was noted that fermentation is caused by cellular organisms, the yeast cells. The question arose then whether fermentation depended on some "vital force" or could also take place outside living cells. Pasteur, who had accumulated many experimental data on the metabolism of yeast and other microorganisms and who had described lactic acid fermentation, citric acid fermentation, and other types of fermentation, considered fermentation an expression of life. Some of his contemporaries (around 1870), however, differed with Pasteur and pointed to pepsin, already known at that time, and the digestion of protein, among other things. Traube wrote: "... ferments[4] are chemical substances, akin to proteins ... they possess definite chemical structure and evoke changes in other substances depending on their specific chemical affinities ...." This opinion is almost completely valid even today.

The arguments centering around Pasteur's ideas were settled finally in 1897, when Buchner prepared a cell-free extract of compressed yeast ("yeast press juice") that could carry out alcoholic fermentation. This fundamental discovery removed the grounds for much useless speculation and greatly stimulated further experimentation. Some of the authors who studied reactions in yeast extracts and comparable extracts of muscle and who contributed so much to our present knowledge are Harden and Young, C. Neuberg, H. von Euler, G. Embden, O. Meyerhof, Parnas, O. Warburg, K. Lohmann, C. and G. Cori. The names of some of these workers have been associated with certain phosphate esters. In honor of the principal workers, the glycolytic degradative pathway is also called the *Embden-Meyerhof pathway* or *Embden-Meyerhof-Parnas pathway*.

Principles of the Glycolytic Breakdown. One essential characteristic of glycolysis is that it can proceed anaerobically and still provide energy (36 kcal is produced per mole of glucose, enough to convert 2 moles of phosphate into energy-rich ATP). Although the net equation

$$C_6H_{12}O_6 = 2\ C_3H_6O_3$$
$$\text{Glucose} = 2\ \text{Lactic acid}$$

corresponds to an isomerization reaction, a dehydrogenation step is nonetheless inserted; the coenzyme is regenerated, however, in a separate hydrogenation reaction.

Whereas in the direct oxidation of glucose the dehydrogenation reaction occurs right in the beginning of the reaction chain, in the Embden-Meyerhof pathway this reaction takes place at a much later step. Four phases may be distinguished:

1. The conversion of hexose to 2 moles of triose phosphate; this occurs at the oxidation level of the carbohydrate and requires ATP for phosphorylation.

2. The dehydrogenation of triose phosphate with NAD to form phosphoglycerate with a concomitant yield of energy, which is conserved partially as chemical energy in the form of ATP (1 mole per mole of triose).

3. The conversion of phosphoglycerate to pyruvate, which raises the phosphate to a high energy level and allows the regeneration of another ATP (from ADP).

4. The metabolism of pyruvate, which may proceed either by aerobic or anaerobic breakdown. Aerobic breakdown leads into the citrate cycle and the respiratory chain; in anaerobic breakdown the coenzyme NAD is regenerated in a hydrogenation step which produces lactate in the muscle and ethyl alcohol in yeast.

[4] At that time the German word "*Ferment*" was often used synonymously with yeast cell (= "formed ferments"), and to avoid these connotations Kühne introduced the term "*enzyme*" (cf. Chapter V,I).

Phosphorylation and Conversion to Triose Phosphate. Glucose circulating in blood must first be absorbed by cells. In most cases, there is a selective transport mechanism that is as yet little understood. Inside the cell, glucose is phosphorylated in the 6-position; ATP donates the phosphate through the catalytic action of one of two possible enzymes, the *hexokinase*,[5] which phosphorylates a number of hexoses at C-6, and *glucokinase*, which has a very pronounced specificity for glucose. Glucokinase is the more important one in the liver; it is also under the control of insulin (cf. Chapter XX,8). The product of either enzymatic reaction is *glucose 6-phosphate*, very generally the metabolically active form of glucose. The intracellular form of storage is glycogen, which through *phosphorolysis* can be transformed to glucose-1-Ⓟ and then by isomerization to glucose-6-Ⓟ without the consumption of ATP (cf. Chapter XVII,6).

Glucose-6-Ⓟ is then isomerized by glucose-phosphate isomerase to *fructose*-6-Ⓟ (making up 30% of the equilibrium mixture). A second kinase, *phosphofructokinase*, phosphorylates fructose-6-Ⓟ at C-1 with the consumption of ATP to produce fructose 1,6-bisphosphate.[6] This phosphate transfer is virtually irreversible and consequently is a "pacemaker reaction." Phosphofructokinase lately has enjoyed special attention because its activity is controlled allosterically by the ADP/ATP system (cf. Chapter XIX,1 and 2).

Fructose 1,6-bisphosphate then is cleaved by *aldolase* in an equilibrium reaction to two trioses, namely *dihydroxyacetone phosphate* (C-1 to C-3) and *glyceraldehyde phosphate* (C-4 to C-6). The equilibrium mixture is composed of 89% hexose and 11% triose (under the conditions of Meyerhof's measurements); condensation, therefore, is the preferred (= exergonic) reaction. The reaction is analogous to the aldol condensation, described in organic chemistry (Section 5; see also Chapter I,2). Catalysis of the reverse reaction is explained by the fact that enzymes always catalyze "up to the equilibrium." For the detailed reaction mechanism of aldolase see Chapter V,9 and Fig. V-7.

Aldolase is found in relatively high concentration in muscle; its molecular weight is 150,000.

The ketotriose and the aldotriose are in equilibrium through the enol form common to both of them; 96% is in the keto form, dihydroxyacetone phosphate. The attainment of equilibrium is accelerated by the enzyme *triosephosphate isomerase* which possesses an astoundingly high turnover number (several hundred thousand molecules per minute). Hence, the small amounts of glyceraldehyde phosphate present are replenished as soon as they are used in the subsequent reaction.

---

[5] "*Kinase*" is the general term for any enzyme that transfers a phosphate group from ATP to some other substance.

[6] Fructose 1,6-bisphosphate formerly was commonly known as fructose 1,6-diphosphate; for the new nomenclature see comment in Section 4.

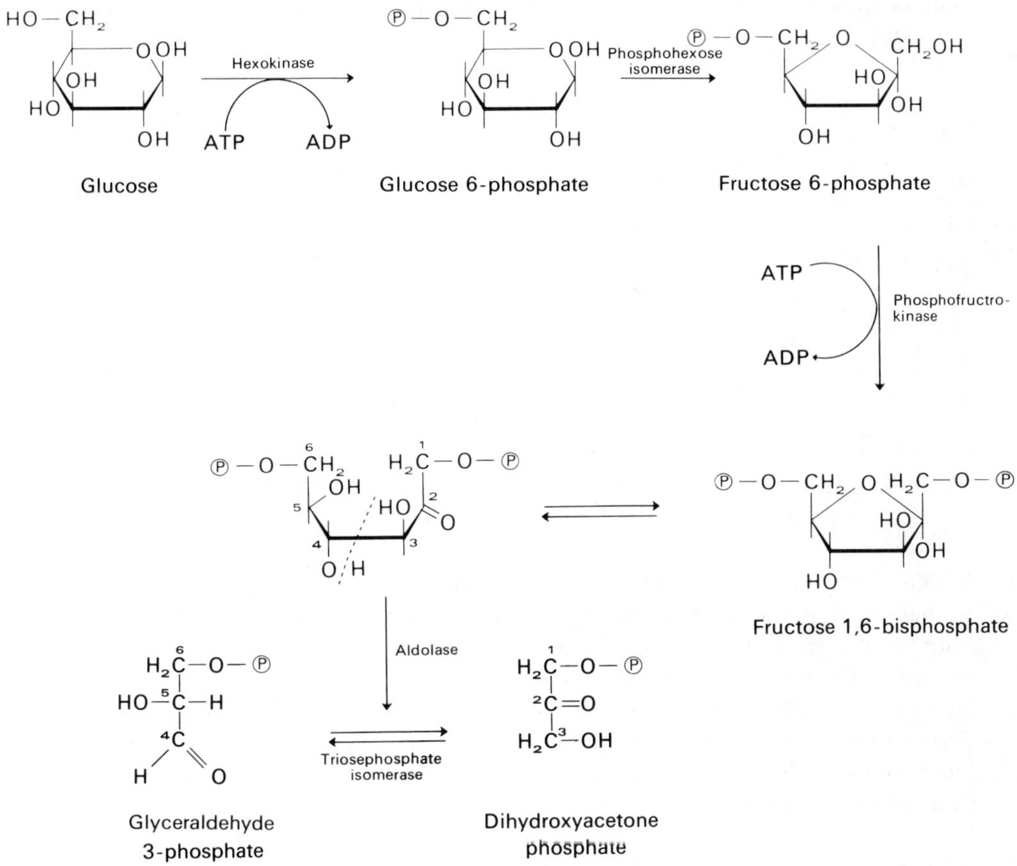

Dehydrogenation of Glyceraldehyde Phosphate. From energetic considerations this is the most important step in the entire Embden-Meyerhof breakdown. The aldehyde group of the triose is dehydrogenated to yield the carboxylic acid. This reaction could proceed in the simplest possible way, since it is strongly exergonic:

$$R{-}CHO + H_2O + NAD^{\oplus} \rightleftharpoons R{-}COOH + NADH + H^{\oplus} \qquad \Delta G^{\circ} = -16\ \text{kcal}$$

However, since alcoholic fermentation and glycolysis are anaerobic, *energy-yielding*, metabolic processes, little would be gained if the reaction ran in this way and all the energy were dissipated as heat. Actually, 7 kcal are stored as chemical energy—in the form of 1 mole of ATP. This is accomplished by conducting the dehydrogenation by way of the following detour mediated by *glyceraldehyde-phosphate dehydrogenase.*

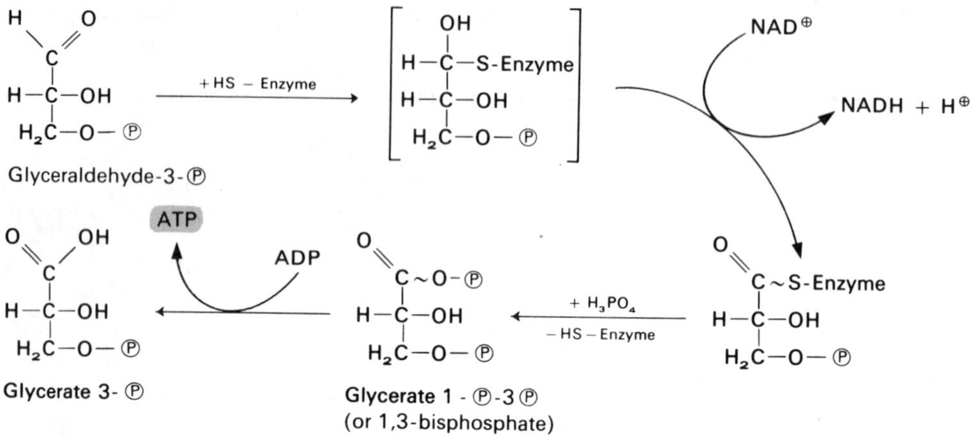

The aldehyde group is added to the SH group of the enzyme; the dehydrogenation is then carried out by transferring hydrogen to NAD. An *acyl-S-enzyme complex* with an energy-rich thioester bond is thus formed. Hydrolysis would release the energy of the thioester as heat; phosphoryolysis, however, conserves it as chemical energy in the acyl phosphate anhydride bond (*3-phosphoglycerate 1-phosphate*). The energy-rich phosphate group is then transferred by the enzyme *phosphoglycerate kinase* onto ADP. The result is *3-phosphoglycerate* and ATP. This type of ATP formation is termed *substrate-linked phosphorylation.*

Formation of Pyruvate. 3-Phosphoglycerate now is rearranged by phosphoglycerate mutase to *2-phosphoglycerate*; the participating coenzyme is 2,3-bisphosphoglycerate. One must visualize that the 3-phosphate is phosphorylated by the enzyme–bisphosphate complex and that the dephosphorylated coenzyme (2-phosphoglycerate) dissociates off the enzyme while the product of phosphorylation becomes the new coenzyme. In the next step water is eliminated by action of enolase; the resulting compound is the phosphorylated enol form of pyruvate called *phosphoenolpyruvate.*

Phosphate is present again in an energy-rich form (namely as the enol ester). It can be transferred by pyruvate kinase to ADP; this transfer affords *pyruvate*, the most important metabolite of both anaerobic and aerobic carbohydrate metabolism, and the second ATP.

The sequence of reactions from phosphoglycerate to pyruvate is equivalent in essence to the cleavage of the phosphate group and its transfer to the adenylic acid system; it is the phosphate which originally—during phosphorylation of the hexose—was supplied by ATP. Thus the ATP invested originally is reclaimed here.

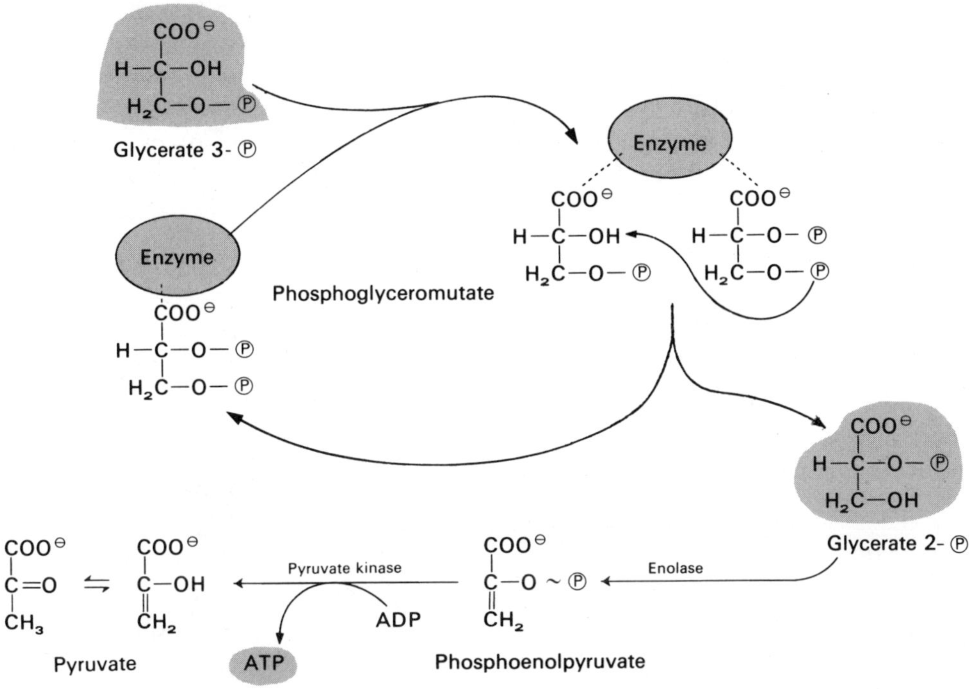

The Pathway of NAD and Pyruvate. In order to formulate a continuous process, the reduced nicotinamide nucleotide NADH must be reoxidized and made available again. The reoxidation could be achieved, e.g., through the respiratory chain, if oxygen were available at the site of carbohydrate breakdown. In the absence of oxygen, however, another solution must be found. In the metabolism of vertebrates, in anaerobic glycolysis of muscle, pyruvate is reduced to lactate; in yeast the reduction is preceded by a decarboxylation to form acetaldehyde, and the product of the reduction is the desired ethyl alcohol.

In glycolysis the chain of reactions is terminated by *lactate dehydrogenase* which reduces pyruvate (the reverse of what the name indicates) and which makes the nicotinamide nucleotide available again in its oxidized form to catalyze the dehydrogenation of more triose.

In summary, one hexose is converted to the bisphosphate with the uptake of 2 ATP; the bisphosphate is cleaved; it traverses a series of equilibrium reactions and releases hydrogen to the nucleotide coenzyme. The breakdown is coupled at two places with the adenylic acid system in order to trap and store chemical energy. The formulas of this scheme are in Fig. XV-3; regulation is discussed in Chapter XIX,2.

The net yield is only 2 ATP per mole of glucose; the conversion to lactate is unable to provide more energy than that. However, the relatively energy-rich substrate *lactate* can be broken down further in the presence of *oxygen*. Lactate is transported from skeletal muscle to the liver, a highly vascular organ well supplied with oxygen.

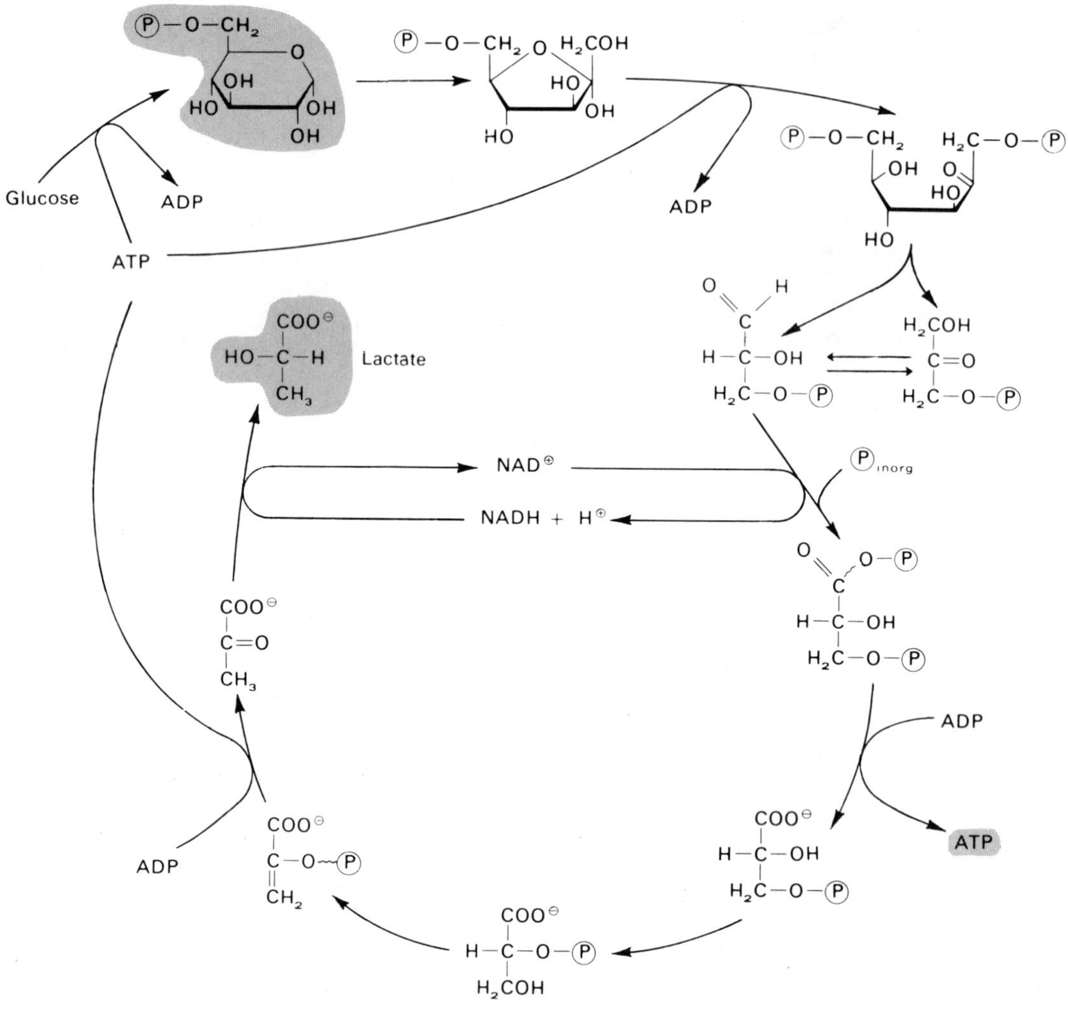

Fig. XV-3. Diagram of anaerobic glycolysis.

In fermentation, as we have pointed out, pyruvate is decarboxylated to form *acetaldehyde*. The enzyme pyruvate decarboxylase requires $Mg^{2\oplus}$ and thiamine pyrophosphate as cofactors. Acetaldehyde is probably bound in its activated form at first, because it can undergo several reactions. Under certain conditions, two molecules can combine to form acetoin. Pyruvate decarboxylase is found only in plants.

In microorganisms acetaldehyde is then reduced by alcohol dehydrogenase and NADH to form *ethanol*. At the same time the coenzyme is returned to its oxidized form. The coenzyme, therefore, is an auxiliary substrate for hydrogen, passing through a cyclic process, just as phosphate when it is removed from ATP and later transferred back to ATP.

8. The Metabolism of Fructose

Cane sugar (sucrose) is composed of glucose and fructose, as will be explained in Chapter XVII,2, and is decomposed to these constituents in the intestines. Thus, fructose is one of the regular components of our diet. In some circumstances, fructose can make up a considerable part of all the carbohydrates ingested.

Fructose phosphates are intermediates in the breakdown of glucose according to the Embden-Meyerhof pathway, but free fructose is broken down in a different fashion. At first it is phosphorylated by a fructokinase and ATP to produce *fructose 1-phosphate*.[7] The phosphate is split by a specific enzyme, 1-phosphofructoaldolase, to *dihydroxyacetone phosphate* and *glyceraldehyde*. This last step corresponds to the aldolase reaction of glycolysis except that the resulting glyceraldehyde is unphosphorylated.

Free glyceraldehyde can now follow a number of different pathways. Most of it is phosphorylated to glyceraldehyde 3-phosphate and thus can enter the glycolytic pathway. Some of it is reduced to glycerol through the action of NADH-dependent alcohol dehydrogenase. Last, some glyceraldehyde can be oxidized directly (un-

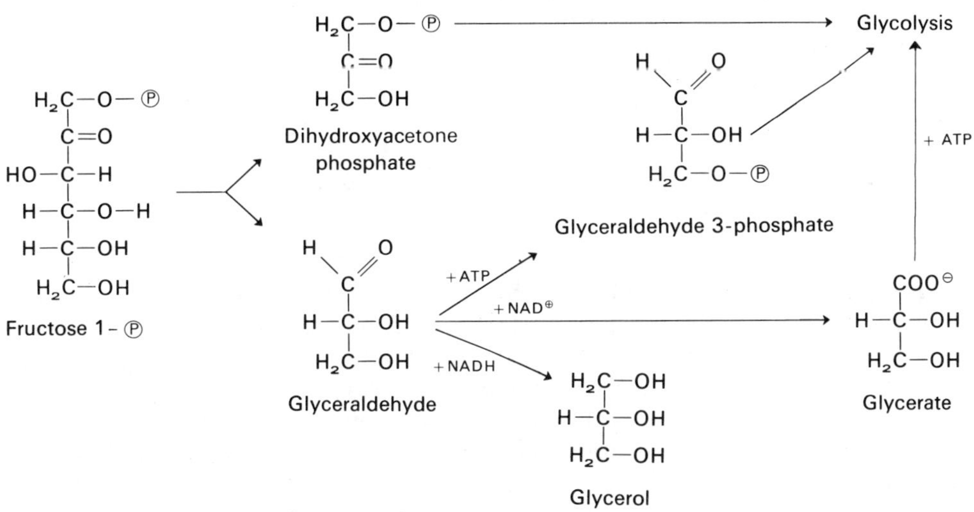

[7] Phosphorylation in the 1-position is favored probably because free fructose occurs predominantly in its pyranose form (the oxygen bridges between C-2 and C-6). The 6-hydroxyl group is thus blocked and cannot be phosphorylated.

phosphorylated) to glycerate, which then is phosphorylated at position 2. The last step is a connection to the Embden-Meyerhof degradative pathway, but without the gain of one ATP.

Dihydroxyacetone can enter the Embden-Meyerhof pathway immediately and be used for the resynthesis of glucose (gluconeogenesis, see Section 10).

9. Aerobic Carbohydrate Breakdown

In general, carbohydrates are broken down aerobically, i.e., with a supply of oxygen. Anaerobic glycolysis seems to be more of a bypass for cells with a poor supply of oxygen. The Embden-Meyerhof pathway is significant primarily because it operates not only under anaerobic conditions, but, up to the step of pyruvate, under aerobic conditions as well—at least to the extent that glucose is not oxidized directly to pentose (Section 6; see also Chapter XIX,1). Reduced nicotinamide-adenine dinucleotide is reoxidized through the respiratory chain, and pyruvate can be metabolized further in a variety of ways.

Oxidative carboxylation is very significant both from a quantitative standpoint and also because it links the different metabolic pathways. The mechanism of this reaction, requiring the participation of lipoate, coenzyme A, and NAD, has already been discussed in detail (Chapters VI,4, VIII,10, and XI,1). Formed on a multi-enzyme complex according to the equation shown below, the end product is "activated acetate" or acetyl-CoA. The greater part of the acetyl-CoA is consumed in the citrate cycle (see Chapter XI), but any excess can be converted to fat (cf. Chapter XII,6), provided that there is sufficient NADPH.

$$H_3C-\overset{\displaystyle O}{\overset{\|}{C}}-COOH + HS\overline{CoA} + NAD^{\oplus} \rightarrow H_3C-\overset{\displaystyle O}{\overset{\|}{C}}\sim S\overline{CoA} + CO_2 + NADH + H^{\oplus}$$

Energy Balance of the Aerobic Carbohydrate Breakdown. Up to the formation of pyruvate 1 mole of ATP and 1 of NADH arise from each triose. Another mole of NAD is reduced during the oxidative decarboxylation of pyruvate to acetyl-CoA. Up to the formation of acetyl-CoA (employing the respiratory chain) $1 + 2 \times 3 = 7$ ATP are stored. Complete oxidation of active acetate in the citrate cycle yields another 12 moles of ATP per triose, i.e., a total of 19 ATP per mole of triose or 38 per mole of glucose.

Carboxylation. Another reaction of pyruvate closely connected with the citrate cycle is the carboxylation to form oxaloacetate. We have to consider two different routes. In the main pathway, activated carbon dioxide is added with the aid of a biotin-containing enzyme, *pyruvate carboxylase.*

$$CH_3-CO-COO^{\ominus} + {}^{\ominus}OOC\sim\text{Biotin enzyme} \rightleftharpoons {}^{\ominus}OOC-CH_2-CO-COO^{\ominus} + \text{Biotin enzyme}$$

The formation of ${}^{\ominus}OOC\sim$biotin enzyme consumes one ATP; the formation of oxaloacetate by this route therefore takes place at the expense of a good deal of energy.

The second CO_2-fixation reaction is a reductive carboxylation (Ochoa). The reaction is catalyzed by the "malic enzyme," or better *decarboxylating malate dehydrogenase*, as follows:

$$CH_3-CO-COOH + CO_2 + NADPH + H^{\oplus} \rightleftharpoons HOOC-CH_2-CHOH-COOH + NADP^{\oplus}$$

Although the enzyme occurs in liver, it nevertheless should be of subordinate significance under normal metabolic conditions since the Michaelis constant for pyruvate and CO_2 is about $10^{-3}M$. In contrast, the biotin-containing pyruvate carboxylase has a K_m for CO_2 of $4.8 \times 10^{-5}M$ and is present at a relatively higher concentration.

The formation of oxaloacetate by one of these two pathways is significant, because it provides the "initial spark" for the citrate cycle, since oxaloacetate is needed as partner for the condensation reaction with acetyl-CoA. Furthermore, oxaloacetate mediates the resynthesis of glucose (see below). It should be pointed out, finally, that pyruvate can be transformed to alanine (Chapter VIII,10) by transamination so that pyruvate also represents a link to the metabolism of protein.

10. Resynthesis of Glucose: Gluconeogenesis

During anaerobic glycolysis a lot of lactate is produced, and it has been known for some time that some of lactate is used for the resynthesis of glucose, though not in the muscle, the place of its origin, but rather in the liver. In theory, the synthesis might be a simple reversal of the Embden-Meyerhof pathway, but the equilibria of some of the reactions are too unfavorable. This is true particularly of the reaction of pyruvate kinase; the formation of pyruvate and ATP from phosphoenolpyruvate and ADP with $\Delta G^0 = -6.2$ kcal is practically irreversible. This may explain the biosynthetic detour over oxaloacetate, which arises from the carboxylation of pyruvate as discussed above (chiefly with the aid of biotin-containing pyruvate carboxylase). Oxaloacetate is phosphorylated much more easily since the enol form is favored. The reaction

$$\text{Oxalocacetate} + \text{GTP} = \text{phosphoenolpyruvate} + CO_2 + \text{GDP}$$

is catalyzed by *phosphoenolpyruvate carboxylase*. The phosphate donor in this case is GTP, and not ATP. The key reactions are summarized below:

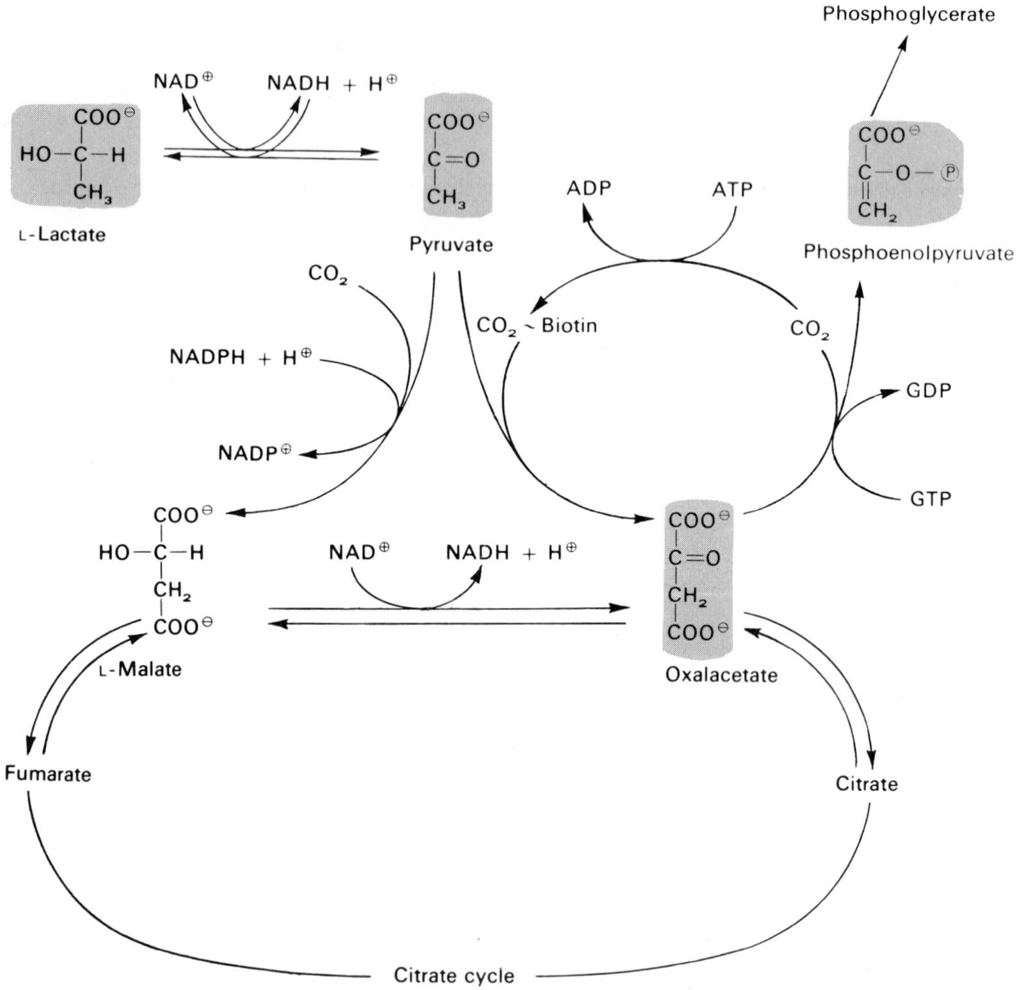

From phosphoenolpyruvate, gluconeogenesis proceeds as a reversal of glycolysis. Enolase catalyzes the addition of water to 2-phosphoglycerate, and the migration of the phosphate group yields 3-phosphoglycerate. The latter becomes phosphorylated with ATP to 3-phosphoglycerate 1-phosphate, which is reduced to triose phosphate by the action of glyceraldehyde phosphate dehydrogenase and NADH as hydrogen donor. Triose isomerase and aldolase then effect conversion to fructose 1,6-bisphosphate, which can be converted to glucose 6-phosphate, and finally to glucose or glycogen (see Chapter XVII,6), the storage form of carbohydrate. The phosphate groups are removed hydrolytically; the transfer back to ATP is precluded for energetic reasons.

Evidently three moles of energy-rich phosphate are required for each triosphosphate: one for the formation of carboxybiotin, a second for the decarboxylating phosphorylation of oxaloacetate (to produce phosphoenolpyruvate), and a third for the reduction of 3-phosphoglycerate. The reversal of the breakdown must, of course, be endergonic. It has long been known that 20–30% of lactate is combusted completely to enable the resynthesis of glucose.

Gluconeogenesis from Amino Acids. The pathway described above has general significance beyond the utilization of lactate. We have mentioned previously that many amino acids can be converted to glucose, provided that they give rise to C_4-dicarboxylic acids. These are participants in the citrate cycle and thus can easily produce oxaloacetate and then phosphoenolpyruvate by Utter's reaction. From that point on, the synthesis of glucose is easy, being simply the reverse of glycolysis. Only the phosphorylation reactions of hexose by ATP are not reversible, but phosphate can be split off the hexose bisphosphate by specific phosphatases. Furthermore, the synthesis of carbohydrates does not end with free glucose but usually with the polymer glycogen. Polymerization, however, proceeds through phosphorylated intermediates.

It may appear peculiar that buildup and breakdown proceed largely via the same route; merely a few key reactions differ. These key reactions are the targets of the cell's regulatory mechanisms which must determine the direction that the processes are to take. These mechanisms will be discussed in detail in Chapter XIX,2. Gluconeogenesis from amino acids (i.e., ultimately from proteins), for example, is strongly stimulated by the hormone cortisol (cf. Chapter XX,2). Such regulation is of decisive importance for the organism's overall metabolism.

The preceding paragraphs undoubtedly have revealed the complicated and diverse nature of carbohydrate metabolism, both on the level of interconversions among the carbohydrates and on that of degradative reactions for the production of energy. Some of the energy is derived anaerobically from substrate-linked phosphorylation; the major part, however, is liberated in the respiratory chain. The situation is further complicated by the obvious fact that carbohydrate metabolism is not an isolated system of reactions, but is closely tied to other pathways and reaction cycles through common intermediates. A separate chapter (Chapter XIX) is devoted to such interrelationships.

BIBLIOGRAPHY

B. Axelrod, Glycolysis, in "Metabolic Pathways," D. M. Greenberg (editor), 3rd edition, Vol. I. Academic Press, New York, 1967.

B. Axelrod, Other Pathways of Carbohydrate Metabolism. *Ibid.*

R. Bentley, Configurational and conformational aspects of carbohydrate biochemistry. *Annu. Rev. Biochem.* **41,** 953–996 (1972).

R. Caputto, H. S. Barra, and F. A. Cumar, Carbohydrate metabolism. *Annu. Rev. Biochem.* **36,** 211–246 (1967): pages 223 to 237 deal with glycolysis and pentose cycle.

CHAPTER XVI

Photosynthesis

1. Importance of Photosynthesis

Photosynthesis, the assimilation of carbon dioxide in green plants, is the most important process on this planet, from both a qualitative and a quantitative standpoint. This is evident if we recall that all forms of life that are unable to photosynthesize depend either directly or indirectly[1] on the assimilation by plants. In fact, present forms of life cannot exist without photosynthesis. The energy for its maintenance is light, which is trapped by chlorophyll, a leaf pigment, and is utilized through a complicated mechanism. The organic substances which arise from this one process serve all the other forms of life as starting materials for the diverse metabolic interconversions, which we have already studied. These substances are ultimately oxidized to CO_2 which can then be incorporated again into new organic compounds by way of photosynthesis.

The equilibrium between CO_2 in the air and organically bound carbon (having remained static for hundreds of thousands or even millions of years) has been disturbed noticeably since the second half of the last century by man's progressing industrialization and civilization. Fixed forms of carbon, fossil fuel (coal, petroleum) and inorganic carbonates, have been mobilized extensively and thus been included in the carbon cycle.[2]

The photosynthetic process in summary appears to be a reversal of the combustion of glucose (light quanta, i.e., radiant energy, are designated by $h\nu$):

$$6\,CO_2 + 6\,H_2O \xrightarrow{h\nu} C_6H_{12}O_6 + 6\,O_2$$

---

[1] Strictly speaking, we have to exclude those few types of bacteria that can fill all of their energy requirements from chemical work alone; they gain all their energy by oxidizing inorganic substances, like sulfur or H_2S.

[2] The extent of this process can be calculated from the content of the radioactive isotope ^{14}C (half-life, 5.5×10^3 years). The same concept is used in "radiocarbon dating" of archaeological or geological objects.

The decisive reaction in the process is not "assimilation of carbon dioxide," as assumed formerly, but rather the utilization of light for driving an endergonic reaction (the cleavage of water) represented by the following equation:

$$2\,H_2O + 2\,NADP^{\oplus} \xrightarrow[\text{chloroplasts}]{h\nu} O_2 + 2\,(NADPH + H^{\oplus})$$

Water is cleaved photolytically, thereby liberating oxygen and transferring hydrogen to the coenzyme $NADP^{\oplus}$. The transfer of hydrogen occurs against an energy gradient, as can be seen from the redox potentials in Table X-1 (cf. Chapter X,3). The requisite energy is derived from absorbed light quanta.

The primary event of all photochemical processes is the absorption of light quanta by special pigment molecules. In the case of photosynthesis this primary event—the true light reaction—is coupled with redox processes. These result on the one hand in the liberation of O_2, on the other both in the formation of NADPH and in the synthesis of ATP from inorganic phosphate and ADP (so-called photophosphorylation).

The reduction of CO_2 to carbohydrate proceeds as a purely biochemical-enzymatic reaction linked to the photosynthetic apparatus only by its requirement for ATP and NADPH.

2. The Photosynthetic Apparatus and Light Reactions

Chloroplasts. The photosynthetic apparatus is located in the chloroplasts, which in higher plants are disk-shaped cell organelles with a diameter of 4–8 μm. The electron microscope reveals the ultrastructure of many membranes forming closed loops called *thylakoids*. They are embedded in the ground substance *stroma*.

Higher plants often contain two morphologically and functionally distinct types of chloroplasts: In one type individual thylakoids pervade the entire interior, whereas in the other type thylakoids are in contact over variable areas and form regular stacks (see Fig. XVI-1). Following the disruption of chloroplasts, these stacks can be isolated by differential centrifugation (see Chapter XVIII). They are also called *grana*. Enzymes of carbohydrate metabolism abound in the stroma; photosynthetic pigments and redox systems are localized in the thylakoid membranes.

Photosynthesis Pigments. In addition to *chlorophyll a* and *b* (for formulas see Chapter IX,6) so-called "accessory pigments," primarily the *carotenoids* in higher plants and in the case of red and blue algae furthermore the noncyclic tetrapyrrole pigments *phycoerythrines* and *phycocyanines*, also belong to this group. The pigments are localized in the thylakoid membranes discussed above and probably are inserted there in an ordered manner in the protein-lipid layers. In the electron microscope, regularly structured units with a diameter of 180 Å have been seen. These *quantasomes* conceivably contain the full complement of pigments and redox catalysts for system I or II.

Botanists have also become quite interested in the phytochrome system, which, however, is not involved in photosynthesis. It participates in phototropisms and similar plant physiological mechanisms. For its structure see the end of Chapter IX,5.

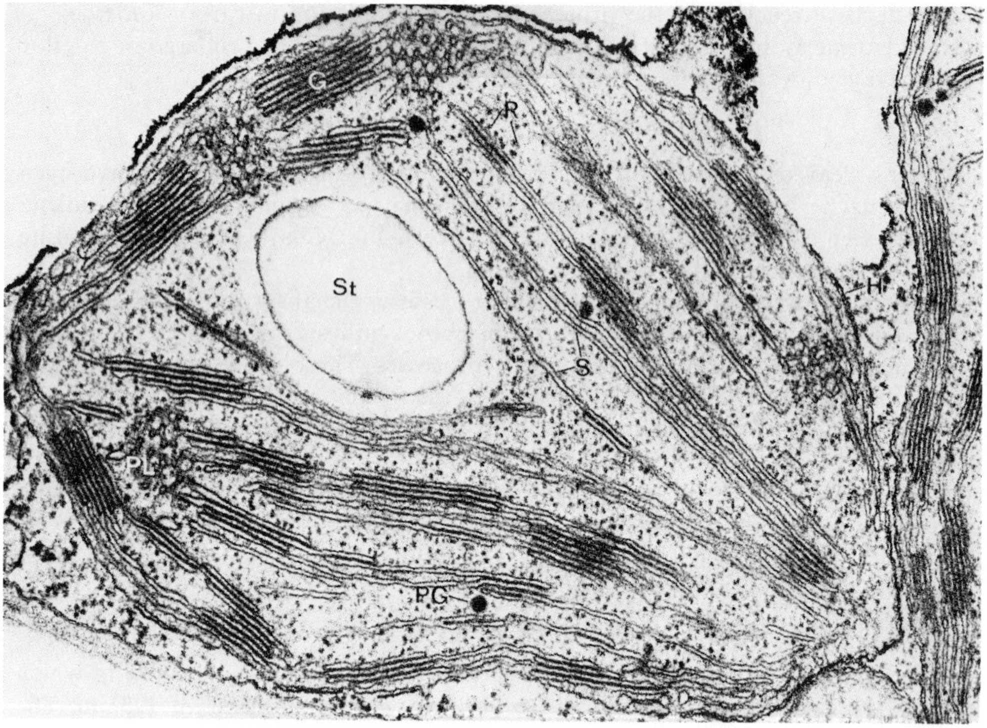

Fig. XVI-1. Electron micrograph of a chloroplast isolated from corn. Long drawn-out stroma thylakoids (S) can be seen in the plastid stroma. They are stacked in the upper left to form grana (G). Below that is a starch granule (St). PG, Plastoglobuli; PL, remnants of prolamellar bodies; R, plastid ribosomes; H, plastid unit membrane. Enlargement ca. 40,000 ×. (Original photo by Prof. Wehrmeyer.)

Higher plants contain two functionally closely related systems of pigments that trap light energy and convert it to chemical energy (= difference of redox potential). They are called "*system I*" and "*system II*," and both contain as chief component chlorophyll *a* and *b*.

System I contains for each 300 to 1000 "ordinary" molecules of chlorophyll *a* one of chlorophyll a_I bound in a special way (among others, linked to cytochrome *f*) and capable of carrying out the photochemical reactions. According to its absorption maximum the pigment is called P_{700}. The other, ordinary chlorophyll *a* molecules can conduct light energy to the P_{700} molecule in the form of "excited states." The process is termed energy conduction. It ensures the utilization of light even when the light itensities are very low.

System II is structured similarly; it contains pigment P_{680} as the photochemically active species and per energy conduction provides excitation energy for a biochemically active molecule of chlorophyll a_{II}.

The two chlorophyll systems function according to the same principle: The absorption of a light quantum elevates an electron to a higher energy level and transfers the electron to an acceptor. Energetically this corresponds to an "uphill transport" of electrons; chemically it is photoreduction of the acceptor. The two systems differ merely in the nature of the acceptor molecule and in their range of redox potential in which they operate (see Fig. XVI-2).

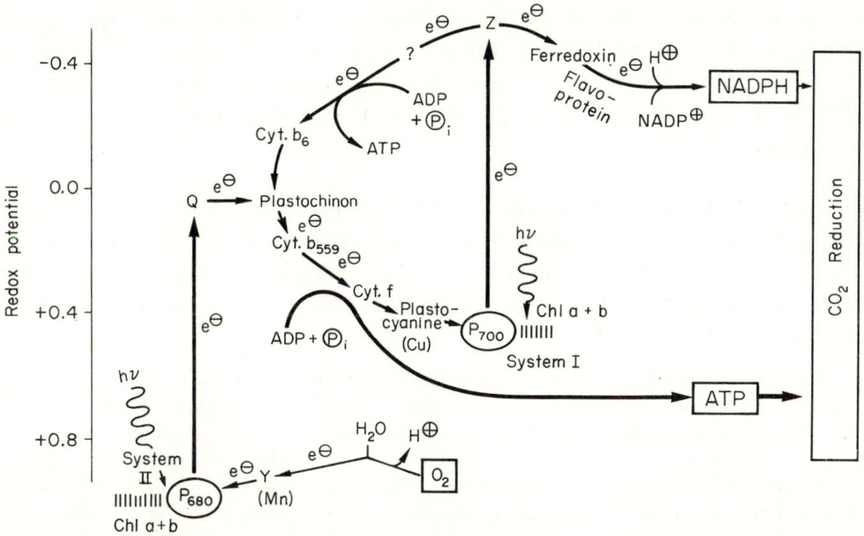

Fig. XVI-2. Light reactions and redox systems of photosynthesis. The diagram is intended to illustrate only the sequence of the redox enzymes. The stoichiometry has not been considered. Further details are in the text.

Photosystem II and the Photolysis of Water. In the ground state, system II has a redox potential of over +1.0 V; in the excited state, about 0 V. The acceptor for the photochemically activated electron of system II is *plastoquinone* (formula in Chapter XIV,9), which changes over to the semiquinone. The electrons are resupplied by an as yet unknown redox system Y with the participation of manganese. Oxidized $Y^{\oplus}$ abstracts electrons from water and oxidizes it to O_2 according to the formal equation:

$$2H_2O \rightarrow O_2 + 4\,H^{\oplus} + 4\,e^{\ominus}$$

This equation constitutes the reverse of oxidation with O_2 (cf. Chapter X,2 and 6).

Photosystem I and the Formation of NADPH. In the ground state, system I has a redox potential of +0.46 V. A quantum of absorbed light energy elevates an electron to a potential more than 1.0 V more negative. From there the electron can transfer to ferredoxin via an as yet unknown redox system Z ("ferredoxin-reducing substance," possibly a pteridine or quinone system). *Ferredoxin* is a protein that contains iron (bound to cysteine and labile sulfur) and has a redox potential of $E_0' = -0.43$ V. Similar nonheme iron proteins have also been discovered in mitochondria and microsomes (cf. Chapter X,4 and 6). Because of its negative redox potential, ferredoxin is able to reduce $NADP^{\oplus}$; the reaction is catalyzed by ferredoxin-NADP oxido-reductase, a flavoprotein. The resulting NADPH constitutes a stable transport metabolite of hydrogen. The entire reaction sequence is seen on the right side of Fig. XVI-2.

The Electron Transport System. The two chlorophyll systems are linked by a chain of redox catalysts, as shown in Fig. XVI-2. The first link in the chain is *plastoquinone* (formula in Chapter XIV,9), the electron acceptor of system II. Other links are *cytochrome b-559* and *cytochrome f*, a cytochrome of type *c* (also called cytochrome c_6) with an $E_0' = +0.37$ V, and *plastocyanine*, a Cu protein of low molecular weight. (The order of cytochrome *f* and plastocyanine is not certain and their position could be reversed.) At the end of the chain is P_{700} which also provides the linkage to photosystem I.

Noncyclic Photophosphorylation. The transport of electrons through the chlorophyll systems and the interposed redox chain is associated with the synthesis of ATP: For each pair of electrons 1 ATP is formed out of ADP and inorganic phosphate. As with respiratory chain phosphorylation, a difference of redox potentials is harnessed for phosphorylation.

The mechanism of the linkage to this phosphorylation is not yet clear. Again in analogy with the respiratory chain, two theories have been advanced, a chemical coupling and a chemiosmotic theory. Certain important experiments, however, seem to favor the chemiosmotic theory, as illustrated in Fig. XVI-3. The primary process is the setting up of a charge difference at the thylakoid membrane through the action of light. The negative charges (electrons) then react via ferredoxin with $H^{\oplus}$ and $NADP^{\oplus}$, whereby the outside grows deficient in $H^{\oplus}$. The inside accumulates a corresponding excess of $H^{\oplus}$ since 2 $HO^{\ominus}$ ions donate their electrons to system II and undergo reaction to yield $H_2O + \frac{1}{2}O_2$. We now have a pH differential which has actually been measured: The interior of the thylakoid at pH 5; the outside at pH 8. It is postulated that this differential is equalized at certain "phosphorylation sites" with the formation of ATP, in principle by a reversal of active transport (see Chapter XXI,3).

This mechanism is well substantiated by numerous measurements by Witt and co-workers and agrees extensively with Mitchell's theory of respiratory chain phosphorylation (see Chapter X,5). The essential difference between the two is that in the thylakoid the charge difference is built up fundamentally by a light reaction whereas in the mitochondrial cristae membrane the pH gradient results from redox reactions.

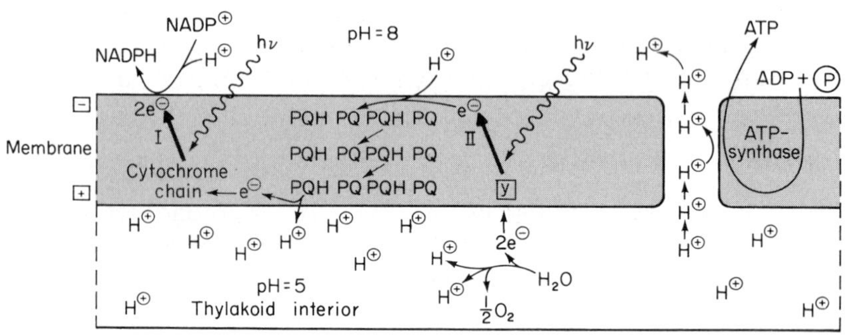

Fig. XVI-3. The chemiosmotic photophosphorylation (according to H. T. Witt, modified). The diagram represents the steady state during illumination. By the photochemical primary process in systems I and II (heavy diagonal arrows) electrons are exported to the outside. They combine with $H^{\oplus}$ and reduce plastoquinone (PQ) or $NADP^{\oplus}$. As a result the outside becomes deficient in $H^{\oplus}$. At the inside of the membranes electrons ($e^{\ominus}$) are abstracted from water (system II) or from PQH via the cytochrome chain and $H^{\oplus}$ ions are released to the interior. Plastoquinone, which oscillates between the quinone, semiquinone, and hydroquinone steps, becomes a conveyor belt for $H^{\oplus}$. The pH gradient arising from this and from the photolysis of water is equalized again at a pore at the right side of the diagram and becomes the motor for phosphorylation of ADP to ATP in the manner of a reversal of an ATP-driven "ion pump."

The analogy between respiratory-chain phosphorylation and photosynthesis is not entirely accidental. Both the mitochondria and the chloroplasts contain highly ordered enzyme systems which permit an exergonic process (the electron transport through a cytochrome chain) to be coupled with the strongly endergonic formation of ATP from ADP and inorganic phosphate. The two reaction chains possess many characteristics in common: presumably the respiratory chain of the mitochondria evolved gradually from the photophosphorylation of the chloroplasts. The only significant additions during this evolution have been cytochrome oxidase and the NADH-dehydrogenase system.

Cyclic Photophosphorylation. Under certain conditions a "short circuit" in electron transport has been observed. Electrons pumped by the photoreaction in system I run through a circular process and form ATP in a similar fashion as photochemical reduction of NADP.

The short circuit between the primary acceptor Z and plastoquinone apparently is catalyzed by a flavoprotein and cytochrome b_6. It is not certain yet whether ATP is formed at this step as shown in Fig. XVI-2. From plastoquinone the electrons run through the electron transport chain to system I producing ATP. The cycle is closed when electrons are driven again by a light reaction to acceptor Z. In summary, no redox reactions take place, only the formation of ATP, i.e., the conversion of light energy to chemical energy. It is as yet an open question to what extent this cyclic photophosphorylation contributes to the energy balance of photosynthesis. Some investigators believe that the cell employs only noncyclic photophosphorylation.

The Hill Reaction. The addition of artificial electron donors and acceptors modifies the photochemical redox chain. Without a detailed description we shall just mention the reaction named after its discoverer Hill. When unphysiological electron acceptors (Hill oxidants), such as ferricyanide or benzoquinone are added, one can see the evolution of oxygen with concomitant reduction of the added oxidant.

Quantum Requirement of Photosynthesis. The quantum requirement is the number of quanta (n) necessary to evolve 1 molecule of O_2 (1 mole O_2 hence requires $n \cdot 6 \cdot 10^{23}$ quanta). According to the scheme in Fig. XVI-2, it takes two quanta to raise 1 electron from the energy level of water to that of NADPH. Since the formation of 1 molecule of O_2 is a four-electron process [cf. equation (4) in Chapter X,2], a total of eight quanta are required. This value has to be increased, of course, if cyclic photophosphorylation proceeds to any appreciable extent because it consumes quanta for the production of ATP without, however, yielding any O_2. Numerous experimenters[3] have reported empirical values of eight to ten quanta, values in excellent agreement with the mechanism described above. The yield of energy, or efficiency, thus would amount to about 30%.

3. Binding of CO_2 and Reduction to Carbohydrate

The other part of photosynthesis, the conversion of CO_2 to carbohydrate, is purely a biochemical change, a dark reaction. Calvin and his associates are credited for shedding light on the very confusing array of these dark reactions.

[3] On the basis of manometric measurements under nonstationary conditions, O. Warburg and his coworkers have found a quantum requirement of 3. This would correspond to an energy yield of about 90%. Warburg even maintains that a single quantum can release 1 molecule of O_2; the additional, required energy would be provided by dark reactions (= respiration). We cannot discuss this hypothesis and possible explanations of the discrepancies any further here.

The use of radioactive carbon ($^{14}CO_2$) made it possible to elucidate the reaction sequence. The old hypotheses postulating formaldehyde as an intermediate were refuted quickly. It took longer, however, to establish the true state of affairs. Some of the intermediate products are exceedingly short-lived; their existence is demonstrated only if the reaction is stopped a few seconds after adding labeled CO_2.

The acceptor for CO_2 is *ribulose bisphosphate*, which arises from *ribulose 5-phosphate* and ATP by the action of a kinase. The enzyme *ribulose bisphosphate carboxylase* catalyzes the addition of CO_2 to ribulose bisphosphate and the cleavage of the labile primary product into *2 moles of phosphoglycerate*. This is the first isolable product of photosynthesis.

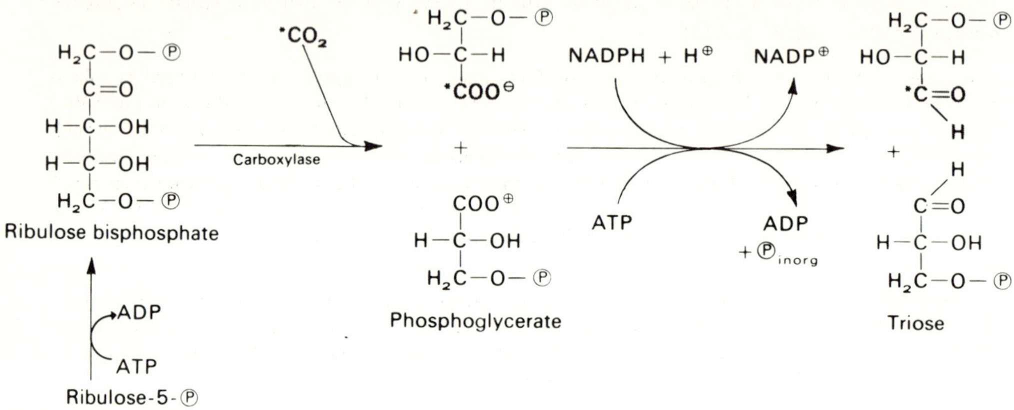

In some plants the fixation of CO_2 proceeds primarily via phosphoenolpyruvate carboxylase. The initial products are oxaloacetate, malate, and aspartate, which arise in the following reaction:

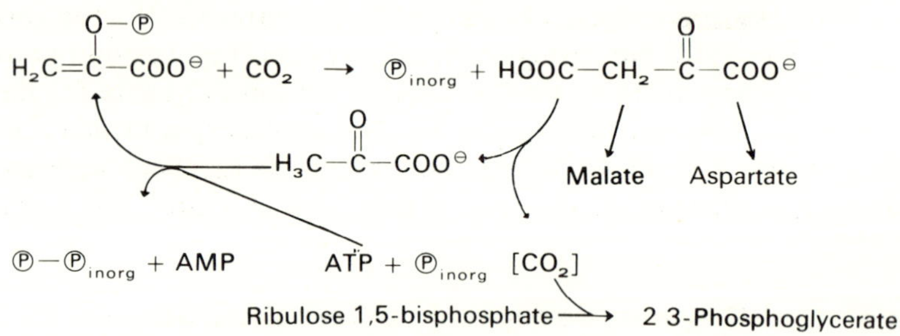

This pathway leading to carbohydrate in the end also involves the carboxylation of ribulose bisphosphate to form 3-phosphoglycerate. The detour, however, permits effective CO_2 fixation even at low CO_2 concentrations.

It is not yet established just how CO_2 initially incorporated into oxaloacetate finally becomes the carboxyl group of 3-phosphoglycerate. Oxaloacetate probably serves as a transport metabolite for CO_2 and by decarboxylation frees CO_2 in the immediate vicinity of ribulose bisphosphate carboxylase. Pyruvate, the decarboxylation product of oxaloacetate, is then converted back to phosphoenolpyruvate by action of the unusual pyruvate-orthophosphate dikinase reaction. This pathway involves two additional moles of energy-rich bonds per mole of CO_2 fixed.

Phosphoglycerate is then reduced to *triose phosphate*. This reaction has been discussed in context of gluconeogenesis (Chapter XV,10). It proceeds as a reversal of the triosephosphate dehydrogenase reaction (Chapter X,7) and requires ATP and, in the case of plants, NADPH.

This energy-consuming reductive step actually fulfills our task. We have reduced carbon from the oxidation level of CO_2 to that of carbohydrate. For the sake of completeness, we must allow this reaction sequence to operate on both molecules of phosphoglycerate. That means we require 2 NADPH and 2 ATP per mole of CO_2. It should be noted that by cleavage of the acceptor molecule into two molecules of a carboxylic acid, the energy required for further reaction is distributed between two molecules and, consequently, is easier to provide.

Formation of Ribulose Bisphosphate. Up to this point the situation has remained relatively simple. Now, however, ribulose bisphosphate has to be regenerated so that it may accept another CO_2. The reactions of the pentose phosphate cycle (cf. diagram XV-6, lower part, read from left to right) with minor modifications serve this purpose.

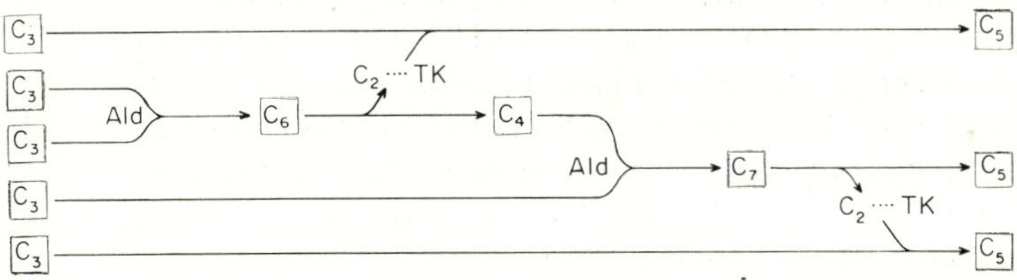

At first two molecules of *triose phosphate* combine to form hexose bisphosphate (fructose-Ⓟ$_2$), which is then hydrolyzed by a specific phosphatase to *fructose*-6-Ⓟ and phosphate. Hexose phosphate and triose phosphate interact in a transketolase reaction to produce *erythrose*-4-Ⓟ and *xylulose*-5-Ⓟ, which then rearranges to *ribulose*-5-Ⓟ (the first pentose molecule). Erythrose phosphate now does not undergo reaction with fructose-6-Ⓟ but rather in a type of aldol condensation combines with triose phosphate to form *sedoheptulose bisphosphate* ($C_4 + C_3 = C_7$), which subsequently is dephosphorylated. The enzyme transketolase then transfers a C_2 fragment from heptulose to another molecule of triose phosphate yielding 2 molecules of pentose (*xylulose*-5-Ⓟ and *ribose*-5-Ⓟ); both must be epimerized to *ribulose*-5-Ⓟ. Having ribulose-5-Ⓟ available, the cycle can commence again; first, phosphorylation with ATP to the bisphosphate, followed by the acceptance of CO_2 by the bisphosphate and production of 2 moles of phosphoglycerate. The following simplified scheme represents an outline and summary of photosynthetic reactions:

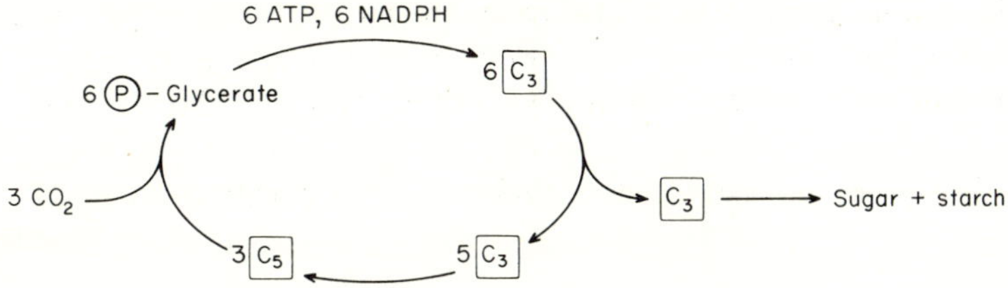

4. Assimilation of Nitrogen

Plants have the general capacity to incorporate inorganic nitrogen into organic compounds. The chief source of nitrogen is *nitrates*. Some specialized plants can also use elemental nitrogen of the air, mainly with the cooperation of symbiotic bacteria in the nodules of the plant root system and other independent, nitrogen-binding microorganisms. Regardless of whether $NO_3^\ominus$ or N_2 is used, the first step of assimilation is reduction to NH_3. The "reduction equivalents" necessary for that are derived from photolysis of water. Assimilation of nitrogen thus falls under the topic of photosynthesis.

Fixation of Nitrogen. Molecular nitrogen N_2 is extraordinarily stable and chemically inert. Although the formation of NH_3 from its elements N_2 and H_2 is strongly exergonic, the energy of activation is very high. Industrial production of ammonia, consequently, has to resort to special catalysts as well as high temperatures and pressures (Haber process).

The biochemical equivalent of the Haber process is the reduction of N_2 according to the equation: $N_2 + 6\,H^\oplus + 6\,e^\ominus = 2\,NH_3$. Although the reaction is exergonic, energy is consumed nevertheless. Electron donor often is $Fe^{2\oplus}$. The enzyme catalyzing this reaction has been called "nitrogenase"; it consists of a molybdenum-iron-sulfur-containing protein and is able to reduce additional substrates, e.g., HCN and acetylene, which becomes ethylene. This latter reaction is frequently employed for the measurement of activity of the enzyme.

The Assimilation of Nitrate. Of critical importance for higher plants is the assimilation of nitrate. It proceeds over two steps. Initially $NO_3^\ominus$ is reduced to $NO_2^\ominus$ by the action of *nitrate reductase*, a molybdenum-containing flavoprotein complex. Hydrogen donor is either NADH, NADPH or ferredoxin.

The reduction of nitrite—probably via the intermediate NH_2OH—results in NH_3 or rather $NH_4^\oplus$. This reduction is catalyzed by an enzyme complex containing ferredoxin, $Cu^\oplus$ (cuprous ion), and flavoprotein and requires ATP. *Nitrite reductase* appears to be closely linked with the phososynthetic apparatus.

Utilization of NH_3. Although the reduction of N_2 or $NO_3^\ominus$ to NH_3 is an important step, free ammonia (or $NH_4^\oplus$) even at low concentrations is a severe cellular poison. In plants it is therefore quickly bound to organic compounds, primarily by the synthesis of glutamine from glutamate and NH_4 with the consumption of ATP (see Chapter VIII,8, last paragraph), which can proceed even at very low concentrations of NH_3. Glutamine donates NH_2 groups for purine synthesis and for one of the two carbamoylphosphate syntheses. The amide nitrogen can also supply the amino group of certain amino acids; glutamate synthase catalyzes the reaction:

$$\text{Glutamine} + \alpha\text{-ketoglutarate} + \text{NADPH} + H^\oplus = 2\text{ glutamate} + \text{NADP}^\oplus$$

Glutamate in a certain sense represents a pool of α-amino nitrogen; almost all other amino acids can be formed from glutamate by transamination, as long as the corresponding carbon skeleton, i.e., the α-keto acid, is available.

Biosynthesis of Essential Amino Acids. Whereas man and other higher mammals depend on food as the source of certain amino acids, plants and most microorganisms can synthesize all amino acids themselves; they are "autotrophic." There is not enough space to present the biosynthetic pathways in plants of all twenty common amino acids. To single out an example, we will, however, discuss the synthesis of the key aromatic amino acids phenylalanine and tyrosine. The diagram shows the formation of an aromatic nucleus:

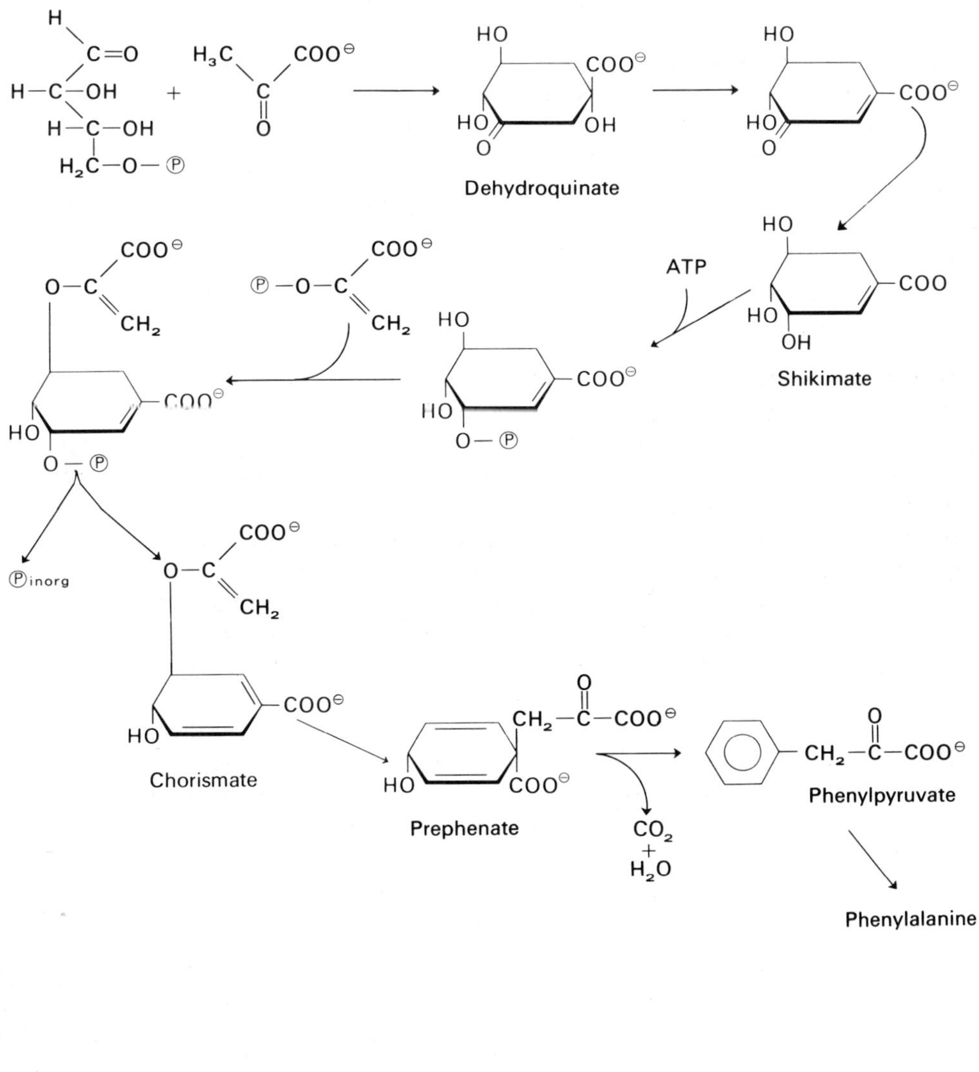

Compounds containing the aromatic benzene ring, in general, are not synthesized by the animal organism.[4] However, in plants and microorganisms such substances can at times be formed in huge amounts, as exemplified by lignin, a major component of woody material.

The first isolable precursor of the aromatic amino acids is *dehydroquinate*, a cyclohexane derivative arising from erythrose 4-phosphate and pyruvate. The initial condensation step is similar to the formation of neuraminate. By elimination of water, dehydroquinate changes to dehydroshikimate and then to *shikimate*, which is phosphorylated in its 5-position. The hydroxyl group in the 3-position then forms an enol ether with phosphoenolpyruvate, which in turn easily rearranges to *prephenate* (the phosphate group is split off in the process). It can undergo various reactions: By the action of *prephenate dehydratase* H_2O and CO_2 are split off resulting in *phenylpyruvate*, a reaction which proceeds spontaneously in an acidic medium. Phenylpyruvate is then transaminated to *phenylalanine*. In another pathway, *prephenate dehydrogenase* dehydrogenates the *p*-hydroxyl group and by loss of CO_2 the ring is aromatized to *p*-hydroxyphenylpyruvate; transamination in this case yields *tyrosine* (not shown in the diagram). In a final pathway, pyruvate can be split off enzymatically with the formation of *p*-hydroxybenzoate.

Tryptophan is formed from anthranilate, which supplies the aromatic ring and the nitrogen for the indole ring.

The biosynthesis of *isoleucine* is discussed in connection with its mechanism of regulation, Chapter XIX,6.

Biosynthesis of Alkaloids. Alkaloids are nitrogen-containing, organic bases; they are products of amino acid metabolism. Plants do not excrete nitrogenous products, mainly because the supply of assimilable N-compounds frequently is the limiting factor of plant growth. With adequate nitrogen supply, several amino acids produced in excess are converted to alkaloids. They appear to be end products of metabolism and are accumulated because there is no suitable mechanism of elimination or excretion. Perhaps some alkaloids are converted back to amino acids under conditions of nitrogen deficiency, or at least the nitrogen of alkaloids is reused for this purpose.

Common precursors of alkaloids are *ornithine*, *lysine*, *phenylalanine*, *tyrosine*, and *nicotinic acid*. Plants build the last compound from C_3 and C_4 fragments rather than from tryptophan.

As a typical example for many other syntheses, we shall briefly see the biogenesis of a few alkaloids of the family of isoquinoline derivatives. *Dopamine*, derived from tyrosine, is the precursor. By undergoing oxidative deamination it can produce *3,4-dihydroxyphenylacetaldehyde*. This aldehyde condenses with dopamine to

[4] An exception is made by the follicular hormone estradiol, which appears to be formed from cholesterol (cf. Chapter XIV,6).

form a benzylisoquinoline derivative, such as *norlaudanosine*, as shown in the diagram.

Norlaudanosine very simply by undergoing methylation gives rise to laudanosine (formula not shown). If the ring is dehydrogenated at the same time, the result is *papaverine*. *Morphine*, the important alkaloid of opium, is also derived from norlaudanosine. In order to bring out this relationship isoquinoline has been rotated around in the right formula. Oxidative ring closure between the two phenyl rings and other conversions finally yield morphine. The diagram also shows *berberin*, which probably arises from norlaudanosine by a ring closure with active formaldehyde.

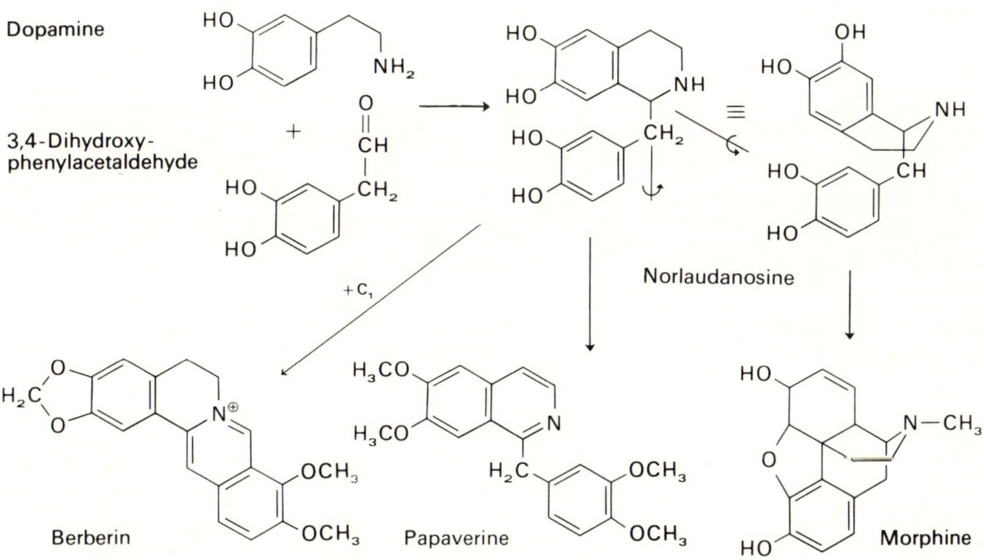

Other alkaloids similarly may be grouped in natural families of compounds. *Ornithine*, for example, is precursor of the pyrrolidine ring (cf. the diagram of formulas in Chapter VIII, end of Section 13), which occurs in *hygrine* and in the bicyclic system of *hyoscyamine*, *atropine*, and *cocaine*. It also leads to the five-membered ring of *nicotine*.

Lysine is the biogenic basis for the piperidine nucleus of *isopelletierine*, *pseudopelletierine*, and *anabasine*. The lobelia alkaloids also belong to this group.

BIBLIOGRAPHY

W. Z. Hassid, Transformation of sugars in plants. *Annu. Rev. Plant Physiol.* **18,** 253 (1967).
E. Leete, Alkaloid biosynthesis. *Annu. Rev. Plant Physiol.* **18,** 179–196 (1967).

S. L. Streicher and R. C. Valentine, Comparative biochemistry of nitrogen fixation. *Annu. Rev. Biochem.* **42,** 279–302 (1973).

I. Zelitch, "Photosynthesis, Photorespiration, and Plant Productivity." Academic Press, New York, 1971.

CHAPTER XVII

Glycosides, Oligosaccharides, Polysaccharides

1. The Glycosidic Bond

The substances named above have the glycosidic bond as a common structural feature. Warming a solution of glucose with methyl alcohol and HCl produces a mixture of two new substances, α- and β-methyl glucoside, with the following structures:

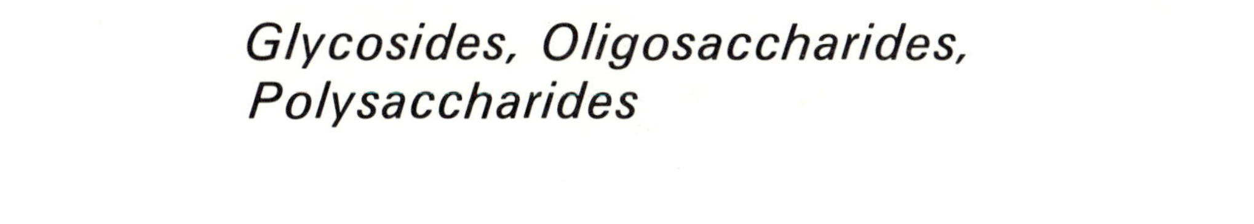

Compounds of this type have the general name of glycosides. They are the acetals (not hemiacetals) of aldehydes (cf. Chapters I,2 and XV,2). These substances are formed from smaller compounds according to our definition (Chapter I,3), and they decompose to their component parts by hydrolysis. "*Glycoside*" is a generic term for this entire class of substances. Those members of the class that are derived from glucose are called *glucosides*; derivatives of galactose are *galactosides*, etc., and the bond between the sugar and some alcohol (or other sugar) is called the glycosidic bond. The method of forming glycosides from sugar and methanol by acid catalysis is similar to that of acetal formation from aldehydes. The living cell, of course, does not employ methanolic hydrochloric acid.

Isomerism between α- and β-glucosides corresponds to the anomerism between α- and β-glucose discussed previously (Chapter XV,2). However, while the two forms of glucose in solution are in equilibrium (mutarotation), the corresponding glucosides cannot interconvert. This is understandable, because glycosides cannot equilibrate by passing through the carbonyl form; the hydroxyl group at C-1 is kept from reaction by the substituent R.

A glycosidic bond can be formed with alcohols, with phenolic hydroxyl groups, and even with carboxylic acids (to give the so-called ester glycosides). In glycosidic natural products the alcoholic (or phenolic) component is frequently called the *aglycon* (the sugarless moiety).

Besides *O*-glycosides there also exist *N*-glycosides, formed by elimination of water between hemiacetal hydroxyl and an HN group. The *N*-glycosides include primarily the nucleosides (Chapter VI,5) and the polynucleotides (nucleic acids, Chapter VII,3), obviously a very important class of compounds.

The *O*-glycosides occur predominantly in the plant kingdom. Several typical examples are presented here:

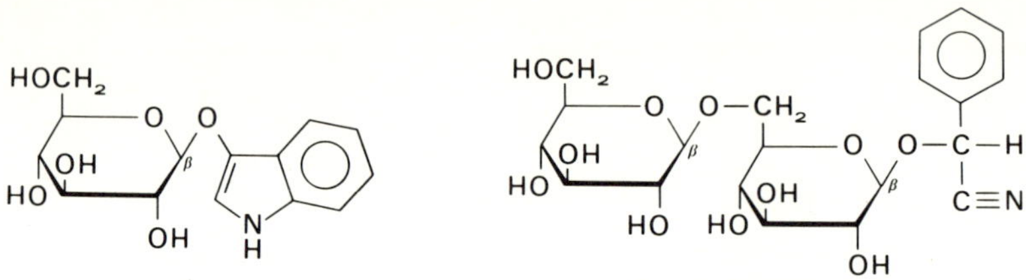

Indican Amygdalin

Plant indican (from *Indigofera* species), a glucoside of indoxyl, at one time important in the isolation of indigo; *amygdalin* from bitter almonds, which upon hydrolysis releases 2 moles of glucose, 1 mole of benzaldehyde, and 1 mole of HCN; *hesperidin*, a flavanone glycoside, formerly assumed to possess vitamin character and last, *streptomycin*, which contains a N-containing inositol (diguanidino-*scyllo*-inositol) as the alcohol, the branched-chain streptose and *N*-methyl-L-glucosamine as the sugars.

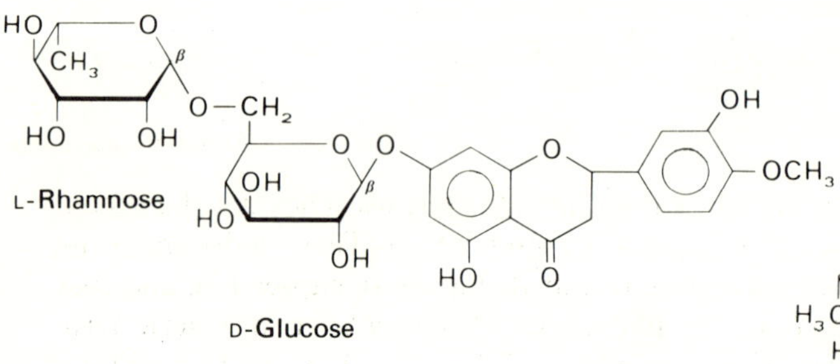

Hesperidin

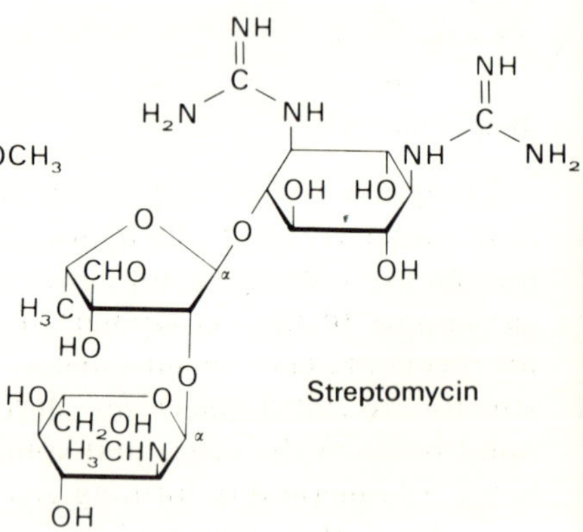

Streptomycin

This group also includes *digitonin* and *strophanthin* (Chapter XIV,3), and the glycolipids discussed in Chapter XIII,4.

Glycosides occur also in lower animals; in insects, e.g., 4-glucosidoprotocatechuic acid. In the mammalian organism they are found rarely and are replaced by the *glucuronides*, which play an important role as detoxication and excretion products of numerous phenols, alcohols, and carboxylic acids. Steroid glucuronides, for example, pregnanediol glucuronide (cf. Chapters XIV,6 and XX,3), are regularly found in the urine.

The glycosidic bond of the glucuronides is completely analogous to that of the plant glycosides described above, but the carbohydrate is glucuronate [for biosynthesis see Section 4(d)].

2. Disaccharides

If sugars can combine with any alcohol to form glycosides, then they should also be able to react with the alcohol group of another sugar molecule. This is indeed possible, and the product arising from two sugar molecules is called a *disaccharide*. Like other glycosides, the disaccharides break down to their components during hydrolysis. A combination of three (or four) units is called a trisaccharide (or tetrasaccharide); *oligosaccharide* is the term for combinations up to about eight carbohydrate groups; beyond that, they are called polysaccharides.

Many chemical characteristics of the monosaccharide (e.g., reducing properties) depend on the free hydroxyl group of the hemiacetal (at C-1 of the aldoses or C-2 of the ketoses). If one of the other hydroxyl groups of the parent sugar is occupied by a glycoside bond, the resulting disaccharide retains the reducing properties of the parent sugar; it still shows mutarotation, and it can form still another glycoside bond with another alcohol. Such oligosaccharides are of the *maltose type* because *maltose* is their prototype (4-α-glucosylglucose).

Disaccharides in which both hemiacetal hydroxyl groups have reacted with one another possess entirely different properties. Both sugar components exist in their full acetal form. Such oligosaccharides are not reducing, show no mutarotation, and form no osazones. Since the simplest natural representative is *trehalose* (1-α-glucosyl-1-α-glucoside), the type of linkage in the nonreducing disaccharides is called the *trehalose type*.

Systematic names indicate the location (number of C atom) of the linkage and the sugar bound at that location. If, for example, β-D-glucose reacts with glucose, all of the following disaccharides are possible products:

1-β-Glucosyl-1-β-glucoside	Isotrehalose (made synthetically)
1-β-Glucosyl-1-α-glucoside	Neotrehalose (made synthetically)
2-β-Glucosyl-glucose	Sophorose (found in plants)
3-β-Glucosyl-glucose	Laminaribiose (found in plants)
4-β-Glucosyl-glucose	Cellobiose (component of cellulose)
6-β-Glucosyl-glucose	Gentiobiose (e.g., in amygdalin)

Furthermore, the corresponding isomers with the α-glycosidic linkage are also possible. The free hydroxyl group of the hemiacetal of a disaccharide cannot be assigned to a configuration: the α- and β-form are in equilibrium. For this reason the formulas show ~OH whenever the steric configuration of the hydroxyl group is unknown.

As explained in Chapter XV,2, the preferred conformation of sugar molecules is the so-called chair form, in which the maximum number of hydroxyl groups is present in the equatorial configuration (in the case of glucose it is the C-1 chair form). Since this manner of representation is more satisfactory for disaccharides we will use it in several examples in this chapter. In Fig. XVII-1, we compare Haworth formulas with chair form formulas of a few important disaccharides.

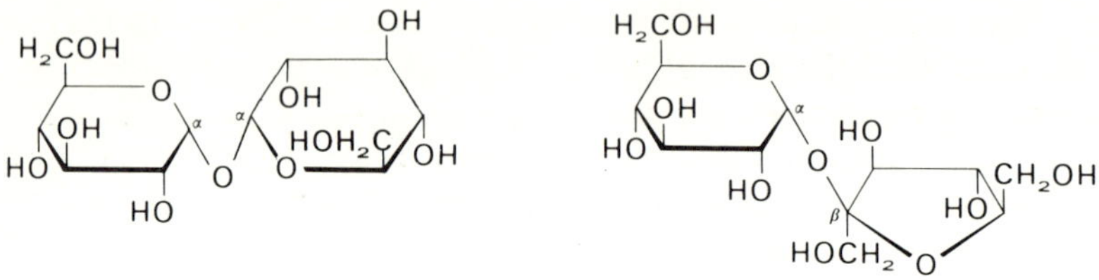

Trehalose Sucrose

Disaccharides of the Trehalose Type. Natural *trehalose* is the α,α′-compound occurring in plants and recently has been identified as the "blood sugar" of insects.

By far the most important compound of the trehalose type is *sucrose.* (α-D-glucopyranosyl-β-D-fructofuranoside), also known as saccharose and cane or beet sugar. It is probably the only foodstuff used in the crystalline form. Although sucrose is widely distributed in the plant kingdom, only sugar cane and sugar beets are used for its industrial preparation. Chemically, it is remarkable that fructose is present in its furanose form, the less stable ring form. For this reason, even dilute acids split cane sugar to glucose and fructose. The process is called inversion because the direction of optical rotation is reversed (cane sugar has $\alpha_D = +66°$; the cleavage mixture, $\alpha_D = -20°$, since fructose is strongly levoratatory). The cleavage mixture called *invert sugar* is the principal component of honey along with sucrose.

Disaccharides of the Maltose Type. ***Maltose*** (4-α-D-glucopyranosyl-D-glucopyranose). A product of the breakdown of starch, maltose occurs in malt (malt sugar). As we have mentioned, it is reducing.

Cellobiose (4-β-glucopyranosyl-glucose) also features the 1:4 linkage of the glucose groups, but in β-linkage. While in maltose (α-linkage) the two six-membered rings are at an angle in the molecular model, the β-linkage imparts a rather planar shape to the molecule; this becomes clearer in the formula in Section 5, in which one ring is turned around.

Isomaltose. This name was given to 6-α-glucosyl-glucose, because it is obtained by hydrolysis of starch. It corresponds to the branch points (cf. Section 5) in starch; the free disaccharide is not important.

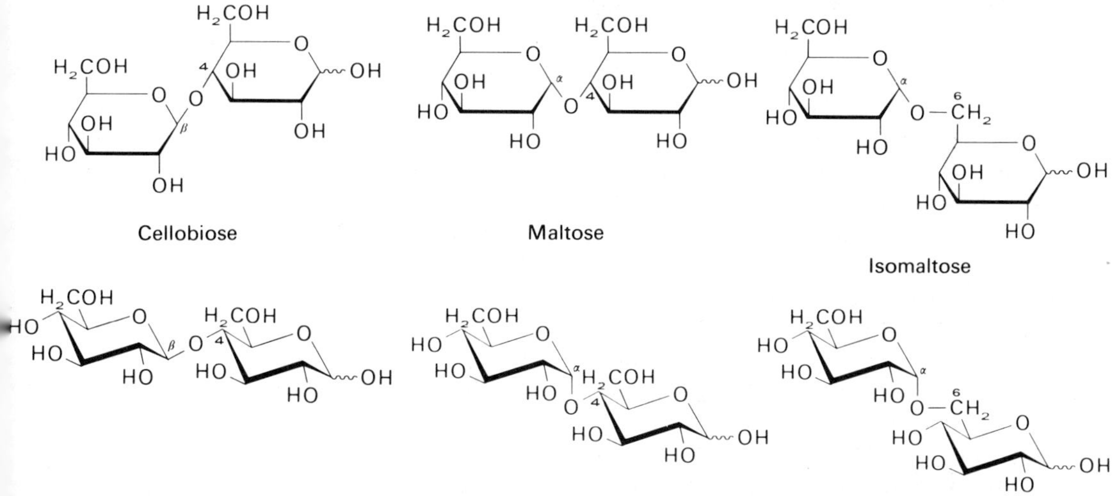

Fig. XVII-1. Disaccharides of the maltose type, represented in the upper part as Haworth formulas and below them in the more realistic chair forms.

Lactose, or milk sugar (4-β-galactosyl-glucose). The most important carbohydrate of the milk of all mammals is lactose. Human milk contains about 6% (and 0.3% of higher oligosaccharides); cow milk, about 4.5%.

Higher Oligosaccharides. These are found in many plants. One example is the trisaccharide *raffinose*, α-D-galactosyl-(1 ⟶ 6)-α-D-glucosyl-(1 ⟶ 2)-β-D-fructofuranoside.[1] Raffinose can be taken as the galactoside of sucrose.

Only small amounts of free oligosaccharides are found in animals. *N*-Acetylneuraminosyl(2 ⟶ 3)-lactose, for example, occurs in human milk. However, in bound form, oligosaccharides are very important as components of glycoproteins, the blood group substances (see Section 8), and gangliosides (see Chapter XIII,4).

3. Enzymatic Cleavage of Oligosaccharides

The hydrolysis of oligosaccharides and glycosides proceeds as shown by the following example:

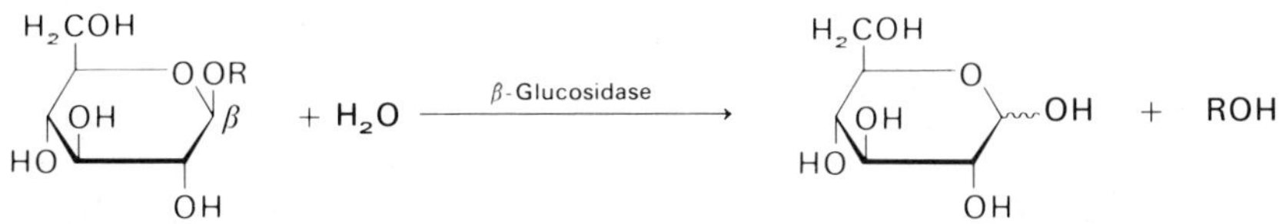

[1] The small arrow shows the direction of the glycosidic bond (always from the hemiacetal hydroxyl to some alcohol or hemiacetal group of the following sugar). The full names of the sugars are often replaced by their standard abbreviations (see Appendix).

This hydrolysis is one of the longest known enzyme-catalyzed reactions. The equilibrium greatly favors cleavage. The enzymes responsible for establishing the equilibrium are called *glycosidases* and are a subgroup of the hydrolases. They are usually group specific, and their specificity is aimed at the nature of the *glycosidically* bound sugar and at the type of glycosidic linkage. Accordingly, we distinguish: α-galactosidases, β-galactosidases, β-fructofuranosidases, etc., α-glucosidases, and β-glucosidases.

"*Emulsin*" from bitter almonds, one of the longest known enzymes, is a mixture of glycosidases whose chief component is β-D-glucosidase. It hydrolyzes both natural and synthetic β-glucosides, e.g., β-methyl glucoside, amygdalin, cellobiose, and others.

α-Glucosidase, formerly called *maltase* because it splits maltose, is frequently found together with starch-hydrolyzing amylase (Section 5). α-Glucosidase can also cleave sucrose. The sucrose-cleaving enzyme of yeast, however, is a β-fructofuranosidase. Since this cleavage involves an inversion of the direction of optical rotation, the enzyme formerly was known as "*invertase*."

Another important group of glycosidases is comprised of the *β-glucuronidases*, which are distributed widely among plants and animals. Substrates are the glucuronides, including not only excretion products but also the *mucoids* (Section 7).

Most glycosidases simultaneously are really also *transglycosidases*, which transfer the glycosidically bound sugar residue to other appropriate molecules with OH groups. Since water is always present in especially high concentrations it is a suitable acceptor molecule; but the transfer to water is equivalent to hydrolysis. It has been observed, however, that galactosidase transfers the galactosyl residue of lactose onto glucose (preferably at the 6-OH) or onto another lactose molecule; in the latter case, a trisaccharide would be formed. *Transglycosidation* is important in the synthesis of polysaccharides by plants (see below).

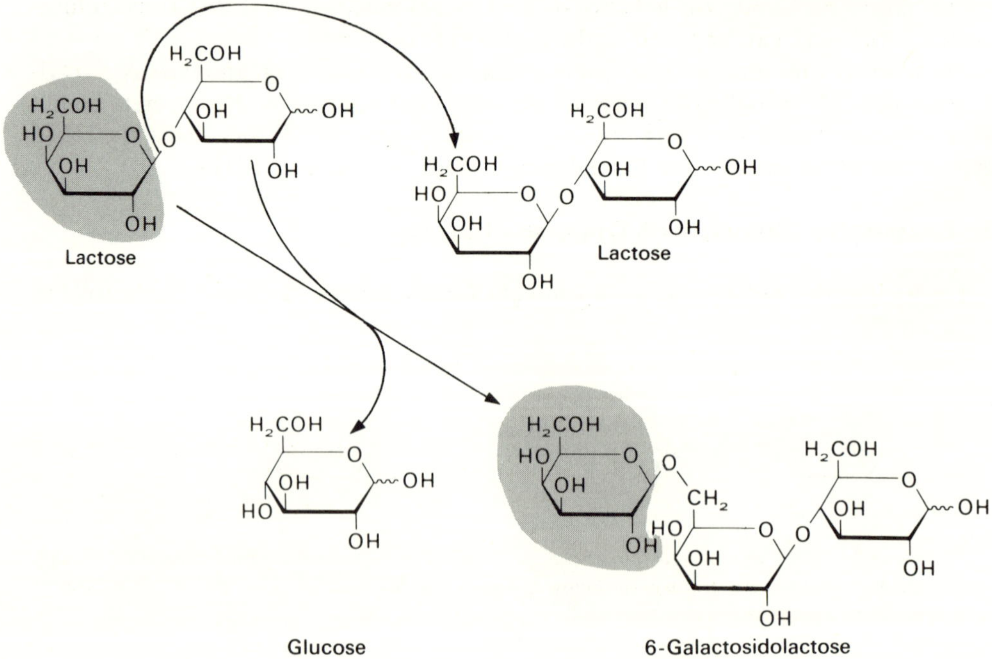

4. Biosynthesis of Glycosides and Oligosaccharides

In principle, glycosidases are also able to promote synthesis. Synthesis from monosaccharides is severely limited by the positive free energy of the reaction; when disaccharides are available, however, longer chains can be formed by transglycosidation. Plants perform transglycosidation to a considerable extent, mainly in the formation of polyfructoside from sucrose (Schlubach). A side product of the reaction is glucose, which can be converted back to sucrose via some detours.

Monosaccharides must first be activated so that they contain an energy-rich bond. The activating group is a nucleoside diphosphate, predominantly uridine diphosphate (coenzyme of glycosidation; cf. Chapter VI,8).

The binding of glucose to the coenzyme is a complicated process. Initially, a kinase with the aid of ATP converts glucose to glucose 6-phosphate, which in turn becomes rearranged to glucose 1-phosphate (see Section 6). The 1-phosphate reacts enzymatically with uridine triphosphate (UTP) to form uridine diphosphate glucose and inorganic pyrophosphate. The phosphate of glucose 1-phosphate is retained in the UDP-glucose bond.

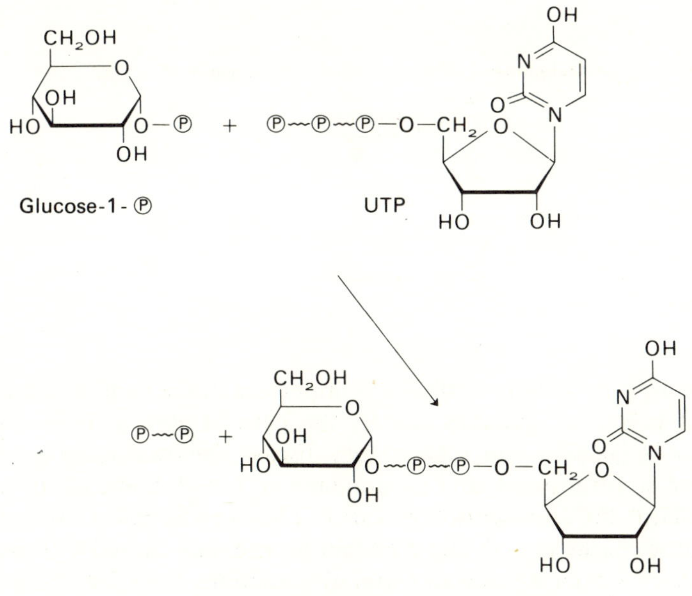

Uridine diphosphate glucose (UDP-Glc)

The glucose molecule activated in this way has various possibilities for further reaction:

1. Epimerization at C-4 to produce the *galactose derivative*, catalyzed by UDP-glucose-4-epimerase (formerly called galactowaldenase). At equilibrium, the ratio is about 1 glucose: 3 galactose. The mechanism of this steric rearrangement is not clear. The supposition is that the rearrangement proceeds via the keto compound. This is supported by the observation that the enzyme contains firmly bound NAD.

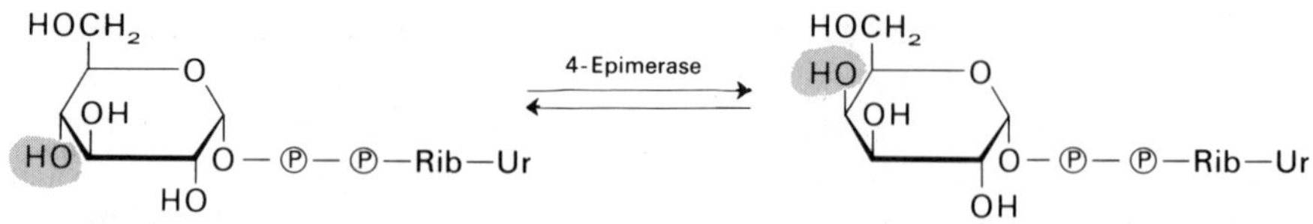

2. ***Exchange between galactose and glucose,*** according to the equation:

UDP-glucose + galactose-1-Ⓟ $\rightleftharpoons$ UDP-galactose + glucose-1-Ⓟ

The enzyme is called *hexose-1-phosphate uridylytransferase* (the systematic name is UDP-glucose:α-D-galactose-1-phosphate uridylytransferase). By this reaction, galactose is bound to the coenzyme. The cooperative action of transferase and 4-epimerase converts galactose-1-Ⓟ to glucose-1-Ⓟ, as illustrated:

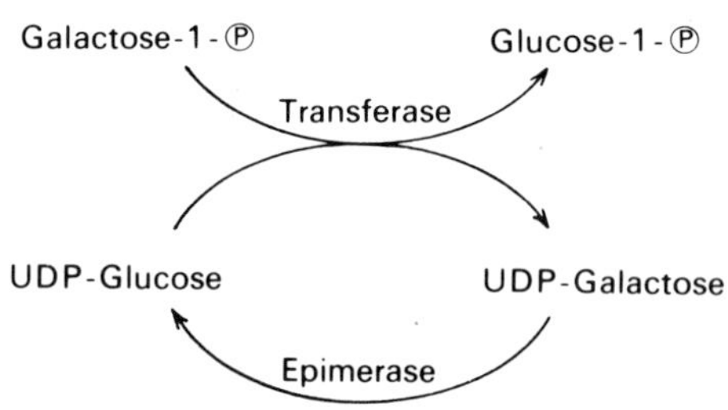

Hexose-1-phosphate uridylytransferase is important in the human organism for the metabolism of galactose; galactose enters the path of glucose breakdown by the above reaction. In *galactosemia*, a hereditary disease, the transferase is lacking. As a result galactose is not utilized, and accumulates in blood. Some of the galactose is reduced to dulcitol, the corresponding hexitol. This is an excellent example of how a change in genetic character is in fact a metabolic anomaly caused by the lack of one enzyme (Kalckar). Under normal nutritional conditions the effects of galactosemia are fatal; complete elimination of galactose (and the milk sugar lactose) from the diet from birth on improves the prognosis considerably.

3. ***Disaccharide synthesis.*** Active glucose (UDP-glucose) can also combine with a second monosaccharide molecule to give a disaccharide. If combined with glucose-6-Ⓟ, trehalose-6-Ⓟ results; if combined with fructofuranose, *sucrose* is formed. The reaction is important especially in plants and illustrates the biosynthesis of cane sugar.

The mammary gland of mammals synthesizes *lactose* in a similar manner. Uridine diphosphate galactose combines with glucose; the galactose group is transferred to the 4-hydroxyl of glucose in β-glycosidic linkage. The product is lactose:

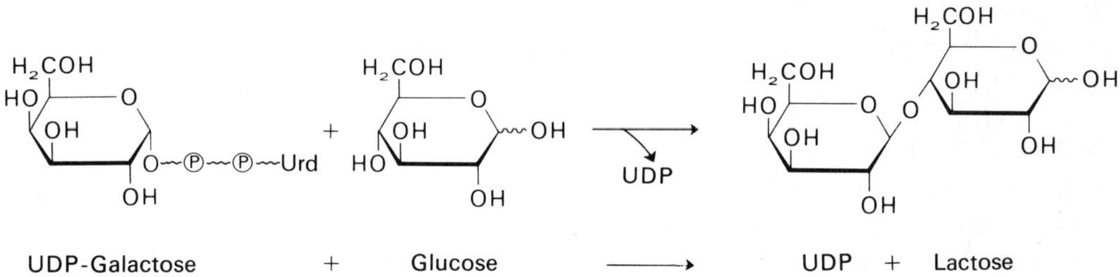

The enzyme system lactose synthetase consists of two separate active peptide chains, the A protein and B protein. The A protein by itself is an enzyme specific for transferring the activated galactose residue onto *N*-acetylglucosamine. The action of the B protein is to alter the specificity of the A protein such that glucose (or glucose-1-Ⓟ) can accept the galactosyl residue. The B protein is identical with α-lactalbumin of milk, whose amino acid sequence is remarkably homologous with lysozyme, a hydrolase that primarily cleaves muraminic acids. Common to both proteins appears to be their affinity for carbohydrate.

4. Glucuronide formation. Uridine diphosphate glucose can be oxidized enzymatically (with NAD as the H acceptor) to give glucuronyl diphosphate uridine, i.e., activated glucuronate.

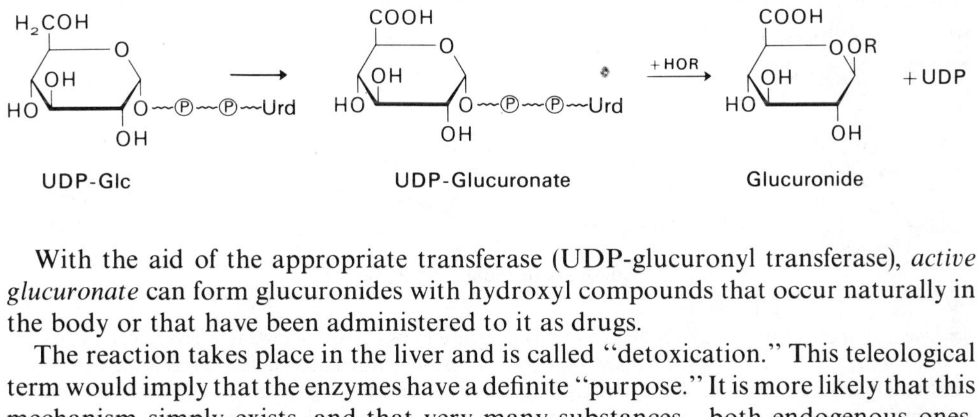

With the aid of the appropriate transferase (UDP-glucuronyl transferase), *active glucuronate* can form glucuronides with hydroxyl compounds that occur naturally in the body or that have been administered to it as drugs.

The reaction takes place in the liver and is called "detoxication." This teleological term would imply that the enzymes have a definite "purpose." It is more likely that this mechanism simply exists, and that very many substances—both endogenous ones, like hormones or bilirubin, and exogenous ones, like drugs and poisons—are bound in this fashion and are excreted, independent of their "toxicity" to and any "intentions" by the organism.

Metabolism of Mannose. This sugar is not activated by uridine diphosphate but by *guanosine diphosphate* instead. There appears to be a certain specificity regarding the activating coenzyme. The reactions otherwise are largely analogous.

5. Polysaccharides: Homoglycans

A combination of very many monosaccharides, linked by glycosidic bonds and analogous to the structure of oligosaccharides, is called a polysaccharide, or *glycan*. Polysaccharides are widely distributed in nature, with a greater variety in plants. Some possess low molecular weights, corresponding to 30–90 monosaccharides; but the majority (especially the more important ones) contain several hundred or even thousands of monosaccharide units.

Polysaccharides have various functions. In plants they serve both as structural support (for example, *cellulose* by far the most abundant polysaccharide) and as storage compounds in seeds and tubers (*starch*). In animals they rarely serve as structural substances (cellulose is found only in tunicates; *chitin* occurs in insects and crabs). *Glycogen* is accumulated by animals as a storage substance.

The chemical structure of polysaccharides permits the following classification: (1) Homoglycans (analogous to homopolymers), which contain only a single kind of monosaccharide component; (2) heteroglycans (analogous to copolymers), which consist of different components (usually only two or three); (3) conjugated compounds (glycoproteins and glycolipids). The "glucans" cellulose, starch, and glycogen are composed of glucose alone; they will be discussed first.

Cellulose. This substance occurs in plants, usually associated with other structural substances (lignin). The cotton fiber and related fibers are almost pure cellulose (seminal threads of the *Gossypium* species). Industrial cellulose is usually obtained from wood and purified in various ways, i.e., it is freed from lignin and other contaminants. In cellulose, the glucose units are linked between C-1 and C-4 by β-glycoside bonds. The simplest component, the repeating unit, is therefore *cellobiose*.

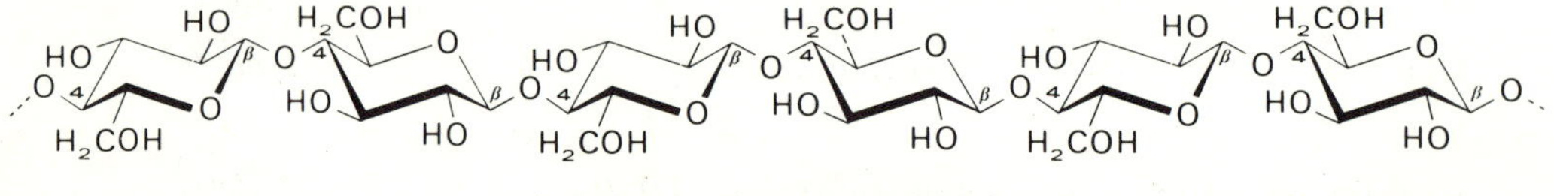

Cellulose formula (segment)

Native cellulose consists of about 8000 to 12,000 glucose units corresponding to a molecular weight of 1.3 to 2.0 million. The chain is a several μm long micelle. The long threadlike molecule is repeatedly folded, and the individual sections are linked by hydrogen bonds. This folding gives rise to elementary fibrils visible in the electron microscope with a diameter of 35 Å. Layers of the fibrils at right angles make up the cell wall. They are cemented together by interlaced "hemicelluloses" (i.e., polyxylans in β-1 → 4 linkage), pectins (see below), the protein extensin, as well as lignin.

Chitin. A structural substance similar to cellulose, chitin is a linear molecule with β-1 → 4 linkages between the individual units, which in this case consists of *N-acetylglucosamine*. The repeating unit of chitin, chitobiose, is shown in the formula below. The molecular size is not known, but probably is large.

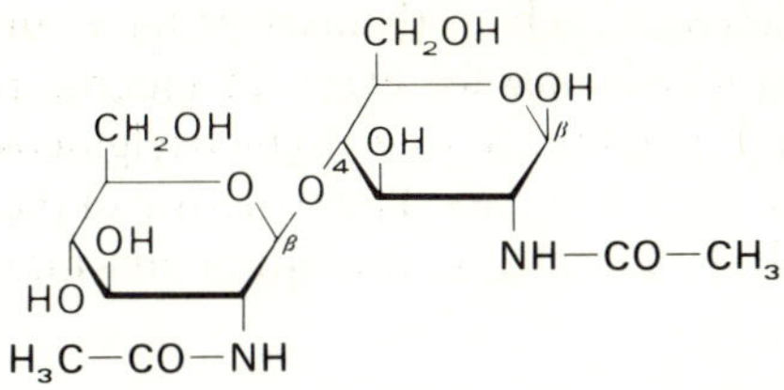

Chitobiose

Chitinase (from the gastric juice of snails or from bacteria) decomposes the polysaccharide to *N*-acetylglucosamine. Chitin is found in fungi, but principally among the arthropods (crabs and insects). The armor of crabs and exoskeleton (or cuticula) of insects consists mostly of chitin and some protein. The excellent mechanical properties of insect cuticula are due to chitin.

Starch. A storage substance of plants, starch is deposited abundantly in seeds (grain) and tubers (potatoes, etc.) in the form of granules. Extraction and fractionation yields two types of compounds.

a. Amylose makes up 20–30% of most native starches and consists of 250–300 glucose residues linked by 1 → 4-α-glycoside bonds. The repeating unit is maltose (see Section 2 and Fig. XVII-1), which is linked to itself in the same way. Because of the α-glycoside bonds the molecule does not appear long and straight, but rather coiled up like a spring (Fig. XVII-2).

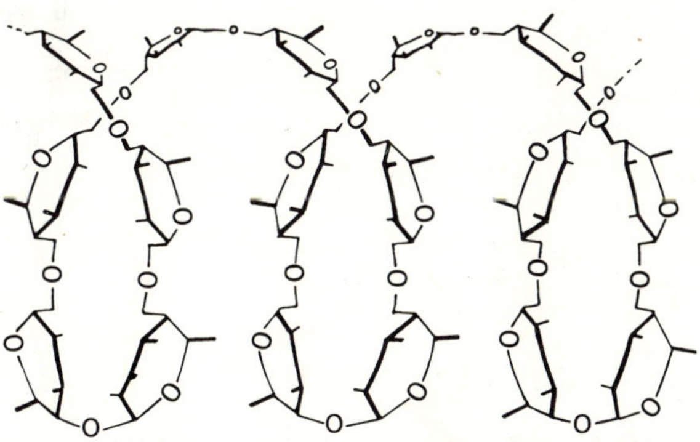

Fig. XVII-2. Amylose.

The molecular shape of amylose explains some of its chemical and biochemical properties. With *iodine*, amylose turns intensely blue. This venerable reaction has been explained only recently. It was found that iodine molecules enter the hollow center of the coil created by the glucose units. In such "enclosure compounds" iodine exhibits changed physical characteristics, such as the strong absorption of light. There are enzymes that slit open this tube and free fragments consisting of 6, 7, or 8 glucose units that are joined together as loops (see also Section 6). These products are "cyclodextrins" or "Schardinger dextrins."

b. Amylopectin, the other component of natural starch, is also built up of glucose units in α-glycosidic linkage but the molecule is branched. Linked to the main chain by 1 → 6 bonds are side chains which themselves have further side chains. On the average, the chain is branched once for every 25 glucose residues. After enzymatic hydrolysis one can isolate isomaltose (6-α-D-glucosylglucose, formula in Section 2) which is derived from the branch points. The structure of the molecule resembles that of glycogen (see Fig. XVII-3). The chains and side chains may in fact be helical instead of linear as shown.

Inulin and Pectins. Besides plant polysaccharides composed of glucose, there are those that are made up of *fructose*, and others. *Inulin* is a polyfructosan consisting of fructose residues in the furanose form with 2 → 1 linkage. The water solubility of inulin is due to the relatively small number of fructose units in the molecule (fewer than 100). Possibly, the chain is terminated by glucose in a sucrose-type bond. Inulin has been employed by physiologists for the determination of the extracellular space of the body (see Chapter XXI,1). *Agar-agar*, a polysaccharide derived from sea algae, consists of D- and L-galactose, predominantly with 1 → 3 bonds, and always contains some amount of sulfuric acid. Agar-agar forms highly viscous gels and is used extensively for the preparation of bacterial culture media. The various plant gums and *pectins* have more complicated structures. Pectins contain the important gel-forming component, polygalacturonate (α-1 → 4 linkage); the carboxyl groups are esterified extensively with methanol.

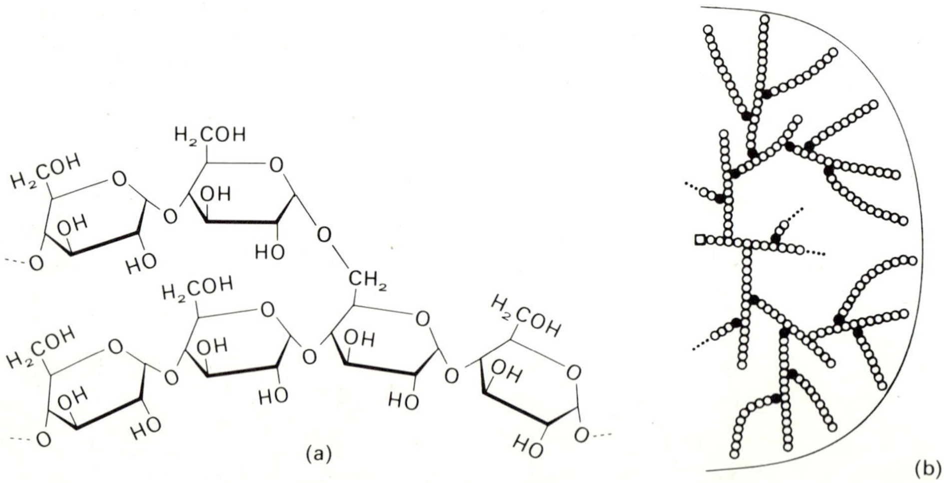

Fig. XVII-3. Molecular structure of glycogen. (a) Diagrammatic representation of a branch point. (b) Diagram of a larger segment of the glycogen molecule (according to Whelan). The little square stands for the lone reducing end group; the solid circles stand for glucose residues existing in 1 → 6 linkage in the molecule. Some of the nonreducing end groups are buried in the interior of the molecule and thus not susceptible to enzymatic attack. The diagram shows, of course, only a segment of the entire macromolecule.

Glycogen, sometimes called "animal starch," occurs in liver, muscle, and many other cells as a storage substance.[2] The chemical structure of glycogen is very similar

[2] Highly polymeric substances are stored better than low molecular ones for two reasons: they are usually insoluble, and even when they are in solution their osmolarity is so low that the accumulation in a cells does not result in hypertonicity.

to that of amylopectin except that glycogen is even more highly branched. In the center of the molecule every third to fifth glucose residue of the main chain bears a side chain in 6 ← 1 linkage. The mean length of the side chains is 10–14 glucose units. For a schematic illustration see Fig. XVII-3.

The molecular weight is very high: for muscle glycogen it has been estimated to be about 10^6; for liver glycogen, about 16×10^6 (corresponding to about 10^5 glucose residues). The glycogens are nevertheless water-soluble; the molecules have the shape of a flattened ellipsoid. Insoluble glycogens are of still higher molecular weight and are bound to protein.

6. Enzymatic Breakdown of Polysaccharides

A distinction must be made between the digestion of polysaccharides, which is hydrolysis, and the metabolism of glycogen in liver and muscle, which follows a phosphorolytic pathway.

Hydrolytic Cleavage. The principal enzymes, the amylases, break starch (and glycogen) down to maltose units. These enzymes are distributed widely. The term amylase is now used as a generic term; α-amylase (dextrinogenic amylase) and β-amylase (saccharogenic amylase) are distinguished.

α-Amylases occur in saliva, pancreas, and malt. In their point of attack they resemble the endopeptidases (Chapter VIII,1); they cleave in the middle of the macromolecule. The initial cleavage products are oligosaccharides of six or seven glucose units. The assumption is that the enzyme attacks the helical structure (Fig. XVII-2) and splits neighboring glycoside bonds that are one whole turn of the helix apart. This also explains the observation that the branch points of the side chains (the 1 → 6 linkages) in amylopectin are skipped: although they present no obstacle to α-amylase, it cannot cleave the 1 → 6 linkages. After prolonged action of the enzyme, the fragments are degraded largely to *maltose*; hydrolysis stops with the disaccharides.

Endoamylase (α-amylase) has a "liquefying" effect, since the viscosity of a colloidal starch solution decreases rapidly and the iodine color reaction disappears without the appearance of reducing sugars. The reducing effect develops later, in a second phase, when the fragments have been broken down further. As indicated by the name of the enzyme α-amylase, maltose is released in the α-form (the optical rotation decreases, however, due to mutarotation).

β-Amylases occur predominantly, if not exclusively, in plants. They are exoglycosidases because they attack the ends and always split off two terminal glucose residues as maltose in its β-form (mutarotation increases). During hydrolysis, a Walden inversion takes place at C-1.

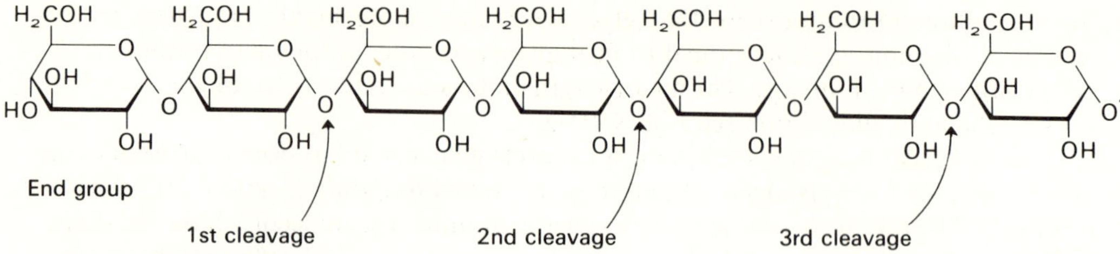

Action of β-amylase

The terminal cleavage explains the early appearance of reducing sugars ("saccharogenic amylase"). Although amylose is hydrolyzed almost completely, only about half of amylopectin is, since the enzymatic reaction stops even before a branch point is reached. Phosphoric acid, an occasional component of starch, also is an inhibitor of the enzyme. Eventually, a "limit dextrin" of relatively high molecular weight is left, and the breakdown continues only if α-amylase or a special $1 \rightarrow 6$-linkage-splitting enzyme is active at the same time.

The end product of the degradation of polysaccharides by α- and β-amylase is *maltose* (and isomaltose); glucose arises to a minor extent by the cleavage of maltotriose into two fragments. In the intestinal tract maltase, or α-glucosidase, splits maltose to glucose and thus ends digestion. α-Glucosidase can also split sucrose.

γ-Amylase recently was discovered in the liver. The enzyme splits glycogen exclusively into glucose units, and no maltose fragments appear. It hydrolyzes both the $1 \rightarrow 4$ and the $1 \rightarrow 6$ bonds and thus can break down glycogen completely.

Cellulases are commonly found in microorganisms, but rarely in animals. The utilization of cellulose almost invariably requires the participation of the microbial flora (e.g., in the rumen of ruminants).

Phosphorolysis of Starch and Glycogen. Within cells, the storage carbohydrate is broken down by a different route, involving phosphorolysis. This holds true both for the breakdown of starch in plant cells and for that of glycogen in muscle and liver.

The term phosphorolysis (coined in analogy to hydrolysis) describes the cleavage of a (glycoside) bond by inorganic phosphoric acid. It may be more correct to conceive of the process as a transfer of the terminal glucose residue onto phosphoric acid (as the acceptor molecule) and to classify the enzyme as a transglycosidase. The net result is the same for the two mechanisms.

It is the action of *glycogen phosphorylase* to remove a glucose residue from the nonreducing end of the polysaccharide and to transfer it to an inorganic phosphate group. The long chain is thus shortened, and glucose-1-Ⓟ is formed. The enzyme can then repeat its action on the new end group of the polysaccharide. The breakdown is halted at the branch points; $1 \rightarrow 6$ linkages are not attacked by phosphorylase, but rather by a hydrolase. The complete degradation of glycogen therefore entails the release of about 10% of free glucose derived from the branch points.

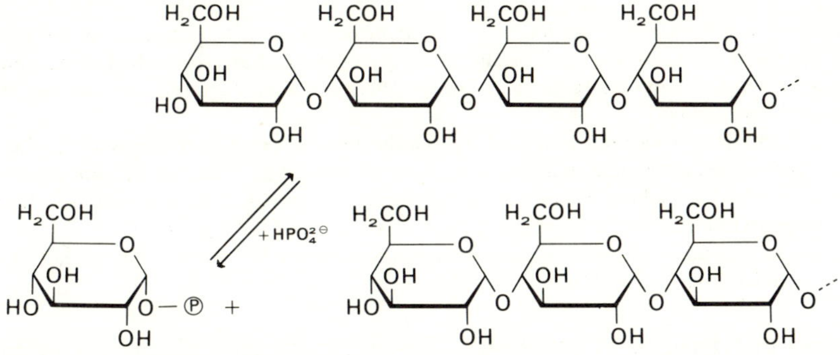

Glucose 1-phosphate

Phosphorolysis of polysaccharides is energetically advantageous to the cell since glucose is broken down further in its phosphorylated form—either through the Embden-Meyerhof pathway or through the pentose phosphate cycle. The cell thereby saves itself the initial phosphorylation step.[3] In the subsequent step, *phosphoglucomutase* changes glucose-1-Ⓟ to glucose-6-Ⓟ. "Cofactor" for this enzyme is glucose 1,6-bisphosphate whose 1-phosphate group is transferred to the 6-position of the substrate glucose-1-Ⓟ. In this way, a new molecule of the cofactor is formed and the cofactor molecule, now converted to glucose-6-Ⓟ, is the product of the reaction. The entire process is analogous to the conversion of glycerate-3-Ⓟ to glycerate-2-Ⓟ. Phosphoglucomutase provides the link to the glycolytic pathway (Chapter XV,7).

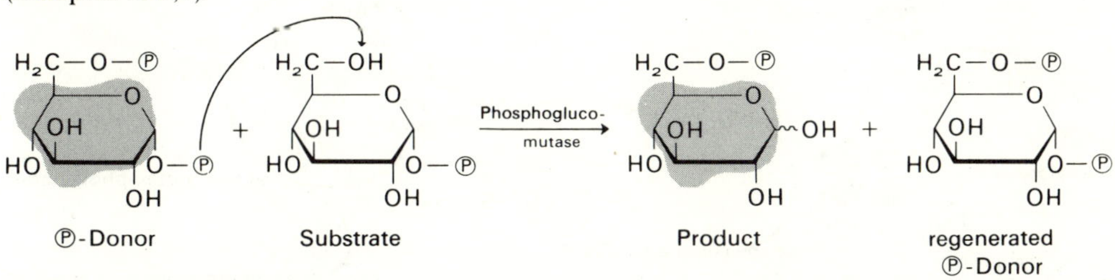

Synthesis of Glycogen. The phosphorylase reaction in vitro is freely reversible. In the cell, however, phosphorylase functions only to break down. The synthesis proceeds via *uridine diphosphate glucose*, i.e., with a more highly activated molecule. The principle of this type of glycoside synthesis has already been discussed (cf. Section 4). A primer molecule (glycogen or dextrin; in the extreme, maltose suffices) is

[3] However, it is fallacious to assume that the degradative pathway of glucose is more efficient energetically via glycogen than the direct breakdown. The formation of glycogen from glucose demands two phosphorylation steps because the synthesis proceeds via UDP-glucose (see below). In other words, only part of the energy expended is actually conserved.

necessary for *glycogen synthetase* to attach further glucose residues in $1 \longrightarrow 4$ linkage.

Some energy is wasted in this reaction, since 2 ATP are consumed in the formation of UDP-glucose from glucose. On the other hand, the equilibrium favors glycogen synthesis (being highly exergonic), and the separation of synthetic and degradative pathways offers the advantage that the two can be regulated independently (see below).

For the formation of natural, branched polysaccharides, a special transglycosidase, the "branching enzyme" (Q-enzyme), is required in addition to glycogen synthetase. The enzyme removes one chain link from the $1 \longrightarrow 4$ linkage and reattaches it at the 6-hydroxyl group. A new branch point is thus created upon which phosphorylase can then act further.

In plants the synthesis proceeds probably without the help of UDP-glucose, but instead with the corresponding adenosine derivative *ADP-glucose*.

Control of Glycogen Metabolism. The hormone epinephrine (= adrenalin) has long been known to rapidly mobilize glucose from glycogen storage. This mobilization first requires the activation of *phosphorylase*, which involves a cascade of enzyme interconversions illustrated in Fig. XVII-4.

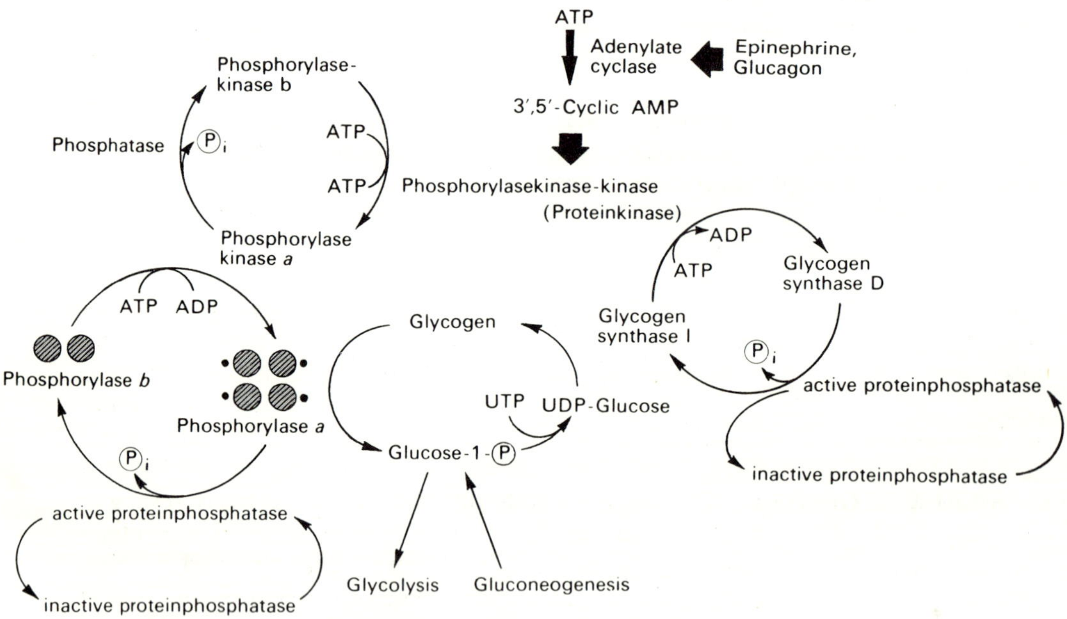

Fig. XVII-4. The regulation of the metabolism of glycogen by its control enzymes. The aggregation of phosphorylase *a* to tetramers shown in the left part of the figure applies only to the muscle enzyme; the activation of adenylate cyclase by glucagon applies only to the liver enzyme. Further explanations in the text.

The phosphorylase exists in two forms: Phosphorylase *b* is relatively inactive; it is converted by the action of *phosphorylase kinase* to active phosphorylase *a*. Enzymes such as the phosphorylase kinase often are called *control enzymes.*

The *mechanism of activation* is the following: A serine residue is phosphorylated by ATP. In the muscle the phosphorylation is associated with the aggregation of 2 phosphorylase *b* molecules (themselves dimers with the molecular weight of (185,000) to the tetramers. In the liver both the active and inactive forms exist as dimers. Reconversion of phosphorylase *a* to *b* is catalyzed by *phosphorylase phosphatase.*

Phosphorylase kinase is itself an interconvertible enzyme. The inactive form is changed by *phosphorylasekinase kinase* (also called *protein kinase*) and ATP to the active form. Protein kinase in turn is *activated* allosterically by cyclic *3'5'-AMP* (formula, Chapter VI,5), which is produced from ATP by *adenylate cyclase.* Finally, adenylate cyclase of the liver is stimulated by both glucagon and epinephrine, that of muscle only by epinephrine (see also Chapter XX,1 and 11).

The action of epinephrine on the mobilization of glycogen reserves therefore proceeds over a series of three control enzymes until, with the activation of phosphorylase, glycogen degradation is finally stimulated. In analogy to the clotting of blood, an amplification effect is achieved.

Protein kinase, however, has an additional function in this scheme of regulation. It also phosphorylates the I-form of glycogen synthetase thus converting it to the less active D-form.

The D-form is active only at high concentrations of the allosteric effector glucose 6-phosphate; in other words, it is dependent on glucose 6-phosphate. The I-form is independent of glucose 6-phosphate and is assumed to be the physiologically active form. The interconversion of I ⟶ D evidently amounts to switching off the synthesis of glycogen, which is appropriate under conditions of glucose mobilization.

If metabolic conditions call for storage of glycogen, glycogen synthetase D is dephosphorylated by activation of protein phosphatase and thus converted to its I-form. The same protein phosphatase also removes the phosphate group from phosphorylase *a* thereby changing the enzyme into its inactive form *b* and halting glycogen breakdown. The control mechanism therefore effectively shifts between two metabolic pathways whose uncontrolled simultaneous operation would be but a waste of UTP.

Glycogen Storage Diseases (Glycogenoses). Several inherited metabolic anomalies consist of the deletion, as a result of mutation, of enzymes involved in glycogen breakdown. These mutations express themselves as different disease syndromes that affect either the liver or muscle tissue. The more prominent diseases will be described here.

In McArdle's disease glycogen phosphorylase is absent from the muscle. As a result the muscle can not utilize its glycogen reserves for energy production (work performance), which is equivalent to profound muscle weakness. Since glycogen breakdown is impaired while its synthesis is unaffected, glycogen accumulates in the muscle tissue. In the disease described by Hers the hepatic phosphorylase is absent and the muscle enzyme is present in normal concentration. This finding is proof that the two enzymes are coded by different genes.

Forbe's disease, on the other hand, is characterized by elevated glycogen storage in both liver *and* muscle. Here amylo-1,6-glucosidase is deleted; it is the enzyme responsible for the hydrolytic cleavage of the 1 ⟶ 6 glycosidic linkage at the branch points of glycogen (see above). The breakdown, therefore, comes to a halt at these points. Since, with continued supply of carbohydrate, glycogen continues to be synthesized in normal fashion and since more and more branch points are being added that can no longer be cleaved, the final result is the deposition of very high molecular weight and highly branched glycogen.

Still another storage disease, named after von Gierke, affects only the liver and can be traced to the deficiency of glucose-6-phosphatase. This enzyme has the important function to liberate the glucose released into blood (see also Chapters XV,10 and XIX,1). Its absence hence engenders extremely low fasting values of blood sugar (below 1 mM). The defect furthermore decelerates glycogen breakdown while synthesis again proceeds unaffected. Increased quantities of glycogen are stored.

7. Heteroglycans

Several polysaccharides are composed not only of simple sugars, but also of derivatives of sugars, such as *amino sugars* and *uronic acids*. Most of these are structural support material for connective tissue or mucous substances of the body. They are also called mucopolysaccharides.

The *mucopolysaccharides* have a common structural feature. They consist of disaccharide units in which a uronic acid is bound glycosidically to the 3-position of an acetylated amino sugar. These disaccharide residues are polymerized by 1 ⟶ 4 bonds to give a linear macromolecule; the uronic and sulfuric acid residues impart to these substances a strongly acidic character. Besides D-*glucuronic acid*, one can also find L-*iduronic acid*, distinguished from glucuronic acid only by the steric configuration at C-5.

Hyaluronic Acid. The least complicated member of the series, hyaluronic acid consists of glucuronate and *N*-acetylglucosamine linked according to the principle just discussed. The formula is

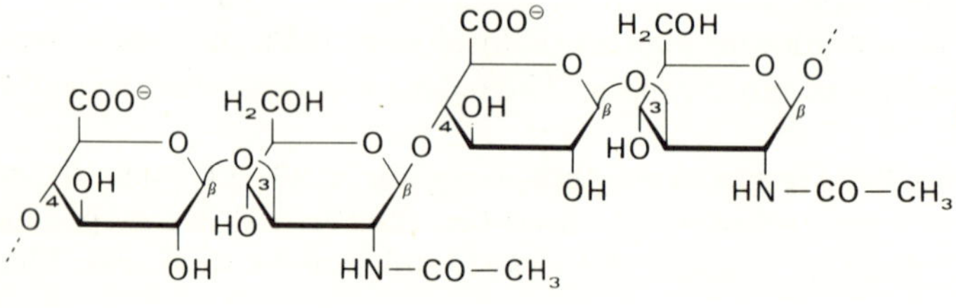

Hyaluronic acid

The molecule is probably not branched, or only slightly so. The molecular weight is very great (several million). Hyaluronic acid is a principal component of the ground substance of connective tissue. It is found, among other places, in the synovial fluid, in the vitreous humor of the eye, and in skin, usually in combination with protein. Frequently it is prepared from umbilical cord.

It is physiologically significant that hyaluronic acid is split swiftly by the *hyaluronidases.* Two enzymatic effects have been distinguished: First, the depolymerizing effect, which causes a rapid drop in viscosity of a hyaluronic acid solution without setting free significant amounts of reducing groups; second, cleavage to smaller fragments (even-numbered oligosaccharides).

Hyaluronidase is the "spreading factor" of skin and connective tissue. The depolymerizing effect permits any foreign bodies, such as ink, colloidal pigments, but also pathogenic bacteria, to penetrate the tissue, since the cementing substance is being dissolved.

Chondroitin Sulfate. Along with hyaluronic acid, chondroitin sulfate forms part of the structure of connective tissue. Cartilage has the highest content of chondroitin. Several polysaccharides of this type have been isolated and designated as chondroitin sulfate A, B, and C. The formula shows *chondroitin sulfate C* (= chondroitin 6-sulfate), composed of glucuronate and *N*-acetylgalactosamine 6-sulfate. In type A, the sulfate is bound in the 4-position, while chondroitin sulfate B (also called dermatan sulfate) contains L-iduronate.

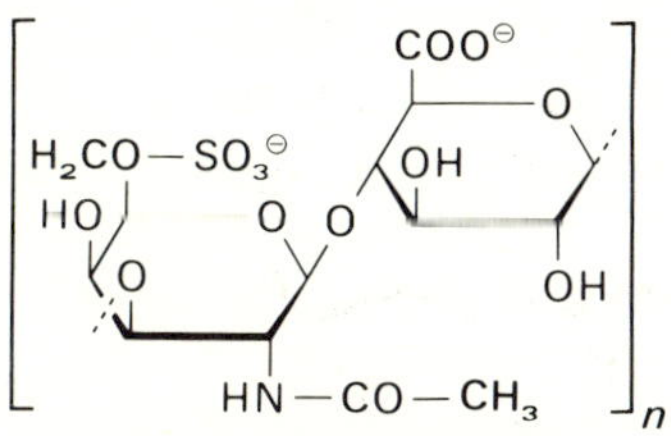

Chondroitin sulfate C

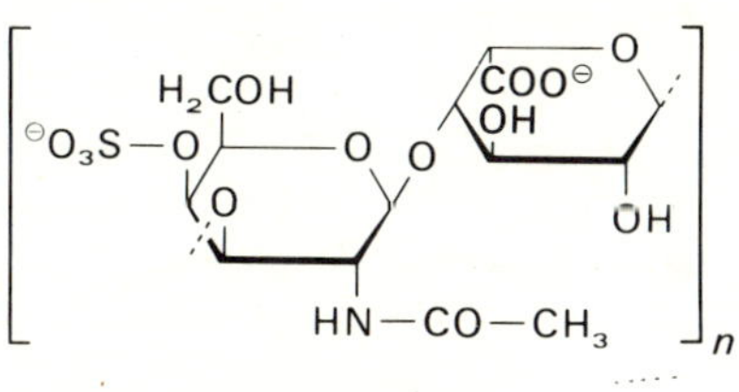

Dermatan sulfate

Chonroitin sulfate is very often associated with collagen, and possibly with other proteins, too. It seems to augment the formation of the fibrillar structure of collagen.

Heparin. A polysaccharide (molecular weight 17,000–20,000), heparin is composed of sulfonylaminoglucose (= glucosamine *N*-sulfate) and the sulfate esters of glucuronic acid. The type of linkage recently was shown to be $1 \longrightarrow 4\text{-}1 \longrightarrow 4$; the structure of heparin thus differs from that of chondroitin sulfate. The sulfate content is very high and corresponds to 4–5 molecules per tetrasaccharide repeating unit. The relative positions of the sulfate residues vary.

Heparin acts as an anticoagulant. It prevents blood clotting by inhibiting the prothrombin-thrombin conversion and thus eliminating the thrombin effect on fibrinogen.

Mast cells contain unusually large amounts of heparin, possibily for the purpose of neutralizing biogenic amines (e.g., histamine; cf. Chapter XX,13).

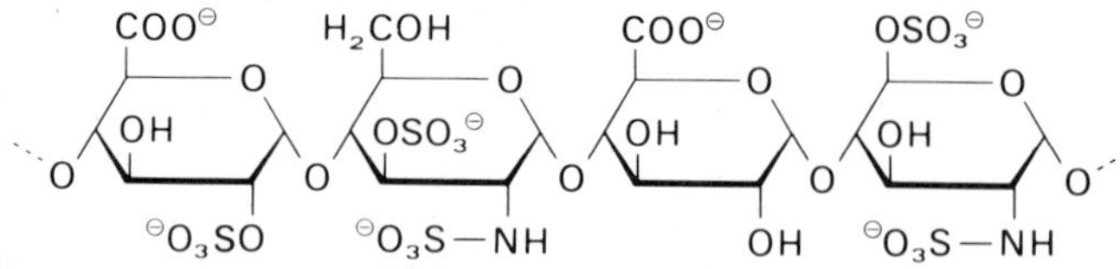

Heparin

Bacterial Cell Wall Structures. These are very large macromolecules of a complex nature; the designation "*mureins*" is being introduced for them. The fundamental building unit is the disaccharide *N*-acetylglucosamine(1 ⟶ 6)*N*-acetylmuramic acid. *Muramic acid* is the 3–*O* ether of glucosamine and lactate. The disaccharide units are linked 1 ⟶ 4-glycosidically to form a polysaccharide. The free carboxyl groups of the lactate residues form short peptide chains (muropeptides) with such amino acids as L-alanine, D-glutamate, L-lysine, and *meso*-diaminopimelate, which at the same time allow cross-linking among different polysaccharide chains. Indeed, it is possible that the entire cell wall is a single bag-shaped macromolecule (Weidel).

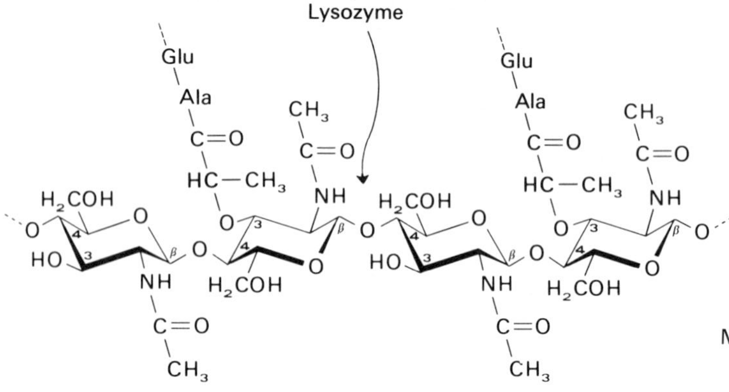

Murein

Glycosides of muramic acid are cleaved specifically by *muramidase* (= lysozyme), an enzyme common in animals. It occurs, among other places, in tears, mucus of the nose, blood plasma, and egg white. For its structure see Chapter IV,5.

Other components of cell walls are *teichoic acids*. They consist of a polyalcohol (glycerol or ribitol) and phosphate. Long chains of molecules are built entirely from phosphate diester bonds; the remaining hydroxyl groups are occupied by *N*-acetylglucosamine residues (in glycosidic linkage) and by D-alanine (in ester linkage). Teichoic acids can be interlocked with muropeptides.

Capsular Substances. Many bacteria are enveloped by a mucus capsular substance of polysaccharide nature that is responsible for serological specificity (cf. Chapter IV,8). O. Wesphal and co-workers have elucidated the structure of the capsular material of enterobacteria—*Salmonella* and *E. coli* bacteria belong to this group. To a highly polymeric chain consisting of heptose and phosphate groups are bound pentasaccharide units consisting of *N*-acetylglucosamine, 2-glucose, and 2 galactose molecules. This basic structure is common to all strains. In addition, side chains of oligosaccharide units are attached to the pentasaccharide in glycosidic linkage. The individual oligosaccharide determines the serologic specificity; it is the so-called O-antigen. Omission mutants are known that lack certain sugars in the oligosaccharide moiety. This alters the antigen specificity. In the well-known mutation from the S to the R type, the terminal unit of the basic pentasaccharide is omitted. As a result, O-specific oligosaccharides are unable to attach. Most of the R types, therefore, are serologically identical, even when originated from completely different S types.

The synthesis of the capsular material is determined genetically. It has been possible to "smuggle" hereditary factors into cells so that the bacteria become transformed from one type to another (transformation of pneumococci, Chapter VII,4).

8. Glycoproteins

Glycoproteins do not have the same structure as the mucopolysaccharides just discussed. The backbone of the macromolecule is a polypeptide chain; in other words, it is a protein molecule. The carbohydrate moiety is comprised of side chains of oligosaccharides. The types of linkage have been demonstrated to be *O*-glycoside bonds with hydroxyl groups of seryl and threonyl residues and *N*-glycosidic bonds with, among others, the amide group of asparaginyl residues. The structure can be typified by the following formula:

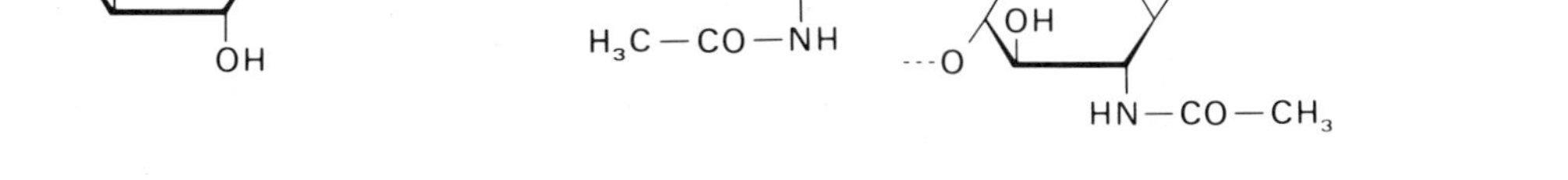

So many proteins contain small amounts of carbohydrate that a carbohydrate content of up to 4% is regarded as "normal." Only proteins exceeding this percentage are classified properly as glycoproteins. Even with this definition almost all plasma proteins, several hypophyseal hormones, and many other proteins are glycoproteins.

Glycoproteins of Blood Plasma. These glycoproteins have properties similar to those of simple proteins. They contain usually 10–25% of carbohydrate. One exception is the *acidic α_1-glycoprotein* (orosomucoid) with the high carbohydrate content of over 40%. Generally only relatively few oligosaccharide groups are bound to the protein chain. The oligosaccharides, however, are rather complex and consist each of 10–15 monosaccharides (including glucose, galactose, mannose, fucose, *N*-acetylhexosamine, and *N*-acetylneuraminate) in branched chains.

Mucoids. Components of the various mucus substances of the body, mucoids have a high content of carbohydrate (over 40%) consisting of large numbers of shorter chains of oligosaccharide. In submaxillary mucin, for example, numerous disaccharides [*N*-acetylneuraminate(2 → 6) *N*-acetylgalactosamine] are linked to seryl and threonyl residues by *O*-glycoside bonds. The abundance of short chains of carbohydrate probably determines the slimy property.

Blood-Group-Specific Substances. These substances occur in erythrocytes. The specific serological reaction with isoagglutinins ruptures red blood cells (= hemolysis), as during transfusion of incompatible blood groups.

The familiar blood groups A, B, and O are characterized by the occurrence of their antibodies in blood (see Chapter IV,10). Many more blood groups can be distinguished, however. The antibodies to substances of the blood groups arise only after blood transfusions and after repeated transfusions can cause disturbances (shock, incompatibility, hemolysis). Blood group characters are hereditary and comprise a very important group of properties in human genetics.

Certain *oligosaccharide end groups* are antigens and thus determine blood group specificity. In erythrocyte membranes the end groups are glycosphingolipid. Body fluids (saliva, gastric mucin, ovarian cysts) contain high-molecular weight glycoproteins with a high proportion of carbohydrate (up to 85%) that bear these groups terminally. The structure is shown in the following formulas:

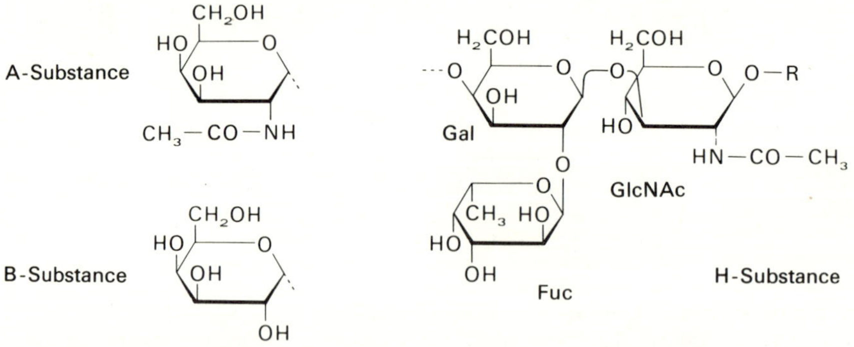

The structure may perhaps be approached from a consideration of the biosynthesis, which proceeds by stepwise attachment of the individual carbohydrate residues that must be supplied in an activated form (Section 4). Specific transferases catalyze the attachment; if they are absent as a result of a gene mutation, then the corresponding sugar is not incorporated. The formula shows that the most "primitive" oligosaccharide is the H substance found in carriers of blood type O. Persons of blood type A, in addition, possess a transferase which attaches N-acetylgalactosamine to the galactose residue of the H oligosaccharide, thereby producing blood group substance A. Carriers of type B have an analogous transferase for galactose. The blood group substance Lewis b (Le^b), formula not shown, features a second fucose, at the 4-position of the glusosamine residue, but no further substituent on galactose besides the fucose. The inherited blood group characters thus can be traced to the presence or absence of certain enzymes, which is in good agreement with our notion of the function of the gene (cf. Chapter VII,7).

The "rhesus factor" should also be mentioned. This blood group property was discovered first in monkeys (*Macaca rhesus*) and later in man. The Rh factor turned out to be a group of different substances (called C, D, and E). The genes C, D, and E, control the formation of the corresponding substances whose appearance in erythrocytes causes the latter to be *Rh positive* (common is the D substance). Erythroblastosis of the newborn may set in when the mother possesses the genetic characteristics c, d, and e but the embryo inherited the characteristic D from the father. If any of the D substance appears in the organism of the mother, then antibodies against D are formed in the maternal serum which may diffuse back into the fetal blood through the placenta and destroy the blood cells there.

BIBLIOGRAPHY

L. Glaser, Bacterial cell surface polysaccharides. *Annu. Rev. Biochem.* **42,** 91–112 (1973).

A. Gottschalk, "Glycoproteins." Elsevier, Amsterdam, 1966.

W. Z. Hassid, Biosynthesis of complex saccharides. *In* "Metabolic Pathways" (D. M. Greenberg, ed.), 3rd ed. Vol. I. Academic Press, New York, 1967.

E. C. Heath, Complex polysaccharides. *Annu. Rev. Biochem.* **40,** 29–56 (1971).

R. W. Jeanloz and E. A. Balazs (eds.), "The Amino Sugars," 2 vols. (in 4 parts), Academic Press, New York, 1965–1973: a comprehensive review of all aspects and ramifications.

W. Pigman and D. Horton, "The Carbohydrates." 2nd ed., 2 Vols. Academic Press, New York, 1970–1972.

CHAPTER **XVIII**

Topochemistry of the Cell

In preceding chapters we have repeatedly touched upon the question of control of metabolism. One critical devise of regulation is the subdivision of the cell into definite submicroscopic structures that fulfill different tasks. The enzymes catalyzing

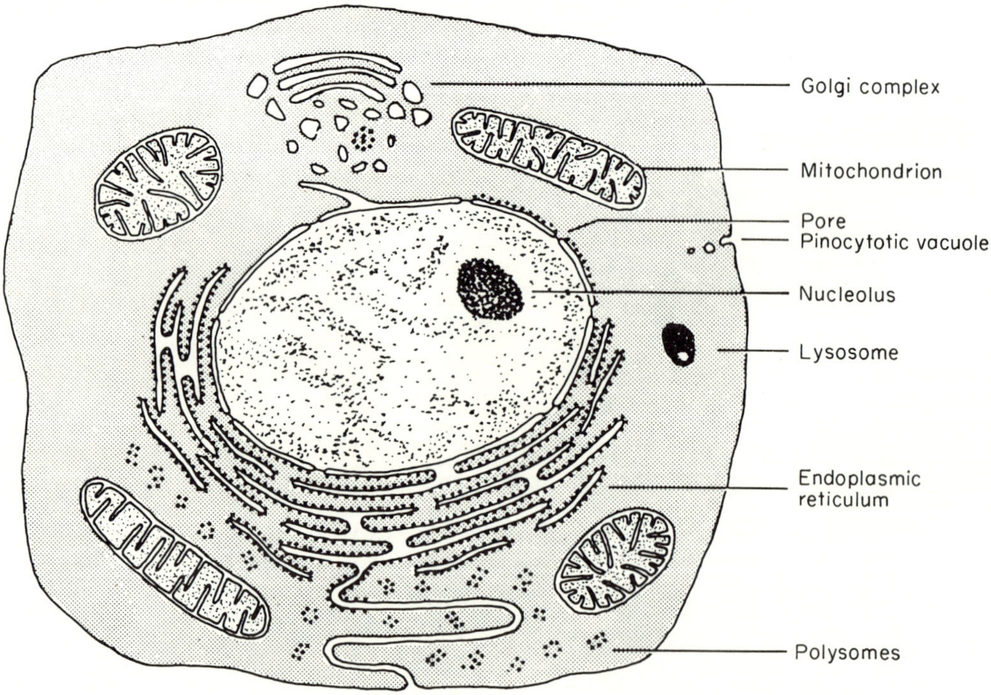

Fig. XVIII-1. Diagram of submicroscopic structures of the cell. The nucleus is continuous with the cytoplasm through some pores. The cytoplasmic space is pervaded largely by the structures of the endoplasmic reticulum, the mitochondria, and the Golgi complex. A few polysomes are shown in the hyaloplasm (according to Vogell).

the various reactions are not all (or even mostly) dissolved in the cell sap. Some are bound to definite structures, others are encapsulated in subcellular particles. Between these structures and the cytosol there exist certain permeability barriers. The modern concept that the cell consists of many different compartments is diametrically opposed to the old idea of the "protoplasm as a colloidal system." The development of electron microscopy of biological objects, particularly with the thin-section technique, finally permitted real insights into the structure of cells. And now the frontier of the science is the establishment of correlations between biochemical function and cellular morphology. A schematic diagram of the organization of the "typical" cell is shown in Fig. XVIII-1.

The electron microscopic visualization reveals very little, of course, about the biochemical function of the cell organelles. The correlation of ultrastructure and biochemical function is under very active investigation. The technique of differential centrifugation permits the isolation from cell homogenates of individual organelles or fractions of organelles. The procedure most commonly employed is shown diagrammatically in Fig. XVIII-2. The fractions obtained in this manner and purified by repeated centrifugation can then be tested separately for enzyme content and metabolic activity.

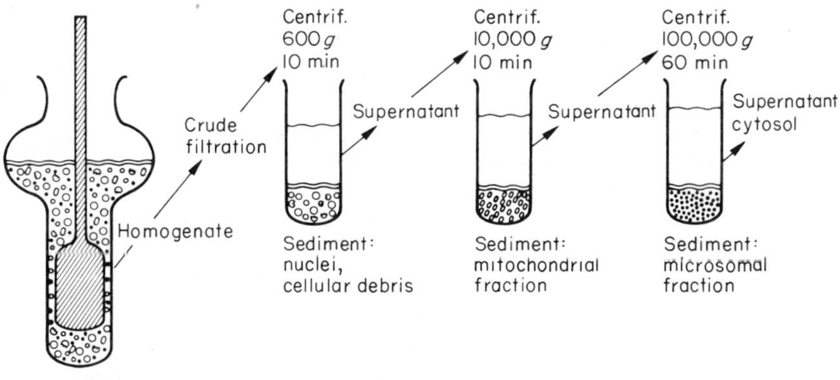

Fig. XVIII-2. Scheme of differential centrifugation. Minced tissue (e.g., liver) is suspended in buffered 0.25 *M* sucrose solution and broken up in the homogenizer (at left) between the rotating Teflon pestle and the glass wall of the vessel; shear forces disrupt the cells. The homogenate is then freed of connective tissue fibers, cellular debris, etc., by crude filtration before centrifugation first at low speed and than at higher speeds. Centrifugal force is expressed in units of gravity (*g*).

To estimate the purity of a fraction or lack of contamination with another fraction, the activity of certain "marker enzymes" has to be determined. Succinate dehydrogenase and cytochrome oxidase, for example, occur only in mitochondria. If the activity of one of these two enzymes is detected in a nuclear fraction then that is a measure of contamination of the fraction with mitochondria.

1. The Nucleus of the Cell

Chromosomes are highly helical structures consisting of *deoxyribonucleic acid* (DNA) and protein. In the customary manner of representation they can be seen only during cell division.

The fine structure of chromosomes can be studied best on giant chromosomes found in resting nuclei of the salivary glands of certain insects (flies, mosquitoes). Highly stained crossbands can be distinguished from weakly stained interbands (see Fig. XVIII-3). The crossbands consist of DNA (shown by Feulgen staining) and basic and nonbasic protein, whereas the interbands contain only nonbasic protein.

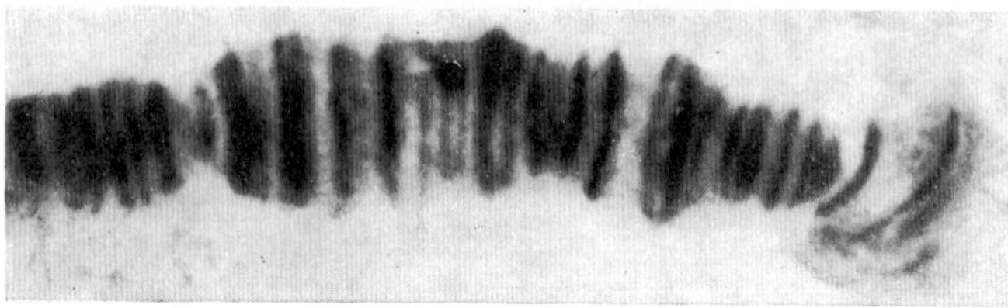

Fig. XVIII-3. Part of the first chromosome of the salivary gland of the mosquito *Chironomus tentans*. The dark crossbands are the gene loci where deoxyribonucleic acids are concentrated. (Light microscopic photograph by Dr. Clever.)

Giant chromosomes are bundles of several thousand individual chromosomes aggregated in strict order. Individual strands (chromonemata) consist probably of a structural protein to which (possibly ring-shaped) giant molecules of DNA are attached. In the resting state, DNA is packed very tightly; in the giant chromosomes it is the substance of the cross bands. The gene map of *Drosophila* has been correlated in part with the pattern of crossbands.

The study of giant chromosomes shows that RNA is synthesized at various times at the gene loci (Beermann). Protein is accumulated simultaneously at the same sites. The transition of an inactive gene to the functional state may be accompanied, in the giant chromosomes, by a *puffing* of the corresponding crossbands (the so-called puffing phenomenon). RNA is being synthesized at the sites of puffing. Usually only a relatively small number of genes is activated. In insects, it was possible to provoke puffing of a certain gene locus by the administration of a hormone (Clever and Karlson). This observation points to one mechanism by which the activity of genes can be controlled by factors from the cytoplasm.

Activity structures resembling puffs have been observed in vertebrates; they are called descriptively *lampbrush chromosomes*. Their DNA is uncoiled extensively into lateral loops. Vigorous RNA synthesis has been demonstrated at these loci. Lampbrush chromosomes have so far been seen only during the development of oocytes.

Euchromatin and Heterochromatin. The giant chromosomes are a special chromosomal form enormously interesting for both the biochemist and cytogeneticist. The nucleus in interphase does not normally reveal any defined structures, but only has stainable "chromatin." On the basis of stainability a distinction is made between eu- and heterochromatin. Structurally the difference may be that heterochromatin consists of highly compact (hyperhelical) DNA, while in euchromatin it is loosened up, or in a state corresponding to the puff regions mentioned above. Euchromatin therefore is more active in terms of RNA synthesis.

Biochemical Function of the Cell Nucleus. The nucleus above all is the site of nucleic acid metabolism including the synthesis of both DNA and RNA. Cellular division must be preceded by DNA synthesis; it occurs in the interphasic nucleus at a very definite time (at the synthetic or S phase.) No DNA synthesis takes place in the subsequent G_2 phase of the cellular cycle (cf. Chapter XXIII,9.)

The nucleus is also the site for the synthesis of nicotinamide-adenine dinucleotide (NAD). For some time this appeared to be a biochemical oddity, but now we appreciate the metabolic role of NAD in the cell nucleus. By contributing the energy held in its pyrophosphate bond, NAD effectuates the linking of two DNA fragments, i.e., it repairs a break in the DNA chain. This reaction is possibly also involved in DNA synthesis during chromosome duplication.

There are three types of RNA syntheses, namely that of mRNA, tRNA, and rRNA (see Chapter VII,3). All three types take place at DNA, but for rRNA (and probably tRNA) there is a special synthetic locus within the nucleus, called the *nucleolus*. Ribosomal proteins conceivably are synthesized there too.

Messenger RNA (mRNA) is synthesized in the nucleus at "chromatin," especially euchromatin. The requisite enzyme *RNA polymerase* is bound relatively tightly to DNA. The synthesis of specific mRNA is the first step of enzyme induction (cf. Chapters VII,7 and XX,1).

2. The Endoplasmic Reticulum

This cellular structure, closely associated with the nucleus, consists of two kinds of components; a *membrane* and small *granula*. With the electron microscope a three-dimensional network of tubules and globules with a diameter of 1000 to 1500 Å can be distinguished. The interior (cisternae of the ergastoplasm) of these structures is encased in a thin membrane. On the outside of the membranes, small granula with a diameter of 150 Å are attached. The granula are the *ribosomes* (cf. Figs. XVIII-1 and 4). The interior appears to be connected with the perinuclear space, and sometimes also with the extracellular space (through pores in the cell surface). The endoplasmic reticulum is very well developed in tissues with active protein synthesis, e.g., in exocrine cells of the pancreas.

The *cisternae* (enclosed spaces) of the endoplasmic reticulum possibly play a part in the exchange of material between the cell and the extracellular fluid. It is quite probable that this exchange of material occurs not only by diffusion and active transport, but also according to the principles of phagocytosis and pinocytosis. A small area of the cellular membrane engulfs the material, forming a vacuole, which is then transported to the interior of the cell. The cellular membrane reforms, the entering vacuole disintegrates within the cell, and its content mixes with the cytoplasm. The process is properly called pinocytosis. In the opposite direction, small portions of the membrane of the endoplasmic reticulum can leave the cell and spill their content to the exterior. It is widely assumed that the *Golgi complex* is part of the endoplasmic reticulum and specialized for this task.

In the process of homogenization the fine membranes of the ergastoplasm are torn apart, and by differential centrifugation a fraction called the *microsomal fraction* is obtained which contains the fragments of this cytoplasmic structure (originally

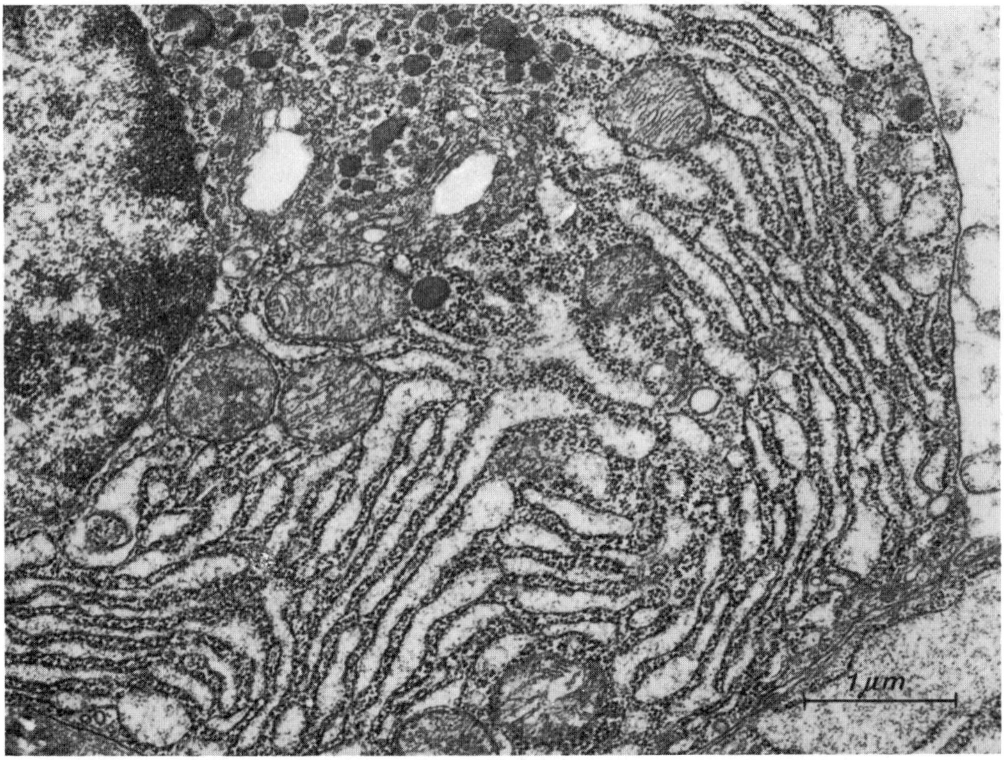

Fig. XVIII-4. Electron microscopic photograph of part of a plasma cell. The widely branched endoplasmic reticulum with adhering ribosomes (small dense bodies) can be seen clearly. On the left, part of a nucleus; next to it (near top), Golgi complex surrounded by a few mitochondria. (Photograph by Dr. Vogell.)

defined as that fraction which can be sedimented in the ultracentifuge at forces of 100,000 *g*). The microsomal fraction is heterogeneous; it comprises *ribosomes*, *lysosomes*, and fragments that reorient themselves to small spherical bodies of non-uniform size and that constitute the true microsomes in the biochemical sense. Such microsomes characteristically contain hydroxylating enzymes and a redox chain instrumental in hydroxylation reactions (see Chapter X,6). The redox chain contains a special cytochrome, a ferredoxin-like ferroprotein, and flavoproteins.

Ribosomes. Also called ribonucleoprotein particles, ribosomes are widely known as the site of protein biosynthesis. Their particle weight is about 3 million; in $Mg^{2\oplus}$-free solutions they dissociate to two subunits of 1 and 2 million particle weight which reaggregate on addition of $Mg^{2\oplus}$. (Bacterial ribosomes are a little smaller: 2.8 million; with corresponding subunits.) Each subunit contains one molecule of RNA with molecular weights of 600,000 and 1,200,000, respectively. Other constituents are

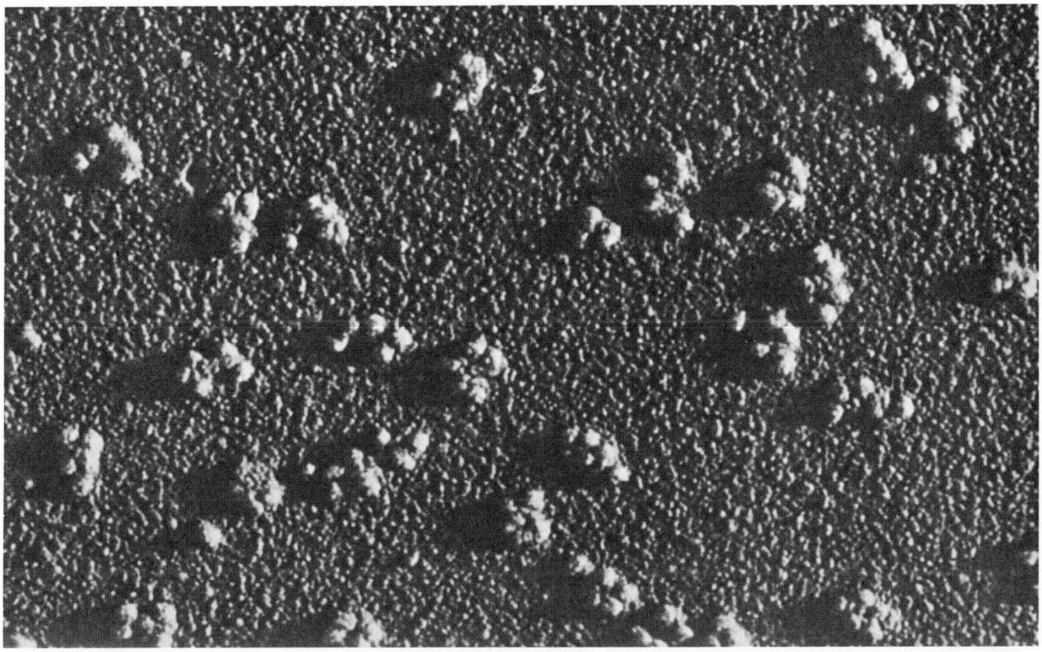

Fig. XVIII-5. Polysomes from reticulocytes. Groups of five or six individual ribosomes are held together by a molecule of messenger RNA. (Photograph by Dr. A. Rich.)

basic proteins and low-molecular weight bases whose function might well be to neutralize the nucleic acids. Ribosomes are not able to synthesize protein until they are charged with messenger RNA and thereby turned into *polysomes* (Fig. XVIII-5). The line-up of ribosomes on the endoplasmic reticulum may facilitate the process. Some individual polysomes, however, are found free in the hyaloplasm. They appear to be responsible for the biosynthesis of protein for the cell's own use, whereas the polysomes attached to the endoplasmic reticulum predominantly seem to form secretory proteins. In the pancreas, for example, these would be digestive enzymes (Chapter XXIII,1); in the liver, proteins of blood plasma.

Lysosomes. Particles of a size between that of microsomes and mitochondria, lysosomes are shaped like grana or vacuoles (de Duve). They contain many hydrolases: phosphatases, cathepsin (i.e., a protease), ribonuclease, β-glucuronidase, and perhaps others (see Fig. XVIII-6). Hydrolases are in a manner packaged there and separated from their substrates. Only when the lysosomal membrane is ruptured can the hydrolases enter the cytoplasm and exercise their specific functions. Conceivably this occurs only under conditions of cytolysis.

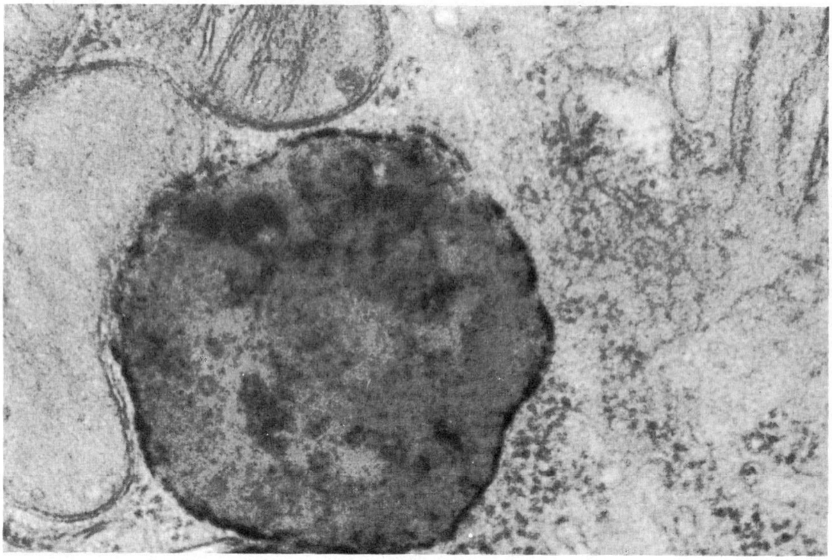

Fig. XVIII-6. Lysosomes. Electron microscopic presentation of acid phosphatase. (Photograph by Professor F. Miller.)

Peroxisomes. These subcellular organelles are rich in catalase, peroxidase, and such flavoproteins that form H_2O_2 (e.g., xanthine oxidase). Electron-microscopists have called them "microbodies." Plants contain a specialized form, the glyoxysomes, that contain the enzymes of the glyoxylate cycle (Chapter XI,4).

3. Mitochondria

Structure. The mitochondria are the most thoroughly studied of the cell components. They can be seen in the light microscope as elongated oval bodies of varying size. A thin section in the electron microscope reveals a membranous interior structure termed *cristae mitochondriales* (instead of cristae, the mitochondria of some tissues have a system of tubes or *tubuli*). Thus two structurally different spaces can be distinguished, the intracristate space and the matrix space. The latter is rich in enzymes, whereas the intracristate space contains lower molecular weight substances in solution. The two spaces are separated by a double membrane consisting of the "structure protein" of the mitochondria and of phospholipids. This membrane, too, is a boundary of variable area subject to, among others, osmotic changes.

Mitochondria swell extensively in a hypotonic medium; they contract again in a hypertonic medium—faster with the addition of ATP. It is possible that the variable degrees of swelling, observed in thin sections of fixed cells, are entirely physiological. Perhaps the exchange of metabolites is thereby facilitated. Mitochondria, nevertheless, do have a system for the active transport primarily of $Ca^{2\oplus}$ and $Mg^{2\oplus}$ ions. The requisite energy evidently is derived directly from energy-rich intermediates of oxidative phosphorylation.

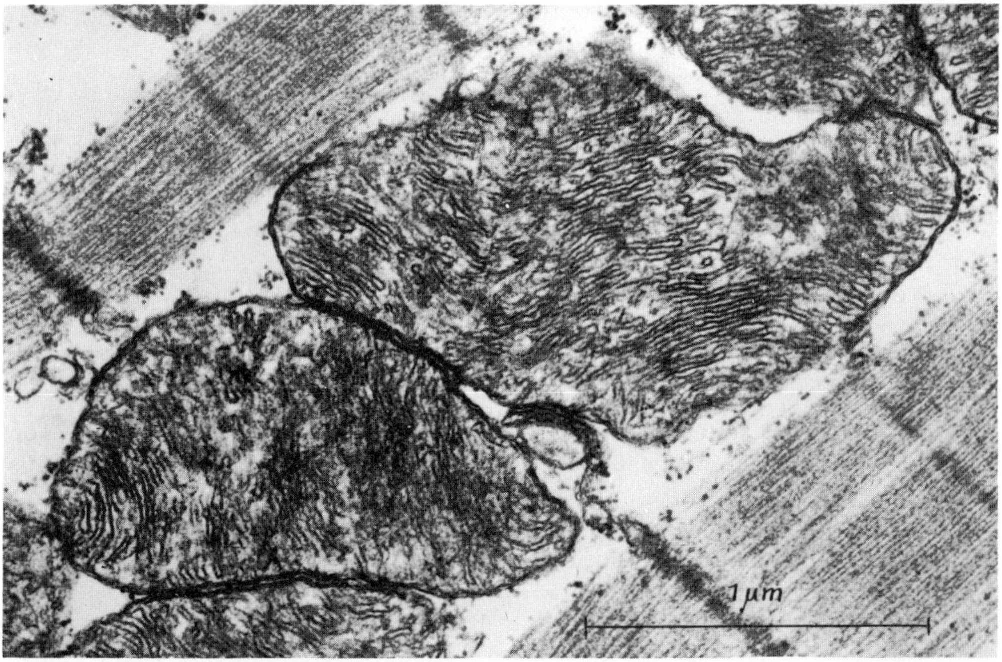

Fig. XVIII-7. Mitochondria of a muscle cell (heart muscle of the rat). The mitochondria are densely packed with "cristae mitochondriales." (Photograph by Dr. Vogell.)

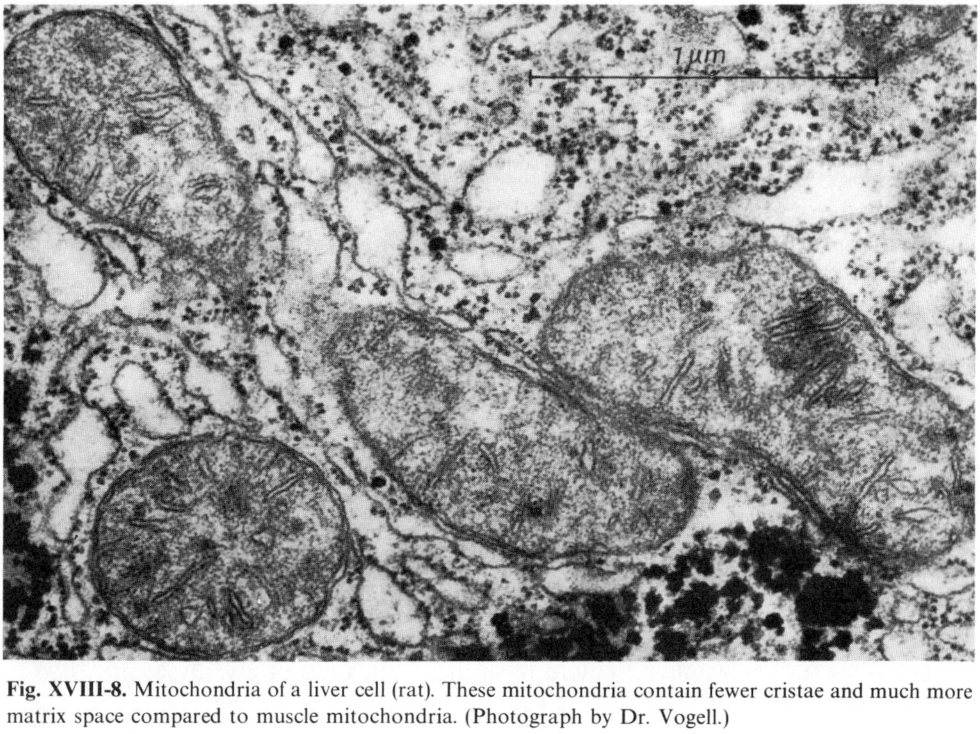

Fig. XVIII-8. Mitochondria of a liver cell (rat). These mitochondria contain fewer cristae and much more matrix space compared to muscle mitochondria. (Photograph by Dr. Vogell.)

The interior structure of mitochondria differs from organ to organ. As a general rule, mitochondria are full of lamellae if their main job is respiration and energy production (as, for example, in aerobically working muscles). However, they have fewer internal membranes (as in liver cells) if they contain many enzymes with synthetic tasks (Palade; Vogell and Klingenberg; see Figs. XVIII-7 and 8).

Mitochondria contain a small amount of deoxyribonucleic acid (DNA) which probably contains the code for the "structure protein," as well as special ribosomes that resemble bacterial ribosomes. This observation suggests that mitochondria have evolved from intracellular symbionts. The capacity for self-reproduction supports the suggestions. Most mitochondrial enzymes, however, are synthesized on cytoplasmic ribosomes and are transferred from there to the mitochondria.

Biochemical Function. Mitochondria contain numerous enzyme systems. The foremost is the respiratory chain. Its localization in the inner membranes and the significance of its structure for the redox chain itself and for respiratory chain phosphorylation has been discussed extensively in Chapter X (near the end of Section 4; also Fig. X-3). Disruption of the membrane structure abolishes oxidative phosphorylation.

Cytochrome oxidase and succinate dehydrogenase are some of the marker enzymes of mitochondria. The remainder of the mitochondrial enzymes exist in the matrix space. This includes primarily enzymes of β-oxidation of fatty acids, of oxidative decarboxylation, and of the citrate cycle. The close proximity of these enzymes to the respiratory chain appears to be extremely meaningful physiologically. The supply of these enzymes with substrate is described below in Section 4 (see also Fig. XVIII-9).

4. The Cytoplasmic Space (Cytosol)

The ground substance or "hyaloplasm" (an expression used in electron microscopy) contains no discernible structures. Possibly it is pervaded by extremely fine fibrils that still defy resolution in the electron microscope. The other term *cytosol* is generally defined as the unstructured, "soluble" portion of the cell. The "supernatant" that remains after centrifugation of the particulate fraction of a homogenate contains all the enzymes and substrates of the cytoplasm.

The content of the endoplasmic reticulum becomes intermixed, however, with the original cytoplasm during the preparation of the soluble fraction in the centrifuge. With present biochemical methodology it is impossible to investigate these two cellular compartments separately. Often, however, especially the interaction between mitochondria and cytoplasm is important, in which case the "soluble fraction" is taken as the "hyaloplasm," undifferentiated electron microscopically.

Enzymes. The soluble fraction of the cell homogenate, the cytosol, contains all the enzymes of glycolysis. Every observation supports the fact that the Embden-Meyerhof pathway operates here. The possibility cannot be excluded, however, that the membranes of the endoplasmic reticulum have many enzymes of glycolysis adsorbed to themselves. If this is the case, then glycolysis is also structure-bound.

Exchange of Substrates between Mitochondria and Cytosol. The inner mitochondrial membrane is an effective permeability barrier which can be passed by few substances. At the same time mitochondria are the location of important biochemical degradative processes, and it is to be expected that the substrates of β-oxidation, citrate cycle, and others penetrate the mitochondria and that NADH having been generated in the cytoplasm is oxidized in the mitochondria. As illustrated in Fig. XVIII-9, there exist regular metabolic cycles for the special purpose of shuttling transport metabolites in and out of mitochondria.

The exchange of coenzyme-bound hydrogen (NADH) takes place via *malate* as auxiliary substrate. The latter enters the mitochondria and releases its hydrogen there to NAD. The resulting oxaloacetate, however, cannot pass through the membrane and must first undergo transamination with glutamate to complete the cycle. The products aspartate and α-ketoglutarate leave the mitochondria and permutate in the cytosol to oxaloacetate and glutamate. Oxaloacetate is reduced to malate, and glutamate migrates back into the mitochondria. The entire shuttle traffic can proceed in either direction: depending on the metabolic situation hydrogen is transported either into or out of mitochondria.

Hydrogen of NADH having been generated on the inside can, of course, enter the respiratory chain where it is oxidized to water. This reaction generates 3 moles ATP by oxidative phosphorylation. ATP is linked with the external ADP-ATP system (see Chapter X,5).

Long-chain fatty acids are channeled into mitochondria as substrate for β-oxidation. As described in Chapter XII,3, this takes place as carnitive derivatives.

The main substrate of mitochondria derived from carbohydrate metabolism is *pyruvate*, which enters mitochondria unhindered and is oxidized there to acetyl-CoA. This oxidative decarboxylation is also a key reaction for the conversion of carbohydrate to fat. Since fatty acid synthesis is an extramitochondrial process, activated acetate has to be brought out again. This is achieved by condensation with oxaloacetate to form *citrate*; citrate leaves the mitochondria and is split again in the cytosol to acetyl-CoA and oxaloacetate by action of ATP-citrate lyase (also known as "citrate cleavage enzyme"). The reaction consumes one mole of ATP placing the equilibrium position of the reaction far on the side of cleavage:

$$\text{Citrate} + \text{ATP} + \text{CoA} \longrightarrow \text{Acetyl-CoA} + \text{oxaloacetate} + \text{ADP} + \text{Ⓟ}_{\text{inorg}}$$

Oxaloacetate is reduced again to malate which in turn passes the mitochondrial membrane. The transport of citrate is coupled to that of malate. For each mole of citrate exported one mole of malate has to be imported into mitochondria, which would otherwise starve for C_4 dicarboxylic acids.

Only the most important transport processes have been described here. In several special metabolic pathways (for example, in the biosynthesis of heme or of steroid hormones) certain individual reaction steps actually take place inside mitochondria with the rest of the pathway being outside. Here, too, starting material is brought in and products are transported out.

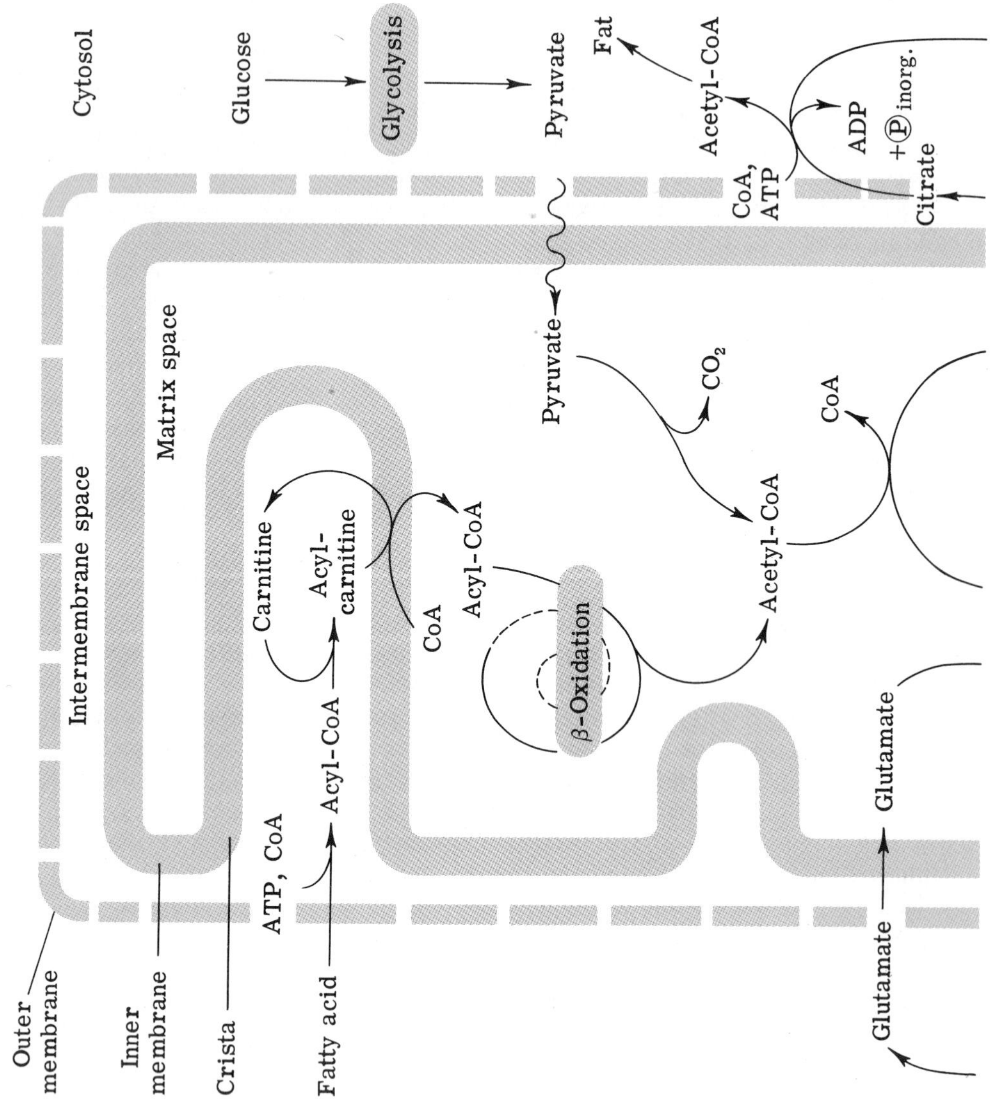

Cytosol
Glucose
Glycolysis
Pyruvate
Fat
Acetyl-CoA
CoA, ATP
ADP
+Ⓟinorg.
Citrate
Intermembrane space
Matrix space
Pyruvate
CO_2
CoA
Acetyl-CoA
Carnitine
Acyl-carnitine
Acyl-CoA
CoA
β-Oxidation
ATP, CoA
Acyl-CoA
Fatty acid
Glutamate
Glutamate
Outer membrane
Inner membrane
Crista

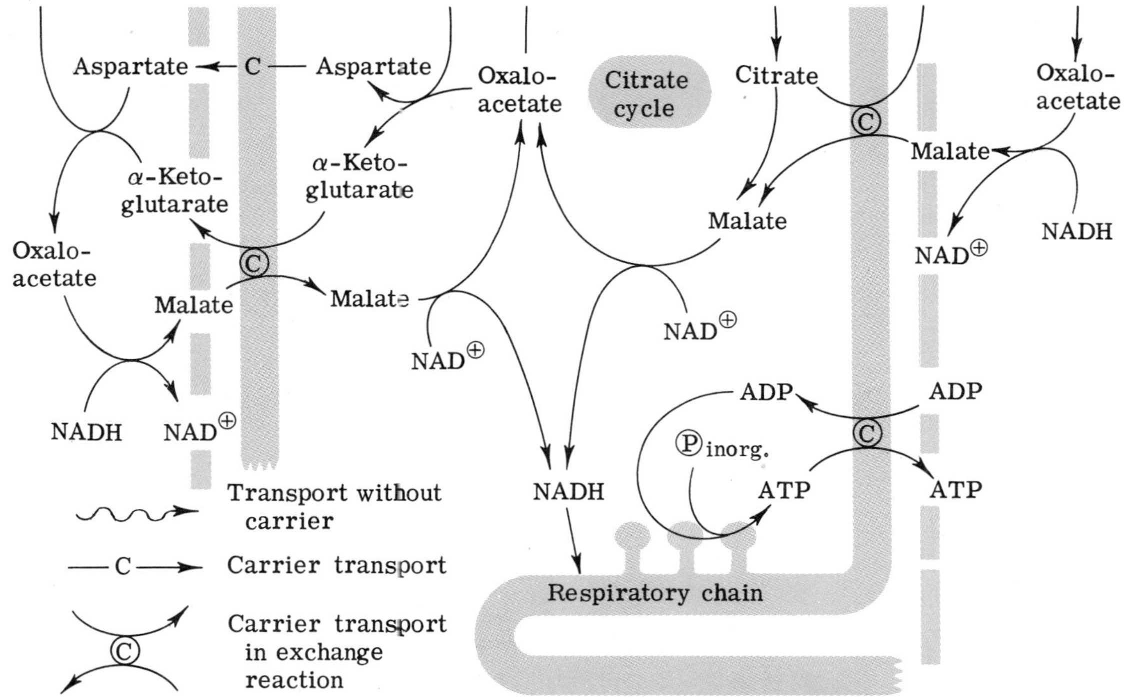

Fig. XVIII-9. Several key metabclic relationships between mitochondria and cytosol. At left the malate-aspartate-glutamate shuttle is depicted; it serves to transport hydrogen on the level of NADH across the mitochondrial membrane, which is impermeable to NADH itself. At right the exportation of acetyl-CoA is shown mediated by citrate and requiring additional ATP. For the exportation of ATP (not shown here) see Chapter X,5.

The foregoing transport processes and associated metabolic steps open the possibility for numerous sites of metabolic control, which will be the subject of the next chapter.

A. D. Bangham, Lipid bilayers and biomembranes. *Annu. Rev. Biochem.* **41,** 753–776 (1972).
G. H. Bourne, "Division of Labor in Cells," 2nd ed. Academic Press, New York, 1970.

CHAPTER XIX

Regulatory Mechanisms and Correlations in Intermediary Metabolism

The two main tasks of intermediary metabolism are: (1) to provide all the intermediary compounds necessary for the synthesis of specific endogenous substances, and (2) to supply all the chemical energy, in form of ATP, that is required for such biosynthesis as well as for the maintenance of complex structures and for special functions, e.g., work performed by muscles. These tasks have to be accomplished moreover under conditions of widely fluctuating food supply both qualitatively and quantitatively. Clearly this cannot be achieved by the random and uncontrolled running of enzymatic reactions.

The answer lies in effective regulation of virtually all key reactions of metabolism and in the phenomenon of self-regulation of all major metabolic pathways and cycles, which adapts the supply of end product to its consumption. The following section will describe the various mechanisms involved in the control of intermediary metabolism while subsequent sections will deal with interrelationships and regulation of each major pathway.

1. Principles of Metabolic Regulation

An uncontrolled side-by-side of enzymatic reactions would lead to the idle running of many metabolic reactions characterized by a waste of energy. This can be illustrated with the following example from the metabolism of carbohydrate.

In the metabolism of glucose the liver cell is faced with two jobs. First, glucose abundantly supplied in the diet must be stored as glycogen. To this end glucose is converted to glucose 6-phosphate (glucokinase reaction), which via glucose-1-phosphate and UDP-glucose is incorporated into glycogen. Second, at times of insufficient glucose supply the cell must synthesize glucose from amino acids (called gluconeogenesis) and transport it to the blood stream. In gluconeogenesis, the initial product is glucose 6-phosphate which then by action of glucose-6-phosphatase yields glucose and inorganic phosphate.

If the two enzymes glucokinase and glucose-6-phosphatase were to exist side-by-side in the cytosol then the two reactions would amount merely to ATP cleavage without any net effect:

$$\text{Glucose} + \text{ATP} \xrightarrow{\text{glucokinase}} \text{Glucose-6-}\text{Ⓟ} + \text{ADP}$$

$$\text{Glucose-6-}\text{Ⓟ} \xrightarrow{\text{phosphatase}} \text{Glucose} + \text{Ⓟ}_{\text{inorg}}$$

$$\text{Sum:}\quad \text{ATP} \longrightarrow \text{ADP} + \text{Ⓟ}_{\text{inorg}}$$

Such wasteful cycles can be avoided only by sensitive regulation of metabolism. In the case illustrated, the regulation is achieved in part by spatial separation or compartmentation of the participating enzymes.

Compartmentation. In the preceding chapter we have seen that the cell is compartmentalized into numerous specialized regions. This is an essential feature for the separate regulation of synthetic and degradative pathways that often proceed via the very same intermediates. The breakdown of fatty acids, for example, proceeds in the mitochondria, while the formation of fatty acids takes place in the cytoplasm on the multienzyme complex. The complex could be regarded as a microcompartment with minimum space. In the example of the preceding paragraph the idling reaction (the ATPase effect) is avoided by the circumstance that glucokinase occurs in the cytosol and glucose-6-phosphatase is bound tightly to the endoplasmic reticulum. Glucose arising from the action of the phosphatase is transported immediately into the cisternae of the reticulum and from there out of the cell.

Regulation by Limiting Metabolites. A single key metabolite of a pathway sometimes can limit the overall turnover of a substrate. A good example is the *control of respiration in mitochondria.* The respiratory chain under normal metabolic conditions is not operating at maximal capacity. The uptake of oxygen by mitochondrial preparations can be raised considerably, e.g., by the addition of uncoupling agents, such as dinitrophenol or dicumarol, which disconnect the process of phosphorylation from electron transport and thus shut off phosphorylation (cf. Chapter X,5). From this and other observations it can be deduced that electron transport is not the limiting element, but rather it is phosphorylation. There is not enough ADP available.

This is a self-regulating mechanism. The more ATP is used up by the cell—for work by the muscle, or for synthesis—the more ADP is released, which is reused by the respiratory chain to drive phosphorylation ahead and to produce more ATP. In this manner, respiration (i.e., the consumption of oxygen) adapts itself to the energy requirements of the cell.

Dynamic Equilibria and Pacemaker Reactions. Ordinarily a dynamic equilibrium encompasses the entire sequence of reactions of a longer degradative pathway. Such a steady state equilibrium is characterized by relatively constant or *stationary concentrations* of the individual metabolites (see Table XIX-1) but also by *changing reaction rates* (the rates depend on changing conditions). In such a system, some reactions serve as *pacemakers*. In the Embden-Meyerhof pathway, for example, it is the phosphofructokinase reaction, i.e., the phosphorylation of fructose 6-phosphate to fructose 1,6-bisphosphate. It is practically irreversible (because it is highly exergonic), whereas the steps before and after are reversible, and the concentrations of the metabolites involved approach equilibrium concentrations. Pacemaker enzymes such as this are key control points of metabolism since any influence on them (activation or inhibition) also involves preceding or succeeding steps. These other steps adapt themselves to the new flow rates at the control points.

TABLE XIX-1

Concentrations in Rat Liver of Some Metabolites of Carbohydrate Breakdown

Metabolite	Conc.[a]	Metabolite	Conc.[a]
Glucose	8.6	3-Phosphoglycerate	0.268
Glucose 6-phosphate	0.370	2-Phosphoglycerate	0.040
Fructose 6-phosphate	0.075	Phosphoenolpyruvate	0.130
Fructose 1,6-bisphosphate	0.022	Pyruvate	0.145
Glycerol 1-phosphate	0.253	Lactate	1.450
Dihydroxyacetone phosphate	0.038		

[a] In micromoles per gram fresh weight. This unit is roughly comparable to the concentration micromoles per milliliter (i.e., m*M*) because the liver contains about 70–80% of water. At the site of enzyme activity the concentration may actually be higher by a factor of 2.

Regulation by Michaelis Kinetics. The steady state concentrations mentioned above (see also Table XIX-1) generally are within the range of Michaelis constants of the enzymes involved. This fact alone constitutes one form of self-regulation of metabolism (albeit not a very effective one). Should the concentration of metabolite rise unduly then the reaction rate is raised as well, as can be seen at a glance from the Michaelis curve (Fig. V-3, Chapter V,7). The metabolite will be converted faster; there will be less of it and the original concentration of the dynamic equilibrium will be approached.

A similar regulative function is frequently seen in *product inhibition* of an enzyme. The product of a reaction may accumulate if it is not transported away fast enough or converted further; but the accumulation of product results in a slowing down of the reaction if the enzyme is inhibited by the product. This phenomenon also contributes to the restoration of the original equilibrium condition.

Allosteric Control. Over key enzymes, allosteric control is one of the most effective mechanisms for regulating metabolism. As explained in Chapter V,8 (near the end), allosteric control is the alteration of enzyme activity by a metabolite that binds to the enzyme protein at a site other than the active center. Enzyme activity is influenced by a change of conformation of the enzyme. Several examples of allosteric control have already been mentioned (cf. citrate cycle, Chapter XI,3) and more will be seen later in this chapter. The example given immediately below also involves this mechanism of metabolic control.

Regulation by Negative Feedback. Feedback in this context may be defined as the influence of an end product on a step at the beginning of the chain of reactions. This type of control mechanism, although more complicated, is disproportionately more effective than any mechanism discussed so far. The principle of feedback control is of fundamental significance for the regulation of metabolism. The tool in this kind of mechanism is *allosteric inhibition* of a key enzyme. Our example will be the well-studied bacterial synthesis of the amino acid isoleucine.

The synthesis of isoleucine starts with threonine and proceeds over a total of five steps. The first step is deamination of threonine to give α-ketobutyrate, which in the next step undergoes condensation with active acetaldehyde. A rearrangement yields the α-keto analog and finally isoleucine itself.

Now the aim is to adapt the production of isoleucine to the requirements. The latter can change radically. Possibly a large amount of isoleucine suddenly becomes available in the nutrient medium. Endogenous synthesis at this point would be superfluous. Moreover, isoleucine can be broken down extensively for the extraction of energy. On the other hand, it is equally possible that isoleucine is completely absent from the medium and endogenous synthesis becomes limiting for protein synthesis and hence for cell growth as a whole.

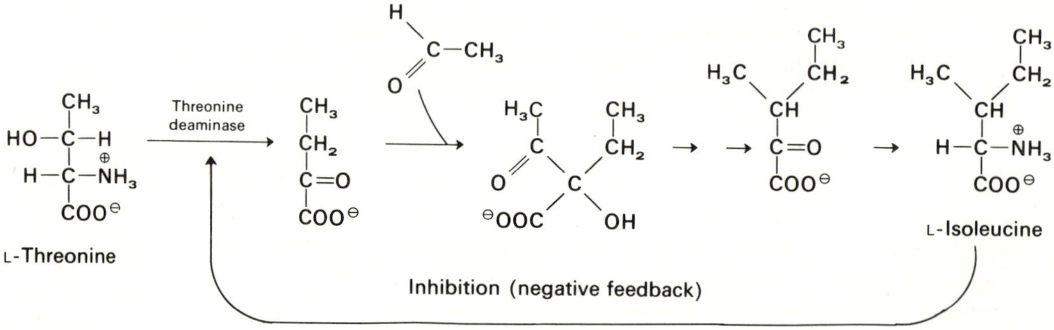

Adaptation to the cell's requirements in this case is effected by the feedback influence of product on the first enzyme of the biosynthetic pathway. Isoleucine acts as *allosteric inhibitor* (for this concept see Chapter V,8) of threonine deaminase. With an abundance of isoleucine the first synthetic step is inhibited and subsequent intermediate products do not have a chance to accumulate. (This points out the essential difference with the mechanism of product inhibition, which achieves the same result, but only for one step. Every intermediate would have to pile up. This is patently uneconomical.) This mechanism is, of course, self-regulatory in that inhibition vanishes with the consumption of end product and synthesis resumes.

Regulation by Enzymatic Modification of Enzymes. The classical example for this control mechanism is glycogen phosphorylase whose inactive form *b* is converted by phosphorylation with ATP into the active form *a*. The conversion is carried out by a special *control enzyme*. A phosphatase can then remove the phosphate group and thus return the enzyme to its original, relatively inactive form (see Chapter XVII,6). Similar to allosteric control, this mechanism is also capable of reversible activation or deactivation of a particular enzyme. Two important differences, however, must be pointed out: The mechanism allows for only two discrete states that are not related by an equilibrium reaction. Second, the new level of enzyme activity is maintained constantly until a new signal reverses the reaction, in our case the dephosphorylation of the phosphorylase.

An increasing number of enzymes has been found to fall under this regulatory mechanism. More will undoubtedly be found when such mechanisms will be looked for systematically.

Regulation by Changed Levels of Enzyme. The amount of catalyst has no effect whatsoever on a true thermodynamic equilibrium. This is not so with a steady state or a dynamic equilibrium; increasing the amount of enzyme (especially a pacemaker enzyme) generally increases the turnover of material. This control mechanism is indeed met with in metabolism. Bacterial systems have been studied particularly well. A number of enzymes are induced by their substrates, i.e., the synthesis of new enzyme commences only with the appearance of substrate (see "Enzyme Induction," Chapter VII,7).

Enzyme induction plays a significant part also in mammals. In some cases enzymes are induced—as with bacteria—by substrates; much more frequently, however, they are induced by *hormones*, which in this manner can affect metabolism. We will return to this in the following chapter under the title "Mechanism of Action of Hormones."

The focus in the following sections will be on the correlation between metabolic chains of reactions. These are presented in one large foldout table included at the back of this book.

2. Metabolism of Carbohydrates

Glucose. In free form, glucose circulates in blood at a normal concentration of 5 m*M*. It is taken up by cells and phosphorylated in the cytosol to glucose 6-phosphate, which also arises *via* glucose 1-phosphate during the mobilization of glycogen reserves by *phosphorylase*. Glucose 6-phosphate is the metabolically active form of glucose and thus can enter a number of different pathways.

For the formation of intracellular glycogen, glucose 6-phosphate has to be converted to glucose 1-phosphate and then activated with UTP to UDP-glucose. This activated form is also employed for the rearrangement of glucose to other hexoses (upper left in the foldout chart) and for the synthesis of diacylglycerols and glucuronides. By reversal of this pathway, *galactose* can enter the degradative pathway.

Pentose Phosphates. From the direct oxidation of glucose 6-phosphate with NADP[1] pentose phosphates are formed. The reaction serves primarily the production of NADPH which is required for the synthesis of fatty acids and cholesterol as well as for hydroxylation reactions. Pentoses can also arise from a nonoxidative pathway from fructose 6-phosphate via the pentose phosphate cycle. The pentoses, of course, are critical building blocks of nucleic acids.

Embden-Meyerhof Degradation. Quantitatively this degradative pathway is the most important one, at least in many organs. It proceeds from fructose 6-phosphate to fructose 1,6-bisphosphate. Fructose-6-phosphate kinase is the controlling enzyme in the Embden-Meyerhof pathway because this phosphorylation is essentially irreversible. The key steps of the breakdown (also called *glycolysis*) are shown on the foldout chart. At the conversion of glyceraldehyde 3-phosphate to 3-phosphoglycerate a substrate chain phosphorylation step is interposed. The end product is *pyruvate*, which anaerobically is transformed to *lactate*, but aerobically to *acetyl-CoA*.

Gluconeogenesis (Reversal of Glycolysis). The starting material for the synthesis of glucose is primarily lactate, which arises during anaerobic glycolysis, and secondarily, a group of amino acids derived from proteins. The key substance for both pathways is phosphoenolpyruvate. For its formation, a detour over *oxaloacetate* is adopted. Oxaloacetate arises directly from pyruvate by the action of a $^{\ominus}$OOC ~ biotin enzyme (i.e., with the consumption of ATP) and is phosphorylated and at the same time decarboxylated to yield phosphoenolpyruvate. The reaction then proceeds along the reverse of the glycolytic pathway, requiring only one additional ATP at the reductive step to form triose phosphate. In the transition fructose 1,6-bisphosphate → fructose 6-phosphate, inorganic phosphate is split off since the transfer of this phosphate to ADP is precluded energetically.

Minor Pathways. Dihydroxyacetone phosphate can be transformed to glycerol phosphate, which is used for the synthesis of fats and glycerophosphatides. Phosphoglycerate can be converted to 3-phosphohydroxypyruvate and further to hydroxypyruvate. This last compound is convertible to active glycolaldehyde (by decarboxylation with thiamine pyrophosphate), and then to glycolate and possibly to glyoxylate, and finally by this route to glycine. Pyruvate can undergo transamination and become alanine; hydroxypyruvate becomes serine.

---

[1] By official recommendation (see Chapter VI,4) the two hydrogen-transferring coenzymes, formerly known as di- and triphosphopyridine nucleotide (abbreviated as DPN and TPN) are to be called nicotinamide-adenine dinucleotide (NAD) and nicotinamide-adenine dinucleotide phosphate (NADP). The new nomenclature offers greater correctness and consistency, among other advantages.

One way to indicate the oxidized and reduced forms of the coenzymes is $NAD^{\oplus}$ (or $NADP^{\oplus}$) and $NADH + H^{\oplus}$ (or $NADPH + H^{\oplus}$), respectively. Generally we have abbreviated this throughout the text and in the foldout chart in the back of this book to NAD (or NADP) for the oxidized and NADH (or NADPH) for the reduced form.

Regulation. Since glucose 6-phosphate can undergo four different reactions some definite control has to be exercised so that depending on the need one or the other pathway is followed. The control operates in the following fashion:

When the *pentose phosphate cycle is active* the intermediates 6-phosphogluconate, sedulose 7-phosphate, and erythrose 4-phosphate inhibit phosphohexose isomerase, i.e., they exert a negative control on the first enzyme of glycolysis. The effect is a limitation on the amount of material passing through the Embden-Meyerhof pathway. In this situation the operation of the pentose phosphate cycle makes available NADPH for the synthesis, for example, of fatty acids if simultaneously there is sufficient ADP. An accumulation of long-chain acyl-CoA derivatives, on the other hand, inhibits glucose-6-phosphate dehydrogenase. This latter control prevents excessive fat synthesis.

After a meal a great deal of glucose reaches the liver and is converted there to glucose 6-phosphate, which is in equilibrium with glucose 1-phosphate. A high level of glucose 6-phosphate activates the enzyme glycogen synthetase, thereby enhancing the formation of glycogen as reserve store.

The key enzyme for *glycolysis* (the Embden-Meyerhof pathway) is fructose-6-phosphate kinase. Even though ATP is a substrate this enzyme is inhibited by higher concentrations of ATP and reactivated by ADP and AMP. Under aerobic conditions more ATP is released by the respiratory chain which then by inhibiting the kinase limits the amount of glucose 6-phosphate flowing through the Embden-Meyerhof pathway. Under anaerobic conditions, ATP is used up relatively quickly by other metabolic reactions concomitantly producing ADP, which in turn activates phosphofructokinase. With that substrate chain phosphorylation is stimulated.

The Pasteur Effect. The regulation of the Embden-Meyerhof pathway by its adaptation to the prevailing levels of ATP provides an explanation also of the phenomenon of metabolic control observed over 100 years ago by Pasteur. With a deficiency of oxygen, many cells, e.g., muscle or yeast cells, derive the necessary energy from the anaerobic breakdown of carbohydrate. As soon as oxygen is made available, however, glucose is broken down aerobically. What one observes is a drastic decline in glucose utilization which is but an adaptation to the energy needs of the cells. (The aerobic breakdown, of course, yields 38 moles of ATP per mole of glucose, in contrast to the 2 moles derived from the anaerobic breakdown.) The limitation of the glucose turnover can now be accounted for largely by the allosteric regulation of phosphofructokinase.

3. Metabolism of Fatty Acids

The Breakdown of Fats (in the chart, upper right). This process begins with hydrolysis to fatty acids and glycerol. The latter is closely related to the trioses. The further breakdown of the fatty acids demands activation first (consumption of ATP and combination with coenzyme A to give the thioester) and then proceeds through dehydrogenation (by a flavoprotein), addition of water, and a second dehydrogenation to the β-keto acid. The whole process is commonly known as β-oxidation. Thioclastic cleavage of the activated β-keto acid releases acetyl-CoA and a shorter fatty acid in its activated form, which then can undergo the same sequence of reactions several times, until the entire carbon chain is broken down to C_2 fragments.

Modifications of the breakdown are necessitated by methyl side chains. Branched-chain fatty acids arise also from the amino acids leucine, isoleucine, and valine (cf. Chapter XII,4). Propionyl-CoA (instead of acetyl-CoA) is formed from isoleucine and valine, and this is converted to succinate by carboxylation and isomerization.

Acetyl-CoA, the Main Crossing Point of the Metabolism of Carbon. Acetyl-CoA, active acetate, arises partly from oxidative decarboxylation of pyruvate and partly from β-oxidation of long-chain fatty acids and several amino acids. It is no exaggeration to call active acetate a metabolite of cardinal importance. By being converted to acetyl-CoA the various intermediary metabolites all enter the final pathway of breakdown, known as the citrate cycle (see below), and are combusted there in the "common terminal pathway."

The energy content of acetyl-CoA, i.e., its free energy of hydrolysis is a little higher than that of ATP. Yet, this reactive thioester cannot be considered to be the universal reservoir of energy. Rather, acetyl-CoA, is the reactive intermediate that provides the C_2 fragment for so many conversions and syntheses.

Syntheses with Active Acetate. Acetyl-CoA is an important starting material for biosynthetic reactions. First, it is available for diverse ester and amide syntheses. These reactions yield such important products as acetylcholine, *N*-acetylglucosamine, etc.

A second important anabolic pathway of acetyl-CoA produces the *isoprenoid lipids* especially the *steroids*. Three molecules of acetyl-CoA condense at first to form a branched-chain compound, hydroxymethylglutaryl-CoA. With a superabundance of acetyl-CoA, such as occurs in some pathological metabolic conditions (like diabetes, cf. Chapter XX,11), acetoacetate can be formed from hydroxymethylglutaryl-CoA (ketogenesis). Normally, however, the reduction of the thioester group of that compound yields mevalonate which is then converted to isopentenyl pyrophosphate with an expenditure of 3 moles of ATP. The subsequent synthesis of squalene and cholesterol does not require any further supply of energy beyond NADPH.

In addition to steroids, the carotenoids and the various terpenoids are formed by analogous routes in plants and microorganisms. The side chains of the lipoid quinones (vitamins E and K, ubiquinone, and plastoquinone) also arise in this fashion in some mammalian organisms.

Formation of Fatty Acids. Acetyl-CoA is the starting material for the biosynthesis of fatty acids and consequently of the neutral fats. This pathway is heavily used in the conversion of carbohydrate to fat.

Superficially the synthetic steps appear to resemble the degradative steps, but in fact they differ in a number of important details. In synthesis malonyl-CoA, and not active acetate, reacts with activated fatty acids for chain elongation yielding β-keto acids which then give rise to saturated fatty acids. Some other differences are: The use of NADPH in the two reductive steps of synthesis and NAD and FAD in the corresponding dehydrogenation steps of β-oxidation. The β-hydroxyacyl derivative is in the D-configuration in the synthetic sequence and in the L-configuration during β-oxidation. The dehydrogenases are specific for these intermediates and hence are different, too. The location of synthesis within the cell is the cytoplasm, whereas β-oxidation generally takes place in the mitochondria. The most striking difference comes from the finding that all the intermediate substrates of the synthesis are complexed with a protein by thiolester bonds (a multienzyme complex in yeast and avian systems; an acyl-carrier protein in bacterial systems), whereas the intermediates of breakdown are all derivatives of CoA. Finally at the C_{16} or C_{18} stage, the fatty acids are released from the complex as CoA derivatives (cf. Chapter XII,6).

The formation of neutral fats and phospholipids out of acyl-CoA compounds requires glycerol phosphate, the alcohol component of fats. This compound is derived from the metabolism of carbohydrates. The phosphatidic acids are intermediates in the synthesis of neutral fat (= triacylglycerol; cf. Chapter XII,6) and phospholipids (= phosphatides; cf. Chapter XIII,3). In the latter case, the phosphatidic acids are activated by CTP and coupled to serine. Decarboxylation and methylation (with adenosylmethionine) finally results in phosphatidylcholine ("lecithin").

Regulation of Fatty Acid Synthesis. The conversion of carbohydrate to depot fat takes place primarily in the adipose tissue whose enzymatic equipment is characterized by high levels of glucose oxidase for the preparation of NADPH.

Acetyl-CoA arises from pyruvate by oxidative decarboxylation in the mitochondria. In order to be transported to the cystosol it is condensed with oxaloacetate to form citrate, which can leave the mitochondria and which can be cleaved again in the cytosol by ATP-citrate lyase (see Chapter XVIII,4 and Fig. XVIII-9).

The cleavage yields acetyl-CoA which can enter fatty acid synthesis, and oxaloacetate which can be reduced again to malate and as such either reenter the mitochondria or be dehydrogenated by malate dehydrogenese (NADP) to form pyruvate and CO_2. This last reaction can yield as much as 35–40% of the total NADPH required for fatty acid synthesis.

The pacemaker enzyme of the biosynthesis of fatty acids is *acetyl-CoA carboxylase*, the biotin-containing enzyme which catalyzes the carboxylation to form malonyl-CoA. It is *activated allosterically by citrate* so that fatty acid synthesis is heightened when sufficient citrate comes from the mitochondria. This is indeed the case at high levels of ATP since isocitrate dehydrogenase, which in the mitochondria competes for citrate, is inhibited allosterically by ATP. Fatty acid synthesis as a result proceeds to a considerable extent only with an abundant supply of ATP.

Acetyl-CoA carboxylase is *inhibited allosterically by long-chain acyl-CoA*, the end product of fatty acid synthesis. This end product inhibition ensures that fatty acid synthesis is shut down under conditions when acyl-CoA can no longer be converted to neutral or depot fat, for example, during starvation.

4. Citrate Cycle and Respiratory Chain

The end products of catabolism are mainly CO_2 and H_2O. Carbon dioxide arises largely from decarboxylation in the citrate cycle; water originates from biological oxidation in the respiratory chain. Since the breakdown products of the three main foodstuffs, carbohydrates, fats, and proteins, all enter these common terminal pathways, it is not surprising that these foodstuffs can replace each other as nutrients.[2]

The citrate cycle is indeed the pivot on which the metabolic processes revolve, with respect to both the terminal oxidation of foodstuffs and the synthetic activities. The citrate cycle requires acetyl-CoA, which is also the universal starting material for the synthesis of endogenous substances. Acetyl-CoA (active acetate) is formed mainly in two reaction sequences: from the breakdown of fatty acids and from the breakdown of carbohydrates.

Citrate cycle. The biological purpose of the cycle is to decompose acetate (which enters the cycle as acetyl-CoA), with the help of 2 H_2O, into CO_2 and [H].

$$CH_3COOH + 2\,H_2O = 2\,CO_2 + 8\,[H]$$

The decomposition commences with an enlargement of the molecule, by the condensation of acetyl-CoA and oxaloacetate to give citrate, from which the cycle is named. Isomerization yields isocitrate; dehydrogenation and decarboxylation produce ketoglutarate. Next, an oxidative decarboxylation with *thiamine pyrophosphate* and *lipoate* is interposed. The product, active succinate (= succinyl-CoA), is converted to succinate with the gain of 1 ATP. Succinate is then dehydrogenated (with a flavoprotein) to form fumarate. The uptake of water and subsequent dehydrogenation complete the cycle by yielding oxaloacetate again. (For the regulation see Chapter XI,3.)

Only 1 ATP is gained from the citrate cycle itself, namely through the oxidative decarboxylation of ketoglutarate. Almost all the carbon dioxide is produced in the cycle. Most of the energy expected from the decomposition of foodstuffs is produced in the *respiratory chain.* The reduced coenzyme NADH is oxidized by the enzymes of the respiratory chain and at the end of the chain by atmospheric oxygen. The formation of water thus occurs in a stepwise fashion which at three places is coupled with the phosphorylation of ADP. Respiratory chain-linked phosphorylation (or oxidative phosphorylation) is the most productive source of chemical free energy.

[2] The phenomenon of interchangeability was discovered by physiologists more than 60 years ago and has been the basis of calculations of the caloric value of food intake (cf. Chapter XXII).

Another significant feature of the citrate cycle is its central location between various degradative and synthetic pathways. Its interrelationships to the metabolism of amino acids are particularly numerous and will be discussed in the following Section. For glucogenic amino acids the branch point leading to phosphoenolpyruvate is important since it is the route of gluconeogenesis. Carbon atoms of the fats also end up in carbohydrates by this route. A net synthesis is not possible, however, since two C atoms are lost along the way. Mammals cannot synthesize sugar from acetyl-CoA.

A second branch point is provided by succinyl-CoA, which can unite with glycine to form β-aminolevulinate; the latter condenses to form porphobilinogen, a key intermediate in the biosynthesis of the hemin ring system (cf. Chapter IX).

Shemin has pointed out that the oxidative deamination of aminolevulinate and the subsequent oxidation of ketoglutarate semialdehyde to give ketoglutarate constitute a second cycle which would effect complete oxidation of glycine. The significance of this is still obscure (see Chapter XI,4).

The glyoxylate cycle (cf. Chapter XI,4) has recently been discovered in certain plants and bacteria. Here, two molecules of acetate are united to form succinate and thus a net synthesis of carbohydrate can be achieved by way of the familiar reactions. This cycle is not shown on the metabolic chart because it seems to be without significance for the mammalian organism.

5. Metabolism of Proteins

Proteins are broken down to amino acids by proteolytic enzymes. The metabolism of the different amino acids is as individual as their structures. Only a few steps can be integrated into one general scheme.

Metabolism of Nitrogen. This proceeds by transamination to glutamate and aspartate and on to urea (other N-containing substances have only subordinate significance for the excretion of nitrogen in man and most other mammals). Urea synthesis itself proceeds by a cyclic process. In this case, energy is consumed (4 moles of ATP per mole of urea) rather than produced. The pathway from ornithine via citrulline or arginine and back to ornithine is shown in the diagram on the foldout chart. The key compound for the first step is *carbamyl phosphate.* (For more information on the individual steps refer to Chapter VIII,8.) The ammonia donors glutamate and aspartate enter the citrate cycle after their deamination, unless they are charged again with amino groups by transamination.

The direct oxidation of amino acids (oxidative deamination) is insignificant compared to transamination. Decarboxylation is important only for some special metabolic processes, e.g., in the formation of certain hormones and tissue hormones (cf. Chapter VIII,5).

The Carbon Metabolism of Amino Acids. Sooner or later amino acids funnel into the citrate cycle as illustrated in Fig. XIX-1. The shortest pathway is that of glutamate, which by oxidative deamination in the mitochondria produces α-ketoglutarate directly. This pathway is traversed as well by glutamine, proline, histidine, and arginine (or ornithine). Aspartate yields oxaloacetate by transamination, but fumarate through the urea cycle. Fumarate (along with acetoacetate) is also derived from the degradation of phenylalanine and tyrosine. Alanine, serine, and cysteine are closely related to pyruvate (and then via oxaloacetate to acetyl-CoA). That the carbon skeletons of amino acids enter the citrate cycle is significant not only for their terminal breakdown but also for gluconeogenesis, which is the net synthesis of carbohydrate such as glucose (cf. Chapter XV,10). The synthesis of amino acids frequently proceeds via transamination of the appropriate α-keto acid. There are, however, a number of essential amino acids, so named because they cannot be formed by higher organisms and instead must be supplied in the diet (see Chapters VIII,6 and XXII,2).

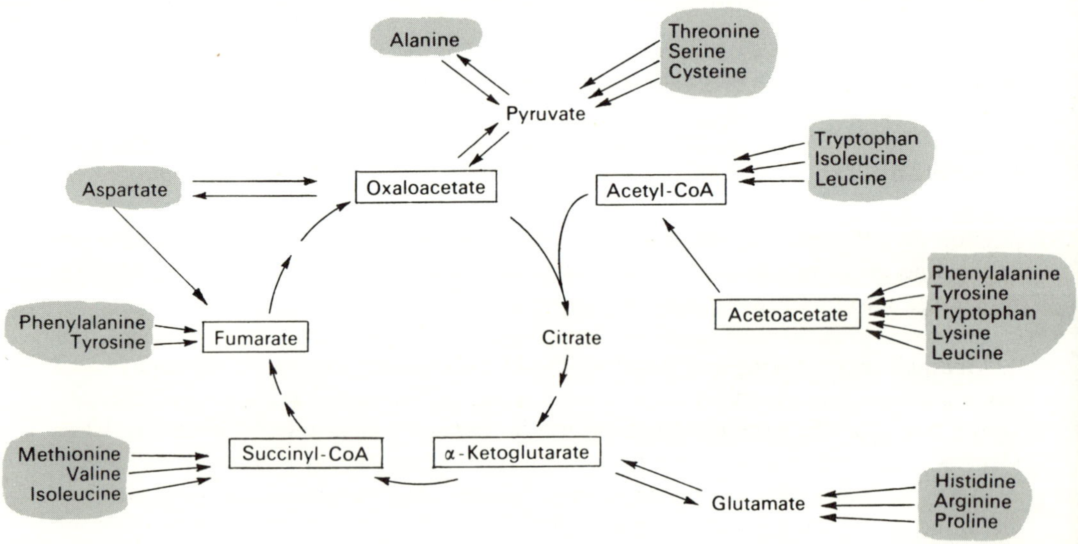

Fig. XIX-1. Influx of carbon skeletons of amino acids into the citrate cycle. Double arrows between amino acid and keto acid indicates direct reversible transamination. The breakdown of amino acids to the intermediates of the citrate cycle is discussed in some detail in Chapter VIII. Oxaloacetate is the key compound of gluconeogenesis. Further explanation in the text.

Some amino acids enter the pathway of fatty acid catabolism after transamination or oxidative deamination. Since they give rise to branched-chain fatty acids, they undergo several changes (cf. Chapter XII,4, and the right side of the foldout chart).

Leucine is broken down to acetoacetate after passing through the step of hydroxymethylglutaryl-CoA. Acetoacetate arises also from phenylalanine and tyrosine as well as from lysine and tryptophan via glutaryl-CoA.

C_1 Metabolism. The metabolism of *glycine* and *serine* should be mentioned briefly because of their numerous relationships with other classes of compounds. An outline has been provided (large diagram of formulas, Chapter VIII,12). By transamination of serine, hydroxypyruvate arises; it can enter the metabolism of carbohydrates. Serine is especially important as a donor of the C_1 fragment. Formaldehyde and formate (in its activated form) originate here; they are utilized primarily in the synthesis of the purine derivatives (to form nucleotides). Methionine supplies another C_1 fragment, the methyl group.

Synthesis of Proteins. From the available pool of free amino acids, proteins are synthesized. The synthesis requires ATP and specific nucleic acids as cofactors and as templates. This is the place where hereditary factors (deoxyribonucleic acids) take effect, since they bear the information for the structure of the proteins (including enzyme proteins). This information is transcribed to messenger RNA which then adheres to ribosomes and serves as the template for the lining up of activated amino acids. Protein synthesis is one of the most important energy-consuming processes.

Metabolism of Nucleic Acids. The biosynthesis of nucleic acid in the cellular nucleus proceeds from nucleoside triphosphates by action of the polymerase in conjunction with DNA which serves as template. In *degradation*, the high-molecular weight substances are first broken down to mononucleotides and then to the components of the latter. The sugar moiety enters carbohydrate metabolism, and the purine bases are oxidized to uric acid and excreted as such.

The *biosynthesis* of purine and pyrimidine bases has been described in Chapter VII,2. It proceeds largely from small fragments closely related to the metabolism of protein. Note that C_1 fragments (active formate) are used extensively. Vitamins are often required in the biosynthesis of nucleotide coenzymes (Chapters VI,2 and XXII,3).

6. The Common Pool of Intermediary Metabolism

The foldout chart and the preceding discussions reveal that the main metabolic pathways are interconnected in many ways. In order to be understood properly, the entire diagram must be looked upon as a dynamic equilibrium (see Chapter V,4). On the one hand, substances are continuously added, and on the other, waste products are excreted. Quantitatively, carbon dioxide is the principal waste product, with around 1 kg being discarded each day (we have emphasized repeatedly, however, that CO_2 can also reenter metabolism).

The metabolic chart also reveals that various foodstuffs and endogenous substances are broken down constantly to produce common intermediates. This is often referred to as forming a "common pool of metabolites."[3] Acetyl-CoA is a very typical example; it is derived from many different sources. The metabolic pool, therefore, has numerous inlets as well as outlets; one of the outlets flows to the citrate cycle, another to fat synthesis, another to isoprenoid synthesis, etc. These "inlets" and "outlets" generally are regulated rather tightly, frequently by allosteric controls.

Molecules are mixed uniformly in such a pool; once submerged in the pool, the origin of a molecule (whether from fat or from carbohydrate) can no longer be ascertained. This explains how radioactively labeled fragments can be spread over an entire organism.

In theory, there exists a metabolic pool of this kind for every substance. Such pools are particularly significant if they consist of metabolites situated at the junctions of metabolic pathways, e.g., acetate, pyruvate, α-ketoglutarate, succinate, and oxaloacetate, or of hydrogen (the reduced coenzymes NADH and NADPH), or of ATP, which in a sense is the common pool for chemical energy.

A clear distinction must be made between size of the metabolic pool and rates of flow into it and out of it. The *steady concentration* of a metabolite, i.e., the size of the pool, can be rather low, although a large amount of material is actually metabolized. Concentrations of metabolites range within the order of magnitude of 10^{-3} mole/liter. Greater concentrations are attained by storage substances. The turnover, in contrast, can be gigantic indeed. It can be calculated, for example (cf. Chapter XXII,1) that the human body produces during 24 hours as much as 70 kg of ATP; its own weight! Of course, ATP is used up as fast as it is produced; it is usually split to ADP and phosphate and can be resynthesized from these components.

BIBLIOGRAPHY

D. M. Greenberg (editor), "Metabolic Pathways." 3rd edition, 6 volumes. Academic Press, New York, 1967–1972. See particularly vol. 5, Metabolic Regulation, H. J. Vogel (editor), 1971.

H. Holzer and W. Duntze, Metabolic regulation by chemical modification of enzymes. *Annu. Rev. Biochem.* **40,** 345–374 (1971).

[3] The term *metabolic pool* is used in different senses. At one time, it may refer to the total amount of a substance present in the whole organism (or in certain organs, such as blood); at another time, it may have the more restricted and original meaning, namely the amount of a substance that is in a state of active turnover. The second is the proper meaning for the expression "the common pool of metabolism." Pool size is usually estimated by administering a radioactive substance and measuring the dilution with unlabeled substance that it undergoes.

CHAPTER **XX**

Hormones

1. Principles of Hormonal Regulation

Humoral or hormonal control means that "peripheral" tissue is influenced by substances with specific effects. The active substances are formed in specialized glands or tissues (endocrine glands) and secreted into the blood stream. This can also be called "*internal secretion*," the substances produced by endocrine (or endosecretory) glands are the *hormones*.

The concept *hormone* has been applied more or less broadly in the past. A hundred years ago, C. Bernard coined the term "internal secretion" to describe the production of glucose by the liver. The example has been dropped, but the term has been kept. The situation is similar with Starling's "secretin," which occasioned the formulation of the term *hormone*. Secretin is not a hormone of the endocrine glands, but rather one of the tissue hormones, which today constitute a separate group. A critical discussion of the hormonal concept is given by Verzar.[1]

The Principle of the Master Gland. Several hypophyseal (=pituitary) hormones act on "peripheral" endocrine glands. The activity of the hypophysis is itself regulated by yet another hormone,[2] so that there evidently exists a hierarchy of hormone glands, which comprises three levels in the case of the hypophyseal–adrenocortical system (see Fig. XX-1). Certain centers of the midbrain elaborate a neurosecretion that travels through the pituitary stalk to the anterior lobe of the hypophysis, where it stimulates the release of the adrenocorticotropic hormone (ACTH). ACTH is itself a glandotropic hormone; its target is the adrenal cortex, which is stimulated to produce increased amounts of the so-called corticosteroids and to release them to the blood circulation. A feedback system controls the master gland, the hypophysis, in

[1] F. Verzar, "Lehrbuch der inneren Sekretion," 609 pp. Liestal, Ars. medici Ludin, Basel, 1948.

[2] The principle of the "master gland" can be detected at rather early stages of phylogenesis; several endocrine glands of insects are guided by the endocrine activity of certain nerve cells (neurosecretion; cf. Sections 9 and 15).

that its activity is adjusted by the amount of circulating peripheral hormone; an increase of the level of corticosteroids inhibits the activity of the hypophysis.

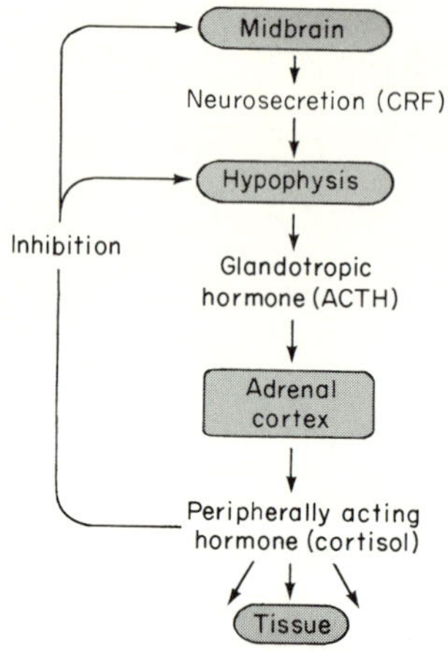

Fig. XX-1. Example of a regulatory loop of hormone production.

Isolation of Hormones. Because hormones always occur in minute amounts, they are difficult to isolate. Often a 100,000- to 1,000,000-fold enrichment of the biological material is required before the active substance can be crystallized. A workable biological assay is needed for the isolation, because initially the hormone can be recognized only by its activity, and the enrichment procedure is pursued by following the rise in activity. The isolation and elucidation of structure of the steroid hormones was performed mainly by Butenandt, Doisy, Kendall, and Reichstein. Thyroxine was isolated by Kendall, and its constitution was determined by Harington. Banting and Best first prepared pure insulin. The hypophyseal hormones have been enriched and finally isolated in pure form chiefly by Evans and Li.

Target Organs and Hormone Receptors. It has long been known that most hormones act only on certain organs and tissues which have been called *target organs* or *target tissues.* Thus, for example, estradiol, the female gonadal hormone, acts on the female reproductive organs, the mammary gland, and on certain centers in the midbrain. Although estradiol circulates in the blood throughout the whole body it acts so selectively only because of the presence of specific receptors for estradiol exclusively in the cells of target tissues. The receptor is a protein extracted from uterine tissue and now purified (Jungblut) that can bind estradiol very tightly. It may have the additional role of transporting the hormone to its site of action, the cell nucleus (see below). Similar receptors have been discovered for the other steroid hormones, namely, progesterone, testosterone, and cortisol. It is quite probable that receptor proteins are generally part of the mechanism of action of hormones.

Additionally, some hormone receptors are fixed components of the cell membranes of target tissues; they are allosteric activators of the adenylate cyclase system.

Mode of Action of Hormones.[3] Another interesting chapter in biochemistry is the manner in which hormones exert their activity. As a chemical substance, a hormone can only have chemical effects; it must influence some chemical processes. The observed "physiological effects" usually appear much later, as a consequence of the primary chemical effect.

Two different primary reactions are recognized today: The activation of the adenylate cyclase system and the control of gene activity. The hormone receptors mentioned above quite probably are involved in both primary reactions.

Adenylate Cyclase and Cyclic AMP as the Second Messenger. Many hormones, particularly the fast-acting ones such as epinephrine, glucagon, and several other peptide hormones, primarily effect in the cell *membrane* a stimulation of the adenylate cyclase system. The system most likely comprises the hormone receptor that imparts the specificity for the corresponding hormone (glucagon, for instance) and the enzyme adenylate cyclase itself. The hormone modifies the receptor protein in such a way that adenylate cyclase is activated allosterically and consequently ATP is converted to 3′,5′-cyclic AMP (formula in Chapter VI,5; see also Fig. XX-2). Cyclic AMP, dubbed by Sutherland as *second messenger*, can then interact with various processes in the cell. It can activate other enzyme systems, as seen in detail in the case of phosphorylase (Chapter XVII,6). Another example would be the stimulation of the synthesis of cortisol in the adrenal cortex by the adrenocorticotropic hormone (Section 9). Cyclic AMP can also influence transport systems of membranes. This action explains the effect of histamine on gastric secretion of HCl and the effect of vasopressin on the reabsorption of water.

A *phosphodiesterase* in the cell rapidly cleaves cyclic AMP to AMP; the action of the second messenger therefore is very brief. Phosphodiesterase is inhibited by methylated xanthines (caffeine, theophylline), and the familiar effects of caffeine at least in part can be explained by the fact that it prolongs the cyclic AMP-mediated effect of epinephrine.

The overall significance of this mode of action appears to be first its *amplification* effect derived from the formation of the "second messenger" and, second, the opportunity of diverse cells and tissues, each with their complement of receptors and cyclic AMP-sensitive enzyme systems, to be influenced by the hormones by the same basic mechanism but in a highly individual direction. There is, however, a concomitant loss of specificity: if several different hormones influence the same cell and have

[3] For a broader treatment, see P. Karlson and C. E. Sekeris, Biochemical mechanisms of hormone action. *Acta Endocrinol.* **53,** 505–518 (1966).

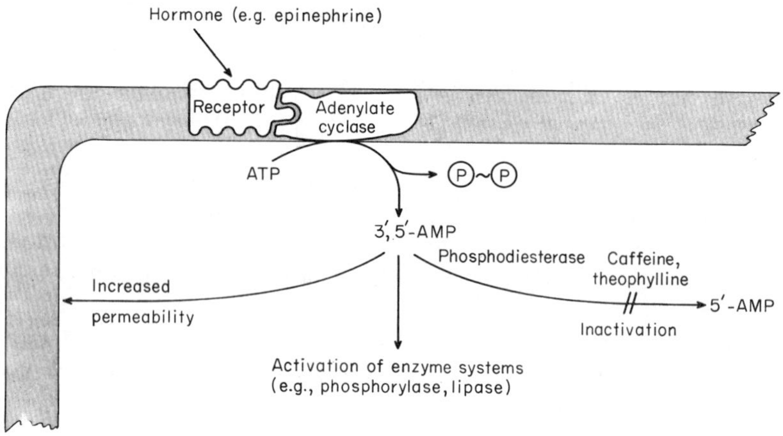

Fig. XX-2. The action of hormones by the adenylate cyclase system. Explanations in the text.

to act through the mediation of the same second messenger, then at best a synergistic or antagonistic effect is achieved and not a differentiated one.

Hormone Action by Gene Activation. Many hormones, especially steroid hormones, induce certain enzymes or other proteins in the target tissue. This mechanism of action may be explained on the basis of activation of certain gene sites discovered by Clever and Karlson (see Chapter XVIII,1). In Fig. XX-3, the assumption is made that the *hormone combines with a repressor molecule* (itself a protein, cf. Chapter VII,7). In this way hitherto repressed DNA is exposed, the synthesis of messenger

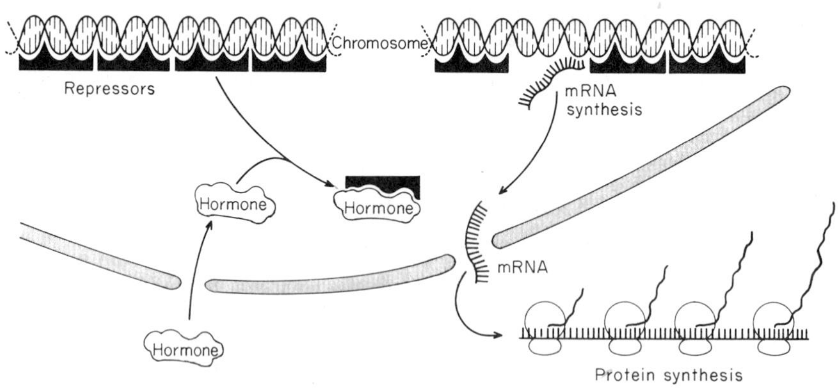

Fig. XX-3. Diagram of the action of hormones by enzyme induction. At left is shown chromosomal DNA covered with repressors. The hormone traps a repressor, and the corresponding gene starts to produce messenger RNA (mRNA), which in turn migrates into the cytoplasm and directs the synthesis of the corresponding protein. (Scheme by the author.)

RNA commences, and eventually enzyme protein is formed at the ribosomes. Evidence in support of this mechanism has been adduced for a number of hormones, particularly steroid hormones (Karlson and Sekeris). The best evidence is the demonstration that incubation of isolated cell nuclei with hormone stimulates the synthesis of RNA.

The scheme of Fig. XX-3 is based on the theory of Jacob and Monod on the control of gene activity in bacterial systems (see Chapter VII,7). Although several facets of the theory have been observed in mammalian systems, expecially the oft-repeated enhancement of RNA synthesis as well as the blockage of hormone action by actinomycin and other inhibitors of RNA biosynthesis, the mechanism of gene activation in higher organisms is probably more complex. The receptor proteins most likely are involved in the mechanism of hormone action in the cell nucleus; indeed, there are in the nucleus specific receptor proteins that differ from those in the cytoplasm. In addition, there is recent evidence that cyclic AMP may take a part even here.

The mechanism of induction in the cellular nucleus and the activation of membrane-bound adenylate cyclase are not necessarily mutually exclusive. Some hormones appear to operate via both mechanisms.

Metabolism of Hormones. Hormones are regulatory substances; they act in trace amounts. Being endogenous substances, they are produced by the organism itself. This is in contrast to vitamins, which also are effective in very small amounts, but which must be supplied from the environment. A few glands (e.g., thyroid) store some of the hormones that they produce; others, however, release their hormones into the circulation at the same rate at which they form them (e.g., the adrenal cortex).

Regulation is, of course, practicable only if the hormones can be removed again from the blood, for otherwise the organism would be flooded with hormones. The study of the biochemical pathways of synthesis and degradation of hormones—and their control—continues to be an intriguing aspect of endocrinology.

Aberrations of the Hormone System. Very often aberrations of the hormone system result in characteristic forms of disease stemming either from an overproduction (frequently caused by tumors of endocrine glands) or from an underproduction of a particular hormone; both abnormalities may result from genetic defects. The topic can only be mentioned here. It is actually clinical endocrinology, an important division of medicine.

Some hormones have been administered therapeutically even in situations where there is no primary lesion of hormone metabolism. A good example is the very successful clinical use of corticosteroids (see below).

Classification of Hormones. According to their chemical constitutions and their patterns of biosynthesis, three groups of hormones can be distinguished: (1) steroid hormones, (2) amino acid-derived hormones, and (3) peptide and protein hormones. Table XX-1 gives a survey of hormones and their effects; it is not intended, however, to be a complete list.

TABLE XX-1

Hormones of the Endocrine Glands of Vertebrates

Hormone	Abbreviation	Gland producing the hormone	Effect
Steroid hormones			
Aldosterone	—	Adrenal cortex	Mineral balance: $Na^{\oplus}$ retention
Cortisol	—	Adrenal cortex	Metabolism: Gluconeogenesis
Progesterone	—	Ovary (corpus luteum)	Secretory phase of the uterine mucosa
Estradiol	E_2	Ovary (follicles)	Proliferation of the uterine muscosa (estrus)
Testosterone	—	Testes (interstitial cells)	Maintenance of the accessory glands of the genital tract and secondary sex characteristics
Amino acid–derived hormones			
Thyroxine	Thx, T-4	Thyroid gland	Increase of basal metabolic rate; development
Epinephrine (adrenalin)	—	Adrenal medulla	Glycogen breakdown
Melatonin	—	Pineal gland	Contraction of melanophores
Peptide and protein hormones			
Parathyroid hormone (parathormone)	—	Parathyroid gland	$Ca^{2\oplus}$ mobilization
Thyrocalcitonin	—	Thyroid gland	Lowering of $Ca^{2\oplus}$ level
Insulin	—	Pancreas	Lowering of blood sugar level
Glucagon	—	Pancreas	Raising of blood sugar level
Relaxin	—	Ovary	Relaxation of pelvic ligaments during pregnancy
Erythropoietin	—	Kidney	Maturation of erythrocytes

In subsequent sections the individual hormones will be discussed with reference to their biogenesis and metabolism, and their physiological effects will briefly be characterized. Finally, a few examples will be given of hormonal regulation in which several hormones cooperate.

A number of hormones are steroids (see Table XX-1). The chemistry of steroids has already been discussed briefly in Chapter XIV,2. With the exception of the estrogens, all steroid hormones have an α,β-unsaturated keto group in ring A.

TABLE XX-1 ***(continued)***

Hormones (all peptides or proteins)	Abbreviation	Site of elaboration	Effect
Corticotropin releasing h.	CRH	Hypothalamus	Release of ACTH
Thyrotropin releasing h.	TRH	Hypothalamus	Release of TSH
Growth hormone-releasing h.	GHRH	Hypothalamus	Release of GH
Gonadotropin releasing h.	LHRH	Hypothalamus	Release of LH and FSH
Prolactin inhibiting h.	PIH	Hypothalamus	Inhibits prolactin release
Ocytocin	—	Neurophypophysis (posterior lobe)	Contraction of uterus
Vasopressin	—	Neurophypophysis (posterior lobe)	Antidiuretic action
Melanotropin (melanocyte-stimulating hormone; intermedin)	MSH	Pars intermedia (middle lobe)	Dilation of melanophores
Somatotropin (growth hormone)	GH	Adenohypophysis (anterior lobe)	Growth and metabolism
Lipotropic hormone	LPH	Adenohypophysis (anterior lobe)	Lipolysis
Corticotropin (adrenocorticotropic hormone)	ACTH	Adenohypophysis (anterior lobe)	Stimulation of the adrenal cortex
Thyrotropin	TSH	Adenohypophysis (anterior lobe)	Stimulation of thyroid gland
Follicle-stimulating hormone	FSH	Adenohypophysis (anterior lobe)	Stimulation of maturation of gametes
Luteinizing hormone (interstitial cell-stimulating hormone)	LH (or ISCH)	Adenohypophysis (anterior lobe)	Stimulation of production of sex hormones
Prolactin (luteomammotropic hormone, luteotropin)	LTH (LMTH)	Adenohypophysis (anterior lobe)	Stimulation of mammary gland and of corpora lutea
Gonadotropin ("human chorionic gonadotropin")	(HCG)	(Placenta)	Similar to LH

2. Adrenocortical Hormones

The adrenal cortex is an absolutely vital organ. It elaborates two chemically related but functionally distinct types of adrenocortical hormones; the *glucocorticoids* (cortisol and corticosterone) and the *mineralocorticoids* (chiefly aldosterone).

All adrenocortical hormones are twenty-one carbon steroids. Besides the unsaturated keto group in position 3, they also have a keto and an alcohol group in the side chain (the so-called ketal group).

The *biogenesis* proceeds from progesterone (cf. Chapter XIV,6). Hydroxylases mediate the introduction of hydroxyl groups into the steroid skeleton in the following order: the 17-position, the 21-position, and the 11β-position. In addition to the 11β-hydroxy compounds corticosterone and cortisol, the 11-keto analogs 11-dehydrocorticosterone and cortisone are also found. The gland, however, elaborates predominantly 11-hydroxy compounds. Out of the multiplicity of adrenocortical steroids found in the gland, only the three compounds cortisol, corticosterone, and aldosterone can be considered true hormones of the adrenal cortex. The formulas of the three hormones are shown again here; the *per diem* production in man is indicated along with the formulas.

In blood the bulk of cortisol is bound to a special protein, *transcortin*.

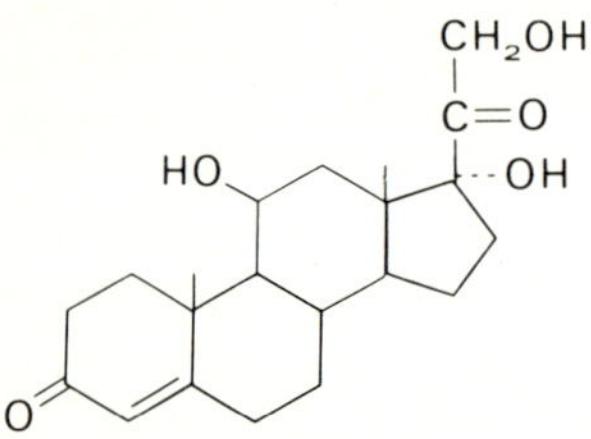

Cortisol

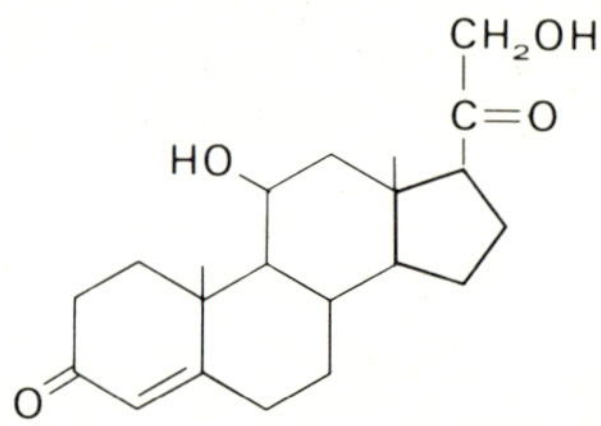

Corticosterone

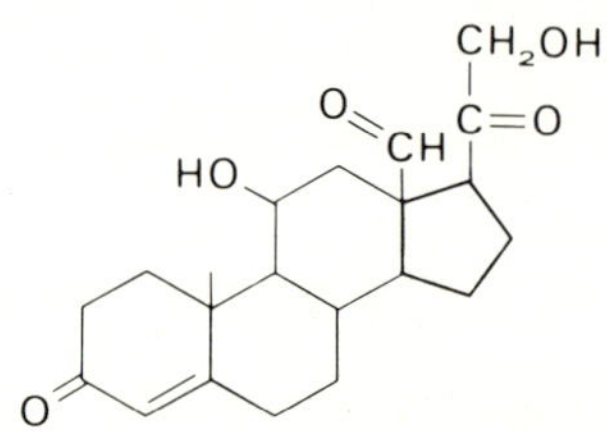

Aldosterone

Daily elaboration: 10–20 mg | 3 mg | 0.3 mg

Although the adrenal cortex contains very little hormone, its synthetic capabilities are extensive. It secretes several times its total content into the blood every minute; the entire amount of corticosteroid hormones circulating in the human organism is renewed completely every 2–3 hours. This synthetic activity is matched by an equally active breakdown in the liver, which proceeds partly to the tetrahydro compounds (saturated ring A, 3α-ol) and partly to the 17-ketosteroids (cf. Chapter XIV,6).

The control of corticosteroid production is effected chiefly by the *adrenocorticotropic hormone* (ACTH) of the hypophysis. A circadian (day–night) rhythm of hormone production has been observed. The secretion of corticosteroids is elevated markedly and rapidly in response to physical or psychic *stress* (see also Section 9, under "Adrenocorticotropic Hormone"). The production of aldosterone is stimulated strongly by *angiotensin II* (cf. Section 13).

Biological Significance of the Adrenal Cortex. The adrenocortical hormones constitute some of the most vital substances. Extirpation of the adrenal gland in experimental animals is followed by most serious disturbances—particularly, the metabolism of electrolytes is unbalanced (increased $Na^{\oplus}$ excretion) and in a few days becomes lethal. The lack of the mineralocorticoid aldosterone is primarily responsible for this since aldosterone causes the retention of $Na^{\oplus}$ and the excretion of $K^{\oplus}$ (cf. Chapter XXI,5).

Besides the *mineralocorticoid* effect on the electrolyte metabolism, the adrenocortical hormones also influence the metabolism of glucose by promoting glycogen formation in the liver, especially from amino acids, and by inhibiting peripheral glucose utilization. This effect is termed the *glucocorticoid* effect and is manifested especially by the 11β-hydroxy compounds. Emphasizing the breakdown of protein, this effect is also called the *catabolic effect*. The mechanism of this hormone action seems to involve the *de novo* formation of more enzymes of amino acid metabolism, e.g., tyrosine α-ketoglutarate transaminase, tryptophan pyrrolase, etc., and enzymes of gluconeogenesis, e.g., pyruvate carboxylase, fructose 1,6-bisphosphatase (cf. "Enzyme Induction," Chapter VII,7). The main human glucocorticosteroid is *cortisol*.

The effect of glucocorticoids on the *metabolism of protein* is catabolic. They inhibit protein biosynthesis and enhance protein breakdown in muscle, bone, and lymphatic organs. (In the liver, in contrast, enzyme induction results in increased RNA and protein biosynthesis.) The accelerated breakdown of protein releases larger amounts of free amino acids to the blood stream which then are utilized by the liver for gluconeogenesis. Nitrogen is excreted in the urine as urea; the organism is in negative nitrogen balance.

The consequence of inhibition of protein biosynthesis in lymphatic organs is the suppression by corticosteroids, especially at higher dosages, of the formation of antibodies. This is called the *immunosuppressive effect*, which is of such critical importance in the postoperative treatment of organ transplantation. The slowed defense reaction is also the basis of the *anti-inflammatory effect* of glucocorticoids.

The adrenocortical hormones therefore have been highly effective drugs in many diseases (arthritis, skin disorders, and inflammation). Modification of the chemical structure, particularly by introduction of fluoro or methyl groups, has provided substances with different action spectra and with much greater effectiveness than the natural hormones. From the multitude of such compounds we present the following two formulas. Both substances have strong anti-inflammatory properties, but relatively low mineralocorticoid effects.

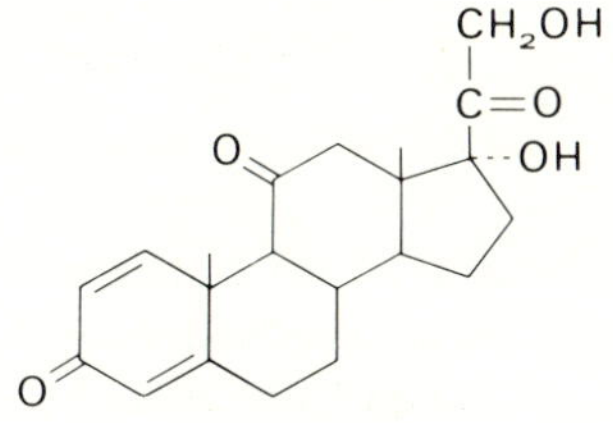

Prednisone

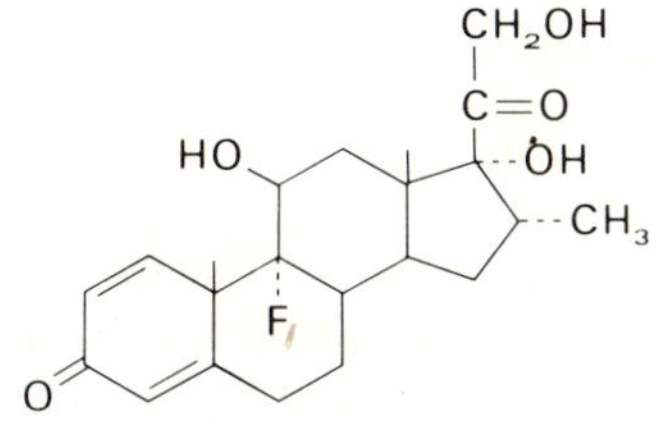

Dexamethasone

3. Gonadal Hormones

The female organism produces two physiologically distinct sex hormones, the estrogens (follicular hormones) and the gestagens (progestins; corpus luteum hormones.) In the male organism, on the other hand, the testicle produces only one type, the androgens.

Prostaglandins. Many authors consider these as hormones of the male genital tract; these active substances, however, are not only produced in the prostate gland, but also in a number of other tissues. Since their biosynthesis starts with certain essential fatty acids we will discuss prostaglandins in that context (Chapter XXII,4). The hormonal role of prostaglandins is not yet clear, although it is currently under intensive investigation.

Male Sex Hormones (Androgenic Hormones). The true testicular hormones are *testosterone* (17β-hydroxyandrost-4-en-3-one) and *androst-4-ene-3,17-dione*, both formed by the interstitial cells of the testicular tissue from 17-hydroxyprogesterone (cf. Chapter XIV,6). At the site of action, however, testosterone is reduced to *dihydrotestosterone* (5α-androstan-17β-ol-3-one), which must be looked upon as the active form of the hormone at least in the prostate gland and in other peripheral target tissues. The actual site of action is the cell nucleus where dihydrotestosterone stimulates RNA synthesis (see end of Section 1 and Fig. XX-3).

Hormones with androgenic activity are produced not only by the testes, but by the adrenal cortex as well, e.g., *androstenedione* and *11β-hydroxyandrostenedione*. Under certain conditions, such as tumor of the adrenal cortex, the androgen production may rise greatly; in women this causes the pathological condition known as virilism.

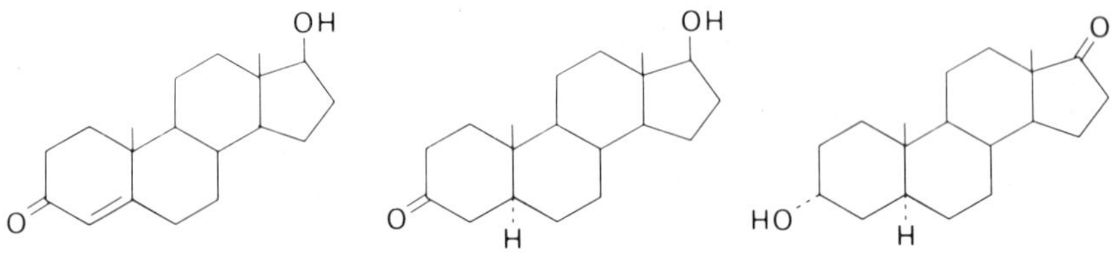

Testosterone 5α-Androstan-17β-ol-3-one Androsterone

In order to be *excreted*, testosterone is converted into androsterone and 5β-androstanolone, both of which can be detected in urine. Androsterone, which retains considerable androgenic activity, was the first androgen to be isolated in pure form (Butenandt). A small (about 1%) and variable fraction of testosterone is excreted as the glucuronide of testosterone.

Biological Function. The effects of castration have long been known and need not be listed here. The reversal of these effects by transplantation is a classical experiment from the beginning of hormone research (Berthold, 1849). The embryonic development of the testicle is determined genetically (by the presence of the Y chromosome).

The subsequent differentiation of the male reproductive organs during prenatal and postnatal development and the formation of secondary male sexual characteristics in puberty depend on the influence of androgens. Even the psychosexual differentiation and the establishment of male behavioral traits during embryonic development take place only in response to androgenic hormones. In the adult male organism the continuous production of the hormone is necessary for the maturation of sperm and the activity of the accessory glands of the genital tract.

Beyond this specific effect on the genital tract, the androgens also have a general effect on metabolism. They promote protein synthesis and elevate nitrogen retention. This *anabolic effect* is also found with several steroids that have been derived by chemical modification of the natural hormone. These derivatives are relatively weakly androgenic; one example is *1-methyl-1-androstenolone.* "Anabolic steroids" are very useful therapeutically.

Antiandrogens. Chemical derivatives of steroids that inhibit the action of androgens at the target organ have been synthesized. The testosterone analog with a $1\alpha{:}2\alpha$-methylene group, i.e., a cyclopropane ring, is very active. The antihormonal effect probably may be explained as competitive inhibition at the target site (the "hormone receptor"), which operates on the same principle as competitive inhibition of enzymes (see Chapter V,8). Antiandrogens are very useful both in research and in clinical application.

Estrogens. These are characterized chemically by their aromatic ring and the phenolic OH group at C-3. The angular methyl group at C-10, present in all other steroids, is absent.

The naturally occurring estrogens are estrone (with a keto group at C-17), estradiol-17β [systematic name; 1,3,5(10)-estratriene-3,17β diol], and estriol-3,16α,17β. The ovarian follicles contain *estradiol*, the most active estrogen.

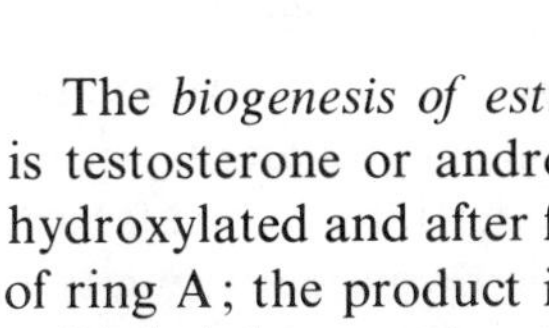

Estrone

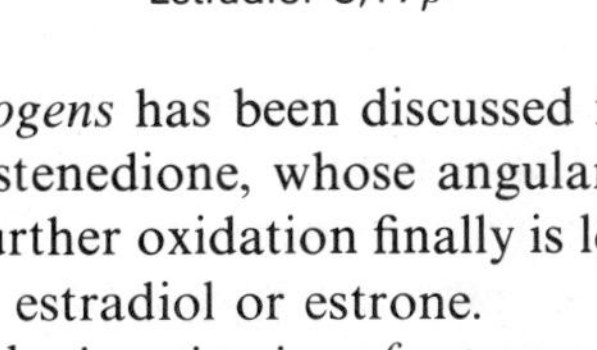

Estradiol-3,17β

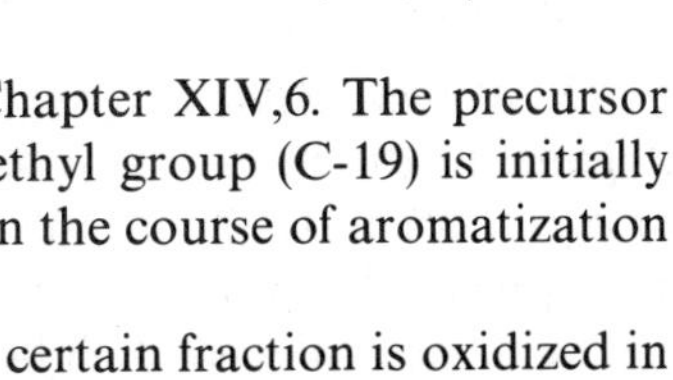

Estriol-3,16α,17β

The *biogenesis of estrogens* has been discussed in Chapter XIV,6. The precursor is testosterone or androstenedione, whose angular methyl group (C-19) is initially hydroxylated and after further oxidation finally is lost in the course of aromatization of ring A; the product is estradiol or estrone.

Little is known about the *inactivation of estrogens.* A certain fraction is oxidized in the liver (higher hydroxylated metabolic products appear in the urine), the bulk, however, is excreted in the urine as estrone sulfate or estriol 16α-glucuronide.

The *mechanism of action* of estradiol has been investigated more than once. Chronologically one of the earliest biochemical effects is the stimulation of RNA and protein synthesis, but the induction of specific enzymes has not been demonstrated. One early physiological effect is the marked uptake of water by uterine tissue; histamine is involved in this phenomenon.

Physiological Effects. Estradiol (the follicular hormone) is responsible primarily for the regular course of the reproductive cycles, the estrous cycles (assay procedures on laboratory animals) or the menstrual cycles (in man and other primates). Estradiol causes uterine mucosa to proliferate and the mammary glands to develop. By feedback mechanism estradiol controls the gonadotropin output of the hypophysis and thus interacts cooperatively in the cycle with progesterone and the hypophyseal hormones (see Section 12). Estradiol has very little pronounced effect on general metabolism; it does lower blood lipids and causes accumulation of subcutaneous fat depots.

Progestins (Corpus luteum hormones, Pregnancy hormones). The principal hormone of the corpus luteum is *progesterone*, a steroid that has already been mentioned as a starting material for the biosynthesis of adrenocortical hormones. It is formed from cholesterol in the corpus luteum and in the placenta by a shortening of the side chain, dehydrogenation of the 3β-OH group, and migration of the double bond from Δ^5 to Δ^4 to bring it in conjugation with the 3-keto group. In addition to progesterone, two *pregnenolones* (20α-OH and 20β-OH) have been found; both are biologically active.

Progesterone is produced only during certain phases of the menstrual cycle and during pregnancy. Excretion products in urine are the glucuronides of *pregnanediol* and some stereoisomers thereof; as in the case of the adrenocortical steroids, inactivation is accomplished by hydrogenation of ring A.

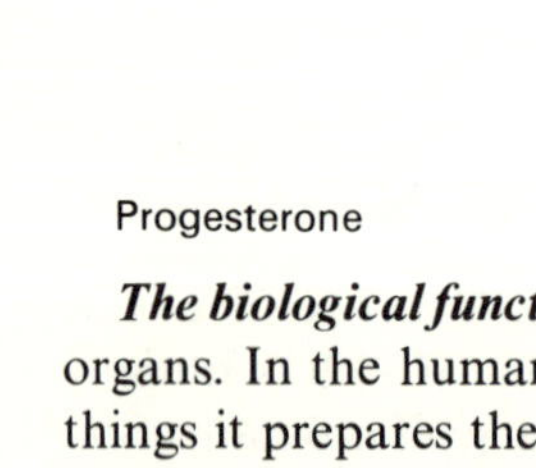

Progesterone

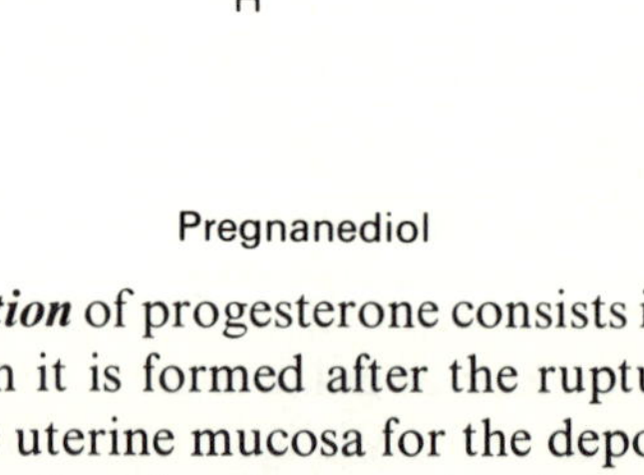

Pregnanediol

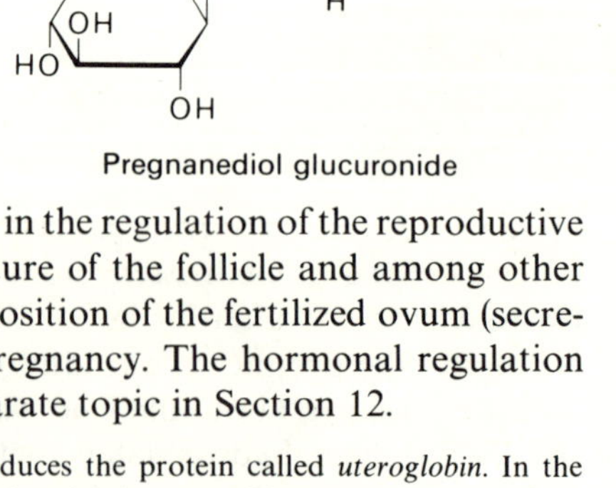

Pregnanediol glucuronide

The biological function of progesterone consists in the regulation of the reproductive organs. In the human it is formed after the rupture of the follicle and among other things it prepares the uterine mucosa for the deposition of the fertilized ovum (secretory phase). Its prime function is to maintain pregnancy. The hormonal regulation of the menstrual cycle will be discussed as a separate topic in Section 12.

In the rabbit uterus, early in pregnancy, progesterone induces the protein called *uteroglobin*. In the oviduct of chicken, *avidin*, one of the proteins of egg white, is induced. It is reasonable to postulate, therefore, that progesterone also operates by the activation of specific genes.

4. Hormones of the Adrenal Medulla

Epinephrine (Adrenaline). One of the hormones of the adrenal medulla, epinephrine was the first hormone to be isolated in pure form. Much later, a second hormone, the closely related *norepinephrine* (noradrenaline), was found in the same gland.

Both hormones are pyrocatechol (*o*-dihydroxybenzene) derivatives, which are easily oxidized. This fact explains the histochemical reaction of the "chromaffin tissue." The hormonal content of the adrenal medulla is relatively high (several milligrams per gram of gland). The two active substances of the adrenal medulla, epinephrine and norepinephrine, are both derivatives of phenylethylamine, which possesses strong pharmacological effects, which we will not discuss further.

Biosynthesis. The biosynthesis of epinephrine begins with tyrosine. The reaction sequence shown here also provides information about the chemical nature of the hormones.

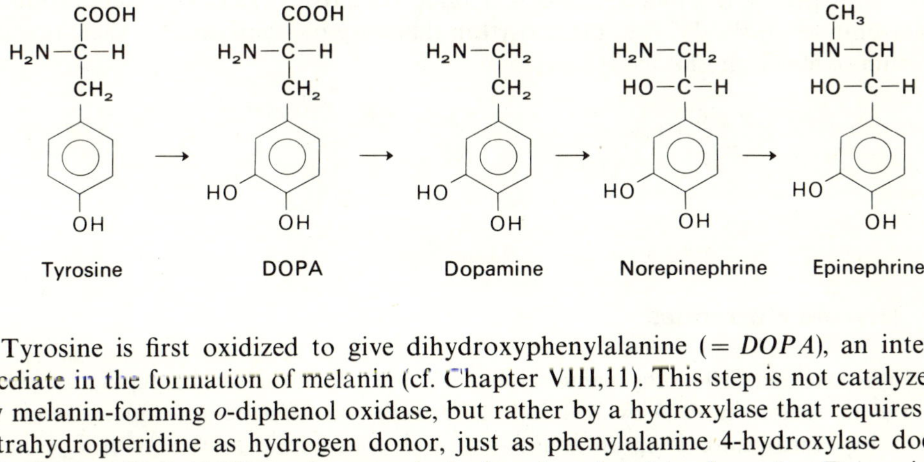

Tyrosine is first oxidized to give dihydroxyphenylalanine (= *DOPA*), an intermediate in the formation of melanin (cf. Chapter VIII,11). This step is not catalyzed by melanin-forming *o*-diphenol oxidase, but rather by a hydroxylase that requires a tetrahydropteridine as hydrogen donor, just as phenylalanine 4-hydroxylase does (cf. Chapter X-6). DOPA subsequently is decarboxylated to *dopamine.* Dopamine already possesses some biological activity; it is one of the "tissue hormones" (cf. Section 13). The next step is hydroxylation of the side chain. The resulting norepinephrine is finally methylated by active methionine.

The first step in the *inactivation of the hormone* is the methylation of the 3-OH group; this is followed by conversions on the side chains, starting with the oxidation of the amino group.

Biological Effect. The main *pharmacological effect of norepinephrine* is on the vascular system. Norepinephrine causes peripheral vasoconstriction and raises the blood pressure, but epinephrine causes vasodilation and vasoconstriction in different regions. The principal *biochemical effect of epinephrine* is to elevate blood glucose by mobilizing glycogen reserves (see Section 11 below). The mechanism involving cyclic AMP was discussed in Chapter XVII,6. In adipose tissue lipase is activated and consequently more free fatty acids are released to the blood.

Epinephrine and norepinephrine are the *neurotransmitters of the sympathetic nervous system* (cf. Chapter XXIII,8). This close relationship is understandable, since the adrenal medulla developed from nerve tissue. Epinephrine thus provides a link between the conventional hormones and the so-called tissue hormones and substances with hormonelike activity. Once again we observe that conceptualization draws arbitrary boundaries where in reality there may be a continuous, boundary-free transition.

5. The Hormone of the Pineal Gland

Melatonin is found in the pineal gland (= epiphysis). As is evident from the formula, it is an indole derivative which probably arises from typtophan and is closely related to serotin (cf. Section 13). Melatonin blanches amphibian skin; it is the melanophore-contracting principle (the antagonist of melanotropin, Section 9). In mammals, melatonin appears to inhibit the secretion of gonadotropic hormones and of ACTH. In many species, the pineal gland is a photoreceptive organ. The elaboration of melatonin follows a day-and-night rhythm that may be significant for the circadian rhythm of many physiological functions.

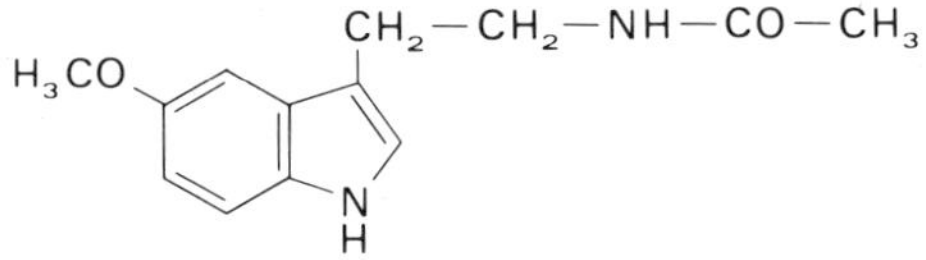

Melatonin

6. Thyroid Hormones

Thyroxine and Triiodothyronine. Thyroxine is one of the earliest recognized hormones (isolation in 1915 by Kendall; determination of structure and synthesis in 1925 by Harington). It is an iodine-containing aromatic amino acid. The non-iodinated parent compound of thyroxine is called *thyronine*. The unusual structural feature is the diphenyl ether group. Iodine atoms may be substituted at the positions 3, 5, 3′, 5′. A few diiodo- and triiodothyronines have physiological significance, especially *3,5,3′-triiodothyronine*, which is about five times as active as *thyroxine* itself.

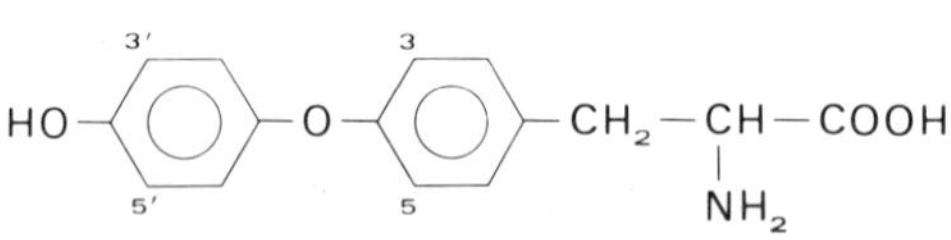

Thyronine

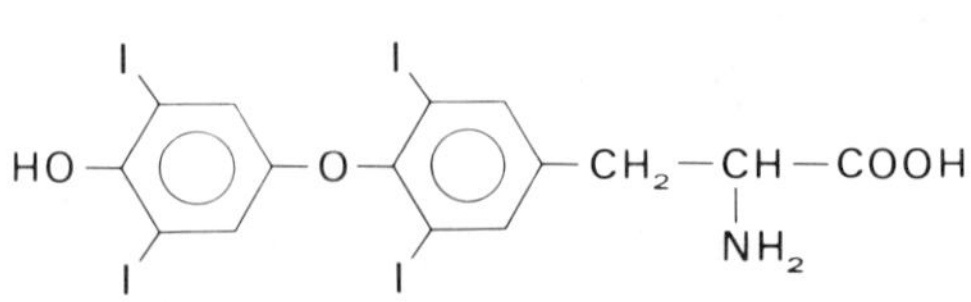

Thyroxine

Biosynthesis. The thyroid gland contains a special protein, *thyroglobulin*, which has many reactive tyrosyl residues that are iodinated first. Iodide ions are taken up from blood (against a concentration gradient), oxidized to elemental iodine, and inserted into the tyrosine radical. The thyronine structure seems to be formed with particular ease; in fact, it arises nonenzymatically from iodinated protein or iodinated tyrosine peptides. The reaction probably proceeds via semiquinone radicals as shown.

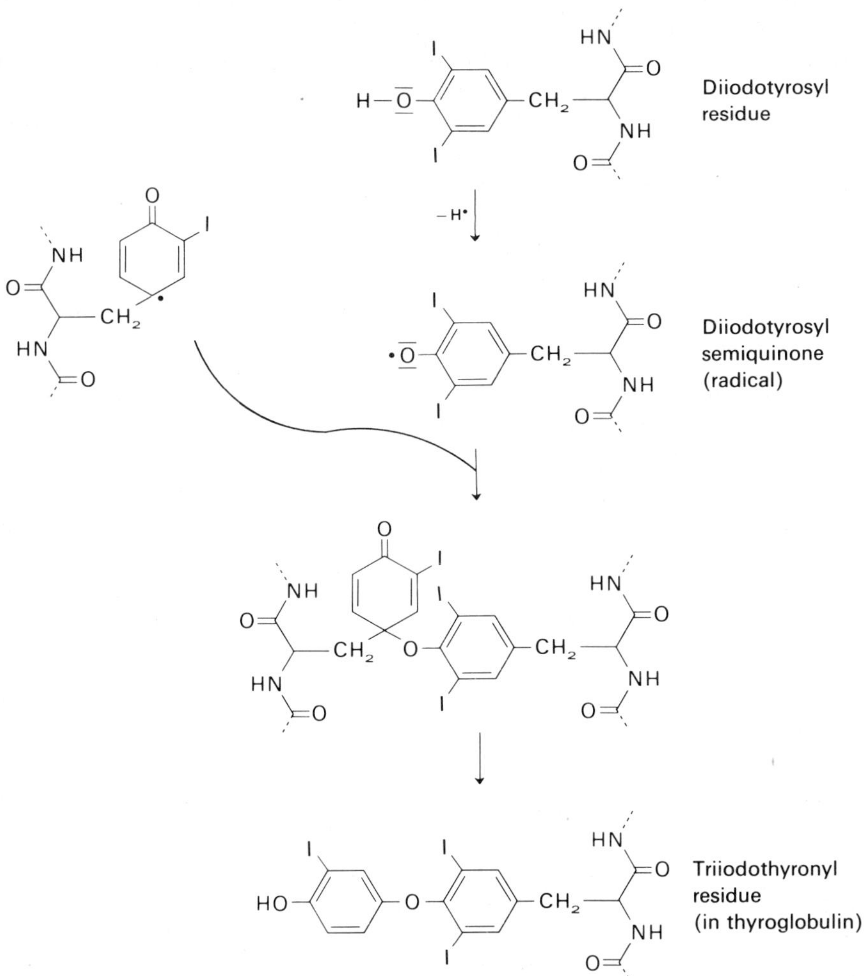

Depending on the requirements of the organism or the stimulation of the gland by the thyrotropic hormone (see Section 9), thyroglobulin is broken down to a lesser or greater extent and thyroxine (or triiodothyronine) is secreted into the blood stream. The secretory rate is 80–100 μg per day. The entire activity of the thyroid gland can be represented schematically, as shown in Fig. XX-4.

Thyroxine and triiodothyronine can be synthesized by the organism only when sufficient iodine is supplied in the diet. Iodine therefore is one of the vital trace elements (for a list of these see Chapter XXII,2) whose function is understood. If too little iodine is present in the diet, especially in drinking water and table salt (as, for example, in remote areas of the Alps), the result is goiter. The tissue of the thyroid gland hypertrophies in order to compensate as well as possible for the underproduction of thyroxine.

Antithyroid Substances. Various substances are able to interfere with the normal process of thyroxine synthesis. First of all, several univalent anions (e.g., thiocyanate, nitrate, perchlorate, iodate) block the active transport of iodine and thus prevent its accumulation. A second important group of antithyroid substances includes the goiterogenic, sulfur-containing compounds found in certain plants (e.g., cabbage). These compounds damage the thyroid gland of rabbits fed exclusively on cabbage. Numerous substances of this type have been synthesized for pharmacological application in hyperthyroidism. Thiouracil is the example chosen in the scheme of Fig. XX-4. Their primary effect is to inhibit the incorporation of iodine into organic compounds; but they also interfere with iodide uptake. Perhaps a specific enzyme, an iodide oxidase, is poisoned. Finally, some substances appear to compete with tyrosine for available iodine, e.g., *p*-aminosalicylate, many sulfonamides, and simple phenols, such as resorcinol.

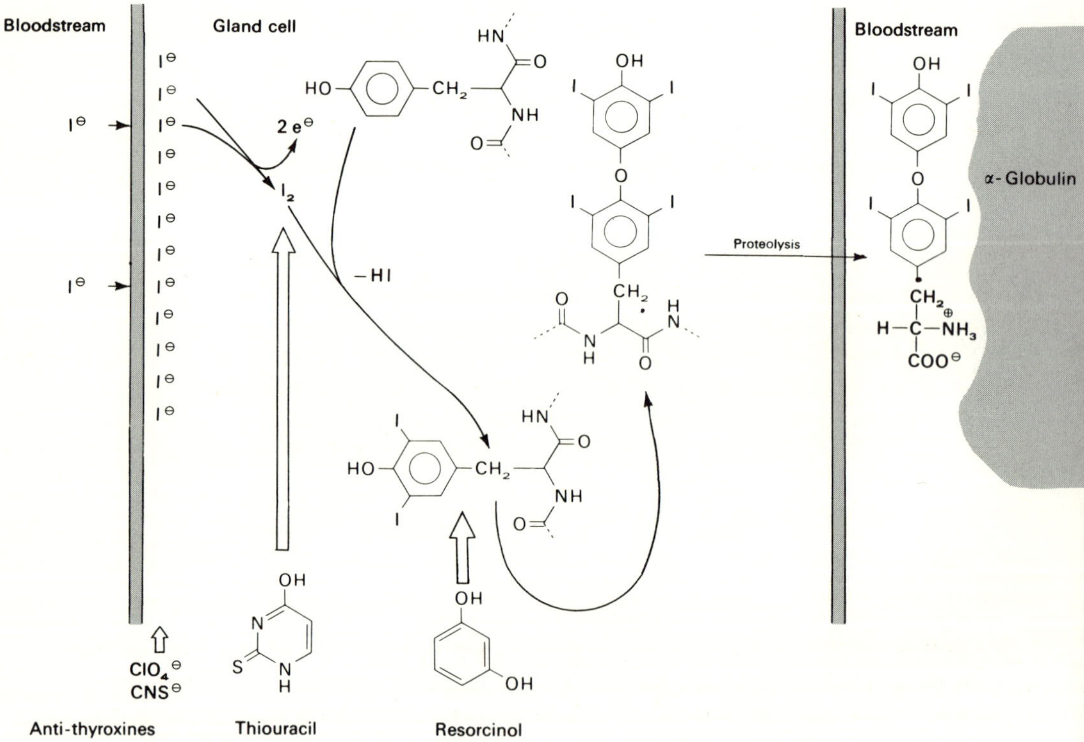

Fig. XX-4. Diagram of thyroid activity. Iodide ions are taken up from blood (left), oxidized in the gland by removal of electrons and bound organically. From the iodinated tyrosine residues in the thyroglobulin molecule arise tri- and tetraiodothyronine, which are freed by proteolysis and released to the bloodstream. At right is the addition compound with α-globulin. At the bottom, the sites of action of the inhibitors (antithyroid substances) are indicated.

Thyroxine Metabolism. Thyroxine or triiodothyronine, elaborated by the gland into the blood, is bound partly to prealbumin and partly to a glycoprotein of the α-globulin group. The clinical chemical determination of this *protein-bound iodide* is of diagonistic value for thyroid function. Another diagnostic tool is the '*radio-iodine test*, where the uptake of the radioactive nucleid ^{131}I by the thyroid gland is measured.

Both thyroxine and triiodothyronine are deaminated when they are catabolized, just as all other amino acids are (either oxidatively or by transamination). The resulting thyropyruvate is decarboxylated to give thyroacetate which still retains hormonal activity. Thyroxine and its catabolic products may also be bound to glucuronate and enter the bile. For the proper iodine balance of the organism it is important that iodine be liberated from iodine-containing catabolites by a specific deiodinase and returned to the thyroid gland as iodide.

Biological Effects of Thyroid Hormone. Thyroid hormone is indispensable for growth and development. Hypofunction or absence of the hormone in the growing organism results in serious developmental disturbances (cretinism). Extirpation of the thyroid gland in animals causes delay in growth and sexual maturation. In amphibia metamorphosis of larvae (tadpoles) is triggered by the thyroid hormone; administration of thyroxine causes premature metamorphosis.

External metamorphosis is accompanied by a shift in metabolism. Tadpoles excrete $NH_4^{\oplus}$ as the end product of nitrogen metabolism; frogs produce urea. At the time of metamorphosis, all enzymes of the urea cycle are formed. Their induction by thyroxine has been demonstrated.

The thyroid gland affects the entire metabolism of the adult organism. Disturbances of thyroid function are reflected in the *basal metabolic rate* (metabolism while resting, cf. Chapter XXII,1). In hyperthyroidism (Basedow's disease) the rate is elevated; with hypofunction (clinical manifestation: myxedema), it is lowered. This phenomenon is used frequently in judging thyroid function. Administration of thyroxine raises the overall metabolic rate and oxygen consumption and simultaneously stimulates protein synthesis. It is possible that the elevated metabolism is a result of increased RNA and protein synthesis. Several enzymes, e.g., α-glycerophosphate dehydrogenase of mitochondria, are induced specifically.

With certain hyperthyroses, even after thyroidectomies, there appears in the serum an immunoglobulin that stimulates the thyroid gland and is called *long-acting thyroid stimulator* (LATS). It may be involved in the pathogenesis of several thyroid diseases.

High dosages of thyroxine are deleterious. Hyperfunction of the gland can lead to the disease thyrotoxicosis, in which thyroxine probably acts as an uncoupling agent of oxidative phosphorylation. The suggestion that this effect be the basic mechanism for the control of metabolism could not be substantiated.

Calcitonin (Thyrocalcitonin). Recently a second hormone has been discovered in the thyroid gland. It controls the $Ca^{2\oplus}$ level in antagonism to parathormone. Calcitonin has been isolated in pure form from pig thyroid glands; it is a polypeptide consisting of 32 amino acids. The sequence is known and has been confirmed by chemical synthesis.

Biological Activity. Calcitonin effects a rapid and brief lowering of the $Ca^{2\oplus}$ level in the blood. Presumably more calcium is deposited in bony substance during the influence of calcitonin (see Chapter XXI,6).

7. The Parathyroid Hormone

The parathyroid glands elaborate a vital hormone now called *parathormone*. It is a polypeptide that has been obtained in pure form very recently. Its molecular weight is 8600. The hormone dissociates under certain conditions to yield smaller peptides that retain some activity. Oxidation inactivates the parathyroid hormone rapidly.

Biological Effect. Parathormone regulates the ionic milieu, especially the $Ca^{2\oplus}$ and phosphate content of tissue and blood. A deficiency of the hormone, e.g., after parathyroidectomy, quickly results in a decrease of the $Ca^{2\oplus}$ level in blood and eventually in tetanic convulsions due to changes in the electrolyte balance in muscular and nervous tissue. Injection of the hormone results in an increase of blood phosphate, citrate, and $Ca^{2\oplus}$ concentration and increased phosphate excretion by the kidney. The hormone exerts an influence on the metabolism of bone and mobilizes the mineral content. This explains the elevation of $Ca^{2\oplus}$ in blood (hypercalcemia). The counter regulator is calcitonin, but parathormone also affects the kidney directly and regulates phosphate excretion by reducing the reabsorption, so that more phosphate is flushed out.

Another substance, chemically related to vitamin D_2 (calciferol), has an effect similar to that of parathormone. This is *dihydrotachysterol* (*AT 10*). The effect is, however, in some aspects more comparable to D-hypervitaminosis. Dihydrotachysterol is used frequently in the condition of tetany caused by a deficiency of the parathyroid hormone, because the peptide hormone is not readily available.

Hyperparathyroidism. Very often hyperparathyroidism is caused by tumors of the parathyroid gland. The effects are a demineralization of bone (osteodystrophy), elevated serum $Ca^{2\oplus}$ level, elevated $Ca^{2\oplus}$ and phosphate excretion, and the formation of phosphate calculi in the kidneys and the urinary tract. *Alkaline phosphatase* (a phosphatase with a pH optimum in the alkaline region), a typical enzyme of the osteoblasts, is elevated in serum in this condition.

8. The Pancreatic Hormones

The pancreas produces digestive enzymes and the two hormones insulin and glucagon. The hormones are synthesized in the islets of Langerhans. This tissue consists of two types of cells, the α-cells which produce glucagon and the β-cells which produce insulin. Both hormones are polypeptides.

Insulin. The chemical structure of insulin has been elucidated completely by the brilliant work of F. Sanger. The amino acid sequence is shown in Chapter III,3. The polypeptide has a molecular weight of 5733 and consists of two chains that are held together by two disulfide bridges. Reduction of the disulfide bonds inactivates the hormone. Insulin easily binds zinc; in the process, double molecules with a molecular weight of around 12,000, or higher aggregates, are formed.

Biosynthesis. Insulin biosynthesis follows the general principles of protein synthesis (cf. Chapter VII,6). Initially a single peptide chain consisting of 84 amino acid residues is synthesized. In this inactive *proinsulin* the disulfide linkages are already in the correct configuration. The scission between amino acid residues 30–31 and 63–64 generates the active hormone:

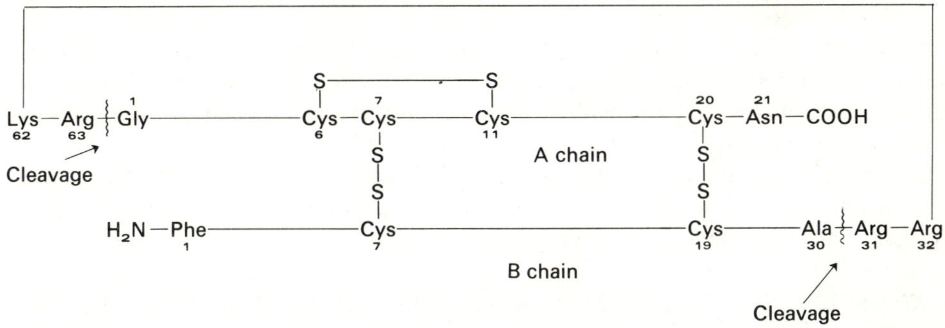

The half-life of active insulin is only about 35 minutes. Inactivation takes place in the liver primarily by enzymatic reduction of the disulfide bonds with the participation of glutathione as reductant.

Biological Effects. Insulin lowers the level of blood glucose very drastically. It increases the permeability of cell membranes for glucose and a few other sugars; at the same time it enhances the breakdown of carbohydrates in the cell, primarily in the *liver*, by inducing the enzyme glucokinase. In *adipose tissue* insulin inhibits lipolysis, i.e., the release of free fatty acids (cf. Section II). Simultaneously the conversion of carbohydrate to fat is enhanced, chiefly by providing NADPH derived from an accelerated pentose phosphate cycle. In the liver the synthesis of fatty acids is also stimulated, whereas the formation of ketone bodies is decreased.

A long-known metabolic disease is *diabetes mellitus*, characterized by elevated blood sugar and urinary excretion of sugar, and often of ketone bodies. Classical diabetes is an insulin deficiency disease. The symptoms can be relieved by insulin injections (being a polypeptide, insulin is destroyed in the gastrointestinal tract and is thus ineffective when administered orally). The treatment is called substitution therapy.

Synthetic pharmaceutical products can also lower blood glucose levels but are not peptides; they are *orally active hypoglycemic* agents. Derivatives of sulfonylurea promote the secretion of insulin from pancreatic β-cells; they thus act by mobilizing endogenous insulin reserves.

Alloxan Diabetes. In experimental animals diabetes can be induced artificially by the administration of alloxan. The results resemble those following pancreatectomy. The β-cells of the insular apparatus are damaged by alloxan, and insulin is no longer produced.

In experimental animals, antibodies against insulin from a different species have been produced, and by means of these antibodies the content of insulin in serum has been determined. In experiments of this type, it was noticed that normal serum contains yet another substance (a protein or polypeptide) with insulinlike activity that does not react with the antibodies. The site of formation and the physiological significance of this substance are not yet known. Its concentration is independent of the level of metabolites, in contrast to that of insulin.

Glucagon. The molecular weight of glucagon is 3500. It consists of one polypeptide chain with 29 amino acid residues, whose sequence is known. The sequence differs greatly from that of insulin. The glucagon peptide is not homologous with insulin but remarkably with secretin (see Section 13).

Glucagon is produced by the α-cells. Its synthesis can be inhibited specifically by the administration of $Co^{2\oplus}$, which creates a condition analogous to alloxan diabetes, but with the opposite symptoms.

Biological Effect. In contrast to insulin, glucagon raises the blood sugar level. It mobilizes the carbohydrate reserves, mostly liver glycogen in two ways, (a) by promoting gluconeogenesis from lactate, and (b) by increasing phosphorylase activity (cf. Chapter XVII,6 and Section 11).

9. Hypophyseal and Hypothalamic Hormones

The Hypophysis (Pituitary Gland). An accessory gland of the brain, the hypophysis consists of two anatomically distinct organs, neurohypophysis and adenohypophysis, and both parts elaborate several different hormones. The pars intermedia (developed to different extents in different animals) produces another hormone. Table XX-1 (Section 1), lists all the different hypophyseal hormones along with their common abbreviations. Many hormones are "master hormones" (Section 1), or *tropic hormones* that act on other endocrine glands and there stimulate the production of the corresponding hormones.

Since the hypophyseal hormones are all peptides, their chemical nature is very similar. Their molecular weights, however, differ considerably; they range from about 1000 to 50,000. Some contain a carbohydrate component and are, therefore, glycoproteins. Unlike other hormones, these peptides and protein hormones possess a certain degree of species specificity; their amino acid sequences may differ from one animal species to the next (cf. Chapter III,3 and below).

Hypothalamic Hormones. The activity of the hypophysis is itself directed to a large extent by the *hypothalamus*. This is where the neurosecretory agents, rather infelicitously[4] called "*releasing factors*," are formed. These agents probably constitute the link between neural and hormonal regulation. Furthermore, they are included in the feedback system: peripheral gland–hypophysis (see Section 1). The hypothalamic releasing hormones will be discussed along with the appropriate hypophyseal hormones.

Neurosecretions. These are hormones or hormonelike substances produced by nerve cells, transported in the axon of a nerve cell, and released into the blood stream by end organs. With the light microscope certain staining techniques can visualize granula in neurosecretory cells. It is not clear what the connection is between granula and hormones; in any case, they are not identical.

Typical *neurosecretory hormones* are *ocytocin*, *vasopressin*, and the *releasing factors* for corticotropin, thyrotropin, and gonadotropins.

The Hormones of the Neurohypophysis. These hormones originate, as mentioned, as neurosecretions. Although the hormones of the posterior lobe differ physiologically, they are very similar chemically. They are oligopeptides consisting of nine amino acid residues with one disulfide bridge. The ring structure arises secondarily by oxidation of the cysteine residues in the molecule. As shown in the formula (Chapter III,3) only two positions (3 and 8) differ; *ocytocin* contains isoleucine in position 3 and leucine in 8; *vasopressin* has phenylalanine and lysine (or arginine) in the corresponding places.[5] *Vasotocin* has been extraced from reptiles, amphibians, and fish; it has isoleucine in position 3 and arginine in 8.

Van Dyke has isolated a protein from the neuropophysis that possesses the activity of both ocytocin and vasopressin (the "Van Dyke protein"). By suitable physical methods, i.e., without cleavage of any peptide bonds, ocytocin and vasopressin can

---

[4] The nomenclature of hypothalamic and hypophyseal hormones is currently being worked out by an international commission.

[5] Compared to the greatly differing physiological effects, the chemical differences at first appear minor. All too often one overlooks the fact that the side chains of amino acids are after all quite different. If one were to compare the complete structural formulas of the peptides, then the differences would indeed appear much greater than they are, for example, among progesterone, corticosterone, and testosterone.

be liberated from the protein. The protein may be considered a storage and transport form of low-molecular weight peptides and can be traced to its neurosecretory origin.

Physiological Effect. *Ocytocin*[6] acts on the smooth muscle of the uterus and enhances contraction; it undoubtedly plays a major role during parturition (by initiating labor). In addition, ocytocin acts on the lactating mammary gland by stimulating the ejection of milk (also through muscle contraction).

Vasopressin, as the name indicates, influences blood pressure; an injection of the substance causes a long-lasting elevation of blood pressure. Furthermore, it excites the smooth musculature of the intestines. In normal physiology, however, the effect on the kidney is most significant. Vasopressin inhibits diuresis, i.e., it helps in the reabsorption of water and consequently affects the concentration of urine. Impaired production of vasopressin results in the syndrome called *diabetes insipidus* in which huge quantities of very dilute urine are excreted; in one documented case, as much as 56 liters per day! A corresponding thirst accompanies such a loss of liquid. Injection of the hormone relieves the symptoms.

Hormones of the Adenohypophysis. Table XX-1 (in Section 1) shows that the anterior lobe produces at least eight different hormones; melanotropin, somatotropin, thyrotropin, corticotropin, luteotropic hormone, follicle-stimulating, and interstitial cell-stimulating hormones. The last three hormones are grouped together under the generic term gonadotropic.

Melanocyte-Stimulating Hormone, or Melanotropin (MSH). MSH is produced in the *pars intermedia* (functionally part of the adenohypophysis). The amino acid sequence of the peptide hormone is known completely (A. B. Lerner; C. H. Li) and greatly resembles that of corticotropin.

Melanotropin promotes the dilation of melanophores in the skin of amphibia and fishes, which darken as a consequence. The hormone can also be extracted from hypophyses of mammals that do not possess melanophores. It can nevertheless cause a darkening of the skin in man (A. B. Lerner). Its biological significance in mammals is not known. Some evidence indicates that it assists in dark adaptation and facilitates the resynthesis of visual purple.

The Growth Hormone, or Somatotropin (GH). This hormone possesses the greatest species specificity, and as isolated from bovine hypophyses is ineffective in man. Most recently, growth hormone has been isolated from human hypophyses (autopsy material) and tested clinically with some success.

All somatotropins studied are proteins. Their molecular weights differ considerably: The hormone from sheep hypophyses has a molecular weight of 48,000; somatotropin from monkeys, 25,000; and human growth hormone (HGH), 21,500. HGH is a polypeptide consisting of 190 amino acids of known sequence with two disulfide bridges (C. H. Li).

---

[6] The name of the hormone is derived from the Greek, meaning "fast birth." An alternate spelling is "oxytocin." The form *ocytocin* is preferred in order to avoid confusion with the prefix oxy = oxygen-containing.

Growth is a complicated process, and the growth hormone must influence a multiplicity of physiological processes. Bone and cartilage growth is stimulated; fat is burned at a higher rate; nitrogen is retained and more protein synthesized; blood sugar is elevated and weight is increased. The weight gain has been used as an assay method for the hormone (along with bone growth). Some effects on metabolism will be discussed later in connection with blood sugar regulation (Section 11). The name "growth hormone" may not be particularly appropriate for this hormone.

The elaboration of somatotropin by the hypophysis is controlled by the *growth hormone-releasing hormone* produced in the hypothalamus. This releasing hormone is produced in increased amounts when the concentration of blood glucose decreases. Since somatotropin elevates the blood glucose level we have another feedback control system. Newly discovered *somatostatin* inhibits the release of growth hormone.

Lipotropic Hormone, Lipotropin (LPH). Li and co-workers recently have isolated from sheep hypophyses a polypeptide consisting of 90 amino acids and possessing high lipotropic activity. The sequence of amino acids is known and turns out to be identical in part with that of ACTH (see below) and with that of human melanotropin. This and other lipotropins found by Li act on adipose tissue, where it releases free fatty acids, which are then transported to the liver and other organs and combusted. Growth hormone has the same activity, which is why the existence of a separate lipotropic hormone has been doubted for so long.

The Thyroid-Stimulating Hormone, or Thyrotropin (TSH). It is one of the glandotropic hormones and appears to be a glycoprotein with a molecular weight of around 30,000. Complete purification has not yet been achieved.

Its *biological effect* is to control the thyroid gland. Thyrotropin stimulates the elaboration of thyroxine. Circulating thyroxine has a feedback effect on the hypophysis, inhibiting the further elaboration of the stimulating hormone (compare with corticotropin, below).

An *exophthalmos-producing substance* has recently been separated from thyrotropin. It is a glycoprotein with the molecular weight of 40,000. The substance causes exophthalmos in *Basedow's disease*; no other biological functions are known.

Thyrotropin-Releasing Hormone. The production of thyrotropin is controlled by a neurosecretion of the hypothalamus. The substance has been extracted from tissue and found to be the tripeptide *pyroglutamyl-histidyl-prolinamide* (for the formula see Chapter III,2).

Adrenocorticotropic Hormone, or Corticotropin (ACTH). ACTH is a polypeptide consisting of 39 amino acids. The sequence is known and reveals certain species-specific differences. The fragment consisting of the first 23 amino acid residues is fully active, and is now produced synthetically for therapeutic application.

The physiological effect of corticotropin consists in the stimulation of the adrenal cortex; the synthesis of adrenocortical hormones is enhanced. The mechanism of action of ACTH involves (a) hydrolysis of cholesteryl esters (the storage form of cholesterol in the adrenal cortex) to provide precursor for hormone production; (b) the stimulation of the mitochrondrial side-chain cleavage of cholesterol (the rate-limiting step); and (c) the stimulation of direct glucose oxidation to provide sufficient NADPH for the various hydroxylation reactions occurring in corticosteroid synthesis. At least some of these mechanisms of action of ACTH entail the stimulation of the adenylate cyclase system (see Section 1).

Corticotropin-Releasing Hormone (CRF). This hypothalamic hormone was the first "releasing factor" whose existence was established. It is a peptide, now extensively purified, but whose structural formula is still not determined unambiguously. It acts on the adenohypophysis by increasing the elaboration of corticotropin especially in response to stress (see below).

Control of Corticotropin Production. The elaboration of corticotropin depends closely on the concentration of circulating adrenocortical hormones. This feedback control loop comprising the hypothalamus, the adenohypophysis, and the adrenal cortex strives for a constant concentration of corticosteroid hormones in the manner shown in Fig. XX-1 (in Section 1). External factors, however, can affect this control loop by shifting its "set point." Selye's group of especially stressful factors (trauma, infections, low temperature, etc., and even psychic factors) result in increased elaboration of corticotropin and consequently increased corticosteroid production (chiefly cortisol). This response constitutes a defensive reaction to stress.

Gonadotropic Hormones. Numerous hormones exert an influence on the activity of the gonads. In contrast to sex hormones, they are not sex specific— At present three effects are distinguished:

1. The *follicle-stimulating effect* promotes the development of the follicles in the ovary and the germinal cells of the testes.
2. The *interstitial cell-stimulating effect* enhances the production of hormones in the interstitial cells (estrone or testosterone), and is identical with the luteinizing effect (conversion of the follicles to the corpus luteum).
3. The *luteotropic effect* stimulates the production of progesterone in the corpora lutea.

For each of these effects, one hormone is elaborated by the hypophysis. Several other gonadotropins are found in *urine*, some of them originating in the hypophysis, and some in the placenta (chorionic gonadotropins). The latter differ chemically from the hypophyseal hormones. Biologically, they commonly have FSH and ICSH activity, although in fluctuating proportion depending on source and method of preparation.

Gonadotropic hormones exhibit a certain species specificity in their chemical constitution. Extracts from human hypophyses are very much more active clinically than animal gonadotropins.

Prolactin, or Lactotropin. This hormone is identical with the *luteotropic hormone* (LTH). It has been purified from sheep hypophyses and found to be a protein with a molecular weight of 23,000; the sequence of its 198 amino acids is known (C. H. Li).

Prolactin stimulates the secretion of milk in the mammary gland. The tissue grows and the production of milk rises. Prolactin releases the breeding instinct in many animals, and thus has a strong psychic effect.

Another characteristic property of the hormone is its luteotropic effect: In animal experiments, prolactin stimulates progesterone production in the corpus luteum. It is not yet known whether prolactin and LTH are identical in all species. For the regulation in the menstrual cycle see Section 12.

Prolactin Inhibiting Hormone (PIH). Whereas the elaboration of other pituitary hormones is stimulated by corresponding hypothalamic "releasing factors," the production of prolactin is *inhibited* by a hypothalamic hormone. Only when the chronic secretion of inhibiting factors is decreased does the inhibition vanish commensurately and the elaboration of prolatin rise temporarily.

Follicle-Stimulating Hormone (FSH). This hormone also varies with its origin. Molecular weights between 25,000 and 36,000 are reported. Besides amino acids, the molecule also contains carbohydrates (galactose, mannose, fucose, hexosamine, and *N*-acetylneuraminate). It is interesting that activity is lost when neuraminate is split off enzymatically.

The follicle-stimulating hormone augments growth and development of the gonads: in the ovary, the follicles enlarge. In the testes, spermatogenesis is stimulated. The hormone is particularly active during the menstrual cycle (see Section 12).

A *FSH-releasing hormone* formed in the hypothalamus controls the elaboration of FSH. It is evidently identical with the LH-releasing hormone (see below).

Luteinizing Hormone, or Interstitial Cell-Stimulating Hormone (LH). This hormone is also a glycoprotein (25% carbohydrate) with a molecular weight of 30,000 (from hog hypophyses). Only a few details are known about its structure.

The interstitial cell-stimulating hormone of the hypophysis controls the production of gonadal hormones. In the testes, Leydig's interstitial cells are stimulated to produce testosterone; in the female organism the follicles are stimulated to produce progesterone and estradiol (see also Section 12).

LH-Releasing Hormone (LHRH). Elaboration of LH is regulated by a hypothalamic neurosecretory hormone, which is a decapeptide whose amino-terminal amino acid is pyroglutamate (pGlu), as was the case with thyrotropin releasing hormone. The complete structure (Schally) of LHRH is as follows:

$$\text{pGlu-His-Trp-Ser-Tyr-Gly-Leu-Arg-Pro-Gly-NH}_2$$

This decapeptide possesses at the same time the activity of a FSH-releasing hormone. The production of LHRH in turn responds to the concentration of circulating sex hormones by negative feedback.

Chorionic Gonadotropins. During pregnancy, much gonadotropic hormone is excreted in urine. This fact is the basis of several pregnancy tests. The hormone is produced in the placenta, not in the hypophysis. *Human chorionic gonadotropin* (HCG) resembles the interstitial cell-stimulating hormone in its effects; it enhances estrone and progesterone production and thus, secondarily, the development of the uterus. Chemically, it resembles the hypophyseal hormones, since it is also a glycoprotein with a molecular weight of 30,000 and a very high content of carbohydrate.

Gonadotropins have also been found in the postmenopausal urine of women; the preparation is called *human menopausal gonadotropin* (HMG). Its effect is that of a follicle-stimulating hormone; it is possibly a derivative of the hypophyseal hormone, because its activity is lower.

10. Further Proteohormones

Relaxin. Relaxin may be considered to be an additional female sex hormone. It is formed probably in the ovaries. Its purification has not yet been achieved, though it appears to be a protein with a molecular weight of about 12,000.

Its physiological effect consists in relaxing and softening the symphysis pubis. Many mammals (e.g., guinea pigs) cannot have a normal birth without relaxin. Relaxin has also been detected in man and should be important in the initial phases of birth.

Erythropoietin. This hormone is formed in the kidneys. It has been concentrated from blood plasma of sheep made anemic and was found to be a glycoprotein (Goldwasser and co-workers).

Erythropoietin accelerates erythropiesis, which is the formation and maturation of red blood cells. It has been known for a long time that prolonged hypoxia (e.g., a stay at high altitudes, cardiac or pulmonary insufficiency, or increased disintegration of erythrocytes) stimulates erythropoiesis. Such control is due to the hormone erythropoietin; at times of insufficient oxygen supply, more hormone is secreted and the bone marrow is incited to greater activity. In particular the maturation of erythrocytes and the synthesis of hemoglobin are stimulated. Presumably the stimulation of messenger RNA is involved.

11. Hormone Regulation of Blood Glucose

The concentration of glucose in human blood ranges normally between 4–5 mmoles per liter (720–900 mg/liter). The plasma level of glucose thus is held extraordinarily constant; this is significant for the nourishment of the individual tissues. The brain, for example, has practically no reserves of oxidizable carbohydrate and depends for its proper function on blood sugar. Moreover, one must consider that sugar is not taken up continuously and steadily, but rather intermittently in batches, with food during meal times. The regulation of blood sugar levels by dietary intake therefore is not possible. Regulation instead must involve such mechanisms as the addition to or mobilization of storage forms of glucose (liver and muscle glycogen), the conversion to fat depot, and gluconeogenesis from the lactate/pyruvate group and from certain amino acids (cf. Chapter XV,10).

The control of the sugar level therefore involves both glucose-producing and glucose-consuming processes; these processed are:

Glucose producing	Glucose consuming
Glycogen breakdown	Glycogen formation
Conversion of galactose and fructose	Oxidation
Gluconeogenesis	Fat formation

Numerous hormones participate in the control of blood sugar level. The entire system of regulation is represented schematically in Fig. XX-5. As with most diagrams, it is somewhat oversimplified.

Elevation of blood glucose is achieved most easily by mobilization of the depot, i.e., by phosphorolysis of liver glycogen. This is the site of attack of *glucagon*; it activates phosphorylase (for the mechanism see Chapter XVII,6). In addition, glucagon stimulates gluconeogenesis from lactate.

Cortisol strongly stimulates the conversion of amino acids to glucose. In case of a diet deficient in carbohydrate, gluconeogenesis from amino acids ensure a constant level of blood glucose. The action of cortisol is twofold: a protein catabolic effect and an inhibitory effect on carbohydrate oxidation.

The ***lowering of blood glucose*** is effected by *insulin*. It acts predominantly peripherally by promoting the influx of glucose into the tissues and its subsequent utilization. In muscle, insulin enhances both glycogen formation and glucose oxidation. In the liver, it inhibits the release of glucose (by inhibiting glycogenolysis and probably also gluconeogenesis). In both the liver and the adipose tissue, insulin stimulates the conversion of carbohydrate to fat. At the same time insulin inhibits the release of free fatty acids from adipose tissue into blood plasma. The effect on the adipose tissue is significant for the understanding of diabetes (see below).

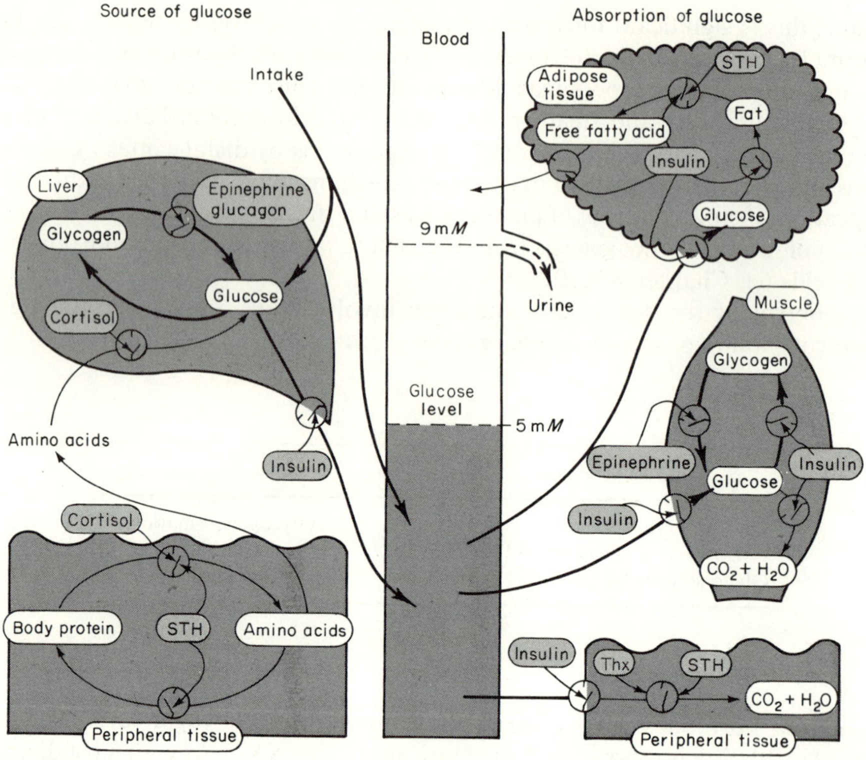

Fig. XX-5. The regulation of blood sugar. The left side shows glucose-yielding processes; the right side, glucose-consuming ones. Hormonal control of these processes is symbolized by the opening or shutting of dampers. STH, somatotropin (or growth hormone); Thx, thyroxine.

Growth Hormone or Somatotropin acts more or less in the opposite direction; it is "diabetogenic." Somatotropin is antagonistic to insulin in that it raises the level of blood sugar, mainly by inhibiting the oxidation of glucose. On the other hand, it generally promotes synthetic processes, especially protein synthesis, and decreases the breakdown of amino acids, which are, of course, indispensable for the synthesis of proteins. Somatotropin thus inhibits gluconeogenesis in favor of accelerated protein production. In summary, both carbohydrate and protein breakdown are suppressed, while fat catabolism is enhanced.

The elaboration of both insulin and growth hormone is controlled directly by the blood level of glucose. A deficiency of blood glucose immediately causes greater production of growth hormone, while the rising blood glucose levels lead to increased secretion of insulin by the β-cells of the pancreas. The pathologically elevated production of growth hormone can cause the exhaustion of the β-cells and consequently hypophyseally dependent diabetes.

Epinephrine finally effects glycogenolysis simultaneously in the liver and the muscle by activating the phosporylase system (cf. Chapter XVII,6).

Although liver glycogen is broken down initially, glycolysis in the muscle is accelerated at the same time. Much of the resultant lactate is reconverted subsequently to liver glycogen, so that paradoxically more glycogen is present in the liver in the end than in the beginning (shift from muscle to liver).

Very generally, it can be stated that the various regulatory mechanisms not only maintain the concentration of blood glucose very constant, but at the same time direct metabolism into different pathways. The guidance of metabolism is the principal function of the adrenal cortex and the hypophysis; the maintenance of a constant level of glucose is primarily the task of insulin. This is reflected in the observation that an elevated content of blood glucose (e.g., by increased food intake) increases the secretion of insulin. A complete feedback loop is in operation here. Accordingly, an insufficiency of the pancreas shows up in the "glucose tolerance test." The action of epinephrine is an emergency reaction which comes into play during drastic lowering of the blood sugar or when unusual circumstances require excessive amounts of sugar. The epinephrine effect is controlled directly by the nervous system.

Diabetes Mellitus. Diabetes is one metabolic disease that has supplied biochemistry with a host of problems. The disease is based on a relative deficiency of insulin (insufficient production of insulin, or increased elaboration of the antagonistic hormones glucagon, somatotropin, and cortisol). A predisposition toward diabetes is inherited, but it takes additional environmental factors to precipitate the disease. The main clinical signs are increased blood glucose levels (possibly accompanied by glucosuria), appearance of "ketone bodies" (= acetone, acetoacetate, and its reduction product β-hydroxybutyrate) in blood and urine with concomitant lowering of alkali reserves in the blood, and an elevation of plasma nonesterified fatty acids.

Because of the conspicuous *hyperglycemia* and *glucosuria*, diabetes has always been considered to be a disturbance of the metabolism of carbohydrates. However, the elevation of blood sugar constitutes in reality a regulatory mechanism: the decreased ability of the periphery to take up glucose from blood is compensated partially by the higher concentration. *Disturbances of fat metabolism* are probably of at least equal significance. During insulin deficiency, ever increasing quantities of free fatty acids pass from the adipose tissue into the blood, and then into the liver where they are broken down to acetyl-CoA by β-oxidation. The large supply of acetyl-CoA then exceeds the capacity of the citrate cycle so that acetyl-CoA is rerouted to the formation of acetoacetyl-CoA and further to free acetoacetate (see Chapter XII,4). This explanation of ketogenesis, still hypothetical in some aspects, can answer only some of the problems of the metabolic disturbances in diabetes.

12. Hormonal Control of the Menstrual Cycle

Cyclic phenomena of physiological activity frequently are regulated hormonally. One example is provided by the reproductive cycle, which differs greatly in different animal species. The estrous cycle of rodents has been of paramount importance in the investigation of the gonadal and gonadotropic hormones, and the main bioassays are performed on rodents (mouse, rat, and rabbit). We will limit ourselves to man, however.

The reproductive cycle of women is characterized by periodic maturation of follicles in the ovaries and by periodic changes of the uterine mucosa. During menstrual bleeding the uterine mucosa is extensively sloughed off. The new cycle is initiated by the hypophysis, which at first elaborates the follicle-stimulating hormone (FSH) and later the interstitial cell-stimulating hormone (or LH). FSH acts on the gonadal cells directly and promotes the maturation of a new follicle. Under the influence of LH the follicle produces estradiol. With the rising concentration of LH, its "luteinizing" effect becomes more prominent. At a certain ratio of LH/FSH, ovulation (rupture of the follicle) occurs and the corpus luteum is developed (cf. Fig. XX-6).

The hypophysis influences mainly the ovaries whereas the processes of the *uterus* are regulated by the gonadal hormones. Estradiol effects production of new mucosa

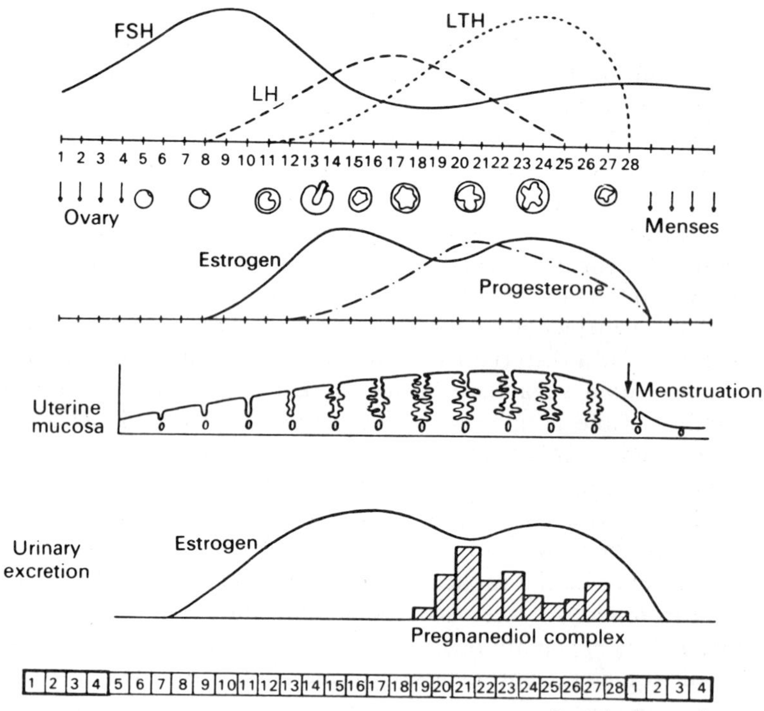

Fig. XX-6. The menstrual cycle. Top: Relative concentrations of the hypophyseal hormones, and below it the activity of the ovaries. Second group of curves: Relative concentrations of the gonadal hormones, and below it the functional state of the uterine muscosa dependent on these concentrations. Excretion of hormone catabolites follows the concentration of the hormones (last curve). After W. Dirscherl *In* "Fermente, Hormone, Vitamine" (R. Ammon and W. Dirscherl, eds.), 3rd ed., Vol. II: "Hormone," p. 169. Georg Thieme, Stuttgart, 1960.

(*proliferation phase*) which ceases shortly before ovulation, and simultaneously exerts an influence back on the hypophysis by inhibiting the elaboration of FSH and enhancing the production of prolactin and LH, which in turn initiates production of progesterone shortly before ovulation. Progesterone promotes transformation of the uterine mucosa to the pregravid state and prepares for the lodging of the fertilized ovum (*secretory phase* of the uterine mucosa). This state persists as long as progesterone is produced.

If the ovum is fertilized, the corpus luteum becomes the corpus luteum graviditatis and gradually accelerates progesterone production, a task later taken over by the placenta.

If the ovum is not fertilized, the corpus luteum atrophies, because production of the luteotropic hormone (LTH) by the hypophysis slows down. In this phase, the stimulating influence of the estrogen on the elaboration of LTH by the hypophysis is presumably absent; progesterone also inhibits production of LH. Regression of the corpus luteum results in decreased hormone production. The secretory mucosa cannot be maintained in the absence of progesterone; it is discarded during menstruation.

With menstrual bleeding the uterine mucosa is returned to the starting point of the cycle. The new maturing follicle produces estradiol, necessary for proliferation, and the process is repeated anew. *The cyclic process depends principally on the alternating stimulation of the ovary by the hypophysis and the feedback control of ovarian hormones on the hypophysis.*

Prevention of Ovulation. Since the adenohypophyseal production of gonadotropins is strongly influenced by circulating steroids, it becomes possible to control the elaboration of gonadotropins by the administration of certain steroids. This mechanism is the basis for hormonal contraception widely used for "birth control." Preparations employed for this purpose contain some orally active progestin (e.g. 17α-ethynyl-19-nortestosterone) that represses adenohypophyseal activity. Since LH is not produced, ovulation is obviated. Such preparations generally contain also an estrogen to stimulate the endometrium. Cessation of medication results in withdrawal bleeding. An artificial cycle can be maintained in this manner.

13. Tissue Hormones

The term *tissue hormone* has been applied to a group of substances that have much in common with the "classical" hormones produced in special glands, but that cannot be included among the hormones in the strict sense.

Certain active compounds of the gastrointestinal tract can be taken as one group of tissue hormones. They are not secreted by discrete glands, but rather are produced by mucosal tissue. They do reach the target organ via the blood stream, however (and for this reason are often classified as true hormones). Another group of agents, histamine, acetylcholine, etc., are produced in many tissues, and their target organ is the site of origin. These substances are hormones only in a very abstract sense. It would indeed by better to introduce a new and neutral name for these primarily pharmacologically active substances.

The Hormones of the Gastrointestinal Tract. These hormones excite the digestive glands. The following can be distinguished:

a. Secretin., the study of which induced Bayliss and Starling to coin the word "*hormone*" (about 1902). The entrance of acidic food pulp (= chyme) initiates the elaboration of secretin by the duodenal mucosa. The hormone stimulates the pancreas to secrete water and bicarbonate and stimulates the flow of bile from liver. Secretin is a polypeptide consisting of 27 amino acid residues, whose sequence is now known and found to be similar to that of glucagon (14 amino acid positions are identical). It seems that the whole molecule is necessary for full activity. The terminal amide is a common structural feature shared by at least four other hormones that act on the flow of body water.

b. Gastrin is formed in the antral mucosa. It stimulates the production of hydrochloric acid in the stomach; at a higher dose level it also stimulates the secretion of pepsin. Two peptides have been isolated from the pig, called gastrin I and II. The sequence of the 17 amino acid residues is known, and the only difference between the two peptides is that the tyrosine in gastrin II is sulfated. The same situation obtains with human gastrin, which differs from porcine gastrin only in one amino acid residue.

c. Cholecystokinin–Pancreozymin is synthesized in the duodenum. The *cholecystokinin effect* is the contraction of the gall bladder whereby bile is expelled. The *pancreozymin effect* is the increased secretion of digestive enzymes by the pancreas. To the best of our present knowledge a single polypeptide of 33 amino acid residues is responsible for both effects (V. Mutt).

The existence of various other substances has been postulated, but they are neither chemically nor physiologically sufficiently characterized to be included here.

Two other hormonelike substances generally regulate blood pressure and the contraction of smooth muscle:

Angiotensin I (formerly angiotonin or hypertensin). A decapeptide, angiotensin I arises by the action of the protease *renin* (from kidneys) from a precursor, a plasma protein of the α_2-globulin fraction called *angiotensinogen.* The further removal of two amino acid residues yields *angiotensin II*, which is probably the true active hormone:

Angiotensin I: H · Asp-Arg-Val-Tyr-Ile-His-Pro-Phe-His-Leu · OH

Angiotensin II: H · Asp-Arg-Val-Tyr-Ile-His-Pro-Phe · OH

Higher doses of angiotensin raise blood pressure. Its physiological role, however, should be seen in the control of adrenocortical activity: Minute amounts of angiotensin II *stimulate the production of aldosterone.*

Bradykinin. (kallidin-9). A nonapeptide, bradykinin arises in similar fashion by the proteolytic action of pancreatic *kallikrein*. The structure of bradykinin is

H · Arg-Pro-Pro-Gly-Phe-Ser-Pro-Phe-Arg · OH

It is a highly effective vasodilating agent and consequently lowers blood pressure. There is evidence that bradkykinin causes functional hyperemia of certain organs. It furthermore effects contraction of smooth musculature.

The following locally acting tissue hormones are all biogenic amines:

Histamine. This hormone arises by the enzymatic decarboxylation of histidine. It is inactivated either by methylation at the imidazole N or via oxidative breakdown by the action of the flavoprotein *diaminoxidase* (see Chapter VIII,5), and then by aldehyde oxidase to imidazolyl acetate.

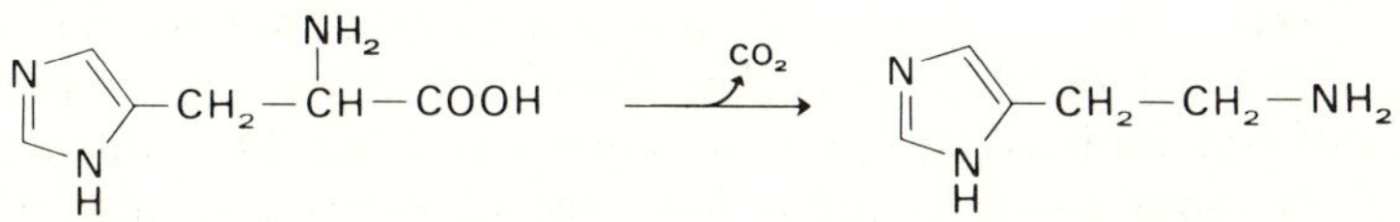

Histidine Histamine

Histamine is widely distributed in the organism. It is particularly abundant in skin and lung tissue, especially in mast cells, but exists there in a bound and inactive form and is released only when needed. Its chief effects are the increase of secretion of gastric juice, the dilation of blood capillaries, and the increase of their permeability. It is suspected that histamine controls local blood circulation in the normal organism. Histamine also participates in the production of allergic reactions. Consequently, substances that inhibit the physiological effects of histamine (antihistamines) are used therapeutically.

Serotonin (5-Hydroxytryptamine). Occurring widely in both animals and plants, serotonin arises from tryptophan by hydroxylation in the 5-position and subsequent decarboxylation (the reverse sequence, proceeding via tryptamine, has been excluded). It is vasoconstricting and, among other effects, migrates from the platelets into the serum during blood clotting. Furthermore, it occurs in the intestinal mucosa, where it promotes peristalsis. Finally, it has been found in the central nervous system, where it serves as *transmitter* at a certain type of synapse.

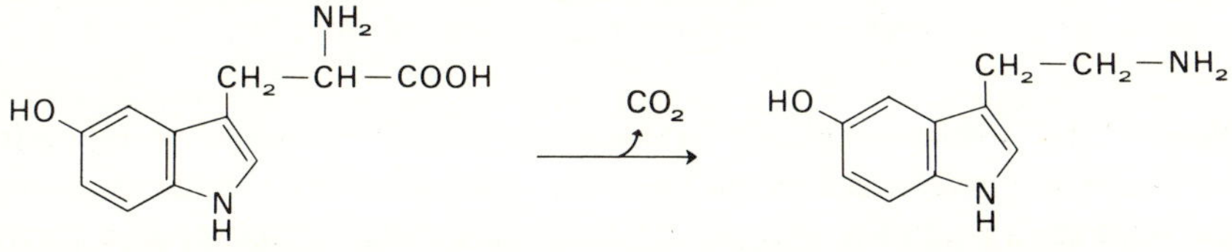

5-Hydroxytryptophan

Serotonin
(5-hydroxytryptamine)

Tyramine. This hormone raises the blood pressure and stimulates smooth musculature (e.g., uterus). More important is *hydroxytyramine* (dopamine) which is formed by hydroxylation of tyrosine and subsequent decarboxylation. On the one hand, hydroxytyramine is the parent substance for the hormones norepinephrine and epinephrine (cf. Section 4), and, on the other hand, it may be another *neurotransmitter*, like norepinephrine, liberated at the ends of sympathetic (= adrenergic) nerves (cf. also Chapter XXIII,8).

Acetylcholine. The *neurotransmitter* substance at most nerve endings (cholinergic nerves; cf. Chapter XIII,8), acetylcholine also lowers peripheral blood pressure.

Phylogenetically acetylcholine is a very old "hormone" occurring in protozoa. It is conceivable that such precursors subsequently evolved into neurotransmitters as well as neurosecretions and the true hormônes.

γ-Aminobutyrate. Our last entry in this group, γ-aminobutyrate, is the decarboxylation product of glutamate. The substance arises mainly in the brain. The function of γ-aminobutyrate is not yet known in detail, but evidently it blocks the synapses. Nervous tissue contains both the decarboxylase that produces γ-aminobutyrate and the transaminase that breaks it down.

14. Hormones of Invertebrates

Hormones are also found among invertebrates. We will single out the hormones of insects from this rather comprehensive area, because they have been investigated most thoroughly.

The juvenile development of insects proceeds through a series of moltings (molting of larvae; among the holometabola also molting of pupae) to result eventually in the adult insect. The most familiar example is the development caterpillar-pupa-butterfly. The process of molting involves three hormones, as shown in Fig. XX-7.

Each shedding is initiated by the activity of neurosecretory cells in the brain; these cells produce a hormone which induces another gland located in the prothorax to elaborate the true molting hormone, *ecdysone*. If ecdysone alone is active, then molting transforms the caterpillar to the pupa and the pupa to the butterfly. Larval shedding (caterpillar to caterpillar) is effected by the elaboration of a second hormone by the *corpora allata* in addition to ecdysone. The second hormone is the so-called *juvenile hormone*, responsible for the development of larval characteristics. For maintaining the caterpillar stage, two hormones are necessary; for development to the pupal stage, or to the butterfly, only one.

Ecdysone. This hormone has been obtained pure and crystalline (Butenandt and Karlson). Surprisingly, the structural formula turned out to be that of a steroid hormone and reveals a certain kinship to cholesterol. Another closely related hormone, 20-hydroxyecdysone, is the molting hormone of crabs; it has now been found in plants as well, but its role there is unknown.

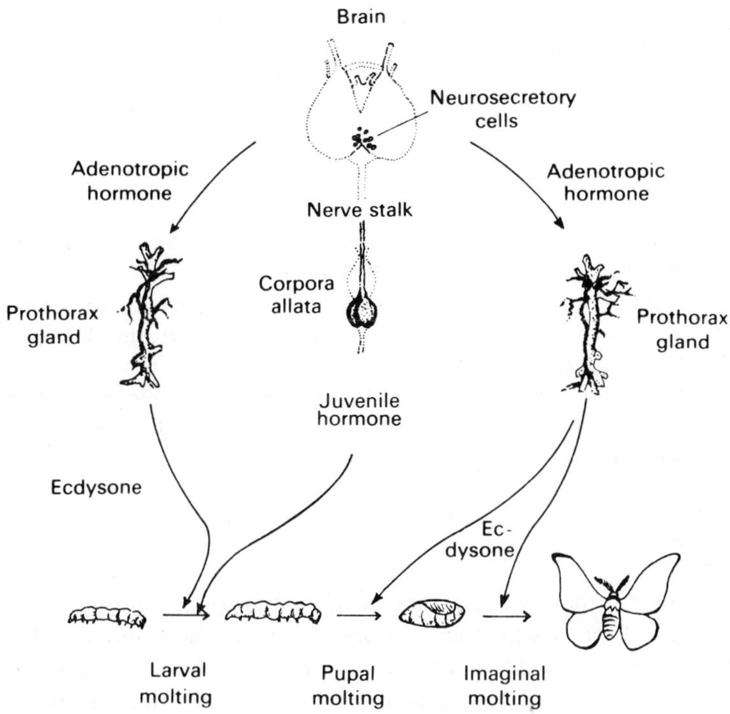

Fig. XX-7. Action of insect hormones. In the upper part, the hormone-producing glands are represented. Below that, the molting stages initiated by the hormones are shown.

In the larvae of flies ecdysone brings about formation of the puparium by acting on the metabolism of tyrosine (Karlson). The first recognizable effect is the activation of certain genes. This discovery was of great significance for our concept of the mechanism of action of hormones in general (cf. Section 1).

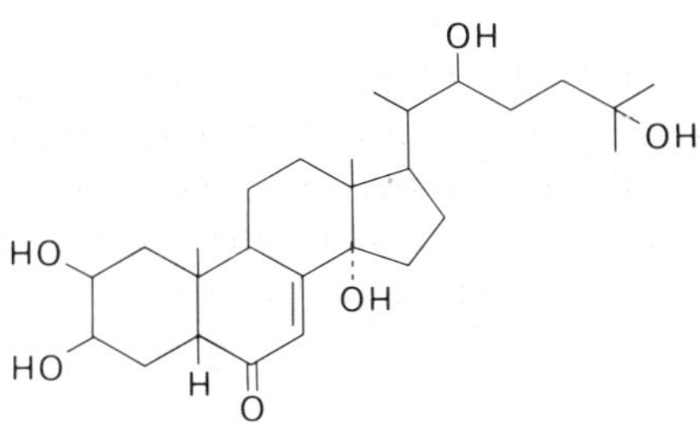

Ecdysone

Juvenile Hormone. Obtained in pure form (Roeller and co-workers), the juvenile hormone has the following structure, substantiated by chemical synthesis:

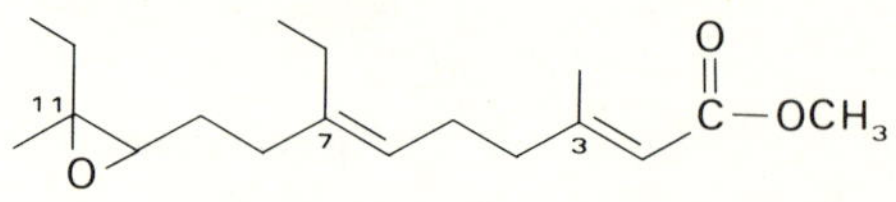

Juvenile hormone

Juvenile hormone can readily be identified as a homolog of an isoprenoid compound, in particular as the methyl ester of a farnesoic acid derivative with an epoxide group and two ethyl groups in place of methyl groups. Various derivatives of farnesenic acid and other plant products, e.g., *juvabion* of balsam fir, have the same activity, but farnesol itself is only feebly active.

Juvenile hormone is important not only for the juvenile development but it acts also in the adult insect as a kind of "gonadotropic hormone" on the maturation of eggs by being the yolk-forming hormone.

15. Pheromones

Active substances that mediate humoral correlations among individuals of a given species have been designated *pheromones* by Karlson and Lüscher. Classical examples are the sex attractants of insects which are emitted by the female and which attract males, often across remarkable distances. In the silk moth this active principle is produced in special scent glands; it has been extracted from the glands and has been purified. The sex attractant of the silk moth is a long-chain alcohol ($C_{16}H_{30}O$) with two double bonds (Butenandt, Hecker, and Stamm).

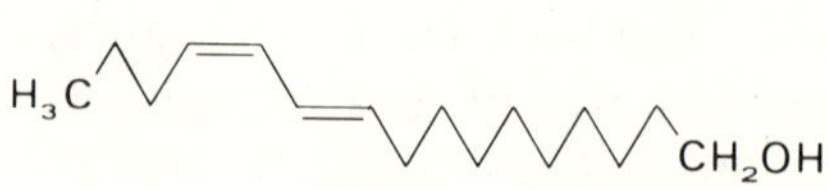

Bombycol, sex attractant of the silk moth

The steric configuration at the double bonds is very critical for biological activity. The pheromone is received by the male with its olfactory sense. Probably only a few molecules per sensory cell suffice to trigger the biological reaction (D. Schneider).

Similar attractants are known for other butterflies, for beetles, and for cockroaches. Pheromones are also used by ants and termites for marking trails and by mammals to stake out territories.

16. Growth Substances of Plants

Growth-promoting substances of plants are also known as *phytohormones*.

Auxin (β-indolylacetic acid). A phytohormone, auxin effects elongation of cells. It is formed in the apex of the plant, especially the germinal shoot, and promotes stretching of the basally located cells. Auxin is also involved in the various tropisms of the plant.

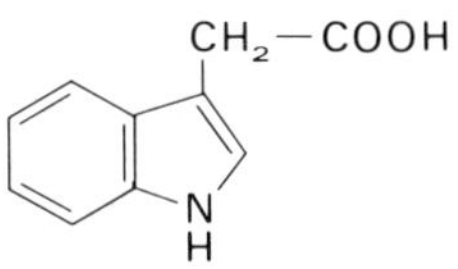

Auxin
(= indolylacetic acid)

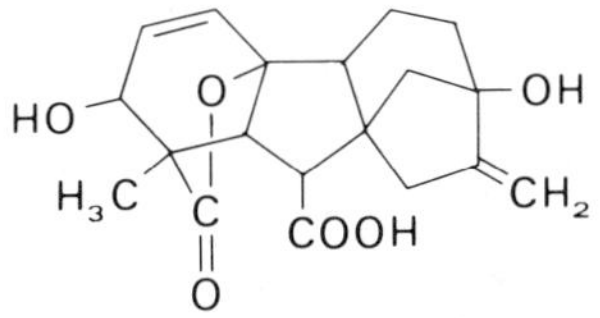

Gibberellic acid

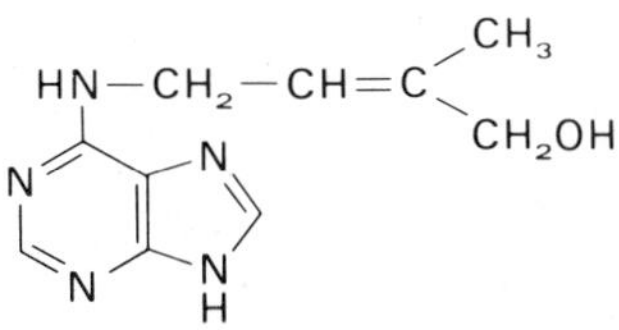

Zeatin

Gibberellic Acid. Isolated initially, gibberellic acid is a product of a phytopathogenic fungus; other gibberellins have also been found in higher plants. They, too, effect cell elongation (particularly of the chief axis) and cause gigantism, but may also stimulate cell division. Gibberellic acid possesses some effects that are also ascribed to the as yet hypothetical "flowering hormone."

Cytokinins. These are plant hormones that stimulate cell division. They are involved mainly in the development of buds where they are antagonists to auxin. The chemical structure is a derivative of adenine with a substitution of the 6-amino group. *Zeatin* has been isolated from maize plants; it contains a hydroxylated isoprene residue.

A pronounced mitotic effect is exhibited by *kinetin* (6-furfurylaminopurine), which has been isolated as a conversion product of deoxyribonucleic acids (derived from animal sources).

A number of "growth substances" of microorganisms are not hormones, but rather vitamins. For nearly every water-soluble vitamin some definite bacterial strain is known that requires this substance as "growth substance" (growth vitamin).

Abscisic Acid. This phytohormone promotes autumnal leaf dropping and ensures that buds remain dormant in the winter. Abscisic acid is an antagonist to other phytohormones. Its chemical structure is related to that of the carotenoids.

BIBLIOGRAPHY

E. B. Astwood (editor), "Recent Progress in Hormone Research" (Proceedings of the Laurentian Hormone Conferences), annual volumes. Academic Press, New York.

S. A. Berson (editor), "Methods in Investigative and Diagnostic Endocrinology." Vol. II, "Peptide Hormones." American Elsevier, New York, 1973.

C. Ezrin, J. O. Godden, R. Volpe, and R. Wilson (editors), "Systematic Endocrinology." Harper and Row, New York, 1973. Clinically oriented.

W. F. Ganong and L. Martini, "Frontiers in Neuroendocrinology, 1973." Oxford University Press, New York, 1973. Up-to-date, excellent reviews on releasing factors and on mechanisms of action of hormones in the brain.

E. V. Jensen and E. R. DeSombre, Mechanism of action of the female sex hormones. *Annu. Rev. Biochem.* **41,** 203–230 (1972).

P. Karlson, Pheromones. *Ergeb. Biol.* **22,** 212–225 (1960).

P. Karlson, Chemistry and biochemistry of insect hormones, *Angew. Chem. Intern. Ed. Engl.* **2,** 175–182 (1963); see also reference in Footnote 3, Section 1 of this chapter.
J. H. Law and F. E. Regnier, Pheromones. *Annu. Rev. Biochem.* **40,** 533–548 (1971).
G. Litwack (editor), "Biochemical Actions of Hormones." 2 vols. Academic Press, New York, 1970, 1972. Vol. II particularly deals with most topics of this chapter.
L. Martini and W. F. Ganong (editors), "Neuroendocrinology," 2 vols. Academic Press, New York, 1967.
V. Mutt and J. E. Jorpes, Contemporary developments in the biochemistry of the gastrointestinal hormones. *Recent Progr. Hormone Res.* **23,** 483–503 (1967).
K. Sláma, Insect juvenile hormone analogues. *Annu. Rev. Biochem.* **40,** 1079–1102. (1971).

CHAPTER XXI

Mineral Metabolism

Inorganic ions and metal complexes have frequently been mentioned in preceding chapters. These substances are also metabolized. They are taken up in food and eliminated again in urine, feces, and sweat.

The metabolism of inorganic ions, or simply "mineral metabolism," differs in one essential point from the metabolism of substances discussed so far. In contrast to proteins, carbohydrates or fats, minerals are neither produced nor consumed in the organism. Their intake from food can be regulated only roughly, if at all. Most animal species, nevertheless, in the course of evolution have developed the ability to keep the concentration of ions constant in the body fluids, thus providing a constant "*milieu interne.*" This is achieved principally by regulating excretion. Several ions have special depots which can be mobilized in periods of insufficient intake.

Salt regulation is of special significance in aquatic animals. Freshwater fish have a much higher concentration of salt in their blood than in their environment. The opposite is true for saltwater fish. Very considerable demands of the salt regulatory mechanism are made by species—such as salmon—which migrate from the sea into rivers and must inhabit regions of widely differing salt content, and yet the salt concentration of the body fluids does not change appreciably with the location of the species.

Mineral metabolism and osmoregulation should interest not only the biologist but also the clinician, because disturbances in the electrolyte balance in man are rather common. Although our discussion will center around the situation in man, we will point out unusual cases of biological interest.

1. Water Balance

Water is by far the most important inorganic component of the organism. In man, 71–73% of the fat-free body mass is made up of water. In some animals (for example, the jellyfish) the water content may be as high as 98%. It is the common solvent and imbibition medium of organic material; it is indispensable for the function of proteins and the metabolic processes in cells—the old maxim still applies, "*corpora non agunt nisi soluta*,"—and, finally, it can participate as a reactant in metabolic conversions. Outside of the cells it becomes the vehicle for transport, as exemplified so well by the blood circulation. Last, it assists in the regulation of temperature. Evaporation is the principal method of conducting heat to the outside and of dissipating it (consult textbooks of physiology).

Distribution of Water. In a discussion of the distribution of water and minerals in the organism, we must distinguish three major compartments: The *intracellular space*, that is, the total space occupied by the fluids within all the cells; the *extracellular space*, which is further subdivided into the water of blood plasma and the interstitial fluids; and the *transcellular space*, which is the content of the intestinal tract. These compartments are not rigidly separated. A considerable amount of fluid is constantly shifted around, for instance, the digestive tract is flooded daily by over 8 liters of a fluid consisting of saliva, gastric juice, and pancreatic secretion. The bulk of the water is reabsorbed by the lower small intestine and the colon and taken up by the blood. Fluid is also exchanged constantly between the vascular system and the interstitial space (cf. Section 2, "The Colloid-Osmotic Pressure"). The blood vessels and the cell membranes are very permeable to water and most dissolved substances, especially to "nutrient substances," such as glucose or amino acids.

Inulin, a polysaccharide of fructose, easily enters the interstitial space but cannot enter the cells and, for this reason, is used to estimate the size of the interstitial space. A certain amount of inulin is injected, and after equilibrium has been reached, the concentration in the blood is determined. From this the total amount of the fluid in which inulin was distributed can be calculated. This amount is equivalent to the "extracellular space"; by subtracting the blood volume from this, one calculates the interstitial space. Substances other than inulin have been used for the determination of these physiological values.

Water Balance. The organism possesses a relatively limited supply of freely available water. Excessive fluctuations in water content must be avoided. Normally the water content is well balanced. The intake of water from beverages, from the water content of solid food, and from the water of oxidation is compensated by the excretion of urine and feces and by perspiration.

There is one item on the intake side which is easily overlooked, and that is *water of oxidation*. As we have seen, the respiratory chain is the most important energy-yielding process. It constantly produces water—in the normally nourished man, about 300 to 400 gm per day. In some organisms the water of oxidation can fill the total requirement for water (desert animals, clothes moths). In these cases, the excretion of water is reduced to a minimum.

In hot climates we must further consider the output of water through perspiration which is aggravated by strenuous exercise. It can amount to several liters per day and must be balanced by an increased intake of fluids.

Regulation of the Water Balance. Water content is balanced chiefly by two mechanisms: thirst, which demands an increased intake of fluids, and the activity of the kidneys, which either conserves water or flushes it out. It is well known that the kidneys produce about 180 liters of primary urine each day. Of this, 178–179 liters are reabsorbed. This function of the kidney is controlled by *vasopressin*, a hormone of the neurohypophysis (cf. Chapter XX,9). In severe cases of *diabetes insipidus* (vasopressin deficiency), reabsorption is greatly diminished and large volumes of urine (20–30 liters per day; even up to 50 liters) are excreted.

Thirst provides the counterbalance in this regulation. Even with excessive perspiration, the fluid balance can be restored by the sensation of thirst, which demands an increased intake of water. However, from a quantitative standpoint this regulation is imperfect and only very approximate. Complete diurnal balance generally is not achieved. Input and output are balanced only over longer periods of time.

2. Water as Solvent

Water is an excellent solvent for organic and inorganic substances. Most salts dissociate completely to their individual ions upon dissolving in water; acids only to a small extent (cf. Chapter I,2, dissociation of carboxylic acids). Organic substances are usually classified as either hydrophilic or hydrophobic. The first group includes amino acids and proteins, nucleic acids, and carbohydrates; the second group covers fats and lipids (cf. Chapters XII–XIV).

The behavior of solutions is described by a series of physical and chemical laws which, of course, apply within living organisms as well. We will briefly discuss these laws (which also hold for nonaqueous solutions), because they form the basis for many vital processes.

Units of Amount and Concentration. In chemistry, the customary unit of amount is the *mole* (1 mole equals the molecular weight expressed in grams) and the unit of concentration is *moles per liter* or *molar* (abbreviated *M*). Since in biochemistry we almost always deal with small amounts and low concentrations, we use units which are smaller by factors of 1000; they are the *millimole* (mmole), for amounts, and the *millimole per liter* or millimolar (m*M*), for concentrations. The next smaller units of concentration are μM (micromolar $= 10^{-6} M$), n*M* (= nanomolar, 10^{-9}), and p*M* (= picomolar, $10^{-12} M$). Concentrations of solutes in tissues are best expressed as micromoles per gram (μmoles/gm) of fresh weight. Care should be taken not to use the abbreviation of concentration m*M* (= millimolar) incorrectly for millimoles, which are amounts.

In the older medical literature, concentrations are often expressed as mg/100 ml: this has been called completely incorrectly "milligram percent," and has been abbreviated as mg %. Reputable scientific journals[1] do not permit this confusing designation; it should be avoided under all circumstances (see also footnote 3, Section 4). If for some reason concentrations are expressed in terms of weight per volume instead of moles per volume, the internationally acceptable designation of *parts per million* should be used; 1 mg per 100 ml corresponds approximately to 10 ppm.

In dealing with electrolytes, it is frequently advantageous to take into account the valence of the ions and to express amounts in *equivalents* or *milliequivalents*. The equivalent weight is the atomic or molecular

[1] For instance, the *Journal of Biological Chemistry*; see "Suggestions to Authors." It is anticipated that the use of molar concentrations will also be required in clinical chemistry.

weight divided by the valence of the ion. In the case of $Ca^{2\oplus}$, for example, 1 milliequivalent equals 40.1 mg divided by 2 (the atomic weight of calcium is 40.1), which is electrochemically equivalent to 1 milliequivalent of $Na^{\oplus}$ or 23 mg. Concentrations are measured by the unit *milliequivalent per liter*, which is abbreviated as meq/liter. We shall make use of this unit.

Osmolarity is a measure of the osmotically active concentration. For nonelectrolytes it is identical with molarity, but for dissociable substances, the molarity must be multiplied by the number of particles (ions) which arise during dissociation (cf. this Section, below). However, this applies only to very dilute solutions. Solutions more concentrated than $10^{-3}M$ deviate from the properties of so-called ideal solutions. To compensate for this fact, *activity factors* have been introduced: The stoichiometric concentration is multiplied by the appropriate correction factor; e.g., for a 0.1*M* NaCl solution the factor is 0.785. The osmotic activity of such a solution is then $(0.1 \times 0.785) \times 2 = 0.157$ osmolar. We multiplied by 2 because we have the two ions $Na^{\oplus}$ and $Cl^{\ominus}$.

The activity factor goes from 1.0 for infinitely dilute solutions to smaller and smaller values with increasing concentration and increasing ionic content.

For calculations in which the interaction among the individual ions is significant, the concept of *ionic concentration* has been introduced. The ionic concentration is arrived at by adding the products of the molar concentration of each ionic species and the *square of its charge*:

$$\Gamma = c_1 \times z_1^2 + c_2 \times z_2^2 + c_3 \times z_3^2 + \cdots + \cdots$$

For a 0.5*M* $CaCl_2$ solution, the following ionic concentration can be calculated, corresponding to the dissociation of

$$CaCl_2 \rightleftharpoons Ca^{2\oplus} + 2\,Cl^{\ominus}$$

$$\Gamma = (0.5 \times 2^2) + (2 \times 0.5 \times 1^2) = 3.0$$

One-half of the ionic concentration is called "*ionic strength* of the solution" and is symbolized with μ; it plays a very important role in the theory of electrolyte solutions.

Diffusion. Free diffusion can be observed in all states of matter, but here we are mostly concerned with diffusion in liquids. If a pure solvent is carefully layered on top of a solution, one may observe a gradual mixing. This process is called diffusion (Fig. XX-1). The dilution of the solution is explained by the spontaneous and exergonic nature of the process, accompanied by a rise of molecular disorder; that is, a considerable increase of entropy.

The velocity at which diffusion proceeds depends on the difference of concentrations and on the diffusion constant D:

$$dn/dt = -D \times q \times dc/dx$$

where dn/dt = transport of substance per unit time, q = cross section, and dc/dx = change of concentration along the x axis (cf. Fig. XXI-1a and b).

For spherical molecules Einstein has evolved a relation between D and the molecular radius r as follows

$$D = \frac{RT}{N_L \times f} = \frac{RT}{N_L \times 6\pi \times \eta \times r}$$

The formula is derived from Stoke's law for the friction of a sphere moving through a liquid of a viscosity η. The friction coefficient f is then equal to $6\,\pi\eta r$. Friction increases as the shape of the molecule deviates from that of the sphere; this is taken into account by the ratio of frictions f/f_0, which is greater than unity. This value is of particular importance in protein chemistry; it permits predictions concerning the asymmetry, or the ratio of axes, of macromolecules.

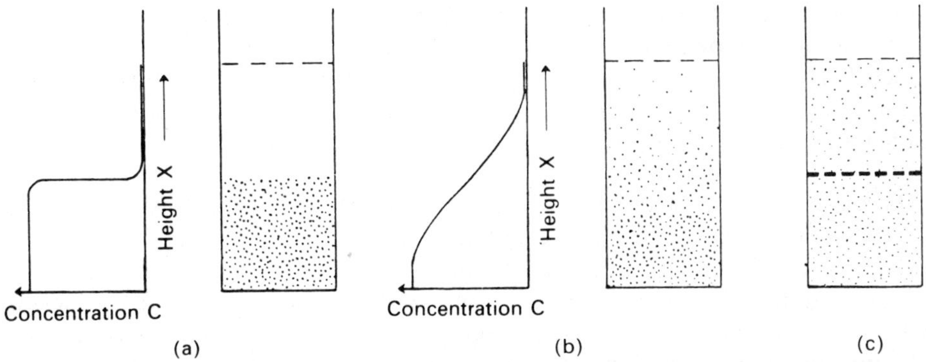

Fig. XXI-1. Diagram illustrating diffusion. In (a) the solution has been covered with a layer of pure solvent; the concentration changes abruptly, as indicated by the concentration curve (at left). In (b) partial mixing has occurred; the concentration curve has flattened out. In (c) a membrane represents restricted diffusion. Since free diffusion takes much less time, complete homogeneity is reached within each compartment long before the two compartments are equalized: the change in concentration is localized at the membrane.

Restricted Diffusion and Facilitated Diffusion. If we insert into the tube (Fig. XXI-1) a membrane through which both solvent and dissolved substance can pass, that is, a completely permeable membrane, then diffusion of dissolved substance (and in exchange for it, solvent) will take place, but at a reduced rate. The rate now depends on the pore size and the thickness of the membrane, which determines the pathway. We may introduce a membrane constant M_K which depends on D, the thickness d, and the effective pore area q. The law of diffusion can then be expressed as $dn/dt = -M_K \times c$ (where c = difference of concentrations).

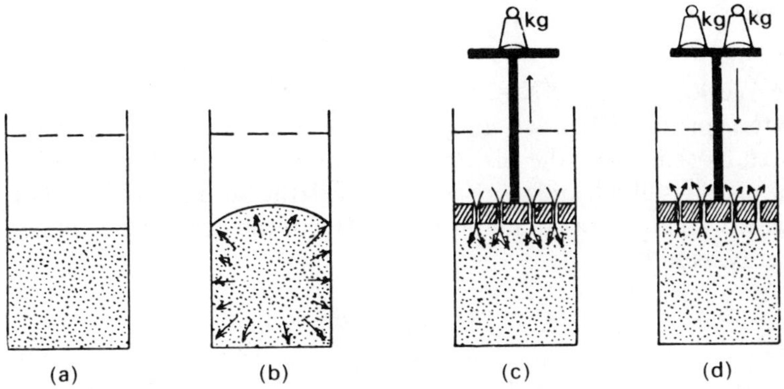

Fig. XXI-2. A diagram of osmosis. In (a) the start of the experiment is shown; the barrier is supposed to be semipermeable. Equilibrium is reached in (b): solvent has migrated to the inside: the pressure has risen and is balanced by mechanical pressure of the membrane. The drawings (c) and (d) illustrate osmotic work. The piston is lifted in (c) letting the solution become dilute (water flows in). In (d) the solution becomes more concentrated when water is pressed out through an expenditure of mechanical work.

In some circumstances, diffusion is very slow because of the very small pore size. Several substances or ionic species may be observed to pentrate more rapidly than expected. This phenomenon is called *facilitated diffusion*. The mechanism resembles that of active transport, although its main criterion, namely the migration of material against a concentration gradient, does not apply here (cf. Section 3).

Osmosis. If we replace the fully permeable membrane of Fig. XX-1 with a semipermeable one, then we obtain the situation represented in Fig. XXI-2. The solute cannot become diluted by diffusion; dilution is achieved by the infiltration of water (solvent) into the lower compartment, whereby the pressure in this area is increased (in the drawing this is indicated by a bulging of the membrane). This pressure can be measured, for example, with a mercury manometer, and the following relationship is observed:

$$\Pi = c \cdot RT$$

or, expressed in words, the osmotic pressure is directly proportional to the concentration c and the absolute temperature T (R = the general gas constant).

The osmotic pressure is an additive effect of the individually dissolved particles. In the case of NaCl, due to electrolytic dissociation, we have in solution the ion $Na^{\oplus}$ along with the $Cl^{\ominus}$ ion; Na_2SO_4 dissociates into the ions $Na^{\oplus}$, $Na^{\oplus}$ and $SO_4{}^{2\ominus}$. With electrolytes, therefore, one must multiply the molar concentration by the number of ions which are expected during solution in order to arrive at the osmotically effective concentration, the *osmolarity*. In addition, with more concentrated solutions, the activity coefficients must be taken into account (see above).

The phenomena of melting point depression, vapor pressure depression, and boiling point elevation are closely related to osmotic pressure. All these effects can be explained by the fact that the solvent loses some of its "chemical potential" upon dissolving a substance. An aqueous solution of the concentration of 1 osmole/liter shows a melting point depression of 1.86° and an osmotic pressure of 22.4 atm (at 0°); or of 25.5 atm (at 37°).

Osmotic Work. Figure XXI-2c represents a cylinder with a freely mobile, but well-insulated, piston which we imagine to be semipermeable. As solvent diffuses into the area closed off by the piston, the volume of the enclosed compartment increases, the piston is lifted, and work is performed (the lifting of a load). The amount of mechanical work can be calculated from the law of the dependency of concentration on free energy as follows:

$$\Delta G = \Pi \times V = RT \times \ln \frac{c_2}{c_1}$$

(Π = osmotic pressure; V = volume increase; c_1 and c_2 = the concentrations before and after dilution).

In principle, osmosis can result either in an increase of concentration (by external expenditure of mechanical work) or in a decrease of concentration. The former—the increase of concentration by pressure—is found in the colloid-osmotic phenomena (see below). The second is far more common, since it corresponds to spontaneously occurring processes. In the organism, osmosis is primarily responsible for shifting fluids among the various fluid compartments; for example, if much salt has penetrated into the interstitial space, then water is removed from the cells by osmosis, whereby the concentrations inside and outside of the cells equilibrate.

The Colloid-Osmotic Pressure. Besides semipermeable membranes which allow passage only for water and not for dissolved substances, there are also those which allow substances of low molecular weight, but not of high molecular weight, to penetrate. It can easily be seen that there may be pores whose diameter is smaller than that of protein molecules. In this case, salts can diffuse freely and give no cause for differences in pressure, while the law of osmosis still applies to macromolecules. Since the number of macromolecules[2] is relatively small, the pressures that result from them commonly are small, too. The organism, however, can develop mechanical pressures greater than that and can press out fluids. In the capillary beds, the arterial pressure is high enough to enable fluids to pass over into the interstitial space; in the venous section, the hyrostatic pressure drops low enough so that the proteins of the blood plasma absorb fluids from the interstitial area.

Ultrafiltration in the glomeruli, too, depends on the reversal of osmotic dilution. The hydrostatic pressure of the capillary system actually pushes fluid into Bowman's capsule. For proteins, a concentration increase is achieved since proteins are not able to pass through.

Donnan Distribution. As discussed extensively in Chapter IV,8, proteins are polyelectrolytes. Most of them exist as anions at physiological pH values. With this, a complicating factor must be introduced in our discussion of colloid osmosis, namely that of electroneutrality: Within the same osmostic cell, the negative charges of proteins hold an equal number of cations (Fig. XXI-3). If the inner compartment has a $Na^{\oplus}$ ion and a $protein^{\ominus}$ ion, and if we add NaCl to the outside, then $Cl^{\ominus}$ is able to migrate to the interior but has to be accompanied by a $Na^{\oplus}$ ion (against a concentration gradient). For this reason, the concentrations assume a relationship which has been expressed by Donnan in the following way:

$$\frac{[Na^{\oplus}]_o}{[Na^{\oplus}]_i} = \frac{[Cl^{\ominus}]_i}{[Cl^{\ominus}]_o}; \qquad [Na^{\oplus}]_o \cdot [Cl^{\ominus}]_o = [Na^{\oplus}]_i \cdot [Cl^{\ominus}]_i$$

From this quantitative relationship the following qualitative effects can be deduced:

1. Protein anions displace equivalent ions to the outside; more $Cl^{\ominus}$ ions will be found outside (and more $Na^{\oplus}$ ions inside). All other diffusible anions present distribute themselves in accordance with $Cl^{\ominus}$; cations, on the other hand, follow the distribution of $Na^{\oplus}$. This holds true also for $H^{\oplus}$; as a consequence, pH changes occur between the inside and the outside.
2. The measured colloid-osmotic pressure of proteins in salt or buffered solutions will be too high because of the unequal distribution of the small ions. This effect can be calculated in advance and taken into account as a so-called "Donnan correction."
3. The Donnan distribution does not result in osmotic equilibrium. The sum of diffusible ions is higher inside than outside and, consequently, the pressure inside must be higher. This is aggravated by the colloid-osmotic pressure of the macromolecules. The cell walls have to adjust to the rising pressure; if this is not

[2] A 6% protein solution (molecular weight of 60,000) is only $0.001M$; the osmotic pressure of this solution therefore is only 0.0255 atm (at 37°).

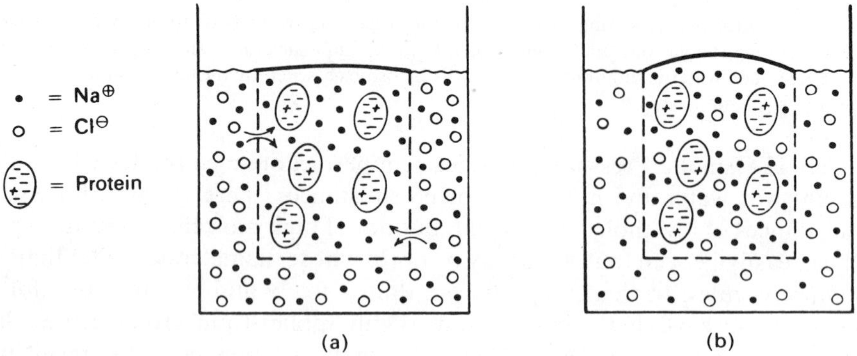

Fig. XXI-3. Donnan distribution. The inner compartment contains a protein solution: protein anions are neutralized by $Na^{\oplus}$ ions. In (a) NaCl has been added to the outer compartment; in (b) Donnan equilibrium has been reached. It can be seen that the inner compartment contains more $Na^{\oplus}$; the outer one, more $Cl^{\ominus}$. Inside, the osmotic pressure is higher.

possible, the cells burst. This is the case, for example, with erythrocytes if they become permeable to cations and become subject to the Donnan laws (osmotic hemolysis).

4. The rearrangement of ions results in an electric potential difference; with negatively charged protein ions, the colloidal compartment becomes positive. The size of the potential difference may be calculated from the laws governing concentration cells, using the following formula (see also Chapter X,3):

$$E = \frac{RT}{nF} \ln \frac{[Na^{\oplus}]_I}{[Na^{\oplus}]_O}$$

All Donnan effects are magnified the smaller the electrolyte concentration and the larger the concentration and net charge of the proteins. However, at the isoelectric point the Donnan effects disappear.

3. Active Transport

The principles of diffusion operate only to effect dilution. By osmosis and the expenditure of mechanical work, a concentration may be achieved; we have seen that the organism makes use of this principle in ultrafiltration and colloid osmosis. The concentration of low molecular weight substances by this mechanism would demand enormous pressures. For this reason, another principle is used, namely *active transport*.

Active transport is a generic term for the transport of substances across membranes *not* operating by osmosis or diffusion. It usually overcomes a concentration gradient. Active transport is found not only across cell membranes, i.e., between the cytoplasmic space and the extracellular space, but also across membranes within the cell such as the mitochondrial membrane.

Active transport mechanisms have been observed for a large number of substances. A variety of ions are transported actively, for example, $Na^{\oplus}$, $K^{\oplus}$, $Ca^{2\oplus}$, and even $H^{\oplus}$ ions (for the production of gastric juice). Among organic substances, the transport of sugars is particularly important (the uptake of glucose into cells), as well as that of amino acids and other low-molecular weight substances which are reabsorbed by the kidney tubules. In several specialized tissues, active transport is a prerequisite for their function, especially in nerves (cf. Chapter XXIII,7). Also in the muscle, excitability cannot be maintained if active transport has come to a standstill (Chapter XXIII,7).

Active transport essentially is a *vectorial process*; it differs from most other metabolic reactions in its directedness. Since it operates against a concentration gradient it is an endergonic process requiring chemical energy. Usually, the cleavage of ATP provides the requisite energy. Indeed $Na^{\oplus}$ and $K^{\oplus}$ ion-activated ATP phosphohydrolase (ATPase) have been located in the membranes and implicated in the transport system. Both the transport phenomenon and the enzyme activity of the ATPase are inhibited by the cardiac glycoside ouabain.

Saturation. The concentration dependence of active transport follows a saturation curve which formally resembles that of Michaelis-Menten kinetics for enzymatic reactions. It is concluded from this observation that proteins are involved in the transport process. Some authors call them permeases, others carriers. Bacterial mutants have been found that have lost the capacity for active transport of certain sugars. This finding is in support of the involvement of a *protein carrier* or *permease*.

The Carrier Model. Numerous models have been devised to explain active transport. Almost all of them postulate the existence of a carrier substance which is supposed to combine with the ion or the molecule that is to be transported—probably in connection with the expenditure of ATP, as diagrammatically represented in Fig. XXI-4. Substance A (for example, an amino acid), in order to be transported, reacts with the energy-rich carrier ~X to form the compound AX, which is split again at the inside of the membrane into A and the energy-poor carrier –X. By some endergonic reaction, which requires ATP or an analogous energy-rich phosphate, ~X is regenerated.

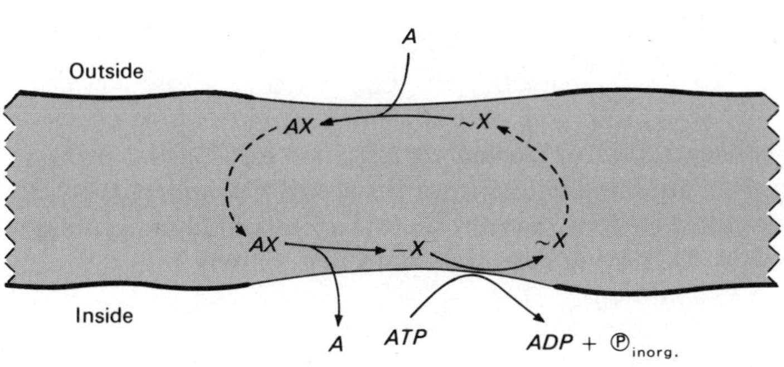

Fig. XXI-4. Diagram of active transport. Substance A diffuses from the outside (top) into the membrane; there combines with carrier X; and then is released to the interior of the cell (bottom). The transport across the membrane is ATP dependent.

The original notion that the carrier migrates within the membrane must be corrected. The migration of a macromolecule in a membrane is quite incongruous with the rapidity of the transport process. Currently discussed are *changes of conformation* of a protein molecule which possibly occupies the entire width of the membrane (70–100 Å) and which would operate on the principle of a revolving or swinging gate. Certain antibiotics provide an interesting model for the selectivity of a carrier with regard to binding $K^{\oplus}$ and $Na^{\oplus}$ ions. Examples are *nonactin* and *valinomycin* (formula at the end of Chapter III), which form clathrates (occlusion compounds) with $K^{\oplus}$, but not with $Na^{\oplus}$, and thus selectively transports $K^{\oplus}$ across the mitochondrial membrane. The *vectorial* nature of transport, however, is not explained by this model.

4. Acid-Base Equilibrium

pH of Blood Plasma. Among the ions whose metabolism we are to consider in this section, the $H^{\oplus}$ ion assumes a special position. The $H^{\oplus}$-ion concentration is rather low, amounting to 4×10^{-8} eq/liter or, in the usual units of the pH scale, the pH value in the extracellular space is 7.4 (pH is explained in Chapter I,2). This value is held remarkably constant; normal deviations lie between 7.35 and 7.45.

"Transcellular fluids" may differ greatly in their pH values: Gastric juice is very acidic (pH 1.5); the content of the small intestine is alkaline with a pH around 8; urine is usually slightly acidic with a pH of 5. Very few data are available about the pH of cells; it is generally somewhat lower than that of extracellular fluids.

Ionic Composition of the Buffer System of Blood. The cations and anions of blood plasma can be compared best in a diagram as shown in Fig. XXI-5.[3] Such a solution can be thought to consist of the bases NaOH, KOH, $Ca(OH)_2$, and $Mg(OH)_2$ (in the amounts indicated) and of the acids, H_3PO_4, H_2SO_4, HCl, and proteins. For the $HCO_3^{\ominus}$ ion, which makes up 27 meq, the acid H_2CO_3 would have to be used, which is in equilibrium with CO_2:

$$H_2CO_3 \rightleftharpoons CO_2 + H_2O$$

The H_2CO_3 concentration in plasma corresponds to a partial pressure of CO_2 of 40 mm of Hg. For attaining equilibrium between CO_2 and H_2CO_3, there exists a special enzyme called *carbonic anhydrase*. It is a Zn-containing protein with a molecular weight of 30,000, having very high molecular activity.

---

[3] The diagram illustrates clearly that meq/liter as the unit of concentration is much to be preferred to the outmoded unit "mg %." First, if expressed in milligram percent, the columns for the cations and anions would not be the same height. Second, the proteins with 7.2 %, i.e., with 7200 mg/100 ml, would stretch the axis intolerably. Last, a completely distorted picture of the relative significance of the various ions would result.

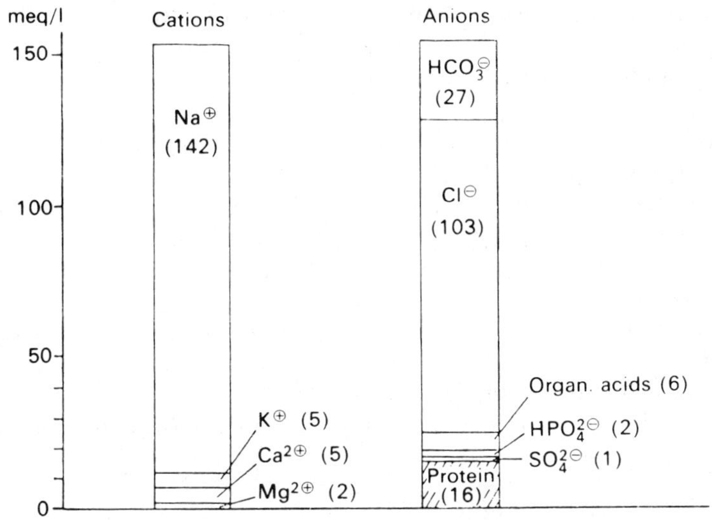

Fig. XXI-5. "Ionograph" of blood plasma. The concentrations of the ions are given in meq/liter.

The system $HCO_3^{\ominus}/H_2CO_3$ is the most important buffer system of the entire organism. At the pH of blood (7.4), $HCO_3^{\ominus}$ and H_2CO_3 are at a ratio of 20:1.

According to the Henderson-Hasselbalch equation (cf. Chapter I,2), the pH of a buffer solution does not depend on absolute amounts, but rather on the *ratio* of concentrations of *undissociated acid to anion.* The pK_1 value of carbonic acid is 6.1. For a nonvolatile acid the ideal buffer region lies around the pK value. But since CO_2 is volatile and carbonic anhydrase rapidly brings CO_2 and H_2CO_3 into equilibrium, the actual concentration of H_2CO_3 depends closely on the partial pressure of CO_2. The great excess of $HCO_3^{\ominus}$ in the buffer system of blood now reveals its advantage; $H^{\oplus}$ ions can be trapped and bound as H_2O while gaseous CO_2 escapes into the atmosphere. The buffer furthermore is practically inexhaustible; any H_2CO_3 lost can be resupplied immediately, since CO_2 is always available.

Two other factors participate in maintaining the pH of blood constant. First, the phosphate system $H_2PO_4^{\ominus}/HPO_4^{2\ominus}$ is an excellent buffer at the pH of blood, since the pK_2 of phosphoric acid is 7. Second, hemoglobin of the erythrocytes also acts as a buffer. When hemoglobin is charged with oxygen, its acid strength goes up, and $H^{\oplus}$ ions are dissociated; as a result, more CO_2 is eliminated in the lungs. In the peripheral tissue, the reverse takes place (*Bohr effect*, cf. also Chapter IX,4).

Alkali Reserve and Acidosis. Since $HCO_3^{\ominus}$ can neutralize the $H^{\oplus}$ ions of a strong acid (entering the blood stream, for example) according to the equation $H^{\oplus} + HCO_3^{\ominus} = H_2CO_3$, the quantity of $HCO_3^{\ominus}$ is often, but somewhat incorrectly, called "*alkali reserve.*" The normal value of the alkali reserve in blood is 25 meq/liter. A situation in which the alkali reserve is diminished greatly is called "*compensated* *acidosis,*" as long as the pH remains normal; if the pH drops, it is called "*uncompensated acidosis.*" Alkalosis refers to the reverse situation and is characterized by a strongly elevated $HCO_3^{\ominus}$ concentration.

Regulation of the Acid-Base Balance. In compensated acidosis the pH can be held constant only if the drop of $HCO_3^{\ominus}$ concentration is accompanied by a proportionate drop of the undissociated acid H_2CO_3, according to the Henderson-Hasselbalch equation. Indeed, the regulatory function of the lungs operates on this principle and consists in lowering the CO_2 tension; accelerated ventilation removes CO_2 faster and keeps its partial pressure low.

The kidneys also participate in the regulation of acid-base balance. By stepping up the acid output into the urine, they can counteract a rise of acid content in the blood. The kidneys accomplish this by exchanging $H^{\oplus}$ for $Na^{\oplus}$ in the tubules, or by eliminating $NH_4^{\oplus}$ in the urine, in principle according to the equation $NH_3 + H^{\oplus} \rightleftharpoons NH_4^{\oplus}$.

A partial failure of kidney function consequently is reflected by the ionic compositions, by the alkali reserve, and possibly by the pH of blood. The regulation by the lungs also may lose it effectiveness and permit corresponding shifts of the alkali reserve (= *respiratory acidosis* or *alkalosis*). Finally, a metabolic defect may cause overproduction of organic acids, as has been known longest in the case of diabetes mellitus. The last situation is termed "*metabolic acidosis.*" There are other pathological shifts of acid–base equilibria and of the electrolyte balance that are not discussed here.

5. Metabolism of Alkali and of Chloride

Sodium and *potassium* are distributed in the organism characteristically unevenly: $Na^{\oplus}$ in extracellular fluids, especially in the interstitial spaces; $K^{\oplus}$ in the cells. The intracellular $K^{\oplus}$ concentration is less than the extracellular $Na^{\oplus}$ concentration; since the cells contain relatively much osmotically active material, osmotic equilibrium would not be ensured otherwise. In some organs (nerve, muscle) this uneven distribution is the basis of their function, i.e., of excitability.

The dietary *sodium intake* fluctuates widely (75–300 meq/day). A large sodium reserve is found in bony tissue; about one-third of the total $Na^{\oplus}$ content is bound in bones and can be mobilized in deficiency states. The *excretion* of $Na^{\oplus}$ in urine is adjusted by the kidneys very closely to the intake. With a low $Na^{\oplus}$ intake more $Na^{\oplus}$ is reabsorbed and *vice versa.* The concentration in the extracellular space thus is maintained very constant at 135–140 meq/liter. This regulation is aided by another effect, the shift of water from the cells to the interstitium, which functions primarily by absorbing the initial shock of a sudden $Na^{\oplus}$ intake.

The reabsorption of NaCl is under control of a hormone of the adrenal cortex; insufficient activity of the gland results in drastic losses of $Na^{\oplus}$ and $Cl^{\ominus}$. The chief "mineral corticoid" is *aldosterone*: it also controls $K^{\oplus}$ excretion, but in the opposite sense. The $Na^{\oplus}/K^{\oplus}$ ratio of the urine, therefore, is an indicator of adrenal cortex function.

With a normal diet, about 100 meq of *potassium* are taken in daily. Not all of this is absorbed; 5–10% is excreted in the feces, and the remainder is eliminated by the kidneys. The reabsorption in the kidney is never complete, not even during periods of extensive $K^{\oplus}$ depletion. $K^{\oplus}$ deficiency shows up, among others, as muscular weakness and lethargy.

Chloride is taken up mainly through table salt. It is found mainly, but not exclusively, in the extracellular space and makes up the bulk of the anions in blood plasma.

Chloride ions are required for the production of HCl in the stomach, where they accompany the $H^{\oplus}$ ions, which are secreted by active transport in the gastric mucosa (see Chapter XXIII,1); the $Cl^{\ominus}$ ions follow the $H^{\oplus}$ ions passively to maintain electroneutrality. Because of the high concentration of $Cl^{\ominus}$ ions in the stomach, persistent vomiting can lead to a state of chloride deficiency.

6. Calcium and Phosphate Metabolism

Calcium is not always available sufficiently in the diet because it is usually absorbed incompletely. Absorbability of dietary $Ca^{2\oplus}$ depends greatly on the other food components. *Oxalate* and *phytin* (inositol hexaphosphate), in particular, impede the absorption in the small intestines by forming slightly soluble salts. *Citrate*, on the other hand, enhances the availability of $Ca^{2\oplus}$, probably by forming soluble complex salts. Last, *vitamin D* greatly improves the absorbability of $Ca^{2\oplus}$; the value of the vitamin in treating rickets is based, at least in part, on this effect.

Only some of the calcium present in blood is in the free, diffusible ionic form. The remainder is bound to proteins. Hence, it is possible to have an overall calcium concentration as high as 5 meq/liter without any precipitation of insoluble $Ca^{2\oplus}$ salts.

Metabolism of Bone. The huge reservoir of calcium contained in the bones plays a unique role in the metabolism of calcium. More than 1 kg is bound there, whereas the daily intake is only around 1 gm. Bound calcium is nevertheless functional: $Ca^{2\oplus}$ salts are essential for the supportive tissue by forming the minerals *hydroxyapatite* and *carbonate apatite*.

Apatites are complex salts with $Ca^{2\oplus}$ as the central atom. Ligands are three molecules of $Ca_3(PO_4)_2$ each occupying two coordination bonds. Associated with the complex cations are the anions $OH^{\ominus}$, $CO_3^{2\ominus}$, $HPO_4^{2\ominus}$ or carboxylate ions of organic acids. These anions are readily exchangeable, and thus the mineral substance of bone has the properties of an ion exchanger.

In spite of its relative insolubility, apatite is remarkably active metabolically. There is a continuous exchange between it and the $Ca^{2\oplus}$ and $HPO_4^{2\ominus}$ ions circulating in blood. This rapid exchange between circulating $Ca^{2\oplus}$ ions and bony substance could be demonstrated only through the use of radioactive isotopes. This exchange is significant insofar as ions related to $Ca^{2\oplus}$, especially *strontium*, can be incorporated as well. In recent decades, large amounts of radioactive strontium (^{90}Sr) have been produced by repeated atmospheric atomic explosions and have been spread in the stratosphere. Rainfall brings it back to the surface of the earth, where it can be taken up by plants and eventually enter food (milk, for instance). Since radioactive strontium is an isotope with a very intense emission, considerable radiation damage (including genetic damage) results from a contamination with it. The danger is increased by the fact that strontium is locally concentrated and held in place in bone tissue.

The formation and breakdown of bone are vital processes carried out by special cells. *Mineralization* occurs in the *osteoblasts*, apparently involving alkaline phosphatase, which liberates inorganic phosphate from phosphate esters and deposits it as calcium phosphate. The enzyme is especially abundant in areas of active ossification, while cartilage is free of it. The *breakdown of bone* in the *osteoclasts* serves to mobilize mineral material. Solubilization of apatite probably is an interphase phenomenon, analogous to its formation.

Regulation of the Calcium Level. Two hormones acting counter to each other control the metabolism of $Ca^{2\oplus}$. They are *parathormone* of the parathyroid gland and *calcitonin* of the thyroid gland. Calcitonin lowers the $Ca^{2\oplus}$ level of blood by promoting the incorporation of calcium in bone. Parathormone, in contrast, activates the osteoclasts and thereby causes a demineralization of bone and a rise of the $Ca^{2\oplus}$ level in blood. It is not yet clear how these minerals are dissolved. *Citrate* probably plays a key role; at least, it occurs in relatively high concentrations in bony tissue. Bone tissue contains citrate synthase, but little aconitase and isocitrate dehydrogenase.

The production of parathormone and calcitonin normally is adjusted by the $Ca^{2\oplus}$ level in blood, a true feedback control mechanism. How this control mechanism with two regulatory links operates is not known in detail. Generally, no "monitor organs" for such concentrations are known and not even a conceptual model has been developed.

Phosphate Metabolism. Phosphate is important not only as a component of bone. We have already seen that the various esters of phosphoric acid are indispensable for many metabolic reactions, for example, in glycolysis or nucleic acid metabolism. The system $ATP \rightleftharpoons ADP + P_i$ is particularly important for innumerable cell functions. The formation of ATP occurs primarily in the respiratory chain.

In man the uptake and elimination of phosphate is generally balanced, with a tendency perhaps to take up too much rather than too little. Renal excretion is governed by parathormone (cf. Chapter XX,7).

7. Iron and Zinc

Iron. As has already been discussed, iron is a component of hemoglobin and cell hemins, both very important biocatalysts. Thus it is a vital element. The human organism contains 4–5 gm, of which $\frac{3}{4}$ is in hemoglobin. The cytochromes and iron-containing flavoproteins sequester only a small fraction of the total iron.

There are specific storage forms for iron. A protein called *apoferritin* can absorb as much as 23% of iron as $(FeOOH)_8 \cdot (FeO \cdot PO_4H_2)$ to become known as *ferritin*. Another storage form, *hemosiderin*, contains up to 35% of iron in the form of a ferrihydroxide gel that can absorb protein and other organic substances. Hemosiderin in the cells forms granula visible by microscope. In iron storage diseases there are increased numbers of these granula. Both storage forms occur primarily in the spleen and liver; ferritin, furthermore, in the intestinal mucosa.

In blood, iron is transported not with ferritin, but with *transferrin* (or siderophilin). It is a β-globulin with a molecular weight of 88,000 that can complex two atoms of $Fe^{3\oplus}$. Iron bound in this manner can only be mobilized and eliminated by reduction to $Fe^{2\oplus}$. *In vivo* this occurs with ascorbate as reducing agent. Iron in this manner enters newly formed hemoglobin in maturing erythrocytes.

The average life span of erythrocytes is about 4 months. One can calculate that every day 8–9 gm of hemoglobin are broken down and must be replaced. The iron freed during the breakdown of hemoglobin is conserved, however, instead of being excreted. In spite of this unusual recycling, iron must be supplied continually to make up for some inevitable losses. But absorption is generally poor and depends on the amount already present in the organism. With ample amounts in storage and a good supply in the diet, the uptake is greatly decreased. With relative iron deficiency, almost all iron is extracted from food. Since the organism has no mechanism for excreting excess iron, the adjustment of absorption of iron is crucial. This situation is unique for iron.

Iron deficiencies caused by insufficient intake are quite common in some parts of the world. Women are affected particularly because they can lose appreciable amounts of iron through menstrual bleeding.

Zinc. The organism contains between 2 and 4 gm of zinc, almost as much as iron. While iron occurs predominantly in blood bound in hemoglobin, zinc is in the cells. Some organs have a conspicuously high content of zinc; the eyes, up to 0.5%.

Zinc combines easily with proteins, e.g., insulin. It is of functional significance in several enzymes, *carbonic anhydrase* and *peptidases*, for instance (Chapter VIII,3), but other wise too little is known about the role of zinc.

BIBLIOGRAPHY

E. E. Bittar, "Cell pH." Butterworth, London and Washington, D.C., 1964. (Much broader in content than indicated by the title.)

C. L. Comar and F. Bronner (editors), "Mineral Metabolism," 3 vols. Academic Press, New York, 1960–1969.

H. W. Davenport, "The ABC of Acid-Base Chemistry," 4th ed. Univ. of Chicago Press, Chicago, Illinois, 1958.

CHAPTER XXII

Nutrition and Vitamins

1. Caloric Value and ATP Yield

Heat of Combustion. Carbohydrates, fats, and proteins can replace each other as nutrients to a large extent. This is not very surprising in view of our discourse on metabolism (Chapter XIX), in which the concept of a "common terminal metabolic pathway" (Netter) through the citrate cycle and the respiratory chain was developed.

Physiologists have chosen the heat of combustion as the proper unit in which to express mutually equivalent amounts of nutrients. According to the "isodynamic law," the various foodstuffs may be substituted for each other as long as their heat of combustion remains the same. For purposes of calculation the following approximate values of physiological heat of combustion are generally used:

Carbohydrate	4.1 kcal/gm
Protein	4.1 kcal/gm
Fat	9.3 kcal/gm

In biochemistry, it is the custom to base all theoretical considerations on free energy, ΔG (Chapter V,2), and not on the heat of reaction ΔH. This applies also to the above calculations, but the error incurred by neglecting changes of entropy, i.e., by using heat of combustion ΔH instead of free energy ΔG, is rather minor (e.g., 4% with glucose).

Production and Consumption of ATP. A far better measure of the utilization of energy would be ATP production, since we have repeatedly described ATP as the driving power of numerous processes. In a sense, ATP is the organism's energy currency, a highly negotiable means of payment. One might easily think that it would be extraordinarily rare, because it is often obtained only by extensive detours, but the amount turned over is not at all small. It is easy to calculate[1] that during 24 hours man produces (and obviously breaks down again) some 70 kg of ATP, an amount equal to his body weight! Most of this (95%) is generated by oxidative phosphorylation in the respiratory chain.

The useful energy made available, expressed in percentage yield of ATP, is practically the same for the oxidation of either fat or carbohydrate. It actually represents the energy yield of the respiratory chain. About 40% of the free energy of oxidation is initially converted to ATP.

Basal Metabolism. For maintenance of body functions, temperature, and thermodynamic nonequilibria (e.g., concentration gradients), the organism requires a definite minimum amount of energy. This is the reason for the so-called *basal metabolism*. Depending on size (body surface), age, and sex, the minimum energy may range between 1400 and 2000 kcal per day and thus is a rather flexible value. Hyperfunction of the thyroid gland elevates basal metabolism, whereas a period of malnutrition will depress it considerably. Added to basal metabolism is the metabolism during activity; this may raise the total energy turnover to 2500 kcal per day for light or sedentary activity, and to 3500 kcal per day or more for heavy labor. Food intake is usually adapted to the requirements so well that body weight does not fluctuate appreciably. Imprudent habits, however, can easily result in undesirable accumulation of fat.

It has been noted that particularly protein ingestion can effect a rise of basal metabolism. In addition to the normal turnover, 15–20% of the overall caloric intake is converted to heat. Rubner called this a "specific-dynamic effect," a rather unfortunate term, for its specificity is obscure and it is debatable whether there is a "dynamic" effect. The explanation for the greater turnover is simply that more protein calories than carbohydrate calories are required to produce the same amount of ATP.

Biological Half-Life. Basal metabolism provides both the material and the energy for *de novo* synthesis and replacement of endogenous substances. Isotope tracer methods showed that practially *all* structural substances of the organism are continuously renewed with the possible exception of DNA, the genetic material. This renewal takes place at different rates. The rate is expressed conveniently in terms of biological half-life, i.e., that time in which one-half of the material present is broken down and renewed (the concept is analogous to the half-life of radioactive substances). For instance, in man the half-life of liver or plasma proteins is about 10 days, whereas the half-life of proteins in skin, skeleton, and muscle is about 158 days.

---

[1] The calculation is based on a turnover of energy of 2500 kcal and a storage of energy (as ATP) of 40%. This means that 1000 kcal of chemical energy are used for ATP synthesis, $1000 \div 7 = 140$ moles $= 140 \times 510$ gm $= 72$ kg ATP can be synthesized. (This is only an approximation value, of course).

The Respiratory Quotient. The end products of metabolism are chiefly CO_2, H_2O, and urea. The ratio

$$\text{R.Q.} = \text{volume of } CO_2 \text{ formed/volume of } O_2 \text{ consumed}$$

called respiratory quotient, allows some conclusions as to the nature of foodstuffs combusted. Pure carbohydrate, oxidized according to the general formula

$$C_6H_{12}O_6 + 6\,O_2 = 6\,CO_2 + 6\,H_2O$$

must have a ratio of $CO_2/O_2 = 1.00$. Similarly it can be shown that the biological oxidation of fats has a respiratory quotient of 0.7, and of proteins, about 0.8. During substantial conversion of carbohydrate to fat, the respiratory quotient rises above 1.00. These quotients are the basis of many calculations in nutritional physiology.

Nutritional Norms. It has been attempted to set up guidelines for sound nutrition; thus a daily intake of 2200 to 2800 kcal is recommended for men and 1600 to 2000 kcal for women. It is difficult, however, to make such general recommendations when the requirement of an individual depends so much on his level of activity. Physically active persons simply require more calories than those with more sedentary habits. The continuous control over body weight is much sounder than nutrition strictly by calorie tables. Very generally the normal feelings of hunger and appetite assure the adaptation of dietary intake to requirements. This physiological control functions surprisingly well; the body weight of the majority of individuals remains quite constant. Bad habits, of course, can override the regulatory mechanism, which results then in obesity. In Europe and North America, malnutrition nowadays has become synonymous with overeating. Many people are overweight and consequently face a prognosis of a shortened life span.

In developing countries, in sharp contrast, hundreds of millions of people suffer from undernourishment. This applies to both the daily caloric intake and the qualitatively inadequate supply of protein and vitamins.

The latter deficiency is more deleterious to health than the inadequate supply of calories. An inadequate supply of an otherwise balanced diet, in which the essential nutrients (see below) and vitamins exist in the proper ratios to each other, may lead to retarded growth, weight loss, or hypoproteinemia, but the absence of a certain vitamin from an otherwise ample diet results in a deficiency disease such as scurvy or beriberi.

2. Essential Food Components

Nitrogen and Protein Needs. For energy production the principal foodstuffs can replace one another extensively; even carbohydrates, usually the main source of calories, are not indispensable, as is shown by the dietary habits of Eskimos. The intake and elimination of the chemical elements must, of course, be balanced. Carbon and hydrogen are contained in all foodstuffs; oxygen is in the air. Nitrogen is derived predominantly from proteins.

Nitrogen balance is the difference between nitrogen taken up in protein and nitrogen excreted as urea. If excretion exceeds the uptake, then the balance is negative and the body suffers a net loss of nitrogen. A certain minimum of protein must be taken in regularly in the diet to maintain nitrogen balance, since a certain amount of protein is broken down continuously and the resultant nitrogen excreted in urea—even during basal metabolism (at rest). The minimum of protein in the diet of a man is around 35–50 gm per day (0.5–0.6 gm/kg body weight). A good diet naturally requires more protein; 70–90 gm of protein per day are recommended. This presumes that the protein is of high quality, i.e., that it is easily digestible and contains the essential amino acids in adequate amounts. An unfavorable ratio of the essential amino acids results in suboptimal utilization of dietary protein for the buildup of body protein and in a greater proportion of amino acids being broken down. Consequently more protein is required in the diet in order to achieve nitrogen balance. Many plant proteins unfortunately are not of high quality in this regard. Protein from wheat, for example, is poor in lysine; proteins of legumes are low on methionine. Animal protein is generally of high quality, with the important exception of gelatin.

Essential Amino Acids. As mentioned in Chapter VIII, 6, several amino acids found in proteins are not synthesized in the animal organism and are therefore indispensable components of the diet. Amino acids essential for man are listed in Table XXII-1:

TABLE XXII-1

Essential Amino Acids[a]

Valine	0.28	Tryptophan	0.04
Leucine	0.39	Methionine	0.07
Isoleucine	0.21	Threonine	0.20
Lysine	0.24	(Cysteine)	0.07
Phenylalanine	0.18		

[a] The values are the daily minimal requirements of an adult for zero nitrogen balance expressed as mmole/kg body weight and with a total protein intake of 0.58 gm/kg body weight (according to Jekat).

The table brings out clearly that the less abundant amino acids Trp and Met are required in proportionately smaller amounts. Cysteine is placed in parentheses because it is not essential in the true sense (for its carbon skeleton), although it is indispensable as source of sulfur.

Growing organisms have higher requirements than mature organisms; furthermore, even some nonessential amino acids (e.g., arginine and histidine) are growth promoting, because these apparently are not synthesized in sufficient quantities during conditions of growth.

An adequate diet of protein and supply of essential amino acids is a grave problem in developing countries. In Asia and Africa many people subsist on a minimum of protein. This problem affects small children more because the growing body requires a relatively higher intake of quality protein. *Kwashiorkor*, a childhood disease common in Africa and Asia, is caused by protein deficiency.

Protein metabolism is influenced furthermore by hormones. Testosterone causes nitrogen retention with normal nutrition (anabolic effect, Chapter XX,3); glucocorticosteroids have a catabolic effect since they promote the conversion of protein to carbohydrate (gluconeogenesis, Chapters XV,10, XX,2, and XX,11).

Essential Fatty Acids. The mammalian organism is not able to synthesize *de novo* higher unsaturated fatty acids of the linoleate type, characterized by the terminal group $CH_3{-}(CH_2)_4{-}CH{=}CH{-}$. Assisted by a mitochrondrial enzyme system, it can merely convert linoleate ($\Delta^{9,12}$-18:2) via γ-linolenate ($\Delta^{6,9,12}$-18:3) to arachidonate ($\Delta^{5,8,11,14}$-20:4; for the formula see Chapter XII,1).

A deficiency of essential fatty acids is manifested by changes in the skin, disturbances in the fluid balance, and reproductive disorders (presumably due to restricted synthesis of prostaglandins, see Section 4, below). The daily requirement for man is about 6–8 gm. Since linoleate is distributed unbiquitously, this level of intake is quite readily achieved and deficiency symptoms are rare. The adult, furthermore, can draw on very considerably body reserves.

Minerals and Trace Elements. The mineral elements necessary for maintenance of the animal organism are; the *anions* phosphate, chloride, iodide, and fluoride; and the *cations* sodium, potassium, calcium, magnesium, iron, zinc, copper, manganese, cobalt, and possibly others.

Among the heavy metals, *iron* and *zinc* occupy the first place; they have been discussed in detail in Chapter XXI,7. Many mineral substances are required only in minute amounts (trace elements). They are widely distributed and usually present in sufficient amounts in the diet. After all, our food is derived exclusively from living material, from plants or animals, and for that reason should contain all the essential elements. In some circumstances, however, deficiency symptoms may develop (the most widely known examples are the endemic goiter due to iodine deficiency, and anemia due to iron deficiency).

3. Vitamins

Vitamins are essential food components, just as the essential amino acids are. The significance of vitamins, however, is much better appreciated by the general public.

The concept *vitamin* has undergone extensive revisions during the history of biochemistry. Hofmeister's definition—vitamins are substances which are indispensable for the growth and maintenance of the animal organism, which occur both in animals and plants and are present only in small amounts in food—is still valid today, but has been interpreted in various ways. Originally it was believed that there was only *one* vitamin, but soon a list had to be started. The substances prepared in pure form turned out to belong to completely different chemical classes of compounds. The categorical phrase "not synthesized in the animal body" became shaky and had to be replaced by more specific statements (such as what vitamin and what animal species). Indeed, for some time it was believed that the distinction between vitamins and hormones was no longer justifiable, although the two classes of active substances differ fundamentally: Hormones are regulatory substances and are elaborated by glands in discrete quantities. Vitamins as accessory nutrients cannot be regulatory substances since their intake in the diet is basically beyond control.

A thorough understanding of the vitamins has been provided by biochemical genetics. Plants and many primitive organisms are capable of synthesizing all necessary substances from the simplest starting materials, i.e., from simple sources of carbon and nitrogen, from minerals, and from energy. The more advanced organisms have lost some of these capabilities. Probably as a result of some mutational changes, the chains of synthesis of some important compounds were interrupted or abolished entirely. The compounds are no longer synthesized by the organism and have to be supplied in the diet. It is of secondary significance in this connection whether or not an essential food component possesses a catalytic function.

Vitamins are ordinarily defined as substances that act in *trace amounts.* Hence, the definition includes only substances with *catalytic functions.* Table XXII-2 reveals indeed that man's daily requirement for any vitamin (except ascorbate) is less than 10 mg.

Biochemical Function of Vitamins. The nature of the catalytic function is now known for most vitamins. They are incorporated as coenzymes or prosthetic groups of enzymes.

Having devoted already an entire chapter (Chapter VI) to this role of vitamins the following remarks can concentrate on the nutritional aspects of vitamins.

Deficiency Diseases. A deficiency of certain vitamins is reflected in characteristic disturbances (deficiency diseases); occasionally only arrested growth is observed. Despite all the progress registered by "vitaminology" during past decades, culminating in the recognition of the role of vitamins as coenzymes, in most cases it cannot be explained why the deficiency appears in terms of specific diseases instead of the expected general disturbances. The unusual sensitivity of the skin toward the lack of various vitamins is very conspicuous, however.

Filling the Requirement of Vitamins. Deficiency diseases appear very rarely under adequate nutritional situations. Such diseases are the consequence of unbalanced nutrition. *Hypovitaminoses* are actually clinically more significant than the avitami-

TABLE XXII-2

Vitamins

Letter	Name	Formula in chapter	Role as coenzyme	Deficiency disease of man	Daily requirement[a] of man
I. Fat-Soluble Vitamins					
A	Retinol	XIV,8	Unknown (opsin)	Night blindness, xerophthalmia	1.5–2.0 mg
D	Calciferol	XIV,4	Unknown	Rachitis	0.025 mg
E	Tocopherol	XIV,9	Unknown	Unknown (muscular dystrophy?)	[5 mg]
K	Phylloquinone	XIV,9	Unknown	Delayed blood clotting	[0.001 mg] own synthesis by intestinal flora
Q	Ubiquinone	XIV,9	In electron transport	Unknown	Probably own synthesis
F	Essential fatty acids	XII,1	—	Debated (dermatitis?)	Unknown
	Lipoic acid (thioctic acid)	VI,4	In oxidative decarboxylation	Unknown	Unknown (own synthesis?)
II. Water-Soluble Vitamins					
B_1	Thiamine	VI,7	Thiamine pyrophosphate	Beriberi (polyneuritis)	0.5–1.0 mg
B_2 complex	Riboflavin	VI,4	Flavin-adenine dinucleotide	Dermatitis	1 mg
B_2 complex	Nicotinamide	VI,4	Nicotinamide-adenine dinucleotide (phosphate)	Pellagra	Own synthesis[b]
B_2 complex	Folic acid	VI,6	Tetrahydrofolate	Megaloblastic anemia	[1–2 mg]
B_2 complex	Pantothenic acid	VI,7	Coenzyme A	Burning foot syndrome	[3–5 mg]
B_6	Pyridoxal	VI,8	Pyridoxal phosphate	Unknown	[2 mg]
B_{12}	Cobalamine	VI,9	In carboxyl shifts (in bound form)	Pernicious anemia	0.001 mg
C	Ascorbic acid	XV,4	In redox systems	Scurvy	75 mg
H	Biotin	VI,6	In carboxylations (enzyme bound)	Very rare ("raw egg white injury," dermatitis?)	[0.25 mg]

[a] Values in brackets indicate estimated amounts.

[b] Pellagra occurs only under the special condition when the body's own synthesis from tryptophan is restricted.

noses. A hypovitaminosis is a state of relative deficiency, without the classical symptoms of disease. This is the usual field of application of vitamin therapy.

The animal organism is able to perform the last synthetic step in the production of some vitamins; i.e., it can transform *provitamins* to the true vitamins. In such cases (e.g., vitamin A), the requirement is filled largely by consumption of provitamins.

Intestinal bacteria contribute considerably toward fulfilling the vitamin requirements listed in Table XXII-2. This can be considered an example of symbiosis. Man's requirement of vitamin K, for example, is filled almost entirely by the bacteria. Cows which received no thiamine or riboflavine still produced milk with a normal content of vitamins. On the other hand, high doses of sulfonamides or antibiotics can liquidate the intestinal flora to such an extent that the source of vitamins stops suddenly and serious avitaminoses may develop unless the diet is corrected.

Hypervitaminoses. These conditions have been described for a few fat-soluble vitamins (A and D). Almost without exception, they are caused by incorrect dosages in vitamin therapy. Inappropriate or unusual diets do not give rise to hypervitaminoses. (One exception: Polar bear liver is considered to be "poisonous" because it contains very large amounts of vitamin A).

Classification of Vitamins. Vitamins are classified as either fat or water soluble. This rather superficial classification is still useful because it indicates the best food source of a vitamin. Table XXII-2 presents the principal vitamins; several factors whose significance is still debated (inositol and choline) have not been included.

4. Fat-Soluble Vitamins

Vitamin A (Retinol). One of the isoprenoid lipids, vitamin A is closely related to *carotene*, its provitamin which the organism can convert to vitamin A (cf. formula in Chapter XIV,8). Two closely related substances, vitamin A_1 and vitamin A_2, are distinguished. Vitamin A alcohol is easily dehydrogenated by the organism to give the aldehyde (by an alcohol dehydrogenase and NAD). Vitamin A aldehydes are components of visual purple (see Chapter XIV,8). The role in the visual process is the only well-established biochemical function of vitamin A.

In animal experiments, *vitamin A deficiency* is manifested first by cessation of growth; hence vitamin A was formerly thought to be a "growth vitamin." Evidently there must be roles for the vitamin other than in the visual process, and it is significant that these other roles can be filled by vitamin A acid. The acid (oxidation product of vitamin A aldehyde) is not reduced by the organism and cannot function in the visual process, but does support growth. There is some evidence that even *in vivo* vitamin A acid is the active molecular species for the general function.

Several deficiency symptoms are known in man. At first, night blindness sets in; later, cornification of the epithelia around the eye (xerophthalmia) may develop.

Vitamin A appears to be required for the normal biosynthesis of mucopolysaccharides. The cornification of the eyes is particularly conspicuous.

Vitamin D. Also called calciferol, vitamin D is related to the steroids. It arises from $\Delta^{5,7}$-unsaturated sterols, the provitamins, by ultraviolet irradiation (cf. the diagram of formulas in Chapter XIV,4). Vitamin D_2 (ergocalciferol) is derived from ergosterol from plant sources; D_3 (cholecalciferol), from Δ^7-*dehydrocholesterol* formed in the human organism itself. With adequate exposure to sunlight this endogenous provitamin is converted in the skin to D_3, removing the need for dietary supplementation.

Cholecalciferol is hydroxylated in the liver to 25-*OH*-cholecalciferol and again in the kidneys to *1,25-dihydroxycholecalciferol.* The latter functions as a true hormone influencing calcium metabolism: It induces a protein in the intestinal mucosa, which is indispensable for the absorption of $Ca^{2\oplus}$ ions by active transport, and it promotes the mineralization of bone (see Chapter XXI,6), at least within the range of physiological dosages. Excessive administration of vitamin D produces a hypervitaminosis, during which calcium is again mobilized out of the bone structure: This reversal of the action on bone is an unsolved puzzle.

A deficiency of vitamin D in childhood leads to rachitis (rickets). The disease is characterized by insufficient deposition of $Ca^{2\oplus}$ in bone as a result of poor absorption of $Ca^{2\oplus}$. For the maintenance of a certain level of serum $Ca^{2\oplus}$ parathormone will mobilize calcium from bone, which then will become soft. Rachitis can, of course, be prevented quite easily today by adequate sun irradiation or by adequate supplies of calciferol in the diet (carefully avoiding overdosages).

Vitamin E (Tocopherol). This vitamin has been isolated from wheat germ oil. There are several closely related tocopherols distinguishable by the number of methyl groups. The formula in Chapter XIV,9 is for common α-tocopherol. It is related to the quinones with the isoprenoid side chain and can easily be oxidized to a quinone (tocoquinone). Tocopherol is used as an antioxidant because of its ready oxidizability. It prevents the spontaneous oxidation of highly unsaturated compounds. This property has been assumed to be the basis of the mechanism of its biological activity, but we still do not really know the biochemical function of this vitamin.

Vitamin E was discovered as the antisterility factor of the female rat. Deficiency in experimental animals results in pregnant females in reabsorption of the fetuses (which observation prompted the naming of the vitamin as tocopherol) and in males in atrophy of the testes and dystrophy of muscles. *In man, a deficiency disease is not known*; some authors, however, suspect that muscular dystropy may be connected with vitamin E deficiency.

Vitamin K (Antihemorrhagic vitamin, Phylloquinone). This vitamin contains a naphthoquinone nucleus and an isoprenoid side chain. Compounds with a phytyl side chain are called *phylloquinone*; those with a difarnesyl side chain, *menaquinone* (Chapter XIV,9). Since these side chains can be synthesized by the mammalian organism, even *menadione* (= 2-methyl-1,4-naphthoquinone) is active as vitamin. The biochemical function formerly (prior to the discovery of ubiquinone) was thought to be its participation in the respiratory chain. Apparently this is the case only in bacteria, where vitamin K indeed replaces ubiquinone.

The symptoms of vitamin K deficiency are a tendency to bleed and *disturbances in blood clotting*. Of the many clotting factors (end of Chapter IV,9), prothrombin is affected primarily; it is produced in insufficient amounts in vitamin K deficiency. The role played by phylloquinone in the hepatic biosynthesis of the clotting proteins is still not entirely clear. In man the deficiency symptoms are rather rare, because the intestinal flora produces enough of the vitamin to contribute materially in the supply.

Dicumarol is an important antagonist of vitamin K. It has found clinical application for decreasing the clotting capacity of blood. High doses are toxic.

Ubiquinone (Coenzyme Q). This can no longer be classified as a vitamin since it is synthesized by the mammalian organism (Chapter XIV,9).

It is now certain that ubiquinone is a redox component in the respiratory chain. Mitochondria contain relatively high proportions of ubiquinone.

Vitamin F. The designation for *essential fatty acids* is vitamin F (cf. Chapter XII,1, and this chapter, Section 2).

Prostaglandins. These are not customarily considered as vitamins (but rather as hormones). They are synthetized in the organism from highly unsaturated fatty acids, but since the latter are essential food components, the whole situation resembles in many ways the conversion of a provitamin to a vitamin. The conversion involves cyclization and oxidation as indicated here:

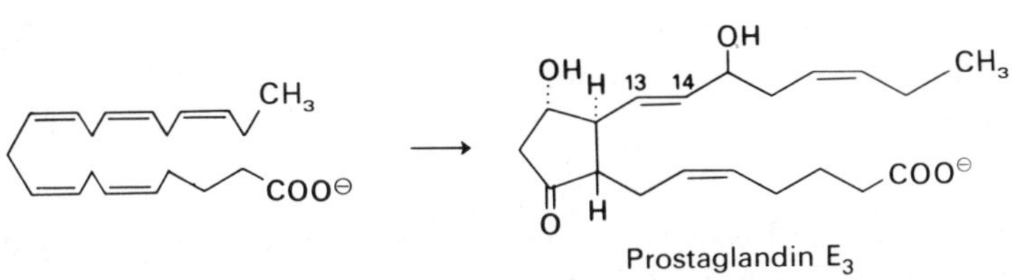

Prostaglandin E_3

The starting material for the example chosen is eicosapentaenoate (C_{20} fatty acid with five double bonds). Analogous starting material could also be acids with four double bonds (i.e., arachidonate) and three double bonds (homo-γ-linolenate). There is a whole family of closely related prostaglandins. They are particularly abundant in seminal fluid and vesicular glands, but can be found in many other tissues as well.

Prostaglandins are vasodilators and stimulate smooth muscles. They also affect adipose tissue by interfering with the fat mobilization initiated by norepinephrine by inhibiting the formation of cyclic AMP. However, in some other tissues prostaglandins can stimulate the synthesis of cyclic AMP. To a certain extent the absence of prostaglandins is responsible for the deficiency symptoms observed in animals that are kept on a diet lacking unsaturated fatty acids.

As already mentioned, prostaglandins and dihydroxycalciferol both behave as *hormones* in the animal organism and therefore might well be discussed in the chapter on hormones. They have however been included here, because their "prohormones," certain polyunsaturated fatty acids (Chapter XII,1) and the calciferols (Chapter XIV,4), respectively, are indeed essential nutrients or vitamins.

5. Water-Soluble Vitamins

Thiamine (Vitamin B_1). One of the longest-known vitamins, the chemical structure of thiamine is somewhat complicated; it contains two heterocyclic rings (a pyrimidine and a thiazole ring, formula in Chapter VI,7) connected at a quaternary N atom.

Thiamine pyrophosphate is the coenzyme of decarboxylases and aldehyde transferases. It plays a key role in *oxidative decarboxylation* of pyruvate (in the breakdown of carbohydrate) and of α-keto glutarate (in the citrate cycle). Man's requirements of thiamine are calculated in conjunction with his caloric intake, since the demand for the coenzyme is apparently higher with a high overall metabolic rate.

Beriberi is the deficiency disease. It appears in individuals on an exclusive diet of polished rice and results in neuritic symptoms and disturbances of heart functions. Chronic alcoholics also often have symptoms of thiamine deficiency ("alcohol polyneuritis"). Some of the deficiency symptoms can be explained on the basis of the role of the vitamin in metabolism.

Vitamin B_2 Complex. The deficiency symptoms—in man, chiefly lesions of the mucosa—were formerly explained by the lack of "vitamin B_2," but now are recognized to be more complex and to result from the absence of a number of factors. One speaks, therefore, of the "vitamin B_2 complex." The individual vitamins of the complex follow.

Riboflavin. An isoalloxazine derivative, its chemical properties have already been discussed in Chapter VI,4. It occurs in bound form in most foods as *flavin mononucleotide* (FMN) or *flavin-adenine dinucleotide* (FAD), or in *flavoproteins*. Milk alone contains free riboflavin.

The biochemical role has already been presented in detail (Chapters VI,4 and X,4). Riboflavin is a component of the "yellow enzymes," which are dehydrogenases. The isoalloxazine ring operates as a reversible redox system.

In experimental animals a deficiency of riboflavin impairs normal growth and causes symptoms of the skin. In man the principal symptoms of ariboflavinosis are dermatitis and inflammations around the mouth. The nutritional supply of the vitamin is, however, adequate in general. Excessive amounts of riboflavin in the body are excreted in urine.

Nicotinamide (Niacinamide, "pellagra-preventive factor"). This is a relatively simple compound (pyridine 3-carboxylic amide, formula in Chapter VI,4). In the mammalian organism, including man, it can arise from tryptophan (cf. Chapter VIII,11). Deficiencies, therefore, develop only under very special circumstances.

Biochemical Role. Nicotinamide functions as a constituent of the *pyridine nucleotides*, which occupy a central role as hydrogen-transferring coenzymes (transport metabolites for hydrogen).

Deficiency of nicotinic acid may cause pellagra (a special form of dermatitis involving the browning of skin), diarrhea, and delirium. Interestingly enough, pellagra can be cured by the administration of tryptophan. This proves that the "vitamin" can be synthesized by the human organism. Pellagra appears endemically only where corn is the main constituent of food. Corn apparently contains a factor that restricts the biosynthesis of the vitamin; furthermore, corn protein contains little tryptophan.

Folic Acid (*Pteroylglutamate*) is a derivative of pteridine; for chemical properties of folic acid see Chapter VI,6. The biochemically active form, *tetrahydrofolate*, is the cofactor for the metabolism of single carbon fragments; it transfers activated formate or activated formaldehyde.

Folate was discovered as a growth substance for microorganisms. The situation is complicated by the fact that some organisms require only one component of the molecule, e.g., *p*-aminobenzoate, and other organisms need the entire folate molecule or even *N*-formyltetrahydrofolate (= folinate or "citrovorum factor," named after *Leuconostoc citrovorum*).

In man, folate affects primarily the composition of blood (megaloblastic anemia, thrombocytopenia). A common cause of deficiency symptoms is a disturbance in the *utilization* of folate rather than inadequate dietary intake. Vitamin B_{12} acts synergistically with folate.

Folate Antagonists. *Aminopterin* has an amino group in position 4 of the pterin nucleus instead of the OH group; it inhibits dihydrofolate reductase and thus the formation of the coenzyme. Aminopterin and its N^{10}-methyl derivative *amethopterin* find therapeutic application in the treatment of leukemia since by inhibiting purine synthesis (Chapter VII,2) they also inhibit cell division.

Sulfonamides inhibit bacterial biosynthesis of folate because they are structural analogs of *p*-aminobenzoate. This is the basis of their bacteriostatic effect, which has been employed in the therapy of infectious diseases. Prolonged oral sulfonamide therapy, however, damages also the normal intestinal bacteria and consequently curtails the source of certain vitamins arising from the intestinal flora.

Panthothenic Acid. Panthothenate consists of *α,γ-dihydroxy-β,β-dimethylbutyrate* and *β-alanine* in peptidic linkage. It can combine with *β*-mercaptoethylamine to give *panthetheine* (formula in Chapter VI,7), which in turn is a component of *coenzyme A* and of the multienzyme complex of fatty acid biosynthesis. The SH group of the mercaptoethylamine is the active group of the coenzyme, which activates acetate and higher fatty acids and thus plays an extremely important role in metabolism.

In various animals, experimental deficiency of pantothenic acid causes various pathological conditions (pellagra in chickens; graying of hair in rats). True deficiency diseases are not known in man. A conjugate of pantothenic acid, pantheteine, is a growth factor for many microorganisms (the "*Lactobacillus bulgaricus* factor").

Vitamin B_6 (Pyridoxol). This is a substituted pyridine. Closely related are pyridoxamine phosphate and pyridoxal phosphate, which are important coenzymes for the metabolism of amino acids (Chapter VIII,4). The generic term covering all these different compounds is Vitamin B_6.

Vitamin B_6 deficiency does not cause a *typical* clinical syndrome. In children epileptic convulsions have occasionally been seen, which might conceivably be traced back to a disturbance of the metabolism of glutamate (the formation of γ-aminobutyrate in the brain). Seborrhealike symptoms have also been noted. In addition, tryptophan catabolism is disturbed which in turn decreases the endogenous *synthesis of nicotinamide*; kynurenine, 3-hydroxykynurenine, and xanthurenic acid appear in urine. Excess vitamin B_6 is oxidized to pyridoxinic acid (like pyridoxal, but with a carboxyl in place of the aldehyde group) and excreted.

Cobalamin (Vitamin B_{12}). By far cobalamin has the most complicated structure of all the vitamins (cf. Chapter VI,9). The ring structure resembles that of the porphyrins, but differs by the absence of one methine group. In hydroxycobalamin the cyanide group is replaced by $-OH$; in the B_{12} coenzymes this place is occupied by a second nucleotide ligand.

Cabalamin is the *anti-pernicious anemia factor*. Minute amounts (a few micrograms) cure pernicious anemia in man. The disease is characterized by a drastic decrease of the erythrocyte count (due to a disturbance in the maturation of red blood cells). The disturbance in maturation may be based on the inability of ribonucleotide reductase (B_{12}-dependent) during coenzyme deficiency to produce sufficient components for DNA synthesis. B_{12} deficiency is manifested biochemically, among other effects, by the appearance in urine of methylmalonate, since its normally occurring conversion to succinate is also dependent on vitamin B_{12}. This appearance has diagnostic value. Pernicious anemia does not arise from a dietary deficiency, but rather is caused by a defect in the *absorption* of the vitamin. Cobalamin (designated the *extrinsic factor* in this connection) can be absorbed by the human organism only in the presence of the *intrinsic factor*, normally formed in the gastric mucosa. The intrinsic factor is a neuraminate-containing glycoprotein with the molecular weight of 60,000. It forms a complex with B_{12} which is absorbed in the intestine. Patients with pernicious anemia lack this factor, and it is this lack which is the underlying cause of vitamin B_{12} deficiency.

Ascorbic Acid (Vitamin C). A derivative of carbohydrates (formula in Chapter XV,4), ascorbate is one of the biochemical redox systems; it can be oxidized (reversibly) to give dehydroascorbate (formulas in Chapter XV,4). Very few oxidoreductases, however, specifically require ascorbate as H_2 donor, e.g., the formation of homogentisate in the metabolism of tyrosine (Chapter VIII,11) and the hydroxylation of proline in collagen (Chapters VIII,13 and XXIII,6). The latter reaction begins to explain the disturbance of the metabolism of connective tissue with vitamin C deficiency. It is noteworthy in this context that the adrenal cortex is well supplied with ascorbate and that the concentration of ascorbate changes with the functional state of this gland.

Ascorbate is not a vitamin for most mammals, because it is biosynthesized by them to fill their own requirements. The biosynthetic pathway is shown in Chapter XV, at the end of Section 5.

The name *ascorbic acid* derives from scurvy, a deficiency disease much feared by mariners through the ages, but readily curable by ascorbic acid. The principal manifestations are capillary lesions, hemorrhages, gingivitis, and loosening of the teeth. Man's requirement for the vitamin exceeds that of other vitamins by one or two orders of magnitude. The reason for this is obscure. Nevertheless, full clinical ascorbate deficiency disease is rare nowadays. So many of our processed foods and beverages are supplemented with synthetic ascorbate, usually to enhance the flavor. Excess ascorbate is excreted by the organism in the urine.

Biotin. This vitamin contains a condensed ring system consisting of two five-membered heterocycles (formula in Chapter VI,6). In the conjugate *biocytin*, the carboxyl group of biotin is linked peptidically with the ε-amino group of lysine. A similar bond holds the vitamin to the protein molecule in the biotin-containing enzymes, which catalyze carboxylations (see Chapter XII,6).

Biotin was discovered as a growth factor of yeast. Experiments with animals also produced deficiency symptoms (dermatitis, loss of hair) when avidin was administered. *Avidin* is a protein isolated from raw egg white that firmly binds biotin, and thus inactivates the vitamin, the complex is not attacked by proteases. Excessive consumption of raw egg white apparently may cause biotin deficiency even in man.

BIBLIOGRAPHY

L. J. Bogert, G. M. Briggs, and D. H. Calloway, "Nutrition and Physical Fitness," 9th edition. Saunders, Philadelphia, 1973.

H. F. DeLuca and J. W. Suttie (editors), "The Fat-Soluble Vitamins." Univ. Wisconsin Press, Madison, 1970.

P. György and W. N. Pearson (editors), "The Vitamins," vol. 7. Academic Press, New York, 1967. About water-soluble vitamins.

R. S. Harris and P. L. Munson (editors), "Vitamins and Hormones," annual volumes. Academic Press, New York.

M. Kleiber, "The Fire of Life." Wiley, New York, 1961.

F. M. Lappé, "Diet for a Small Planet." Ballantine Books, New York, 1971. Excellent treatment on protein and amino acid requirements and utilization.

C. H. Robinson, "Normal and Therapeutic Nutrition," 14th edition. Macmillan, New York, 1972.

R. H. Wasserman and R. A. Corradino, Metabolic role of vitamins A and D. *Annu. Rev. Biochem.* **40,** 501–532 (1971).

R. H. Wasserman and A. N. Taylor, Metabolic roles of vitamins D, E, and K. *Annu. Rev. Biochem.* **41,** 179–202 (1972).

CHAPTER XXIII

Special Biochemical Functions of Certain Organs

Just as certain special structures perform particular biochemical functions within the cell (cf. Chapter XVIII), so do certain organs perform definite functions within the organism. Since this division of labor is observed very easily, and need not be determined laboriously from quantitative measurements and from electron optical pictures, it has been known for a much longer time.

Individual organs very often are characterized by a specialized set of enzymes. In some cases certain enzymes are absent from an organ so that other reactions are found only in that organ; in other cases a characteristic pattern of isoenzymes is present. Metabolism is adapted in this way to the specialized requirements and tasks of the organ. The causes and mechanisms of this *biochemical differentiation* are not known. The primary tasks of the individual organs can be characterized as follows:

Organ	Function
Stomach, gut, kidney, lung	Digestion, absorption, excretion, gas exchange
Blood	Transport
Liver	Metabolism
Muscle	Conversion of chemical into mechanical energy (locomotion)
Bone and connective tissue	Support
Nervous system	Reception and conduction of stimuli; neural correlation
Endocrine glands	Regulation of metabolism and development; humoral correlation

The disciplines of physiology and biochemistry overlap considerably in the study of these functions. We can leave the discussion of the function of the organs largely to the physiologists and will limit ourselves to the biochemical basis of the functions, especially those of the digestive tract and of the nerves and muscles.

1. The Digestive Tract

Food is digested by hydrolases which catalyze the cleavage of high-molecular weight foodstuffs (proteins, starch, neutral fat) to low molecular weight, largely water-soluble components (amino acids, simple sugars, glycerol, fatty acids). These hydrolytic products are more easily absorbed than the polymeric compounds. Hydrolysis commences in the oral cavity with the action of saliva, which contains *amylase* and, in particular, attacks starches. Hydrolysis continues in the stomach and gut, where the proteases and lipases predominate. The various enzymes have been discussed in connection with their substrates (cf. Chapters VIII,1, XII,2, and XVII,6); the mechanical events during digestion are treated in textbooks of physiology. In addition to the enzymes the secretions of the stomach contain *mucin* (a mixture of mucoproteins and mucopolysaccharides); the slippery mucus ensures a smooth passage of the food pulp.

Stomach. The secretion of *hydrochloric acid* into the stomach is a remarkable accomplishment of the *gastric mucosa.* The acid is surprisingly concentrated (pH around 1.5). Such a low pH is bactericidal and thus provides an effective protection against infections. The chief cells of the stomach produce *pepsinogen,* which is activated to *pepsin* by the acidity (cf. Chapter VIII,2), and also a mixture of mucoproteins, which includes the *intrinsic factor* (cf. Chapter XXII,5).

The generation of acidity entails active transport of $H^{\oplus}$ ions against a concentration gradient. Osmotic work is performed and depends on the presence of ATP (i.e., lack of oxygen brings the process to a halt). The $H^{\oplus}$ ions are derived from H_2O (which of course dissociates spontaneously to $H^{\oplus}$ and $OH^{\ominus}$). They are transported against a concentration gradient of 1:1,000,000 (from pH 7 to pH 1). The $OH^{\ominus}$ ions, which would shift the pH toward the alkaline side and would aggravate the osmotic work, are buffered by carbon dioxide: $HO^{\ominus} + CO_2 = HCO_3^{\ominus}$. The enzyme *carbonic anhydrase* is involved in this last reaction and ensures the rapid attainment of equilibrium between CO_2 and H_2O, or rather CO_2 and $OH^{\ominus}$.

Pancreas. The most important site of production of the digestive enzymes: proteinases, peptidases, lipases, nucleases, etc. (cf. Chapters VII,9, VIII,1, and XII,2) is the pancreas. The amount of enzymes produced is considerable. The pancreas is one of the tissues with the highest rate of protein synthesis. Proteinases are secreted as enzyme precursors called zymogens, which then in the intestine are activated by limited proteolysis (Chapter VIII,2). The pancreatic juice is also rich in $HCO_3^{\ominus}$ and consequently is alkaline. It enters the intestine and first neutralizes the acidity coming from the stomach. When acidic food pulp passes the pylorus, the latter stimulates the pancreas to elaborate more of its digestive enzymes. The secretion of pancreatic juice is regulated by tissue hormones (Secretin, Chapter XX,13).

Gall Bladder. Bile is formed in the liver and contains as its principal component the bile acids (Chapters XIV,5), which aid the digestion and absorption of fat; bile also contains cholesterol and bile pigments, which are the end products of hemoglobin catabolism. Some of the components are reabsorbed by the intestine and are passed through the blood and liver back to the gall bladder (*enterohepatic circulation*). Some of the bile pigments invariably enter the colon and impart the normal brown color to the feces.

Small Intestine. The digestion (hydrolysis) of food continues in the small intestine. The peptides that arose in the stomach by the action of pepsin are degraded further by the pancreatic enzymes. Some of the cleavage of oligopeptides actually takes place intracellularly in the mucosa cells. The hydrolysis of starch and other carbohydrates, which began in the oral cavity by salivary amylase, continues in the small intestine and is terminated there. Finally, fats, phosphatides, and nucleic acids are also cleaved.

Mucosal cells furthermore are a locus of synthesis: The cleavage products of the fats (free fatty acids and monoacylglycerols) are reused to build up triacylglycerols (triglycerides) and transported off by the lymphatic system.

Absorption. The cleavage products formed by the action of digestive enzymes are initially dissolved in the digestive juices—of which more than 8 liters per day may be produced. From this solution, the small intestine absorbs the low-molecular weight substances and water; the colon absorbs chiefly water. A minor part of the absorption follows the laws of osmosis and diffusion (passive transport), although most of it proceeds by active transport. The most thoroughly studied systems are for amino acids and simple sugars. Different amino acids compete for the site of active transport. Most of the absorbed substances reach the liver through the portal vein.

2. The Liver

Only a small portion of the absorbed hydrolysis product is "combusted" in the liver for the immediate production of free chemical energy. The larger portion is utilized for the synthesis of endogenous substances (glycogen, proteins, some neutral fats, as well as phosphatides and cholesterol). Modern concepts no longer permit a distinction between structural metabolites and fuel metabolites; the same substances serve both purposes. The structural components of the organism are being broken down continuously, used for the production of energy, and replaced by newly formed substances.[1]

The liver is the organ where very many of these interconversions and synthetic processes occur; if we were to enumerate all the biochemical functions that are localized in the liver, the greater part of this book would have to be reprinted here. The chief functions will be summarized here; for details we refer to the appropriate chapters.

[1] It is not easy to see the biochemical necessity for this turnover, which requires considerable expenditure of free energy, but it is observed in all phyla of the animal kingdom and appears to be advantageous to the species in the evolutionary process of natural selection.

Metabolism of Carbohydrate. This is one of the main functions of the liver. As discussed in Chapter XX,11, the glucose level of blood is regulated precisely by several hormones. The main target for these hormones is the liver, which is the executive organ and as such builds up glycogen as the storage form of excess carbohydrate. During acute glucose deficiency the liver can withdraw some of the stored glucose and bring it into the blood. This task of the liver is facilitated by the fact that liver cells are freely permeable to glucose, in contrast to other organs where this permeability is controlled by insulin. As a result, there is constant exchange between glycogen and free glucose. The half-life of rat liver glycogen is only 1 day (it is 3.6 days for muscle glycogen).

Another important task of the liver is the synthesis of glucose (gluconeogenesis), both from lactate (arising from muscle metabolism) and from the breakdown products of proteins. The carbon skeletons of the so-called glucogenic amino acids are utilized for gluconeogenesis.

The metabolism of other sugars also takes place in the liver. From a nutritional standpoint the breakdown of fructose (largely derived from the disaccharide sucrose) and galactose (from the milk sugar lactose) are important. For the purpose of detoxication reactions the oxidation of glucose to glucuronate is important. Most mammals are able to convert glucuronate to ascorbate. Man is an exception and must ingest it in his diet as "vitamin C." The interconversions of hexoses to pentoses, and the reverse, via the pentose phosphate cycle also take place in the liver, but are not restricted to this organ.

The breakdown of ethyl alcohol, which takes place primarily in the liver, can also be discussed in the context of carbohydrate metabolism, because even in the absence of ethanol ingestion there is always naturally present a very small amount of ethanol (about 10 μM), derived from pyruvate decarboxylation and thus from the normal metabolism of carbohydrate. Ethanol is dehydrogenated by alcohol dehydrogenase and NAD to acetaldehyde and then further oxidized to acetate (either by xanthine oxidase or aldehyde dehydrogenase). In addition, the liver can also perform a direct oxidation of ethanol by a microsomal peroxidase system, which is also responsible for the oxidation of methanol.

Detoxication Reactions. In the liver a vast array of conversion reactions take place that affect endogenous as well as exogenous substances. Frequently these conversions lead to accelerated excretion, and thus result in "detoxication."[2] One example, the detoxication of ammonia in the urea cycle, was discussed in detail in Chapter VIII,8.

There are three main types of detoxication reactions: conjugation, oxidation, and reductions.

The most important ***conjugations*** are with *glucuronate* (involving UDP-glucuronate, Chapter XVII,4) or *sulfate* (involving phosphoadenosyl phosphosulfate, Chapter VI,6). Furthermore, aromatic carboxylic acids (benzoate, salicylate) generally are linked to *glycine*; the reaction proceeds via CoA derivatives:

$$C_6H_5\text{—}COO^{\ominus} \xrightarrow{\text{ATP, CoA}} C_6H_5\text{—}CO\sim\overline{SCoA} \xrightarrow{\text{Glycine}} C_6H_5CO\text{—}NH\text{—}CH_2\text{—}COO^{\ominus}$$

Conjugated bile acids, e.g., glycocholate, are formed in the same manner.

[2] It can happen that the toxicity of a foreign substance is actually enhanced by just such mechanisms. This observation would seem to argue against any inferred teleological intent of these reactions.

Aromatic amines become *acetylated* with acetyl-CoA. These acetyl derivatives very often are less soluble than the free amines. *Methylations* at N or phenolic hydroxyl groups have also been observed, e.g., in the inactivation of norepinephrine or epinephrine.

Mercapturates are conjugates such as found in thioethers with *N*-acetylcysteine. Glutathione is the donor of the cysteine residue. This reaction is used to "detoxify" aromatic hydrocarbons (e.g., anthracene).

The important ***oxidation*** reactions are *hydroxylations* that occur in the endoplasmic reticulum ("microsomes"). The mechanism is described in Chapter X,6. The enzyme system of hydroxylation can be *induced* by appropriate substrates including many drugs (for enzyme induction, see Chapter VII,7). In addition to the introduction of hydroxyl groups in aromatic ring systems, one also finds hydroxylations of $-NH-CH_3 \rightarrow -NH-CH_2OH$, followed by the cleavage of CH_2O (oxidative demethylation).

Reduction is important for the inactivation of hormones (steroid hormones, insulin) and foreign substances, for example, aromatic nitrocompounds are also reduced.

The Metabolism of Fat. This process involves an interplay between liver and adipose tissue. When the metabolic condition of an organism calls for fat mobilization, then the adipose tissue releases free fatty acids (obtained from the hydrolysis of stored triglycerides), which are taken up by the liver and subjected to β-oxidation. An over-supply of fatty acids causes the formation of ketone bodies. The liver is the only tissue capable of this. In the reverse direction—depending on the overall state of metabolism—the liver can also synthesize fatty acids and transport them to the blood stream as phosphatides. The adipose tissue reuses these to synthesize and store its own "neutral fat." (The adipose tissue, incidentally, is also quite capable of fatty acid synthesis.) Last, the bulk of cholesterol is produced in the liver; some of the cholesterol is further converted there to the bile acids.

The Metabolism of Protein. The amino acids freed by intestinal digestion reach the liver through the portal vein and are metabolized. In part they are reused as such for the synthesis of plasma proteins (albumins, some globulins, and the diverse clotting factors). The remainder of the amino acids is broken down. Nitrogen ends up in urea and is transported to the kidneys. The carbon skeletons largely enter gluconeogenesis.

Liver Function Tests. Many of the above named functions can be employed for diagnostic purposes. The metabolism of carbohydrates, for example, can be tested in a "galactose tolerance test." Healthy individuals are able to metabolize within a certain time period a large amount of orally administered galactose. In an individual with a defective liver, galactose appears in the urine or remains in the blood for an extended period of time. Similarly, the determination of serum proteins and of the clotting factors permits conclusions about the liver's capacity for protein biosynthesis.

Many liver function tests involve detoxication reactions, i.e., they test the capacity of the liver to transform foreign substances into derivatives of glycine or glucuronate or other excretionary forms. The excretion in the bile of certain pigments (sulfobromophthalein) can be used to test liver function. Finally, damage to the liver is indicated when numerous enzymes appear in the blood, such as, glutamate-oxaloacetate transaminase, glutamate-pyruvate transaminase, glutamate dehydrogenase, isocitrate dehydrogenase, and lactate dehydrogenase (the liver contains chiefly isoenzyme 5). The determination of enzymes in serum, therefore, is of considerable diagnostic value.

3. Blood

Blood consists of cellular elements and of plasma. Its function is to transport material.

Blood Plasma. If clotting is prevented by the addition of certain anticoagulants (see Chapter IV,10) blood plasma can be studied. If blood (or plasma) is allowed to clot and the clot is centrifuged off, serum is obtained, which still contains nearly all the proteins of blood. The properties of these and other proteins are discussed in Chapter IV,10. Under pathological conditions, proteins from cells of diseased organs—even enzyme proteins—are found in blood. This is the basis of diagnosis by serum enzymes.

Blood also contains many low-molecular weight substances. The content of N-containing, low-molecular weight substances or the nonprotein nitrogen content (or "residual nitrogen," mainly urea, some amino acids, uric acid, creatine, creatinine) is an indicator of kidney function; the concentration is usually low because these substances are excreted. The glucose content is regulated by hormones and generally is very constant (cf. Chapter XX,11). Lactate appears in blood as a product of glucose breakdown. The concentration is low at rest (below 2.5 m*M*) but may increase to 15 m*M* with strenuous muscular exertion. For the mineral content and the buffer capacity of blood see Chapter XXI,4.

Erythrocytes. The most numerous cellular components of blood are erythrocytes; there are about 5 million per cubic millimeter. In addition, there are leukocytes and platelets. The morphology of blood obviously has diagnostic value, but cannot be discussed here. Erythrocytes contain hemoglobin, the red blood pigment, whose function (oxygen transport) was explained in Chapter IX,4.

The metabolism of erythrocytes has several peculiarities. Since erythrocytes are without a nucleus, they have lost the capacity for the biosynthesis of nucleic acid and protein. Further, the citrate cycle and the respiratory chain are also almost completely absent. The metabolism of carbohydrate is limited to glycolysis and direct glucose oxidation (oxidative pentose phosphate cycle). In this way NADPH is formed which is required for the reduction of methemoglobin to hemoglobin (Chapter IX,4) as well as for the maintenance of the reduced form of glutathione. Glutathione appears to be present in erythrocytes in relatively large concentrations to protect various SH proteins from oxidation. Individuals with an inborn deficiency of glucose-6-phosphate dehydrogenase cannot form sufficient NADPH for this protective function and therefore often suffer hemolytic crises. Such crises are precipitated by apparently harmless external factors, such as the ingestion of anti-malaria drugs or the consumption of certain kinds of beans; the latter gave the disease the name favism.

The Lung. Erythrocytes are charged with oxygen in the lungs. Of all the gases of the air, only oxygen is absorbed and bound reversibly to hemoglobin. In areas of low oxygen tension, i.e., in peripheral tissues and organs with very active oxidative metabolism, oxygen is again released[3] and is reduced to H_2O by the respiratory chain. Carbon dioxide arises from decarboxylation reactions usually proceeding at the same place, i.e., in the identical cell. Carbon dioxide is transported in the blood stream to the lungs and exhaled, if it does not reenter metabolism (cf. CO_2 fixation, Chapter VI,6, "Biotin," and XV,9).

4. Kidney and Urine

The kidney is an organ for both excretion and metabolism. It contains relatively high concentrations of amine oxidases and amino acid oxidases, as well as glutaminase (Chapter VIII,8), i.e., enzymes involved in the production of free ammonia. It serves to capture $H^{\oplus}$ ions by forming $NH_4^{\oplus}$ and therefore helps to prevent acidosis. The ammonium ions of urine stem from this reaction. The enzymes of oxidative metabolism (citrate cycle, respiratory chain) are also present—as in all cells. The kidney consumes relatively large proportions of oxygen and produces much ATP, which it needs for its excretionary activity.

Kidney function comprises three distinct processes localized in histologically different areas:

1. Ultrafiltration of plasma.
2. Reabsorption of water and dissolved substances back into the blood stream; this process entails active transport and consequently consumes energy.
3. Secretion of certain other substances into urine, also by an active transport mechanism.

Ultrafiltration. This term refers to the filtration of low-molecular weight substances with the retention of high-molecular weight ones, particularly the proteins. Blood proteins are retained in the glomeruli and normally do not enter primary urine at all. Protein in urine is invariably a sign of kidney damage or other diseases.

Reabsorption. Water is the substance reabsorbed in the greatest amount. It is an open question whether water molecules are themselves transported actively or transported in conjunction with the active transport of dissolved substances (see Chapter XXI,3).

[3] A very intriguing aspect of comparative biochemistry is the correlation of dissociation curves of hemoglobin (as a function of decreasing oxygen tension) with the habitat and general life style of the corresponding animal species.

The concentration of primary urine to about 1/100 of its original volume demands osmotic work. Calculations show, however, that this osmotic work is almost negligible compared with the total energy production of the kidney. The highly selective process of secretion apparently requires all the additional expenditure of energy.

Most of the low-molecular weight substances that are flushed into the primary urine during ultrafiltration are reabsorbed to a large extent. This is true even for such typical excretion products as urea and uric acid, but especially for free amino acids and glucose, which are reabsorbed completely so long as the blood sugar level remains normal. If the level exceeds 9 m*M*, some of the sugar is excreted in the urine. Since reabsorption is active transport (Chapter XXI,3), the enzymatic apparatus of active transport can no longer cope with the excessive amounts delivered by the blood. Similar conditions prevail for other substances; the capacity for reabsorption (formerly called "kidney threshold") differs widely for various substances.

The Secretory Activity of the Renal Tubules. This also constitutes an active transport of substances from the bloodstream into the urine. Ammonium ions originating from metabolic processes in kidney tissue enter urine in this fashion; the $NH_4^{\oplus}$ concentration is very low in blood. Furthermore, $H^{\oplus}$ ions are secreted actively; urine, consequently, is appreciably more acidic (pH 5–6.5) than blood (pH 7.4). Finally, foreign substances, such as penicillin and other drugs, are also removed from blood by active transport.

Hormonal Regulation of Kidney Function. Three different hormones act on the kidney: the adrenocortical hormone *aldosterone* (mineralocorticoid effect, cf. Chapter XX,2), the parathyroid hormone or *parathormone* (cf. Chapter XX,7), and the anti-diuretic hormone *vasopressin* (cf. Chapter XX,9) of the hypophyseal posterior lobe.

Aldosterone restricts the excretion of $Na^{\oplus}$ while stimulating the excretion of $K^{\oplus}$. Parathormone also acts on the mineral balance by promoting phosphate excretion through the kidney. The most important hormone undoubtedly is vasopressin, which is responsible primarily for the reabsorption of water. In its absence, enormous volumes of very dilute urine are excreted (Chapter XX,9).

The kidneys are not only target tissue, but have an *endocrine function* themselves since they are the site of production of erythropoietin, dihydroxycalciferol, and renin which, in the blood, liberates angiotensin (Chapter X,13) which in turn regulates the production of aldosterone. The renin–angiotensin system constitutes an important feedback mechanism.

Urine. The composition and the amount of urine per day fluctuate with the intake of liquids. Data on the daily excretion of various substances, therefore, are always given on the basis of "24-hour urine." Urea is the chief waste product. The excretion of N-containing compounds generally is of prime importance (carbon in food is largely exhaled as CO_2). The amount of urea excreted can be correlated directly with the amount of protein catabolized: 1 gm urea nitrogen $\cong$ 6.25 gm protein.

Other nitrogen-containing excretion products are uric acid, creatinine (arising from creatine phosphate or creatine by ring closure), the $NH_4^{\oplus}$ already mentioned, and various other components occurring in small amounts.

The significance of excreting inorganic salts (especially the alkali salts of chloride, sulfate, phosphate) has been mentioned. Certain combinations of ions result in precipitation of slightly soluble salts in the urine or in the passage ways (kidney or bladder stones). The most common precipitates are calcium oxalate, calcium phosphate, magnesium ammonium phosphate, and salts of uric acid.

Pathological Components of Urine. In various diseases one often finds anomalous components in the urine; for this reason urine analysis has always had great diagnostic value. It is even more valuable today, since refined quantitative techniques can measure changes in concentrations of normal components and detect traces of hormone metabolites (cf. Chapter XIV,6). Discussion of these anomalous conditions is the subject of pathological physiology and of diagnostics; only a very few substances will be pointed out.

Proteins. The appearance of proteins in the urine usually indicates kidney damage. Albumin (with the lowest molecular weight) is excreted most readily. The Bence-Jones proteins appear with certain tumors (plasmocytomas); they have such a low molecular weight that they pass through the glomerular filter even with normal kidney function. They have become conspicuous by their peculiar solubility properties (they precipitate at 50° and dissolve again at 80°). They are subunits of γ-globulins (Chapter IV,10, "Immunoglobulins").

Sugar. Glucose excretion either is due to a high intake of carbohydrates (alimentary glucosuria) or is symptomatic of diabetes mellitus. In many severe cases "ketone bodies" (acetone, acetoacetate, β-hydroxybutyrate) are excreted along with sugar.

Galactosuria and pentosuria, if not of alimentary origin, are caused by hereditary defects.

Pigments. Bile pigments appear with hepatitis; they are derived from the catabolism of hemoglobin (Chapter IX,5). Uro- and coproporphyrins may enter the urine in cases of poisoning or of genetic defects in the biosynthesis of the hemin pigment.

5. Other Excretion Products

Quantitatively *carbon dioxide* is the most important waste product. It is eliminated by the lungs. The establishment of the equilibrium between dissolved and gaseous carbon dioxide is catalyzed by carbonic anhydrase, a Zn-containing enzyme. Furthermore, hemoglobin becomes a stronger acid by addition of oxygen, i.e., it releases $H^{\oplus}$ ions, and thus enhances the evolution of CO_2 from $HCO_3^{\ominus}$:

$$HCO_3^{\ominus} + H^{\oplus} \rightleftharpoons H_2O + CO_2$$

(Compare Chapter XXI,4 for the significance of this reaction for the acid-base relationship.)

Feces consist largely of mucus substances from the gastrointestinal tract and remnants of intestinal bacteria. Their odor is due to indole and skatole; their color is due to the bile pigments (Chapter IX,5).

Both water and salts are excreted with *sweat*. It contains about 0.4% NaCl, but only traces of organic substances.

6. Connective and Supportive Tissue

The connective and supportive tissues make up a variety of structures with widely differing mechanical functions. They can be rigid or flexible; they can be elastic or inelastic. This variety of mechanical properties derives from a differential content of macromolecules with very diverse properties. The macromolecules chiefly responsible for the characteristic properties of these tissues are the proteins *keratin*, *collagen*, and *elastin* and the mucopolysaccharides *chondroitin sulfates* and *hyaluronic acid*.

Keratin. An important component of skin, hair, and nails, keratin is present in the α-helix conformation and forms fibrillar structures which are made rigid by numerous disulfide bridges. The most outstanding mechanical property of keratin is its great elasticity founded in part on the transition of α-helix to pleated-sheet structure.

Collagen. As previously discussed extensively in Chapter IV,4, the structural feature of the triple helix of collagen depends mostly on the high content of glycine, proline, and hydroxyproline. Collagen possesses a high resistance to stretching and is, among others, the main component of tendons. Moreover, it enters the composition of skin, bone, and connective tissue fibers.

Collagen is formed in fibroblasts. According to the general principle of protein biosynthesis, first a peptide chain consisting of about 1000 amino acid residues called protocollagen is formed. Certain hydroxylases with the aid of ascorbate then hydroxylate some of the proline and lysine residues. Three such peptide chains (they may be slightly different) then aggregate to form a triple helix (see Chapter IV,4) called tropocollagen. Next, certain carbohydrate residues are added in glycosidic linkage, namely, galactose followed by glucose, with the result that carbohydrate groups of the type Glc-α(1 → 2)Gal-β-(protein) are formed. This last reaction completes the intracellular synthesis of tropocollagen. The rodlike protein molecules are then secreted into the extracellular space where numerous tropocollagen rods coagulate to form collagen fibrils. They are finally cross-linked by covalent bonds (presumably ester bonds). This is a slow and enduring process continuing into advanced age and is partly responsible for the aging of connective tissue.

Elastin. This is a protein which forms elastic and resilient fibers. It contains up to 95% hydrophobic amino acids among which are glycine, proline, and leucine. The individual peptide chains are cross-linked in a characteristic manner: the lysine residues of different chains are oxidized in part to aldehyde groups and condensed to form a substituted pyridine ring, as shown formally in the following diagram:

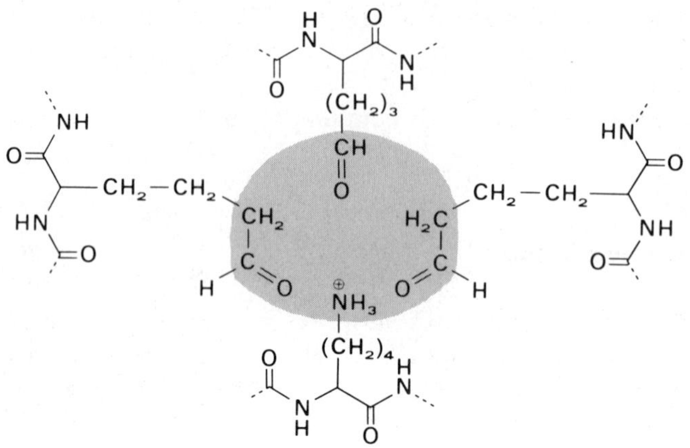

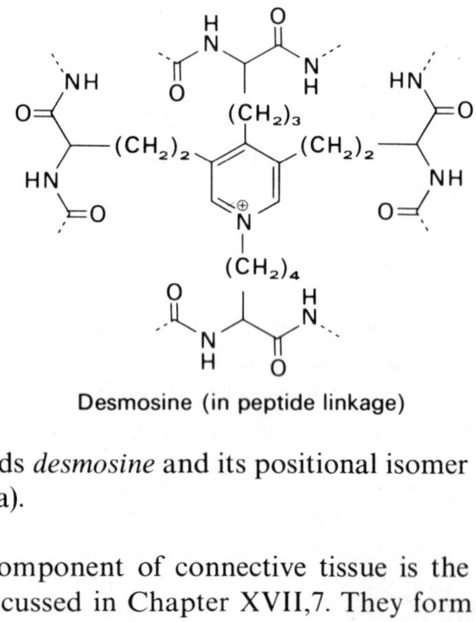

Desmosine (in peptide linkage)

Following hydrolysis the unusual amino acids *desmosine* and its positional isomer called *isodemosine* can be isolated (see formula).

Mucopolysaccharides. The second main component of connective tissue is the mucopolysaccharides. Their structure was discussed in Chapter XVII,7. They form gels in aqueous media and thus bind a great deal of water. Such gels are then made rigid by the incorporation of protein fibers. It is characteristic that *chondroitin sulfate*, a comparatively rigid molecule, is found especially in cartilage, while the fluid in bone joints contains predominantly *hyaluronic acid*, whose gels have lubricating and shock-absorbing properties. The preeminent property of bones is the capacity to bear pressure: the bone skeleton, of course, has to support the entire body weight. This property is achieved primarily by the incorporation of calcium salts into a collagen network.

The structure of bone contains the long-known crystalline hydroxyapatite as well as amorphous calcium phosphate. Mineralization (incorporation of calcium salts) proceeds directly on collagen molecules under the influence of osteoblasts. Several theories offer mechanisms for this kind of directed incorporation, none of which have been proved at this time, however. One such theory assumes a primary pyrophosphorylation of side chains of the collagen molecule, while another ascribes a decisive role to the enzyme alkaline phosphatase. In principle, all collagen can ossify, but mineralization may be prevented in soft connective tissue by chondroitin sulfate.

The reservoir of calcium in bones can be mobilized when necessary through the activity of osteoclasts (cf. also Chapter XXI,6).

7. Biochemistry of Muscles

Muscles convert chemical energy to mechanical energy. The source of chemical energy is ATP; it arises in the muscle by oxidation of foodstuffs.

Enzymes of Muscle. Different types of muscles are equipped with different enzymes. Muscles that must perform a large amount of work for only a short time (e.g., leap muscles) usually fill their energy needs by anaerobic glycolysis. Thus, skeletal muscles contain the necessary enzymes in high concentrations, over 70% of soluble proteins. Other muscles perform work continuously, e.g., the heart muscle or the flight muscle of insects. These muscles operate aerobically primarily by "burn ing" fat. They contain a large number of mitochondria well stocked with the enzymes of the respiratory chain (see Figs. XVIII-7 and 8 in Chapter XVIII,3). They are also equipped with a special oxygen carrier, *myoglobin,* which binds oxygen molecules reversibly, as hemoglobin does, but with greater affinity (cf. the dissociation curve, Fig. IX-2).

Energy-Rich Phosphates. The immediate fuel to power the muscle is adenosine triphosphate, the hydrolysis of which provides the energy for contraction. Creatine phosphate, another energy-rich phosphate in equilibrium with ATP, is the *energy reserve*:

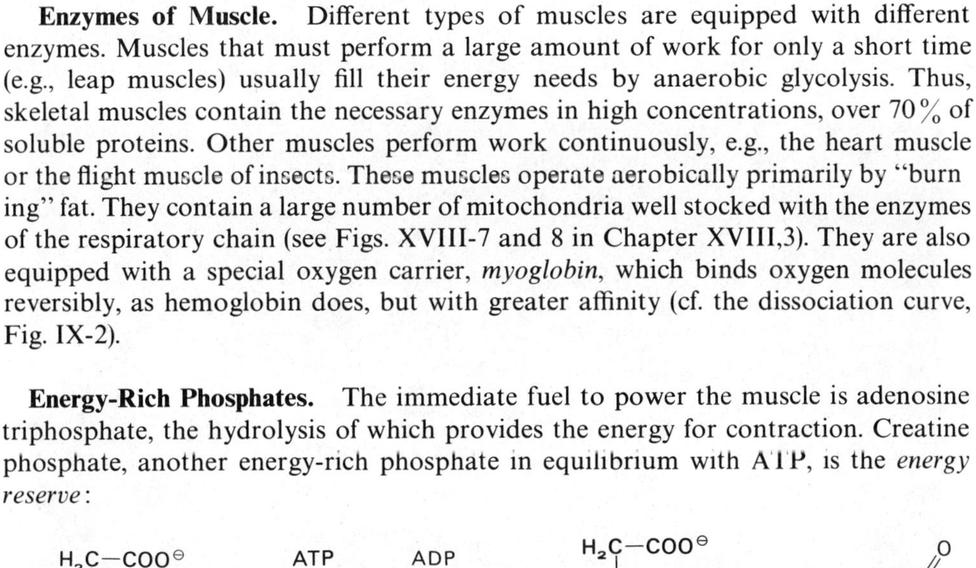

Creatine phosphate is always readily available to regenerate ATP. This system has an advantage over a simple excess of ATP in that an unfavorable ATP/ADP ratio due to ATP hydrolysis is prevented; otherwise this would limit the work capacity of ATP.

Creatine phosphate spontaneously (nonenzymatically) rearranges to form creatinine phosphate, which is normally present in blood and is excreted by the kidneys as such. The amount of creatinine excreted each day is proportional to muscle mass and thus a useful "personal constant" of man.

The free energy of ADP cleavage can also be used. Muscle and other tissues contain the enzyme *myokinase*, or *adenylate kinase*, which catalyzes the following reaction:

$$2\,ADP \rightleftharpoons ATP + AMP$$

At equilibrium, the three phosphates are present at roughly equal concentrations. By this reaction ATP is formed at the expense of ADP, or AMP is phosphorylated to give ADP, which can in turn be phosphorylated further, in the respiratory chain, for example.

Ultrastructure of the Muscle. The usual cell components are of little significance in striated muscles. The cells are filled with myofibrils which produce the familiar striped pattern. Electron optical studies reveal that the A-bands consist of thick, threadlike molecules, presumably myosin. The space between these molecules is occupied by thin filaments which continue into the isotropic area (I-band) and end at

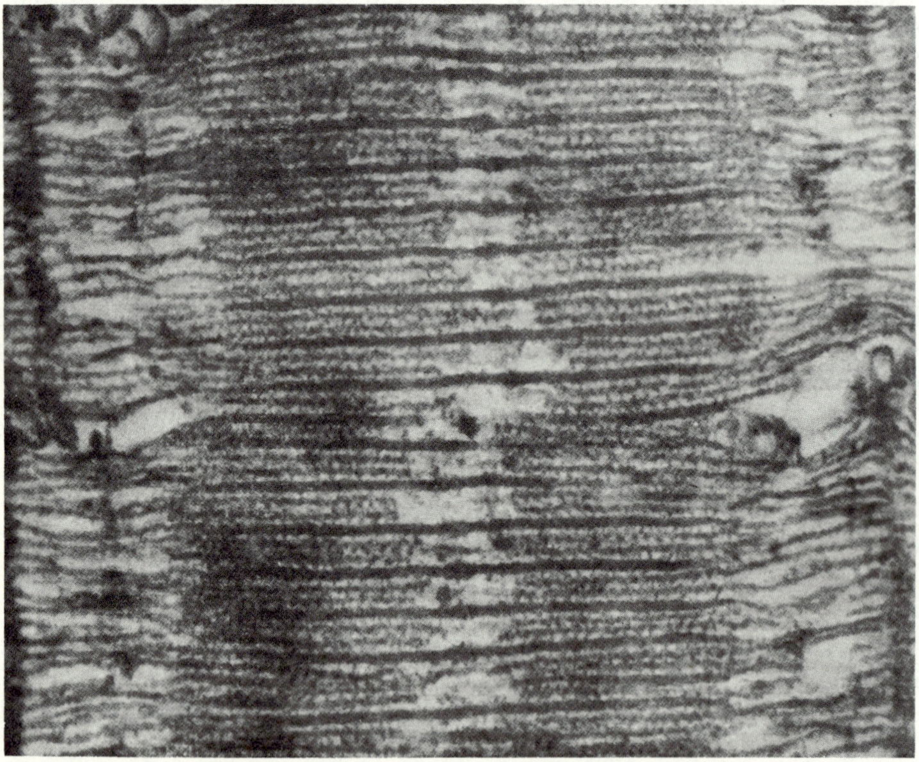

Fig. XXIII-1. Electron microscopic photograph of muscle (thin section; cf. schematic drawing in Fig. XXIII-2). A large number of small cross links can be distinguished between the thin filaments, which are linked together (near the edges of the photograph) in the Z-line, and the thick lines. Enlargement: × 120,000. [From H. E. Huxley, *in* "Molecular Biology" (D. Nachmansohn, ed.). Academic Press, New York, 1960.]

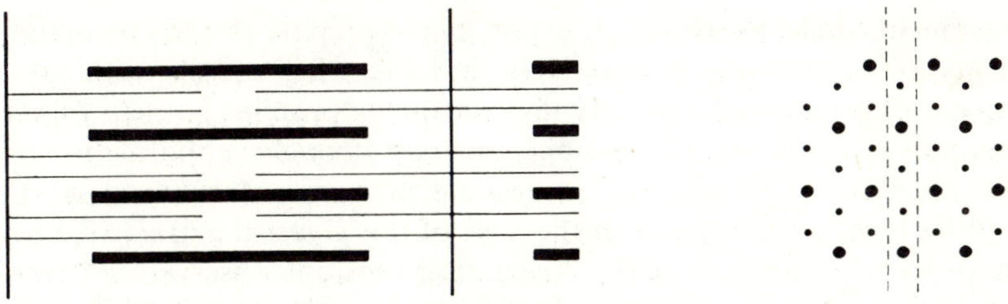

Fig. XXIII-2. Schematic representation of myofibrils. The heavy lines represent myosin filaments; the light horizontal lines, the actin filaments. The vertical lines are the Z-lines. At right, a cross section is shown. It reveals the hexagonal symmetry of the arrangement of the filaments. The direction of the cut of the thin section is indicated by the broken lines. Compare with the electron optical picture in Fig. XXIII-1.

the Z-line. The latter consists of actin and tropomyosin (cf. Figs. XXIII-1 and 2). According to recent investigations by A. F. Huxley and H. E. Huxley, the thin fibrils slide past the myosin fibrils during contraction so that the I-bands eventually disappear and the Z-line approaches the end of the myosin fibrils.

Contractile Proteins of the Muscle Fibrils. The fibrils of muscle contain, besides the insoluble and little-studied structural proteins, also a few soluble, fibrillar proteins, which participate actively in muscle contraction: they are myosin, actin, and tropomyosin.

The protein ***myosin*** has the solubility properties of a globulin. Its X-ray diagram reveals it to be a fibrous protein of the α-keratin type (structure: α-helix). The complex with a molecular weight of 500,000 consists of about 15% of smaller proteins and of two unusually long peptide chains that consist of about 2000 amino acid residues and that are synthesized at particularly large polysomes. The complex is a rodlike structure of about 1400 Å in length and 20 Å in diameter, with a globular, flattened head of 200 × 50 Å in which is localized the ATPase-activity of myosin. Brief treatment with trypsin splits the complex into two fragments, called *L*- and *H-meromyosin*.

Actin is less soluble than myosin. It occurs in two forms. Globular *G-actin* has a molecular weight of 46,000. The addition of salts converts it to the polymeric *F-actin*, which consists of two twisted strands of linearly aggregated globular actin molecules; they also exist in this form in muscle. In solution, actin and myosin combine easily to give *actomyosin*. Under certain conditions, one can isolate native actomyosin from muscle. Actomyosin has pronounced ATPase activity enhanced by $Ca^{2\oplus}$ and $Mg^{2\oplus}$.

Tropomyosin with a molecular weight of 70,000 is a strongly asymmetrical molecule (450 × 20 Å) that readily forms polymeric fibers. Together with actin it participates in the structure of myofibrils especially of smooth musculature.

Another protein of the thin filaments is *troponin* which combines with actin and tropomyosin and represents an important control protein. It inhibits the "ATPase" of actomyosin. This inhibition is abolished by $Ca^{2\oplus}$ ions.

Contraction of Model Systems. It is possible to prepare threads from the isolated proteins myosin and actin. Alternatively, by extracting muscle with glycerol, all soluble material is removed and only the contractile proteins are left. Either system contracts upon addition of ATP; this phenomenon strongly supports the assumption that ATP reacts directly with the contractile proteins. Mechanical work can be performed by these systems, and in the case of the glycerol extract this work is of the same order of magnitude as the mechanical tensions observed *in vivo* (4 kg per square centimeter of cross section; H. H. Weber and co-workers). Contraction is not accompanied by changes in secondary structure of the proteins; the α-helix is retained as the predominating structural feature. Again, the same applies to the living muscle.

ATP is important not only for contraction, but also for *relaxation.* It affects the elastic properties of the muscle and renders it stretchable (H. H. Weber). The softening effect (plasticity effect) is clearly differentiated from the contracting effect. Pyrophosphate and various nucleotide triphosphates also possess this softening effect, but to a less pronounced degree. Probably the interaction between actin and myosin is weakened and actomyosin dissociates to its components.

The softening effect of ATP must be distinguished from the physiological *relaxing factor* of the muscle. The relaxing factor can be isolated as small grana with a diameter of 700 to 1000 Å. The grana are actually fragments of the sarcotubular apparatus. They contain an enzyme system that catalyzes the active transport of calcium ions; 1 ATP is required for each $Ca^{2\oplus}$ ion transported inside. (See also below.)

Contraction of the Living Muscle. Today every theory of muscle contraction must be based on the observation by electron optical methods that during contraction actin and myosin filaments slide into each other in telescopic fashion. This process is accompanied by ATP cleavage. It is believed now that the resting muscle contains a system in which the ATPase effect of myosin is largely inhibited by the lack of $Ca^{2\oplus}$ and by *troponin.* The excitation of the motoric nerve leads to a depolarization of muscle membrane and consequent release of $Ca^{2\oplus}$ ions from *sarcotubuli.* $Ca^{2\oplus}$ concentration jumps from $10^{-7}\,M$ to $10^{-5}\,M$. Troponin binds $Ca^{2\oplus}$, thereby stopping the inhibition of ATPase and promoting ATP cleavage, accompanied by a strong interaction between the myosin and actin filament. In this way linkages are formed between the filaments coming from the globular heads of myosin and making connection with the actin filament; these are visible in the electron microscope. It is assumed that the formation of these linkages with the consumption of ATP is involved in the development of mechanical force and, consequently, in the contraction of muscle. It has not yet been possible to develop a model for this interaction and linkage formation, or for the development of mechanical force.

The end of the twitch and subsequent relaxation of the muscle is brought about by the activity of the sarcotubular apparatus. As soon as the motoric nerve is no longer excited, a $Ca^{2\oplus}$ ion pump is started which removes $Ca^{2\oplus}$ from the cytoplasmic space. With that, troponin can inhibit again the actomyosin ATPase. The cleavage of ATP and the development of mechanical force as a result cease. Second, the grana probably also produce a labile relaxing factor which splits the bonds between actin and myosin and thus enables the muscle to be stretched passively back to its resting position. *Rigor mortis* essentially is caused by the absence of this relaxing factor, leaving the strong bonds between myosin and actin intact.

8. The Biochemistry of the Nervous System

The nervous system serves the neural coordination of the different organs and their functions, the uptake of information (stimuli) from the environment, and the utilization of such information in thought processes. The physiological function of the nervous system as well as any discussion of the bioelectric phenomena are in the domain of physiology. Here we shall only concern ourselves with several biochemical aspects of this function.

Most biochemical investigations have been conducted with the central nervous system. The model system for the conduction of excitation in peripheral nerves has been the giant axon of the squid. If in the following discussion we speak of "nerve cells" or "nervous tissue," one has to remember that we are really dealing with a totality of morphologically and functionally quite different elements. We will not be able to go into any great detail and have to limit ourselves to the discussion of a few selected fundamental phenomena.

Metabolism of the Nervous System. This process is highly dependent on the supply of oxygen. The human brain, for example, consumes approximately 20% of the total oxygen taken up by the organism, although it comprises only 2% of the total body weight. The energy metabolism is essentially the metabolism of glucose, which is "combusted" via the Embden-Meyerhof degradative pathway and the citrate cycle. With a deficiency of glucose, β-hydroxybutyrate can take its place to a certain extent. Fat, however, it not combusted by the brain, and the total respiratory quotient equals 1.0.

The brain also vigorously metabolizes amino acids. The concentration of amino acids is six to ten times higher than in blood plasma. Besides glutamine, glutamate, and aspartate, one also finds *N*-acetylaspartate, γ-aminobutyrate, and cystathionine (cf. Chapter VIII,12). Glutamate and γ-aminobutyrate are intermediate products in a side pathway of the citrate cycle, which we have discussed in Chapter XI,4. The side pathway proceeds from α-ketoglutarate via glutamate, γ-aminobutyrate to succinate semialdehyde, and further to succinate.

The brain is amply supplied with blood, and yet there exists a permeability barrier between blood plasma and the brain commonly called the *blood brain barrier*. It is generally impermeable to proteins, as well as to most low-molecular weight substances. Glucose and amino acids, however, are actively transported into the brain.

Nervous tissue is unusually rich in *lipids*. This is related to the fact that neurons are surrounded by a relatively thick myelin sheath which contains a large amount of sphingomyelin lipids (cf. also Chapter XIII,4).

The neurons are the functional units of the nervous tissue. They are electrically excitable to a very high degree and are specialized to conduct this excitation.

Bioelectric Potentials of the Nerve. In the resting state, the interior of the nerve, i.e., the axoplasm, has a potential of -60 mV with respect to the sourrounding liquid. This membrane potential is established by an appropriate distribution of ions: the $Na^{\oplus}$ ion concentration is very low inside, but relatively high on the outside, and the permeability of the membrane for $Na^{\oplus}$ ions is very small. The interior contains more $K^{\oplus}$ and $Cl^{\ominus}$. This typical nonequilibrium is maintained by a continuous supply of energy; $Na^{\oplus}$ ions diffusing into the interior are transported back out. Very graphically, one speaks of a "sodium pump powered by ATP." This apt expression, however, does not provide a scientific explanation of the phenomenon. The mechanism of active transport has yet to be explained.

ATP is derived primarily from oxidative metabolism (oxidative phosphorylation in the respiratory chain).

Excitation of the Nerve. Permeability is momentarily changed when the nerve is excited. $Na^{\oplus}$ ions flow in (passively, following the concentration gradient), and the polarization is reversed. As a result, the interior becomes positively charged, $K^{\oplus}$ ions then flow out. The state of increased permeability lasts only a few milliseconds; it is followed by the original state of low permeability, during which the $Na^{\oplus}$ ions distribute themselves (and later are pumped out again). Finally, the resting potential is re-established.

The shock of the action potential induces a drop of the resulting potential at an adjacent location (in myelated nerves, at the next ring), resulting in excitation accompanied by a repetition of ion migrations ($Na^{\oplus}$ inflow, $K^{\oplus}$ outflow). The excitation travels in this manner along the nerve to the synapse or to the end plate.

Cause of the Permeability Change. The ionic theory of nerve conduction, which is well supported by measurements of the concentration of ions, assumes changes of permeability to be the *cause* of the excitation. Nachmansohn hypothesizes that acetylcholine is essentially involved in the sudden development of permeability.

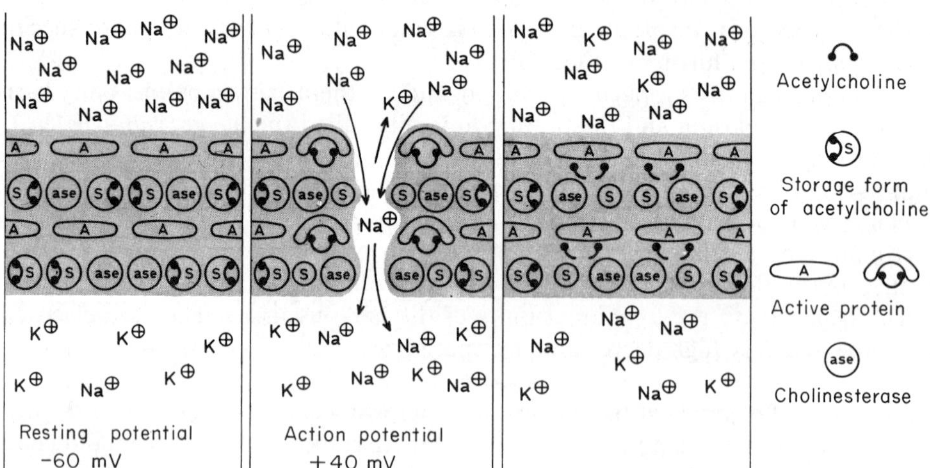

Fig. XXIII-3. Illustration of Nachmansohn's hypothesis of permeability changes in nerves. The diagram shows three phases: left, the resting state; middle, the influx of $Na^{\oplus}$ with concomitant reversal of potential; right, the state after hydrolysis of actylcholine. *S*, *A*, and *ase* represent the proteins. Further explanations in the text.

Electrical stimulation liberates acetylcholine from its storage form (probably bound to a protein) and it then forms complexes with an active protein. One can conceive that the protein layers become permeable by changes in their tertiary structure and allow $Na^{\oplus}$ ions to stream into the interior of the nerve cell. This condition, however, does not last. Acetylcholine is split by cholinesterase to choline and acetate; the protein reassumes its original configuration, so that the "holes" are plugged again, and the very brief changes in permeability are ended (see the diagram, Fig. XIII-3). The great rapidity of the processes—complete permeability lasts less than 1 millisecond—can be explained satisfactorily by the very high turnover number of the enzyme choline-esterase, which amounts to about 3 million. This means that each enzyme molecule can split 50 molecules of acetylcholine per millisecond.

For complete restitution of the original state, the resynthesis of acetylcholine (from choline and acetyl-CoA, which can come from metabolism or intermediary activation of acetate by ATP) and the removal of $Na^{\oplus}$ by the "sodium pump" are required.

Nachmansohn's hypotheses are not yet proved, although they are supported by considerable experimental evidence. Electrophysiologists generally reject the hypotheses without offering another explanation of the changes of permeability. Further work must be performed in order to elucidate these phenomena of molecular biology.

Neurotransmitters. The transmission of excitation at nerve ends (synapses or motor end plates) takes place via a chemical pathway. Active substances are released which are now called neurotransmitters.

According to the nature of these active agents, we distinguish between *adrenergic* and *cholinergic* nerves. Adrenergic (postganglionic sympathetic) fibers elaborate *norepinephrine* (formula and biosynthesis in Chapter XX,4); this compound also possesses the activity of the "accelerating substance" (Loewi). At the motor end plates and at many synapses ("cholinergic nerves"), *acetylcholine* is released, which is identical with Loewi's "vagus substance." Acetylcholine is cleaved by a special enzyme acetylcholine esterase, thus ensuring the rapid disappearance of the effect.

In the central nervous system additional substances also participate in the transmission of stimuli. Glutamate probably is the predominant fast-acting excitatory neurotransmitter of the brain, while acetylcholine is a slow-acting, but longer lasting transmitter. γ-Aminobutyrate, on the other hand, exhibits inhibitory action, probably by increasing membrane permeability to chloride ions. Both inhibitory and excitatory action have been ascribed to dopamine (which replaces norepinephrine) and to serotonin. Here, probably a neuraminate-containing glycoprotein or glycolipid plays the role of receptor substance (Wesemann).

Neurotransmitters are stored in the synapses inside little sacs (synaptic vesicles). With an electric signal arriving at the synapse and coming through the axon, the transmitters are released into the synaptic cleft and there excite the postsynaptic membrane. From there, the excitation is conducted along. Inhibitory synapses reduce the excitability of postsynaptic neurons by excitory synapses.

The Biochemistry of Memory. One important function of the central nervous system is the integration and the storage of information. While integration proceeds primarily through a network of various neurons, the *storage of information* is a chemical phenomenon. The manner in which information—which is present initially merely as nerve impulses—is converted to "molecular information" and how it is recalled from storage, i.e., reconverted to electrical excitation, is still unknown.

Under the influence of recent discoveries of RNA and DNA as information macromolecules, it has been postulated that in the brain, too, information storage involves nucleic acids. Indeed, it has been observed that the learning process is associated with an absolute increase of RNA content and with a change of base composition. It is not clear, however, if this is merely concomitant with brain activity or if it is the actual form of information storage of the brain. No molecular mechanism has yet been advanced that could possibly account for the synthesis of RNA patterned after the information matrix of nervous excitation. The biochemical mechanism of recall of information from nucleic acid storage back into the form of electric excitation is equally unclear.

The frequently cited reports that the feeding or injection of certain RNA fractions would transfer memory information in chemical form is exceedingly controversial and by no means can be taken as substantiated.

Nucleic acids store information by a linear arrangement of their structural elements (bases). One could also conceive of storage of information in the form of a two-dimensional pattern of molecules, for example, on membranes, which would influence the permeability of individual membrane segments in either a positive or negative sense. The conceptual model of an information molecule is certainly not the only possible model, although for some time now it has preoccupied biochemical research in this field.

The problems raised in this section certainly belong to the most difficult ones in biochemistry. It will require much more work to explore the molecular bases of memory and thought processes.

9. Development, Differentiation, and Malignant Degeneration

Preceding chapters have considered the specific biochemical tasks of individual organs of a whole organism, for example, man. The various morphologically, physiologically, and biochemically distinct organs all develop from a single cell, the fertilized gamete. Consequently, all cells of the entire organism, however different their function and morphology may be, fundamentally contain the same genetic information. With that, the question arises of how such a variety, such morphological and biochemical differentiation can come about.

Closely connected with the question of differentiation is that of *malignant degeneration*, which many authors look upon as dedifferentiation. Thus, we come to the problem of cancer which is of such theoretical as well as practical importance in medicine.

Description of the morphological development and of differentiation is subject of *embryology*, while analysis of causes of these processes are subject of *developmental physiology*. In the course of the analysis of causes, developmental physiology very quickly meets genetic and molecular–biological–biochemical problems. This part of biochemistry could also be called *biochemistry of morphogenesis*. In the following paragraphs, we shall deal with these questions.

Molecular Morphogenesis. Morphogenesis means formation of structure. The simplest example of structure formation is found in the protein molecules themselves. From a peptide chain that is in first approximation simply a threadlike molecule, a particular spatial structure is generated spontaneously (Chapter IV,5) solely through the action of physicochemical forces predetermined by the amino acid sequence. Additional "instruction" is not required for this. It could be looked upon as the beginning of morphogenesis. The next step would be protein complexes that are composed of subunits; a simple example is hemoglobin. More complicated still, but operating on the same principle, are multienzyme complexes and structure-forming elements such as collagen.

This principle of "*self-assembly*" can be pursued further. The assembly of bacteriophages from individual subunits is still possible in vitro. Mutants of the more complicated phage T4 could be isolated, which at higher culture temperatures lose the capacity to form certain parts of the phage, such as the head capsule, the tail rod or tail fibers. From such phase cultures it is possible to isolate the still intact parts, for example, mutant heads and from another mutant tail rods. If such isolated parts are mixed together, then the appropriate pieces find each other and aggregate to form complete infectious phages (Wood and Edgard).

The *biological membrane* also appears to be a system whose components (lipid and protein) can aggregate by themselves on the basis of their chemical affinities, i.e., without any external direction, to form a *structure of least energy*. This is concluded from the observation that, for example, after homogenization of cells the fragments of the endoplasmic reticulum can be found again as "microsomes"; the fragments have closed up to form little vesicles. A similar situation obtains with the inner mitochondrial membrane (Chapter X,4).

Molecular Biological Aspects of Cell Differentiation. The differentiation of cells of various tissues is not only expressed in their morphological structure, but also in their different biochemical functions. The latter calls for the equipment of cells with special sets of enzymes. Thus, adipose tissue contains the enzymes of fat metabolism in particularly high activity. Endocrine glands (for example, adrenal cortex) specifically are equipped for the biosynthesis of the corresponding hormone. The enzymes of gluconeogenesis occur only in the cells of liver and kidney. Liver and muscle cells both contain glycogen phosphorylase, but the enzymes in the two tissues are, nevertheless, genetically distinct. The number of examples could easily be multiplied.

Current concepts can explain this differentiated equipment of various parts of the organism with sets of enzymes by postulating that of the total genetic information only a certain amount typical for the kind of cell and for the tissue is actually expressed, i.e., is utilized for the synthesis of RNA and protein.

In the liver parenchymal cells not more than 10–15% of the total genome is transcribed for normal function. A small part of the rest of the information can be activated at any time (enzyme induction, compare Chapters VII,7 and XXI). The largest part, however, remains permanently excluded from transcription. The mechanism for this is still quite unclear. Most probably DNA itself is not changed because when isolated in pure form, DNA can be transcribed completely. *Histones* are now widely considered to be responsible for blocking DNA.

Embryonic Inducers. In the preceding paragraphs we have discussed morphological and biochemical differences between cells that differentiate into specialized tissues. But how does such specialization of cells in organs in the course of embryonic and postembryonic development come about?

Classical transplantation experiments of Spemann and his school have demonstrated that a certain region in the amphibian embryo can induce differentiation of organ anlagen in the adjacent ectoderm. This inductive effect is mediated by substances, two of which have recently been isolated and purified: A *neural factor* that induces, e.g., nervous tissue and eyes, and a *mesodermal factor* that induces the development of muscle tissue, cord, and kidney anlagen (Tiedemann). The cooperation of both factors is necessary for the proper development.

The interaction of substances in developing tissue is not limited to the formation of the very first organ anlagen. Determining substances have also been found in later development. In cord tissue a *chondrogenic factor* has been demonstrated that induces the formation of cartilage. This substance has also been greatly enriched (Zilliken). For the growth and regeneration of nerves a *nerve growth factor* is required; it is a protein with a molecular weight of 60,000 (R. Levi-Montalcini).

The embryonic inducers most probably act on the cell nucleus and activate certain genes which then give rise to cell-specific substances.

Specificity of Cell Surface. Cells from different organs differ in their outer cell membrane. This difference can be demonstrated with immunological methods. According to current concepts, the antigenic character is expressed primarily in the carbohydrate groups of the glycoproteins and the glyolipids that exist on the cell surface. The structures on the surface are important for contact formation of the cells and for the cohesion of cells of a certain organ. During embryonic development the formation of specific-surface structures is the first step toward differentiation; it is critical for the formation of organ anlagen.

Cell Division. Development is associated with a tremendous multiplication of cells resulting from mitotic cell division. We will not discuss the process of mitosis here, but merely recall that each cell division is preceded by a doubling of the genetic material (DNA) taking place during the synthesis phase (S phase). The life cycle of the cell can be subdivided into the following sections: mitosis—G_1 phase—S phase—G_2 phase—mitosis. The G_1 and G_2 phases may be of very different duration.

During the synthesis phase (and during mitosis), transcription of genetic information is suppressed extensively. The synthesis of messenger RNA and its translation, i.e., protein biosynthesis, proceeds mainly during the G_1 and G_2 phases.

During early embryonic development little genetic information is read off. Protein synthesis takes place largely on preformed messenger RNA with a long half-life. Only in a later stage of development does new synthesis of RNA take place for normal development.

If an organ has achieved its final size then cell division halts. Further cells are formed only to replace dead cells.[4] What determines the final size of the organ is largely not known. It is known that the partial removal of the liver in experimental animals results in regeneration of this organ. The cells begin to divide again until the "proper" size of the liver is regained. We do not know yet with certainty what the signal is that controls the division of cells. Recently certain factors have been postulated called *chalones*. They appear to be proteins or glycoproteins that suppress mitosis.

Malignant Degeneration. Cancer cells (*tumor cells*) are somatic cells which have changed in such a fashion that they no longer respond to the mechanisms controlling growth and cell division; they grow autonomously. The conversion of tumor cells in general is irreversible and it is passed on somatically to daughter cells.[5] The uninhibited proliferation of tumor cells results in a displacement of normal body cells and in growth into the healthy tissue (*invasion*). Tumor cells from the primary tumor can spread throughout the body and settle down in other parts of the organism and multiply there (*metastases*).

In the course of carcinogenesis, tumor cells lose more and more of the tissue- and organspecific properties which the original cells have developed in the course of differentiation. This dedifferentiated state resembles in many ways that of embryonic cells. Another similarity is the occasionally observed high rate of cell division of tumor cells. However, dedifferentiation is not an obligatory condition for tumor cells. It is not the cause but rather the result or concomitant phenomenon of malignant degeneration.

Defense Mechanisms. The conversion of body cells to tumor cells probably occurs much more often than the final formation of tumors. The organism is capable, however, under normal circumstances to recognize the malignant, degenerate cell as foreign and by use of its immune system to neutralize it. There is good evidence that in the course of dedifferentiation cell-bound antigens appear which differ from those of normal body cells. These antigens can cause an immune response which is followed by the rejection of the tumor, similar to the rejection of an unaccepted transplanted organ. If for some reason this immune defense mechanism does not operate, then the tumor continues to grow and to proliferate unchecked.

The Metabolism of Tumor Cells. In the hopes of finding some characteristic differences between tumor and normal cells, the metabolism of the former has been studied extensively. The result thus far is rather meager. Tumor-specific metabolic

---

[4] The replacement of cells in tissues that constantly slough off (skin, mucous membrane, etc.) is appreciable. In such tissue, cell division is readily seen.

[5] Somatic inheritance is the designation for information transmittal of somatic mutation, i.e., mutations taking place in the genetic material of a soma cell. The gametes are not affected by this.

pathways, in the sense that perhaps bacteria have a characteristic metabolism, have not been found. Many tumor cells no longer are able to synthesize asparagine and require it as a growth factor, but even that is not universally true. Frequently, one finds quantitative differences between tumor cells and normal cells, but liver tumors in animal experiments are known that show all the properties of liver cancer cells but in no way differ from normal liver cells in their metabolism (*minimal deviation tumors*).

One characteristic of many tumor cells is the production of lactate from glucose even under conditions of adequate oxygen supply. This has been designated as *aerobic glycolysis*. One of the causes for this appears to be the deficient capacity of the mitochondria of tumor cells to reoxidize NADH that has arisen in the cytoplasm with the aid of suitable transport metabolites (for example, α-glycerol phosphate). Warburg took impaired cell respiration and the resultant aerobic glycolysis as the true cause of malignant degeneration and destructive growth of cancer cells. It is very doubtful, however, if this frequently stated property of aerobic glycolysis is the cause of degeneration or merely an associated phenomenon. It is additionally a hallmark of embryonic undifferentiated cells and of several otherwise "normal" cells (for example, the retina cells) of adult organisms. The minimal deviation liver tumors mentioned above, on the other hand, do not show this impaired respiration.

Causes of Carcinogenesis. A number of physical, chemical, or biological factors can initiate, or at least favor, the conversion of normal body cells to tumor cells. These factors (called *carcinogens*) include ultraviolet light or X-ray irradiation, as well as a whole series of chemical substances, and finally *oncogenic viruses*.

The action of carcinogenic substances, of whom hundreds are at present known (see also Fig. XXIII-4), has been studied exceedingly thoroughly over several decades. In this group belong certain aromatic hydrocarbons with numerous fused ring systems such as methylcholanthrene, benzanthracene, benzpyrene, and their methylated derivatives which cause primarily cancer of the *skin*; it also includes certain aromatic amines, for example *p*-dimethylaminoazobenzene or 2-acetamidofluorene, and even such simple substances as dimethylnitrosamine which in the liver initiate the formation of *hepatomas*. 2-Naphthylamine and certain other aromatic amines are responsible for the formation of cancer of the *bladder*. This list of compounds reveals a certain *organotropism* of carcinogenesis. We can be exposed to many of these compounds in industrial plants. The cancers in industrial workers caused by tar, aniline, and other materials originally have led to the discovery of carcinogenic substances. It is also certain now that carcinogenic factors occur in the tar of tobacco generated during smoking. Finally, certain plants and plant products contain natural chemical carcinogens. Particularly interesting in this connection is the group of aflatoxins, products of the mold *Aspergillus flavus* which often attacks a variety of foods. In this manner carcinogens can appear in foods.

The mechanism of carcinogenesis has been studied primarily in rats and mice. A particularly clear model is the development of tumors in mouse skin by carcinogenic hydrocarbons. Here one can distinguish different stages of the conversion of normal cells to tumor cells. The first stage called *tumor initiation* leads to *potential* tumor cells,

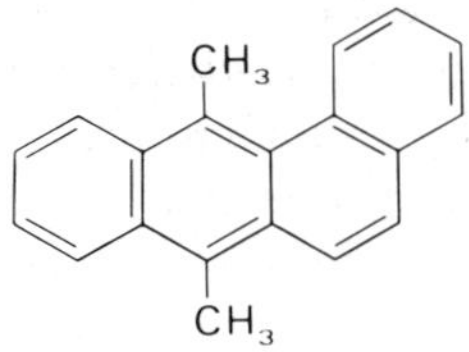

7,12-Dimethylbenzanthracene

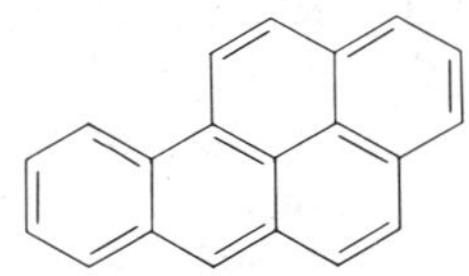

Benzpyrene

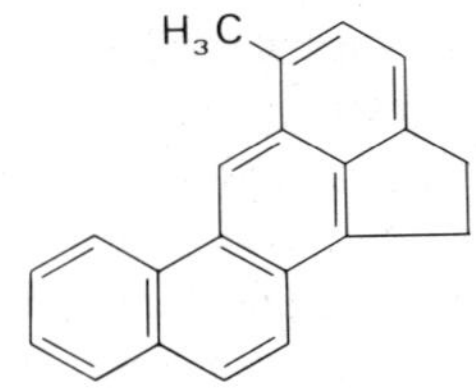

3-Methylcholanthrene

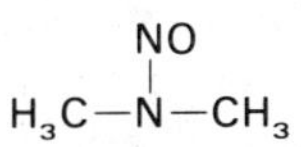

Dimethylnitrosamine

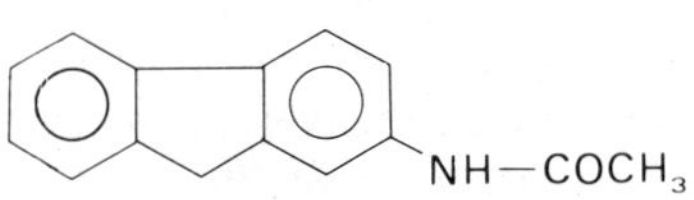

2-Acetamidofluorene

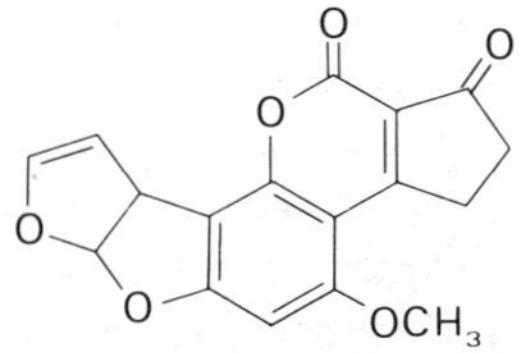

Aflatoxin B_1

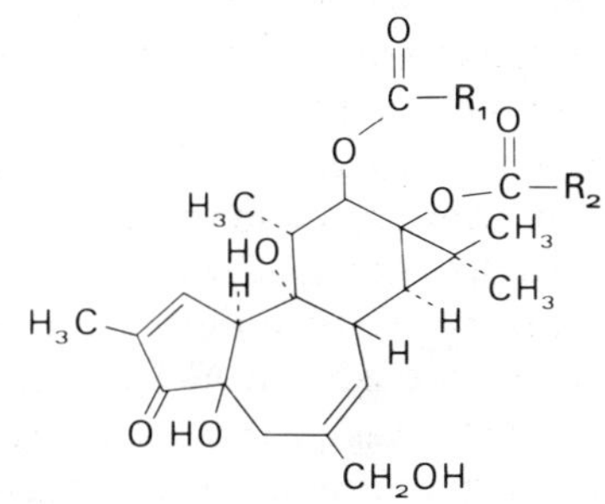

Phorbol esters

Fig. XXIII-4. Several chemical carcinogens. 3-Methylcholanthrene possesses certain structural kinship to steroids. Phorbol esters are not carcinogens, but rather cocarcinogens.

and its formation is probably irreversible. This stage is inherited somatically, as defined earlier in this section, i.e., it is transmitted to subsequent generations of cells. From these potential tumor cells there develop—after variable periods of latency—initially benign tumors, later malignant carcinoma. This second phase, the *development of tumors*, can be accelerated greatly ("tumor promotion") by certain substances which by themselves are not carcinogenic. Such compounds are called *cocarcinogens*. The prototype is croton oil, a plant oil which when applied to the skin by itself causes inflammation. When animals are first treated with a carcinogenic hydrocarbon followed by the application of croton oil they rapidly develop cancers of the skin. The active component of croton oil has now been identified as an ester of a novel diterpene called *phorbol* (Hecker; see formula in Fig. XXIII-4).

Another rather plausible model is tumor genesis by *virus induction*. Experimental animals are known to have a number of tumors that are caused by viruses. The Rous sarcoma virus of chickens and the Bittner virus, which causes mammary carcinoma in mice and is transmitted in the mother's milk, both are RNA viruses. DNA viruses are the SV40 of monkey kidneys and the Epstein-Barr virus from human hemopoetic cells; the latter is capable of transforming cultured cells to tumor cells. A virus-caused tumor disease in man is Burkitt lymphoma which is endemic in Africa.

The formation of virus tumors probably follows the mechanism of so-called *transforming infection* (see Chapter VII,9). In this case, the viral DNA is incorporated into the genome of the cell and duplicated in each cell division in a normal way. With RNA viruses, apparently a complementary DNA molecule is produced by reversed transcription which then behaves just as the viral DNA did. In cell cultures, this kind of infection rapidly leads to the formation of tumor cells. Latent infections are observed *in vivo* where additional unknown factors have to be present in order to activate the virus and to initiate malignant growth of the cells.

Hypothesis of Carcinogenesis. There has been no dearth of attempts at organizing the voluminous experimental material on carcinogenesis to a comprehensive theory. Warburg's idea has already been mentioned. K. H. Bauer has proposed the hypothesis that the conversion to tumor cells depends on somatic mutations (see above). Such a mutation could be initiated by radiation or by carcinogenic substances and would affect the genetic material of the cellular nucleus. On the other hand, Hübner postulates that the transformation to a cancerous cell depends on the action of a virus which is distributed ubiquitously. Normally the virus remains latent but can be activated by environmental factors and then can develop its full transforming properties.

A synthesis of these various hypotheses has been attempted by E. Hecker in his "Loss of Information Hypothesis." He proposes that cancer is caused by a decrement of the information content of normally differentiated cells. The result would be that cells no longer respond to the regulating action of other cells and begin to grow autonomously and continue to divide. This loss of information may result from somatic mutation (released by chemical or physical carcinogenesis) or by transformation (in the case of oncogenic viruses). However, the loss can also occur as a result of irreversible change of intra- or extracellular regulatory mechanisms, for example, the loss of regulatory proteins. This concept leaves the possibility that different types of tumors arise by different mechanisms so that in each individual case it must be determined which mechanism operates in that particular type of tumor.

The problem of cancer has confronted chemistry and molecular biology with very fundamental questions. Many laboratories of the world are working on a solution. Our knowledge at this point is highly fragmentary, although a large number of hypotheses have attempted to bridge these gaps. One can only hope that experimental investigation will provide answers to some of our questions; this would be of great practical and theoretical significance.

10. Summary

Hopkins has defined life as a dynamic equilibrium in a polyphasic system. Enzymology has elucidated the numerous reactions involved in the dynamic steady state equilibrium and now endeavors to understand their regulation. Work on the ultrastructure has shown us the many compartments of the "polyphasic system." Both methods have contributed importantly to our understanding of the activity of cells and organs. The methods also have demonstrated their power, and many problems will yet be solved by them.

A scientific explanation of biological phenomena can be expected to come only from the study of molecules. In this endeavor morphology, physiology, and biochemistry meet; a clear distinction among these overlapping disciplines has ceased to exist.

BIBLIOGRAPHY

E. E. Bittar, "Cell pH." Butterworth, London and Washington, D.C., 1964. (Contains two chapters on skeletal muscle.)

W. Eckhart, Oncogenic viruses. *Annu. Rev. Biochem.* **41,** 503–516 (1972).

Z. W. Hall, Release of neurotransmitters and their interaction with receptors. *Annu. Rev. Biochem.* **41,** 925–952 (1972).

H. E. Huxley, Muscle 1972: Progress and problems. *Cold Spring Harbor Symp. Quant. Biol.* **37,** 689–693 (1973); the whole issue is devoted to "The Mechanism of Muscle Contraction."

M. S. Legator, Environmental mutagenesis and repair. *Annu. Rev. Biochem.* **42,** 683–708 (1973).

W. J. Rutter, R. L. Pictet, and P. W. Morris, Toward molecular mechanisms of developmental processes. *Annu. Rev. Biochem.* **42,** 601–646 (1973).

Appendix

List of Abbreviations Frequently Used in Biochemical Literature

Introduction. Abbreviations are used in the current biochemical literature to an alarming extent. Some publications are rendered almost unintelligible by the numerous abbreviations, and often this is the vehicle for introducing "lab" jargon into the literature. We have intentionally been conservative about the use of abbreviations and should like to suggest that abbreviations be used cautiously. Abbreviations of metabolites are mostly superfluous. Names of enzymes should never be abbreviated, an oft ignored rule of the Enzyme Commission of the International Union of Biochemistry.

Abbreviations are justified in formulas of proteins and other high-molecular weight substances whose individual components are indicated by such symbols. In a sense, this is nothing but an extension of the use of chemical symbols. Monomers are abbreviated by lowercase letters; they are listed in the Systematic List, below.

Though we discourage the use of abbreviations in general, we provide a list here because they do appear frequently in the literature, especially in original contributions.

1. Alphabetic List

A (Ado)	Adenosine
Acetyl-CoA	Acetyl coenzyme A, "active acetate"
ACTH	Adrenocorticotropic hormone
*ADH	Alcohol dehydrogenase
ADP	Adenosine diphosphate
AMP	Adenosine monophosphate
AT-10	Dehydrotachysterol (antitetany compound 10)
ATP	Adenosine triphosphate
ATPase	Adenosine triphosphatase
*BAL	2,3-Dimercaptopropanol (British antilewisite)
C (Cyd)	Cytidine
CDP	Cytidine diphosphate

* Not recommended as abbreviations.

*ChE	Cholinesterase
CM	Carboxymethyl (e.g., CM-cellulose)
CMP	Cytidine monophosphate
CoA	Coenzyme A ("A" stands for acyl activation)
CTP	Cytidine triphosphate
DEAE	Diethylaminoethyl (e.g., DEAE-cellulose)
DFP	Diisopropylfluorophosphate
DNA	Deoxyribonucleic acid
DNase	Deoxyribonuclease
DNP	Dinitrophenyl- (substituted amino acid, e.g.)
DNS	Desoxyribonucleinsäure (German)
DOC	Deoxycorticosterone
DOCA	Deoxycorticosterone acetate
DOPA	Dihydroxyphenylalanine
*DPN, $DPN^{\oplus}$	Diphosphopyridine nucleotide (oxidized)
*DPNH, $DPNH_2$	Diphosphopyridine nucleotide (reduced)
e, or $e^{\ominus}$	Electron
E.C. (with number)	Enzyme code number; official classification (cf. Chapter V,10)
EDTA	Ethylenediaminetetraacetic acid
ESR	Electron spin resonance
FAD	Flavin-adenine dinucleotide
$FADH_2$	Flavin-adenine dinucleotide, reduced form
Fd	Ferredoxin
*FDP	Fructose diphosphate
FMN	Flavin mononucleotide
fp	Flavoprotein
FSH	Follicle-stimulating hormone
G (Guo)	Guanosine
*G-6-P (= Glc-6-P)	Glucose 6-phosphate
*GDH	Glucose dehydrogenase
GDP	Guanosine diphosphate
GMP	Guanosine monophosphate
*GOT	Glutamate-oxaloacetate transaminase
α-GP	α-Glycerophosphate (glycerol 1-phosphate)
*GPT	Glutamate-pyruvate transaminase
GTP	Guanosine triphosphate
GSH	Glutathione (reduced form)
GSSG	Glutathione (oxidized form)
Hb	Hemoglobin
HbO_2	Oxyhemoglobin
HCG	Human chorionic gonadotropin
*HDP	Hexose diphosphate
HMG	Human menopausal gonadotropin
*HMP	Hexose monophosphate
ICSH	Insterstitial cell-stimulating hormone
*IDH	Isocitrate dehydrogenase
IDP	Inosine diphosphate
IMP	Inosine monophosphate
IP	Isoelectric point
ITP	Inosine triphosphate

* Not recommended as abbreviations.

*α-KG	α-Ketoglutaric acid (oxoglutarate)
*LDH	Lactate dehydrogenase
LH	Luteinizing hormone
LTH	Luteotropic hormone
MSH	Melanocyte-stimulating hormone
NAD, $NAD^{\oplus}$	Nicotinamide-adenine dinucleotide, formerly DPN (oxidized)
NADH, $NADH_2$	Nicotinamide-adenine dinucleotide (reduced)
NADP, $NADP^{\oplus}$	Nicotinamide-adenine dinucletotide phosphate, formerly TPN (oxidized)
NADPH, $NADPH_2$	Nicotinamide-adenine dinucleotide phosphate (reduced)
NMN	Nicotinamide mononucleotide
NMR	Nuclear magnetic resonance (method)
OAA	Oxaloacetic acid
PAL	Pyridoxal phosphate
PAMP	Pyridoxamine phosphate
PCMB	*p*-Chloromercuribenzoate
PEP	Phosphoenolpyruvate
P_i	Inorganic phosphate
PLP	Pyridoxal phosphate
PMP	Pyridoxamine phosphate
PMSG	Pregnant mare serum gonadotropin
PP	Pyrophosphate (inorganic)
RNA	Ribonucleic acid
RNase	Ribonuclease
RNS	Ribonucleinsäure (German)
R.Q.	Respiratory quotient
STH	Somatotropic hormone (growth hormone)
T	Thymidine (in DNA)
*TCA	Trichloroacetic acid
Thx	Thyroxine
*TPN, $TPN^{\oplus}$	Triphosphopyridine nucleotide (oxidized)
*TPNH, $TPNH_2$	Triphosphopyridine nucleotide (reduced)
*TPP	Thiamine pyrophosphate
Tris	Tris(hydroxymethyl)aminomethane
TSH	Thyrotropic hormone
TTC	Triphenyl tetrazolium chloride
U (Urd)	Uridine
UDP	Uridine diphosphate
UDP-Gal	Uridine diphosphate galactose
UDP-Glc (UDP-G)	Uridine diphosphate glucose
UMP	Uridine monophosphate
UTP	Uridine triphosphate
UV	Ultraviolet (light)

2. Systematic List

(a) Amino Acids[1]

Ala	(A)	Alanine	Asn (Asp-NH_2)	(N)	Asparagine
Arg	(R)	Arginine	Asp	(D)	Aspartic acid

[1] The letters in parentheses are the one-letter abbreviations which are to be restricted to computer analysis or special graphical presentations (cf. the structure of lysozyme on the foldout chart).

Asx	(B)	Asn or Asp	Leu	(L)	Leucine
Cys	(C)	Cysteine	Lys	(K)	Lysine
Gln	(Q)	Glutamine	Met	(M)	Methionine
Glu	(E)	Glutamic acid	Phe	(F)	Phenylalanine
Glx	(Z)	Gln or Glu	Pro	(P)	Proline
Gly	(G)	Glycine	Ser	(S)	Serine
His	(H)	Histidine	Thr	(T)	Threonine
Hyl		Hydroxylysine	Trp	(W)	Tryptophan
Hyp		Hydroxyproline	Tyr	(Y)	Tyrosine
Ile	(I)	Isoleucine	Val	(V)	Valine

(b) Nucleic Acids and Their Components

DNA	Deoxyribonucleic acid
RNA	Ribonucleic acid
cRNA	Complementary RNA
mRNA	messenger RNA
mtRNA	mitochondrial RNA
nRNA	nuclear RNA
tRNA	transfer RNA
$tRNA^{Ala}$	alanine-specific transfer RNA

Base		Nucleoside[a]		Nucleoside mono-, di-, and triposphate[b]
Adenine	Ade	Adenosine	A(Ado)	AMP, ADP, ATP
Cytosine	Cyt	Cytidine	C(Cyd)	CMP, CDP, CTP
Guanine	Gua	Guanosine	G(Guo)	GMP, GDP, GTP
Hypoxanthine	Hyp	Inosine	I(Ino)	IMP, IDP, ITP
Orotate	Oro	Orotidine	—(Ord)	—, —, —
Thymine	Thy	Thymine riboside	T(Tho)	TMP, TDP, TTP
Uracil	Ura	Uridine	U(Urd)	UMP, UDP, UTP
		Pseudouridine	$\psi(\psi$rd)	

[a] Names and abbreviations stand for *ribo*nucleosides. Deoxyribonucleosides are designated by the prefixed letter d: dA = deoxyadenosine. This also applies to nucleotides: dATP = deoxyadenosine triphosphate.

[b] These abbreviations stand for 5′-phosphates or 5′-polyphosphates. In case of 3′-phosphates, the number 3 is included in the abbreviation: A-3-P = adenosine 3-phosphate. In ambiguous situations, the 5′-phosphates can be designated in the same way: A-5-P_2 = adenosine 5-pyrophosphate (= ADP).

With the use of single-letter abbreviations in cases of oligonucleotides, phosphate groups are abbreviated by a lower case "p" (in front of the capital letter means 5′ position; after the capital letter means 3′ position). Thus pApUp stands for a dinucleotide in which adenine has a 5′-phosphate group and uridine in addition a free 3′-phosphate group. p! or p > signifies the 2,3-cyclic phosphate. In longer sequences, "p" is replaced by a hyphen.

Detailed rules have been published in *Hoppe-Seyler's Z. Physiol. Chem.* **351,** 1055–1063 (1970).

(c) **Sugars**

Rib	Ribose	Glc	Glucose[2]
dRib	Deoxyribose	Gal	Galactose
Rul	Ribulose	Man	Mannose
Xyl	Xylose	Fru	Fructose
Xul	Xylulose		

(d) **Coenzymes:** see Table VI-1 (Chapter VI,3).
(e) **Hormones:** see Table XX-1 (Chapter XX,1).
(f) **Vitamins:** see Table XXII-2 (Chapter XXII,3).

Chronological Table of Important Biochemical Discoveries

1773	Discovery of urea (Rouelle)
1779	Preparation of glycerol from olive oil (Scheele)
1783	Discovery of extracellular digestion of proteins (Spallanzani)
1789	Respiration as oxidation (Lavoisier)
1815	Elucidation of the overall reaction of alcoholic fermentation (Gay-Lussac)
1815	Discovery of optical activity (Biot)
1828	Synthesis of urea from ammonium cyanate: The first synthesis of an organic compound occurring in a living organism (Wöhler)
1837	Investigation of the enzymatic cleavage of amygdalin with almond emulsion (Liebig and Wöhler)
1838–39	Establishment of fermentation as a catalytic process (Berzelius and Liebig).
1847	Diastatic degradation of starch to maltose (Dubrunfaut)
1857	The vitalists' theory of fermentation (essentially dependent on living cells) by Pasteur
1869	Discovery of nucleic acid (Miescher)
1890	First crystallized protein: Egg albumin (Hofmeister)
1893	Classification of enzymes as catalysts in the sense of physical chemistry (Ostwald)
1897	Discovery of cell-free fermentation (E. and H. Buchner)
1902	Formulation of proteins as polypeptides (Hofmeister, E. Fischer)
1903	Isolation of the first hormone: Epinephrine (Takamine, Aldrich)
1905	Role of phosphoric acid in fermentation (Harden and Young)
1905	Discovery of β-oxidation (Knoop)
1912	First scheme of fermentation (Neuberg)
1912	Theory of dehydrogenation in biological oxidations (Wieland)
1922	Isoprene rule as a building principle of numerous natural products (Ruzicka)
1925	Rediscovery of cell hemins (cytochromes) (Keilin)
1925	Formulation of sugars as pyranoses (Haworth)
1926	Isolation of the first vitamin: Thiamine (Jansen and Donath)
1926	First crystallized enzyme: Urease (Sumner)
1925–30	Determination of molecular weights of many proteins with the ultracentrifuge (Svedberg)
1929	Discovery of "labile" phosphate: Adenosine triphosphate (Lohmann: Fiske and Subarrow)
1929	Characterization of the "Atmungsferment" as a hemin compound (Warburg)
1929–34	Isolation of many steroid hormones (Butenandt, Doisy, Laqueur, Reichstein)

[2] Glucose may not be abbreviated as Glu, since this symbol has already been reserved for glutamic acid. The often used abbreviation G should also be avoided because of the possible confusion with guanine (= G). Sugar phosphates may be abbreviated in the same way: Glc-6-P = glucose 6-phosphate, and Fru-1-P-6-P = fructose 1,6-bisphosphate.

1932 Discovery of the yellow enzyme (Warburg; Theorell)
1933 Discovery of the ornithine or urea cycle (Krebs and Henseleit)
1935 New scheme of glycolysis and alcoholic fermentation (Embden-Meyerhof)
1935–36 Discovery of the connection between vitamins and coenzymes. Isolation of the pyridine and flavin coenzymes (v. Euler, Theorell, Warburg)
1935 Threonine as a new, and the first essential, amino acid (Rose)
1935 Isolation of the first crystallized virus: Tobacco mosaic virus (Stanley)
1937 Formulation of the citric acid cycle (Krebs; Knoop and Martius)
1938 Discovery of transamination (Braunstein and Kritzmann)
1939–44 Isolation and structural determination of the first antibiotic: Penicillin (Fleming, Florey, Chain, Johnson)
1944 Mechanism of action of gene factors through enzymes (Butenandt, Kühn)
1944 Isolation of the pneumococcal transformation factor and its characterization as deoxyribonucleic acid (Avery)
1948 Introduction of the centrifugation method for the isolation of cell fragments (Schneider and Hoogeboom; Potter)
1951 Elucidation of "active" acetate (Lipmann; Lynen)
1952 Helical structure of proteins (Pauling)
1953 Structural determination of insulin (Sanger)
1953 Discovery of the pentose phosphate cycle in glucose breakdown (Horecker; Dickens)
1954 Helical model of nucleic acids (Watson and Crick)
1958 Proof of the infectivity of pure virus nucleic acids (Gierer and Schramm)
1958 Isopentenyl pyrophosphate as active isoprene (Lynen)
1960 First mapping of the three-dimensional structure of a protein (Kendrew; Perutz)
1960 Complete sequence of hemoglobin (Braunitzer and co-workers)
1961 Decoding of the base code of nucleic acids (Nirenberg; Ochoa)
1961 Model for the regulation of gene activity (Jacob and Monod)
1963 Allosteric inhibition of enzymes (Changeux, Jacob, and Monod)
1965 First determination of the base sequence of a nucleic acid (Holley and co-workers)
1965 Three-dimensional model of an enzyme (lysozyme) (Phillips and co-workers)
1966 Complete dictionary of nucleotide triplets (Nirenberg; Khorana)
1969 First synthesis of an enzyme, ribonuclease (Gutte and Merrifield; Denkewalter and Hirschmann)

SUBJECT INDEX

D

E

F

H

J

K

M

O

P

Q

R

S

T

V

W

X

Y

Z

A 5
B 6
C 7
D 8
E 9
F 0
G 1
H 2
I 3
J 4